STATISTICAL PROPERTIES OF NUCLEI

STATISTICAL PROPERTIES OF NUCLEI

Proceedings of the International Conference on Statistical Properties
of Nuclei, held at Albany, New York, August 23-27, 1971

Edited by J. B. Garg

Professor of Physics
State University of New York at Albany

Ⴔ PLENUM PRESS • NEW YORK—LONDON • 1972

Library of Congress Catalog Card Number 75-182409

ISBN 0-306-30576-3

© 1972 Plenum Press
A Division of Plenum Publishing Corporation
227 West 17th Street, New York, N.Y. 10011

United Kingdom edition published by Plenum Press, London
A Division of Plenum Publishing Company, Ltd.
Davis House (4th Floor), 8 Scrubs Lane, Harlesden, NW10 6SE, London, England

Printed in the United States of America

DEDICATION

This book is dedicated to the memory of

CHARLES E. PORTER (1927-1964)

and

ROBERT G. THOMAS (1923-1956)

who in their short life-spans contributed greatly to the understanding
of the Statistical Properties of Nuclei.

FOREWORD

An International Conference on the "Statistical Properties of Nuclei" was held from August 23 to August 27, 1971, at the State University of New York at Albany campus. The purpose of the conference was to review the current status of the experimental and theoretical aspects of resonance reaction theories, statistics of resonance parameters such as level spacings, neutron, fission, radiative and reaction widths, level densities, fluctuations in cross sections, strength functions and its relation to the optical model, intermediate structure in particle and photon induced reactions, and statistical aspects of the decay of the compound nucleus. The conference was held under the auspices of the International Union of Pure and Applied Physics.

The organization of the conference was greatly facilitated by the financial support received from the International Union of Pure and Applied Physics, The National Science Foundation and the U.S. Atomic Energy Commission and the generous use of the physical facilities and other audio-visual services provided by the State University of New York. It is with great pleasure that I thank all these agencies for their kind support.

The idea of this conference arose more than two years ago when going through the large accumulation of experimental data of the past five years from high resolution neutron and proton resonance spectroscopy and the corresponding theoretical developments on the statistical aspects of resonances, I felt the great need for a general review of the current understanding of the inter-relation between the data and the theoretical predictions, by the experts actively working in the field.

The last international conference, where statistical properties of resonances were discussed was held at Amsterdam in 1965 entitled as "Nuclear Structure Study With Neutrons." The organizing committee felt it impelling to enlarge the scope of this conference by including borderline areas as well.

The conference was attended by 207 delegates from 20 different countries. Every effort was made to provide enough time to delegates for informal discussions, including the time available during the excursion to the beautiful Lake George on one afternoon. The

conference was concluded with a panel discussion on the last day
with participants summarizing their points of view on topics of
their interest and emphasizing the areas of interest for further
investigation.

I would like to take this opportunity of thanking all the
members of the organizing committee as well as the foreign advisors
who so generously gave their valuable advice and time to make the
conference successful in meeting its objective, and Mrs. Fiona Burde
for her valuable help in making the local arrangements as the
secretary for the conference.

                                        J. B. Garg
                                        Conference Chairman

EDITOR'S PREFACE

This volume of the proceedings of the Conference contains the texts of all the talks orally presented during the Conference. The only exceptions are the manuscripts submitted by S. Suchoruchkin and V. Popov who were invited to present papers orally but were unable to attend the Conference at the last minute.

The delegates were asked to submit one-page contributions of their work to the Conference by June 9, 1971. In all, about 105 contributions were received. From these, only 30 contributed papers were selected for oral presentations due to lack of time. The authors of these papers were asked to submit more detailed versions of the paper or group of papers for publication in the proceedings. All the contributed papers submitted to the Conference were, however, printed at SUNYA in a book form, a copy of which was made available to each registered delegate at the beginning of the Conference. The editor regrets that due to the extra cost and size of the proceedings, all the contributed papers could not be printed, but only a list is given at the end of these proceedings.

A lot of conscientious work was done by the session chairmen and scientific secretaries in going through the recorded versions of the discussions, comments, etc., made during the talks. I would like to express my sincere appreciation to these scientists for their contributions.

We gratefully acknowledge the help of some of the delegates who provided a written version of their comments (which was helpful in clarifying some of the words and expressions in the transcripts). In spite of this, the editor is quite aware that there may still be present errors in spellings, words, and even phrases not attributed to a speaker who failed to give his name and affiliation and could not be later identified by his recorded voice. The editor, however, bears the responsibility for all errors, omissions and misinterpretations.

The editor had felt from the beginning the great need of shortening the time gap between the end of the Conference and the publication of its proceedings. In order to accomplish this objective, certain steps had to be taken which were fraught with great difficulties. Firstly, the invited speakers were asked to submit

their manuscripts about two weeks before the beginning of the
Conference, and I would like to take this opportunity to extend my
warm appreciation to all those people who responded to this request
(even at the expense of their own inconvenience in their earlier
scheduled programs).

Secondly, for the publication of the proceedings the photo-
offset process was adopted in which the final copy is produced by
the photocopy of the original typed manuscript. Hence the errors
overlooked by the authors are reproduced in the proceedings. The
editor has tried to make corrections in the original manuscripts
whenever such corrections were requested by the authors during the
Conference or communicated in writing after the submission of the
original manuscript.

The last session of the round-table discussion was entirely
reproduced from the transcript of the recorded talks with minor
changes in English and various phrases. This method was adopted
in order to preserve the informal atmosphere of their presentation
and the purpose of such discussions. In the opinion of the editor,
the value of such prompt reactions to questions and answers cannot
be substituted by the carefully written versions of the questions,
answers and comments.

Finally, we wish to acknowledge with gratitude the help of
many persons, particularly Ellen Kelly, Cindy Francis, Pat Gardner
and Virginia Davey, who did most of the typing of the recorded
transcripts and did the proof readings.

LIST OF PARTICIPANTS

ABOV, I., Inst. for Theoretical & Experimental Physics, Moscow,
U.S.S.R.
ADLER, F.T., Illinois University at Urbana, U.S.A.
ALLEN, B.J., Oak Ridge National Laboratory, Tenn., U.S.A.
ALTER, H., Atomics International, Canoga Park, Calif., U.S.A.
AUCHAMPAUGH, G., Los Alamos Scientific Laboratory, N.Mexico, U.S.A.
BABA, H., Japan Atomic Energy Research Institute, Tokai, Japan
BAKHRU, H., State University of New York, Albany, N.Y., U.S.A.
BARTHOLOMEW, G.A., Chalk River Nuclear Laboratory, Ontario, Canada
BEARSE, R.C., Kansas University, Lawrence, U.S.A.
BEER, M., Mathematical Application Group, Inc., White Plains, N.Y.
U.S.A.
BEG, K., Columbia University, New York, N.Y., U.S.A.
BEN DAVID, G., Soreq Nuclear Research Center, Rehovoth, Israel
BENENSON, R.E., State University of New York, Albany, N.Y., U.S.A.
BENEZET, L.T., State University of New York, Albany, N.Y., U.S.A.
BERMAN, B.L., Lawrence Radiation Laboratory, Livermore,Calif.,U.S.A.
BHAT, M.R., Brookhaven National Laboratory, L.I., N.Y., U.S.A.
BILPUCH, E., Duke University, Durham, N. Carolina, U.S.A.
BIMBOT, R., Laboratoire de Chimie Nucléaire at Orsay, France
BIZZETI, P.G., University at Firenze, Italy
BIZZETI-SONA, A.M., University at Firenze, Italy
BLANN, M., University of Rochester, N.Y., U.S.A.
BLOCH, C., C.E.N. de Saclay, Gif-sur-Yvette, France
BLOCK, R.C., Rensselaer Polytechnic Institute, Troy, N.Y., U.S.A.
BOHIGAS, O., Institut Physique Nucléaire at Orsay, France
BÖHNING, M., Technische Universität at München, Germany
BOLLINGER, L., Argonne National Laboratory, Illinois, U.S.A.
BORMANN, M., Universität Hamburg, Germany
BOUCHARD, C.A., Laval University, Quebec City, Canada
BOYER, E.R., State University of New York, U.S.A.
CAREW, J., Knolls Atomic Power Laboratory, Schenectady, N.Y., U.S.A.
CAMARDA, H., Columbia University, New York, N.Y., U.S.A.
CHANG, F.S., University of Rochester, N.Y., U.S.A.
CHATTERJEE, A., Saha Institute of Nuclear Physics, Calcutta, India
CHI, B.E., State University of New York at Albany, N.Y., U.S.A.

CHOUDHURY, D.C., Polytechnic Institute of Brooklyn, N.Y., U.S.A.
CLARK, G.J., A.E.R.E. at Harwell, Berkshire, England
CLARK, D., Cornell University, Ithaca, N.Y., U.S.A.
CLELAND, M., Radiation Dynamics, L.I., N.Y., U.S.A.
CHRIEN, R.E., Brookhaven National Laboratory, L.I., N.Y., U.S.A.
CLINE, C.K., University of Rochester, N.Y., U.S.A.
COCEVA, C., CNEN Centro di Calcolo, Bologna, Italy
COHEN, B.L., University of Pittsburgh, Pennsylvania, U.S.A.
COLE, G.W., Brookhaven National Laboratory, L.I., N.Y., U.S.A.
COMMUNEAU, F., Commissariat a l'Énergie Atomique, Paris, France
CUE, N., State University of New York at Albany, N.Y., U.S.A.
DEMEYER, A., Institut de Physique Nucléaire at Lyon, France
DEPOMMIER, P.H., Université de Montreal, Canada
DIVADEENAM, M., Duke University, Durham, N.Carolina, U.S.A.
DRENTJE, A.G., University of Groningen, The Netherlands
DROESSLER, E.G., State University of New York at Albany, N.Y.,U.S.A.
DROZDOV, S., I.V. Kurchatov Atomic Energy Inst., Moscow, U.S.S.R.
EARLE, E.D., Chalk River Nuclear Laboratory, Ontario, Canada
EBERHARD, K.A., Universität München, Germany
EHRLICH, R., Knolls Atomic Power Laboratory, Schenectady,N.Y.,U.S.A.
EJIRI, H., Osaka University, Japan
ELWYN, A.J., Argonne National Laboratory, Illinois, U.S.A.
EYAL, Y., Columbia University, New York, N.Y., U.S.A.
FALLIEROS, S., Brown University, Providence, R.I., U.S.A.
FESHBACH, H., Massachusetts Institute of Technology, Cambridge,USA.
FELVINCI, J.P., Columbia University, New York, N.Y., U.S.A.
FLEURY, A., Université of Bordeaux, France
FLORES, J., Institudo de Fisica, Mexico City, Mexico
FRAENKEL, Z., Weizmann Institute, Rehovot, Israel
FRANCIS, N., Knolls Atomic Power Lab., Schenectady, N.Y., U.S.A.
FRENCH, J.B., University of Rochester, N. Y., U.S.A.
FRICKE, M.P., Gulf Radiation Technology, San Diego, Calif.,U.S.A.
FUBINI, A., CNEN at Casaccia, Rome, Italy
FULTZ, S.C., Lawrence Radiation Laboratory, Livermore,Calif.,U.S.A.
GAERTTNER, E.R., Rennselaer Polytechnic Inst., Troy, N.Y., U.S.A.
GARBER, D.I., Brookhaven National Laboratory, L.I., N.Y., U.S.A.
GARDNER, D.G., Lawrence Radiation Lab., Livermore, Calif., U.S.A.
GARG, J.B., State University of New York at Albany, N.Y., U.S.A.
GIBBS, W.R., Los Alamos Scientific Laboratory, N.Mexico, U.S.A.
GIGLI, A., Universita di Pavia, Italy
GILAT, J., Soreq Nuclear Research Center, Yavne Israel
GOLDHABER, G., Brookhaven National Laboratory, L.I., N.Y., U.S.A.
GORODETZKY, S., Institut de Recherches Nucléaire, Strasbourg,France
GRANT, I.S., Manchester University, England
GREEN, J.M., R & D Associates, Santa Monica, Calif., U.S.A.
HACKEN, G., Columbia University, New York, N.Y., U.S.A.
HALL, W., Chemtree Corp., N.Y., U.S.A.
HALPERN, I., Washington University, Seattle, Wash., U.S.A.
HANSEN, P.G., CERN, Geneva, Switzerland

HARVEY, J.A., Oak Ridge National Laboratory, Tenn., U.S.A.
HAUSMAN, H.J., Ohio State University, Columbus, U.S.A.
HAVENS, W.W., Jr., Columbia University, New York, N.Y., U.S.A.
HILLMAN, M., Brookhaven National Laboratory, L.I., N.Y., U.S.A.
HOCKENBURY, R.W., Rensselaer Polytechnic Institute, Troy,N.Y.,U.S.A.
HUBBARD, H.W., R & D Associates, Santa Monica, Calif., U.S.A.
HUIZENGA, J.R., University of Rochester, N.Y., U.S.A.
HUMBLET, J., Liège Université, Belgium
IORI, I., Universita di Milano, Italy
JACKSON, H.E., Jr., Argonne National Laboratory, Illinois, U.S.A.
JOSHI, V., Duke University, Durham, N. Carolina, U.S.A.
KAHN, P.B., State University of New York at Stony Brook, N.Y.,U.S.A.
KAUSHAL, N.N., Rensselaer Polytechnic Inst., Troy, N. Y., U.S.A.
KAPOOR, S.S., Bhabha Atomic Research Centre, Bombay, India
KEDDY, R.J., University of Witwatersrand, Johannesburg, So. Africa
KEYWORTH, G.A., Los Alamos Scientific Laboratory, N.Mexico, U.S.A.
KINSEY, R.R., Brookhaven National Laboratory, L.I., N.Y., U.S.A.
KIROUAC, G., Knolls Atomic Power Lab., Schenectady, N.Y., U.S.A.
KLEMA, E.D., Tufts University, Medford, Massachusetts, U.S.A.
KOSHEL, R.D., Ohio University, Athens, Ohio, U.S.A.
KRAPPE, H.J., Hahn-Meitner Institute, Berlin, Germany
KRIEGER, T.J., Brookhaven National Laboratory, L.I., N.Y., U.S.A.
KRISHNAIAH, P.R., Aero-Space Research Laboratory, W.P.A.F.B.,
                                        Dayton, Ohio, U.S.A.
LANE, A.M., A.E.R.E., Harwell, Berkshire, England
LE BEYEC, Y., Laboratoire de Chimie Nucléaire at Orsay, France
LE COUTEUR, K.J., Australian National University, Canberra,Australia
LEWIS, F.H., Jr., Lawrence Radiation Lab., Livermore, Calif., U.S.A.
LEUNG, T., Aero-Space Research Lab., WPAFB, Dayton, Ohio, U.S.A.
LIOU, H.I., Columbia University, New York, N.Y., U.S.A.
LOBOV, G., Inst. for Theoretical & Experimental Physics, Moscow,
                                                     U.S.S.R.
LONE, M.A., Chalk River Nuclear Laboratory, Ontario, Canada
LUBITZ, C.R., Knolls Atomic Power Lab., Schenectady, N.Y., U.S.A.
LUSHCHIKOV, V., Joint Institute for Nuclear Research, Moscow,U.S.S.R.
MacDONALD, W.M., University of Maryland, College Park, U.S.A.
MacFARLANE, R.D., Texas A&M University, College Station, U.S.A.
MacKELLER, A., University of Kentucky, Lexington, U.S.A.
MANI, G.S., Manchester University, England
MAHAUX, C., University of Liège, Belgium
MALIK, S.S., University of Rhode Island, Kingston, U.S.A.
MARTINOT, M., Commissariat a l'Énergie Atomique, Paris, France
MARUYAMA, M., Rutgers University, New Brunswick, N.J., U.S.A.
MATHUR, S.C., Lowell Technological Institute, Mass., U.S.A.
MAURENZIG, P., University at Firenze, Italy
McCROSSON, F.J., DuPont Company, Savanah River Lab., S.C., U.S.A.
MEDICUS, H.A., Rensselaer Polytechnic Inst., Troy, N.Y., U.S.A.
MEHTA, M.L., CEN de Saclay, Gif-sur-Yvette, France
MEKJIAN, A.Z., Rutgers University, New Brunswick, N.J., U.S.A.

MELKONIAN, E., Columbia University, New York, N.Y., U.S.A.
MELLO, P.A., Instituto de Fisica, Mexico City, Mexico
MICHAUDON, A., CEN de Saclay, Gif-sur-Yvette, France
MILLER, J.M., Columbia University, New York, N.Y., U.S.A.
MITCHELL, G.E., North Carolina State University, Raleigh, U.S.A.
MOLDAUER, P.A., Argonne National Laboratory, Illinois, U.S.A.
MONAHAN, J.E., Argonne National Laboratory, Illinois, U.S.A.
MOORE, M., Los Alamos Scientific Laboratory, N. Mexico, U.S.A.
MOREH, R., Nuclear Research Centre-Negev, Beer Sheva, Israel
MORETTO, L., Lawrence Radiation Lab., Livermore, Calif., U.S.A.
MOSES, J.D., Duke University, Durham, N. Carolina, U.S.A.
MUGHABGHAB, S.F., Brookhaven National Laboratory, L.I., N.Y., U.S.A.
MUSGROVE, A.R., A.A.E.C. Research Establishment, Sutherland,
                                                              Australia
NAGARAJAN, M.A., Université de Liège, Belgium
NAMENSON, A.I., U.S. Naval Research Lab., Washington, D.C., U.S.A.
NEWSON, H.W., Duke University, Durham, N. Carolina, U.S.A.
NEWSTEAD, C.M., CEN de Saclay, Gif-sur-Yvette, France
NORDHEIM, L.W., Gulf General Atomic Co., San Diego,Calif., U.S.A.
OLSON, W.H., McGill University, Montreal, Canada
PARIKH, J.C., Physical Research Lab at Ahmedabad, Gujerat, India
PATRICK, B.H., A.E.R.E. at Harwell, Berkshire, England
PAULSEN, A., CBNM at Geel, Belgium
PEREZ, R.B., Oak Ridge National Laboratory, Tenn., U.S.A.
PINELLI, T., Universita di Pavia, Italy
PIRAGINO, G., Istituto di Fisica at Torino, Italy
POLIKANOV, S.M., Joint Institute for Nuclear Research, Moscow,USSR
PORILE, N.T., Purdue University, Lafayette, Indiana, U.S.A.
POSTMA, H., Natuurkundig Laboratorium at Groningen, The Netherlands
PREISS, I.L., Rensselaer Polytechnic Institute, Troy,N.Y., U.S.A.
RAHN. F., Columbia University, New York, N. Y., U.S.A.
RAINWATER, J., Columbia University, New York, N.Y., U.S.A.
RAMAVATARAM, K., Université Laval, Quebec, Canada
RAMAVATARAM, S., Université Laval, Quebec, Canada
RATCLIFF, K.F., State University of New York at Albany,N.Y.,U.S.A.
REYNOLDS, J.T., Knolls Atomic Power Lab., Schenectady, N.Y., U.S.A.
RICHTER, A., Max-Planck-Institut, Heidelberg, Germany
RIMAWI, K., Jordan University, Jordan
RODNEY, W.S., National Science Foundation, Washington, D.C., U.S.A.
ROGOSA, G., U.S. Atomic Energy Commission, Washington, D.C., U.S.A.
ROSLER, H., Oak Ridge National Laboratory, Tenn., U.S.A.
ROSENZWEIG, N., S.U.N.Y. at Albany and Argonne National Lab.,U.S.A.
RUDDY, F., State University of New York at Stony Brook,N.Y.,U.S.A.
SAXON, D.S., University of California at Los Angeles, U.S.A.
SCHENTER, R., Wadco Corporation, Richland, Washington, U.S.A.
SCHRACK, R.A., National Bureau of Standards, Washington, D.C.,U.S.A.
SCHOLZ-LI, A., S.U.N.Y. at Albany, N.Y. and R.P.I., Troy,N.Y.,U.S.A.
SEMLER, T.T., N.A.S.A. at Cleveland, Ohio
SEVGEN, A., Max-Planck-Institut at Heidelberg, Germany

SHELDON, E., Lowell Technological Institute, Mass., U.S.A.
SINCLAIR, D., Queens University, Kingston, Ontario, Canada
SINGH, U.N., State University of New York at Albany, N.Y., U.S.A.
SLAGOWITZ, M.B., Columbia University, New York, N.Y., U.S.A.
SODNOM, N., Joint Institute for Nuclear Research, Moscow,U.S.S.R.
SPERBER, D., Rensselaer Polytechnic Inst., Troy, N.Y., U.S.A.
STEHN, J.R., Brookhaven National Laboratory, L.I., N.Y., U.S.A.
SPRINZAK, A., Purdue University, Lafayette, Indiana, U.S.A.
SUMBAEV, O.I., Joffe Physical & Technical Inst.,Leningrad, U.S.S.R.
TAKEKOSHI, E., Japan Atomic Energy Research Inst., Tokai, Japan
TEMMER, G.M., Rutgers University, New Brunswick, N.J., U.S.A.
TRIPATHI, K.C., Atomic Energy Research Institute, Seoul, Korea
TSUKADA, K., Japan Atomic Energy Research Inst., Tokai, Japan
TURINSKY, P.J., Rensselaer Polytechnic Inst., Troy, N.Y., U.S.A.
UHL, M., Institut für Radium Kernphysik, Wien, Austria
UPPULURI, V.R.R., Oak Ridge National Lab., Tenn., U.S.A.
VAZ, L.C., University of Rochester, N.Y., U.S.A.
VONACH, H., Technische Universität München, Germany
WAGNER, R., University of Basel, Switzerland
WASSON. O., Brookhaven National Laboratory, L.I., N.Y., U.S.A.
WATSON, J.W., Knolls Atomic Power Lab., Schenectady, N.Y., U.S.A.
WENESER, J., Brookhaven National Laboratory, L.I., N.Y., U.S.A.
WERY, M., Centre de Recherches Nucléaires, Strasbourg, France
WIGNER, E., Princeton University, N. Jersey, U.S.A.
WILLIAMS, F., University of Rochester, N.Y., U.S.A.
WONG, S.S., University of Toronto, Ontario, Canada
WURM, P., Max-Planck-Institut at Heidelberg, Germany
WYNCHANK, S., Brooklyn College, N. Y., U.S.A.
YOSHIDA, S., Rutgers University, New Brunswick, N.J., U.S.A.
YU-WEN, Y., Purdue University, Lafayette, Indiana, U.S.A.
ZEBELMAN, A.M., Columbia University, New York, N.Y., U.S.A.

CONTENTS

I

INTRODUCTORY SESSION

II

ANALYSIS OF RESONANCE REACTION EXPERIMENTS

III

STATISTICS OF RESONANCE PARAMETERS (PART I)

IV

STATISTICS OF RESONANCE PARAMETERS (PART II)

V

AVERAGE RESONANCE PARAMETERS

VI

NUCLEAR LEVEL DENSITY

VII

DECAY OF THE COMPOUND NUCLEUS

VIII

LIMITATIONS OF THE COMPOUND NUCLEUS

XI

CONCLUSION OF THE CONFERENCE

Round Table Discussion of Selected Topics

STATISTICAL PROPERTIES
OF NUCLEI

BANQUET ADDRESS

Ernest L. Boyer, Chancellor

State University of New York

It is both a privilege and a pleasure for me to be with you
this evening and to bring you greetings on behalf of the State
University of New York. The University is indeed honored to
have been permitted to serve as host for this distinguished
gathering of international scholars and scientists. For us it is
a particularly happy coincidence that the conference is being
held on our Albany campus, which this year established a nuclear
research laboratory, housing the State University's second
nuclear accelerator.

With your indulgence, I should like to say a word or two
about the State University of New York. Since I have been
Chancellor for a little less than a year, I know you will not
think it immodest of me to speak with pride of the University's
achievements. In doing so, I am really paying tribute to the
dedication and superb vision of my predecessors and to all those
who, over the years, have given so unstintingly of their energy
and talents in the service of this great institution.

We are a very young University, having come into being only
twenty-three years ago. Some of our campuses, however, predate
the establishment of the University itself. Indeed, the State
University at Albany, created in 1844 as a teachers college, has
the distinction of being our oldest institution. Along with
thirty-one other State supported colleges and technical institutes,
it became part of the State University of New York in 1948. At
that time, the total enrollment at all of our campuses combined
was approximately 28,300. Today the University enrolls 321,000
students on sixty-nine campuses spread over 48,000 square miles
throughout the State of New York.

Our newest institution--Empire State College--is a "college

without a campus," and will accept its first students next month. It is designed to meet the needs of those students who cannot or should not reside full time on a conventional campus. This college will add one more component to our already large network of public higher education.

While the University's growth has been dramatic, size alone, of course, is not a hallmark of success, nor is it the criterion on which we should be judged. Our goals, rather have been to develop standards of excellence in every aspect of the University's operations, to provide a diversity of educational opportunity to meet the needs of the society we serve, and to work unceasingly toward the elimination of any barriers to higher education which may prevent our young people from realizing their full potential as human beings.

To be sure, these are ambitious goals and their attainment is not without difficulty. But to settle for less would be a betrayal of the faith and trust the citizens of New York have placed in us. Not all of our goals have been achieved, of course. We have had our share of problems. We have experienced periods of anguish, campus disruptions, budget cutbacks. But a new decade has begun, and we must now reevaluate our priorities and explore new ways to carry on our educational work to match the changing times. The State University of New York does not stand alone in this need for reappraisal. I am sure it is a concern shared by all universities throughout the world today.

In all of this, the continuing importance of scientific research cannot be overemphasized. Despite the economic and financial crises we face, at this critical time in history we need more--not less--creative scientific endeavor.

We hear a good deal these days about the cause and effect relationship between scientific discoveries and world problems. Our technological advances have been breathtaking. Science has placed at our disposal marvelous inventions to aid us in understanding nature and the world around us, to alleviate our suffering, to lessen our drudgery, to help us lead better lives. And what have we done with them? We need only look to the field of ecology to find the answer. We have polluted the air we breathe; we have contaminated our rivers, lakes and streams; we have wantonly destroyed our natural resources--all of the things that are essential to our survival on this earth.

And so, we must recognize the deep, interlocking relationship between science and human progress. We must call upon science now to teach us how to deal intelligently with the problems we ourselves have created. We must learn to live as civilized human beings within the restraints of nature--restraints we can no longer ignore with impunity.

Within this context, I want to emphasize the special importance of nuclear research. President Nixon has said that clean energy is one of America's greatest needs. And nuclear energy is still our greatest hope as a reliable, enduring source of clean energy. There are deadly diseases still unconquered, and we have placed our faith in biomedical and radiation studies in the search for a breakthrough in this area.

Low energy physics, as you well know, stands at the juncture of pure and applied science. It has a tremendous capacity to help both our environmental and world health problems. But applied science cannot make any significant progress without the basic data on such essential matters as the behavior of the nucleus itself and its interaction with other particle matter in the universe.

More than ever before, there is a vital need for nuclear research to focus on the humane applications of science. To do so is not only in the best interests of the scientific community itself, but it is vital to the future of the human race.

Further, I must tell you how immensely gratifying it has been for me to observe the spirit of international cooperation and goodwill that has been displayed throughout this conference. It is my firm conviction that our hopes for human progress--indeed our very survival--rest in the worldwide community of learning. Increasingly, the dilemmas of our time are shared by people the world over. They leap over oceans and cross frontiers, oblivious to such matters as visas or customs inspections.

For example, there is the relentless, desperate struggle for food, water, shelter, and at least a minimum standard of health and medical care . . . there is the soaring population growth in countries already barely able to subsist . . . there is the harsh fact of prejudice and discrimination based on ethnic, racial, or religious differences. And, most disheartening of all, there are the stubborn streaks of poison deep within man's own nature which continue to erupt into conflict, aggression, and war.

In an age marked not only by shrinking travel times but also a persistent armaments race, these critical problems are, in the most immediate and literal sense, the concern of all of us. We are, first and foremost, human beings, dependent upon one another, struggling together, living and dying, on a single globe.

"No man is an island, entire of itself . . . Any man's death diminishes me, because I am involved in mankind." This inter-connectedness of all human life is even more true today than it was when John Donne penned his famous lines almost two hundred years ago. And we are only beginning to adjust--emotionally, politically, and educationally--to this awesome fact.

In my comments about the worldwide community of scholarship, it was not my intention to exaggerate its present strength. It is still a fragile fabric, too often strained and torn by the stresses and conflicts which divide our world. I recognize, too, that there are many people today--even within the academic community--who seem to be turning their backs on international problems and international cooperation.

It is true that each university, like each individual, has
its own particular heritage, its special accomplishments, its
unique contributions to make. I would not alter this uniqueness
even if I could, for I cherish institutional diversity just as I
cherish the differences in people.

Nevertheless, the fact remains that the community of learning
does exist. Year by year it is gaining in strength and self-
awareness, and to me it represents a hopeful sign on an often
dark and menacing horizon. The world's universities are united
by a common history, a shared tradition. And scientists and
scholars everywhere are inextricably bound together in a relentless search for truth, a desire to know, to cut through the
stifling miasma of ignorance that dims man's vision and impedes
his progress.

But I would not have you accuse me of being another Dr.
Pangloss. I do not suggest that higher education possesses the
panaceas for all of the world's ills, or the magic formulas that
will instantly resolve these urgent issues. Nor can I accept
without reservation H.G. Wells' statement that human history has
become more and more a race between education and catastrophe.
In spite of education, the threat of catasrophe is even more
ominous than it was in Wells' day--and we are no longer quite so
certain that education will save us.

I do believe, however, that the university--with its tradition
of research and its willingness to explore the mysteries of the
universe, to confront the difficult questions--is the institution
to which society has the right to turn with hope and expectancy
as it grapples with the menaces that press in from all sides. I
believe, too, in the importance of the world of scholarship and
the community of learning as a path to peace and progress.

These, then, are some of the reasons why I welcome you with
a special joy. Though widely separated geographically, we are
co-workers in the same great enterprise. We are united by a
common heritage; we share common concerns. Let us, together,
move ahead with hope and courage toward our common goals.

I thank you for joining us here at the State University of
New York, and I wish you well in all your future endeavors.

INTRODUCTORY SESSION

Chairman   - J. B. Garg, State University of New York
                       at Albany
Scientific Secretary - A. Li-Scholz, State University of New York
                       at Albany

WELCOME ADDRESS

LOUIS T. BENEZET

President

State University of New York at Albany

My greetings from the State University of New York at Albany
come to you first as scientists of the world. There are many ways
we might caption this unusually distinguished group; but "scientists"
will surely cover it.

As students in the classical high schools used to hear, Julius
Caesar was once faced with a mutiny by his legionnaires who were
tired of the long wars far from home. They wanted their pay and
dismissal. "I will give it to them," said Caesar coldly. The men
grew increasingly uneasy. When it came time to part, they begged
him for some final word from their leader. Caesar mounted upon a
chariot, faced the departing throng and spoke to them one loud,
biting word. "Citizens!" he said. The men broke down, "We are
not citizens," they cried; "we are professional soldiers, and we
will fight for you to the last."

My well-known story from the past shows the danger of trying
to caption any group. Yet at this gathering I can think of no
higher praise than to address you as Scientists from one who himself
is not a scientist but who is commissioned to coordinate their work
and to try to see to their needs.

Currently we are in a trough of public understanding about
the role of basic science in the future of society. No one has
stated the problem better than Dr. Philip Handler of the National
Academy of Science in a recent article. There is some likelihood
that the reports of this international conference on statistical
properties of nuclei may give to the general public the impression
that it is one of those remote abstract exercises in which a university engages at the expense of the suffering taxpayer.

Support for science, especially basic science such as nuclear
physics, has taken a drop in the public's priorities ever since the
excitement over the first orbiting satellite brought us the National

3

Defense Education Act in 1958; it has become more difficult to
persuade Congress to support funds for basic research. The growth
in the budget of the National Science Foundation has been heavily
in the direction of research applied to immediate social needs:
the rescue of our environment, the reversal of decay in our cities,
the growing threat of civil strife from our racial divisions. No
one could fault the urgency of such missions. As one university
president, I have pressed for both curriculum and research directed
to such ends.

        Yet I think I have learned enough from my colleagues over the
years in the basic arts and sciences to become convinced that they
too speak to the requirements of the human race. It is only a
question of the time span ahead.

        I don't happen to believe that the most legitimate research in
science is that for which no one can possibly find any practical
use. Nor do I think history has proved that statement to be correct.
What comes out of this conference in the next five days could affect
the course of technology and humanity itself five, ten, or twenty-
five years from now.

        While the scientist is motivated by legitimate ambitions of
his own to push out the frontiers of knowledge wherever it may
lead, and even though James Watson in The Double Helix told the
story of the ego involvement that plays its part, the true scientist
is a man with broad conscience about the human condition. The
Association of Atomic Scientists under the leadership of men like
our own SUNYA colleague, Dr. Eugene Rabinowitch, has testified to
this conscience.

        It is not only correct, therefore; it is proper for me to wish
your conference success in the longest reaches of that term. The
eminence of its participants makes it certain that you will have
that success.

        More informally, let me welcome you to this new university
campus. We are going into our tenth year as a University Center
and into our sixth as occupants of this particular location. The
growth in our own scientific activities is typical of what has
happened throughout the State University of New York of which we
are one of four University Centers. Admittedly we are quite modest
by the standards of great universities. But we are on our way.
The Dynamitron accelerator which will be dedicated this afternoon
(and which I suspect from the current status of State revenues may
be our last major item of scientific research equipment for a while)
is the most recent evidence of our commitment to the march of knowledge.

        That to be sure is not all we are here for. We are a teaching
university. We enroll over 14,000 students in over 50 curricula.
We are here to educate not only future scientists but the enlight-
ened citizen of the future, the boy and girl from disadvantaged
educational background, and the adult who needs to bring his learning
up to date from ten or even forty years back. We engage in public
service; 800 of our students last spring were busy in volunteer

community service projects in welfare and humanitarian agencies throughout the Albany region. We train professionals through graduate schools in Public Affairs, Social Welfare, Business Administration, Education, Library Science, Criminal Justice, and of course, the arts and sciences. In undergraduate teaching we are putting together course sequences in the awareness and control of environmental problems. We are seeking through courses of solid content to give identity to the cultural backgrounds and historic contributions of our chief ethnic groups represented on campus through courses in Afro-American Studies, Judaic Studies, and now Puerto Rican Studies. We are coming to recognize the study content of what it means to be a woman in a predominantly male culture.

The University sponsors, as you would expect, a round of activities in public lectures, topical conferences, concerts, plays, and art exhibits.

In a word we are, I trust, on our way to serving as a center of learning with multiple missions in the tradition of the great mid-Western State Universities from which some of you come and which most of you know favorably in one context or another.

May I close with a personal welcome for you to enjoy, insofar as your schedule makes possible, the natural beauties of this region. Your Lake George excursion will take you to a place which still boasts some of the most uncontaminated fresh water in the United States as well as a beautiful setting that will help restore your soul if your soul should be in need of restoring by that time in the program.

The Albany region is one of the oldest settlements in America and including, of course, those of the American Indians who have claims to make of their own. Our permanent installations here go back 300 years which even by European standards is longer than yesterday. In that spirit may I conclude with a letter addressed by Governor Rockefeller.

Dear Dr. Garg:

It affords me much pleasure to send cordial greetings to all present at the International Conference on the Statistical Properties of Nuclei.

I am also happy to extend a warm welcome to all distinguished delegates from all parts of the world to this important conference.

The members of your organization merit wide appreciation for their contribution to this invaluable work.

Sincerely,

/s/   Nelson A. Rockefeller

Our best wishes then go to each and all of you. We shall look forward with high anticipation to the proceedings of this conference.

My thanks go to Professor J.B. Garg for his organizing genius and to his committee who have made this distinguished gathering an actuality at the State University at Albany in 1971.

INTRODUCTORY TALK

E. Wigner

Princeton University

Princeton, New Jersey

1. Introduction and Specification of our Subject

I have worked, in the course of time, on sufficiently many
books so that I should have realized that the Introduction is the
part of the book which one writes last. If there will be any
merit in my Introductory Talk today, this will be due mainly to
the help which I received from the committee which organized this
symposium and from the participants who submitted papers - papers
short enough so that I could read them. Incidentally, their rather
large number indicates the magnitude of the interest in our subject, and even a quick perusal of the papers submitted shows the
variety of the problems tackled.

Perhaps the first thing we should admit is that our subject is
not precisely defined. In statistical mechanics, the subject of
which _is_ clearly defined, we are interested in the time averaged
properties of systems large or small, which have a definite energy
and, parenthetically, which are at rest and not in rotation. The
quasi-ergodic theorem, if valid, assures us that these time
averaged properties depend only on the energy and are independent
of the other initial conditions. Thus the problem is quite
clearly defined. On the other hand, if we wish to specify the
subject of Statistical Properties of Nuclei, we must say that we
consider situations in which we are not interested in as detailed
a picture of the nucleus as one is generally interested in physics
but tries to find properties and rules which are reasonably simple,
and we believe very interesting, and which are very general,
shared by most all nuclei under appropriate conditions. The ap-

propriate condition is, in practice, principally a rather high
energy even though, as long as we remain nuclear physicists, we
do not go to extremes in this regard, and stay away from energy
regions in which the creation of the particle physicists's
particles become an important process.

To how much detail our interest extends is, of course, a
question which different physicists will answer differently. The
analogue states give herefor an example. These are members of
an isotopic spin multiplet the extreme member of which, that is
the member with the largest neutron to proton ratio, is either the
normal state, or a state of low excitation, of the nucleus which
has one more neutron, and one less proton, than the nucleus with
which we are concerned. The member of the isotopic spin multiplet in which we are interested has, however, a rather high
excitation energy – of the order of 9 MeV in the region of Fe.
Since the state in question has a wave function resembling that of
a low energy state, it has high transition probabilities to the
states which truly have low energies. The point which I am trying
to make is that one may be interested in the explanation of these
high transition probabilities, or not be interested in it and consider such a high transition probability as part of the fluctuations of the transition probabilities. In the latter case, the
excessive transition probability is a subject of the statistical
theory of nuclei, in the former case it is not – for most of us it
is not. However, if we go to even heavier nuclei, the analogue
state is not a single state but is dissolved into a reasonably
large number of states nearby, the isotopic spin of these states
having been before the dissolution of that state, that of the
normal state. In such a case, there is a group of states, rather
than only one, with rather large transition probabilities to low
lying states and one is more inclined to consider these large
transition probabilities as fluctuations and hence subjects of the
statistical theory. Instead of the analogue states, I could have
quoted Feshbach's doorway states which, in heavy nuclei, create
a situation qualitatively similar to that of analogue states. Both
examples illustrate the fuzziness of the boundary of the statistical theory – this boundary depends on our personal interest for
details.

2.   The Statistical Theories' Ensembles of Hamiltonians

Some of the very interesting laws of statistical mechanics can
be derived from the simple postulate that the equations of motion
have a Hamiltonian form. The entropy theorem and the equipartition theorem are in this category. A great deal of other work in
statistical mechanics is based on a reasonably detailed knowledge
of the Hamiltonian which is, in most practical instances, known.

There are, to my knowledge, no theorems in the statistical theory
of nuclei which would have as general a basis as the entropy or
equipartition theorems and what is worse, we do not know the
nuclear Hamiltonian. Further, some relevant properties of the
Hamiltonian are quite complicated. The spectrum of the Hamiltonian has a lower bound but extends, on the positive side, to
infinity. Near the lower bound, the spectrum is discrete but
there is a threshold at which a continuous spectrum sets in. The
character, that is multiplicity, of the continuous spectrum changes,
however, at every threshold drastically. These are properties of
the Hamiltonian which we know; as was mentioned before, we do not
know its exact form. It is not surprising, therefore, that one of
the chief quandaries of the statistical theory concerns the properties of the Hamiltonian to be made use of.

The Hamiltonian is, of course, a matrix in Hilbert space,
and the most natural set of properties to be made use of are
those shared by practically all matrices, or rather self-adjoint
matrices, in Hilbert space. This then leads to the concept of
ensembles of self adjoint, or of real symmetric, matrices in
Hilbert space that is a definition of the measure for such matrices
in Hilbert space so that the concept of "vast majority of all
self adjoint matrices" or of " practically all self adjoint
matrices" be mathematically defined.

The definition of such a measure seems to be strongly facilitated by the fact that, once we restrict ourselves to a single
angular momentum and zero spatial momentum, there seems to be no
coordinate system in Hilbert space which plays a preferred role,
except the one in which the Hamiltonian is diagonal. Since we
do not wish to define the measure of matrices with respect to that
coordinate system, it is natural to demand that the measure be
invariant with respect to unitary transformations. This leads to
a measure which can be an arbitrary function of the invariants of
the matrix, multiplied with the differentials of the independent
components of the matrix elements. If one further demands, as
has been done in their well-known paper by Porter and Rosenzweig,
that the probabilities of the independent components of the
matrix elements be independent of each other, one obtains, essentially, the Wishart distribution. More precisely, the ensemble obtained is such that the number of matrices within unit interval of
the independent components of the matrix elements M_{ik} is proportional to

$$P = \exp(\alpha \Sigma M_{ii} - \beta |M_{ik}|^2) \qquad (1)$$

α and β being arbitrary constants. (See Fig. 1)

The trouble with this distribution is that practically none
of its matrices has characteristics similar to those observed for
actual Hamiltonians. In particular, and I find this most decisive,
the density of the characteristic values of most matrices, as
function of energy has, in the neighborhood of the lower bound
(which can be arbitrarily adjusted by a proper choice of α and β)
a negative second derivative, practically right from the start.
This has been most conclusively demonstrated by B. Bronk. The
actual density, of course, has a positive second derivative with
respect to energy. Since the ensemble with the density (1),
which is essentially indentical with the Wishart ensemble (long
known to the mathematicians) does not give the actual density distribution of the characteristic values, it is natural to question
all other consequences derived theoretically by means thereof.
This has been done most eloquently by Uhlenbeck. As a result,
Dyson who, along with Gaudin-Mehta, developed most consequences of
the Wishart model,has recently decisively turned away from it.

The aforementioned facts have been known for a long time
but attention has been focussed on them only in recent years, but
in recent years attention has been focussed on them increasingly.
No true solution has been found to date but we do have at least
two proposals to which I wish to add another one. Let me, however,
mention first the others.

(a) <u>Return to the Independent Particle Model</u>. A statistical
theorem in nuclear physics was, of course, first postulated by
Weisskopf. However, just about the same time, Bethe proposed an
expression for the density of energy levels in nuclei, and this
expression was based on an extreme independent particle model, on
the model of the degenerate Fermi gas. As we know, this model
gives a fair picture of the density of levels, as function of
energy. In particular, except for the very lowest energy region,
the second derivative of the density with respect to the energy is
positive.

Recently, the independent particle model, or an approximation
thereto, was applied to the region in which we usually expect a
statistical theory to be valid. The original suggestion to do
this is due to French and Wong and a very interesting comparison
with experimental data is due to Bohigas and Flores. The principal
difference between the model considered by these authors, and the
Wishart model considered earlier, is that a great many matrix
elements of the new model's Hamiltonian are 0. These include the
matrix elements which connect independent particle states which
differ in more than two orbits. If I understood them right, French
and Wong mention a case in which less than 10 percent of all the

matrix elements of a 50 by 50 matrix are different from 0. In
the usual ensemble, that is the Wishart ensemble, the probability
of the vanishing, or even approximate vanishing, of so many matrix
elements is practically zero.

I hope it will not be resented if I say a few words about
the conclusions of the new theory which, incidentally, will be
reviewed in more detail by its authors. The first conclusion is
that the density of the levels, as function of the energy, is a
Gaussian rather than the semicircle characteristic of the Wishart
and related ensembles. This is, of course, very different from
the actual density but it does have the property that, at least at
its low energy end, its second derivative is positive. Hence, the
actual density distribution can be obtained by an infinite succession of such distributions, corresponding to configurations
with increasingly highly excited single particle states. The
second conclusion,due to Bohigas and Flores, is that the distance
of the second and further neighbors is subject to much larger
fluctuations than is the case for the matrices of the Wishart
ensemble. Bohigas and Flores find an experimental confirmation of
this greater fluctuation in the spectrum of Th.

The number of zeros in the French Hamiltonians is so great
that I was worried, at first, that the argument given by Von
Neumann and myself for the repulsion of the levels, that is for
zero probability of coincidence,becomes invalid. This is not the
case; I could prove that as long as there is any, no matter how
indirect, way to get from one diagonal element to every other,
going through non-vanishing off-diagonal elements, the changes
for the coincidence of two characteristic values vanish in the
same sense as if all non-diagonal elements had a chance for not
being zero.

Nevertheless, and in spite of my full realization of the
significance of the new picture, I find it difficult to fully
reconcile myself thereto. The independent particle picture has
surely very little validity when the spacing between the energy
levels is many thousands times smaller than the spacing of the
single particle levels. Furthermore, even though the probability
of zero spacing remains zero also in the new picture, the spacing
between two levels which belong to drastically different configurations will have a very good probability of being very small, and in
disagreement with the present experimental evidence. The experiment supports in the energy region to which we usually apply the
statistical model the distribution between the nearest neighbors
which applies for the Wishart ensemble. As to the discrepancy with
respect to the spacing distribution of second and more distant

neighbors which was found by Bohigas and Flores for Th, a recent
paper of Garg, Rainwater and Havens, if I understand it right,
finds the opposite kind of disagreement in the spectra of Ti, Fe
and Ni. I hope I will be corrected if I misunderstood that paper.

Surely, I do not want to claim unrestricted validity for the
Wishart model. As I emphasized before, it gives an absurd level
density. However, I believe that the French-Wong model also has
restricted validity - perhaps restricted to lower energies than we
like to think of when applying purely statistical considerations -
and I believe the proponents of the model realize its limitations
fully as well as I do.

Let me go over, next, to another model which may replace that
of Wishart, that is to another ensemble.

(b) Brownian Motion Toward Prescribed Level Density.
Unfortunately, this time I do not have to apologize to the origina-
tor of the idea for discussing his work: Dr. Dyson does not attend
our meeting. You may recall his 1962 paper in which he pictures
the matrix elements as particles in Brownian motion. Needless to
say, this is a picture only, but a very attractive one. Dyson's
particles ,representing the matrix elements, execute Brownian motion
on a line and the statistical distribution of their position on
that line represents the statistical distribution of the value of
the matrix element which corresponds to the Brownian particle in
question. The particles are subject to two agents. One of them
is an elastic force pulling each toward a zero point. The other
agent provides the irregular momenta which correspond, in the case
of the Brownian particles, to the irregular momenta imparted by the
thermal motion of the atoms of the medium in which they are sus-
pended. These irregular momenta prevent the Brownian particles from
settling down at the equilibrium point of the elastic force, that
is at zero. The different particles do not interact so that there
is no statistical correlation between their positions, i.e.,
between the different matrix elements. In equilibrium, the
Brownian particles will have a Gaussian (normal) distribution
about 0 - the ensemble of matrices is, in the stationary state, the
Wishart ensemble.

So far this is only a picture to illustrate the Wishart
ensemble. Next, however, Dyson derived an equation for the motion
of the characteristic values of the matrices of his ensemble, and
this was the truly surprising contribution. The characteristic
values also obey an equation - the Smoluchowski equation - which
allows us to picture them also as Brownian particles. The term
representing the irregular momenta provided by the temperature
motion of the medium is essentially the same as in the case of

particles representing the matrix elements. The restraining force,
which is in the case of the matrix elements   proportional to
their value, is, however, for the particles representing the
characteristic values augmented by another force. This has the
same form as the electrostatic repulsion between two point
charges in two-dimensional space, i.e., is inversely proportional
to the distance between them. This term provides the repulsion
between the characteristic values so that the distributions of
these are not independent of each other - as they cannot be
because of the well known effect of level repulsion. For the distribution of the characteristic values Dyson obtains, naturally,
Wishart's formula which does express the level repulsion. The
1962 paper of Dyson, though it contains many interesting results
of detail, and also a new derivation of Wishart's results, gives
the Wishart distributions as the distributions of the matrix
elements and of the characteristic values. It does consider the
approach to this distribution if the distribution was different
to begin with - the approach as it would take place if the
Brownian motion picture corresponded to reality.

Dyson's new article proposes a significant departure from the
Wishart picture. He does not refer any further to the matrix
elements as Brownian particles but only to the characteristic
values. These are subject, as in the article which I just quoted,
to the irregular temperature motion of the medium as well as to
the electrostatic-like repulsion by the other characteristic values.
However, the elastic force acting on them is replaced by the
requirement of a force causing their density to become a definite
function of the energy - presumably the experimentally observed
function.

Dyson mentions that proposals similar to his were made also
by Leff, Fox and Kahn, and Mehta. What has been accomplished?
It has been proved that one can find a Brownian motion model for
any level density expression. However, this is perhaps of only
incidental interest. Much more important is that Dyson has
adduced evidence that the local properties of the level distribution,
the probability of a definite spacing between neighbors, and
between second and further neighbors, is the same as for the
Wishart ensemble in the region of the same overall density. Dyson
does not claim to have proved this rigorously but he did make it
very plausible. In this conclusion, his theory is in less conflict
with the theory which we discussed before than with that of French and
Wong, and Bohigas and Flores.

Dyson does not define explicitly the matrix ensemble of which
his Brownian particles are characteristic values. The fact that
he imposes the requirement of a definite density of these char-

acteristic values as function of their magnitudes shows, however,
that the ensemble is not invariant under unitary or real orthogonal
transformations. This is quite reasonable: surely, the ensemble
of Hamiltonians need not be invariant under all transformations.
Nevertheless, it may be of some interest to consider ensembles
which are so invariant. If they are to give a density of
characteristic values different from the semicircle law, the matrix
elements çannot be independent from each other; there must be
correlations between them.

(c) <u>Invariant Ensembles of Positive Definite Hamiltonians.</u>
It has been mentioned before that the actual Hamiltonians all
have a lower but no upper bound. This suggests using positive
definite matrices for the ensemble of self adjoint matrices
characteristic of Hamiltonians since a lower bound, different
from zero, can be obtained from such matrices by adding to them the
unit matrix with a suitable coefficient. Such an addition does not
change the relevant properties of the matrices.

Up to this point the distinction between ensembles of real
symmetric matrices and of hermitean, that is in general complex,
matrices was not clearly brought out - the discussion could be held
sufficiently general to cover both cases. This will not be true in
the present instance and we shall concentrate on the real symmetric case. The simplest form for a positive definite hermitean
matrix is $m^\dagger m$ where m can be any matrix with, in general,
complex coefficients. There are two similarly simple forms for a
real symmetric matrix : $H = r^T r$ is one, r being an arbitrary real
matrix, and

$$H = m^\dagger m + m^\dagger m^* \tag{2}$$

the other, where m is again an arbitrary, in general complex,
matrix. It is tempting to use for r the usual, real, non-symmetric, Wishart ensemble, or if one adopts (2), for m the complex
again non-symmetric similar ensemble.

Since both of the aforementioned Wishart ensembles are invariant under orthogonal transformations, $r \to R^{-1}rR$ and $m \to R^{-1}mR$
where R is a real orthogonal matrix, this will be true also of
the two ensembles, $r^T r$ and (2). We, therefore, have to choose
between them. Now we know from Ginibre's work that the density
of the characteristic values of the Wishart ensemble of real
matrices r is finite at 0. One can conclude from this that the
density of the characteristic values of the ensemble of matrices
$r^T r$ is in fact infinite at 0 (A. T. James has considered such
ensembles). On the other hand, if the H of (2) has a characteristic value 0, one easily sees that its characteristic vector ψ
satisfies both equations

$$m\psi = 0 \text{ and } m^*\psi = 0 \tag{3}$$

so that it can be assumed to be real. This imposes a number of
conditions on m which increases with its dimensions so that, in
the limit of very many dimensional matrices, - and the Hamiltonian
is infinite dimensional - the density of the characteristic values
as function of the energy will be tangent to the energy axis in a
very high order. This means that many of the derivatives of the
density of characteristic values will be positive - the essential
condition which the original Wishart distribution failed to
satisfy.

It seems to me, therefore, that along with the suggestions of
French and Wong, and of Dyson, the properties of the ensemble of
matrices given by (2) may be worth exploring. This does not have
the physical basis which recommends the proposal of French and
Wong, it does not have the flexibility of Dyson's proposal even
though this could be built in. This would be, though, at the expense
of its mathematical simplicity which is the principal element that
renders it to me attractive.

Before closing this subject, let me recall what I said about
our general subject at the start, that it is not clearly defined.
It is quite possible that, from the point of view of one view
of our subject the first, from another the second, and possibly
from a third point of view the third ensemble will appear most
attractive.

3. Comparison of the Statistical Theories with Experiment

My discussion so far was very much the discussion of a theoretical physicist who is fascinated by the theoretical problems,
particularly if they have some mathematical attraction. I would
like, though, to say a few words about the comparison with experiment which, even we theoreticians must admit, is after all the
ultimate touchstone of our ideas.

As the program of our meeting indicates, our subject can be
divided, at least roughly, into two categories: the investigation
of the average properties, such as level density, average transition probability is the first and, for practical applications,
probably the more important subject. The other subject, which was
in the foreground of the preceding discussion because it appears
more interesting to the theoretician, concerns the statistical
distribution of these quantities, of the spacings between levels,
of the transition probabilities. It is interesting how different
our understanding of these subjects is.

Let me begin with the average properties. As to the level

densities, we do have a fair understanding from first principles, if not a complete one. Our Session 6 will be devoted to this subject. The situation is different with respect to transition probabilities. The average transition probability with respect to particle emission can easily be given in terms of what is called the strength function. The strength function, however, is the fraction of the states of the reaction's product nuclei which is present in the states of the compound nucleus within unit energy range. I cannot help but feel that it should be possible to estimate this fraction, at least in crude approximation, from first principles. However, no full scale effort in this direction is known to me. A similar remark could be made with respect to γ-ray emission though the problem appears more difficult in this case. In summary, our knowledge of the average partial width of the levels in the statistical region is less satisfactory from the theoretical than from the experimental point of view.

Let us now look at the distribution of spacings and of transition probabilities about their average values. As for the spacings they seem to agree rather well with the law first deduced by Gaudin and Mehta from the Wishart ensemble, which was, as mentioned before, generalized by Dyson. I mentioned, as exception, the Th spectrum, and there are other surprising situations. However, everyone who occasionally plays with cards knows that the truly surprising hand is the one which does not show any surprising feature. There may be one exception from the general agreement which I should mention. According to theory, levels with different total angular momenta J should show no correlations, there should be no level repulsion between them. In disagreement with this, Garg, in 1964, claimed a strong repulsion, and this claim was reaffirmed, even though in a much milder form, in some later publications. It is true that, according to Dyson, it should be very difficult to distinguish between a weak and a strong repulsion. However, the only source of weak repulsion that I could think of originates via the interaction with the electronic shells. This renders the J value of the nucleus itself an unprecise quantum number. If this is the reason for the repulsion, it should depend on the chemical structure and should, in particular, disappear in the gas phase, at least if the molecules of the gas contain only one atom of the element the spectrum of which is investigated and if the electronic state carries no angular momentum. It may be that there are some experimental data on this question with which I am unfamiliar. If not, some experimental work would be useful because the phenomenon, if it is real, is truly puzzling.

The area for the comparison between experiment and theory

which remains to be discussed concerns the distribution of the
magnitudes of the transition probabilities around their average.
For the neutron widths, and presumably for the widths of particle
emissions which lead to definite states of the product nuclei, the
simplest possible distribution, that proposed by Porter and
Thomas, seems to be, on the whole, well confirmed. There are some
exceptions which may be handled either by an improvement of the
statistical theory, or by searching for their causes as does the
doorway theory of Feshbach, or the isotopic spin theory. This
ambiguity of the borderlines of the statistical theory was discussed in the Introduction and there is no need to repeat what
was said there.

    Let me mention, instead, the very great progress that was made
in the treatment of reactions in which the state of the final
products is not unique. The fission process is a prime example for
this. Before the work of Bohr, Mottelson, and perhaps some others,
one could have thought, either, that the final state of the fission
process is uniquely defined, i.e., that there is only one fission
channel in configuration space. One also could have thought that
there are as many fission channels as are possible pairs of fission
products. What we have learned from Bohr, Mottelson and others is
that neither of these two extreme views is correct; the number of
fission channels is greater than one but very small. In the case
of fission phenomena, statistical considerations permit the interpretation of experimental data, to support and to direct the theory,
very effectively. I would not be a true theorist if I did not wish
that further experiments be performed in this area. It would be
interesting, it seems to me, to ascertain whether the different
fission channels lead to states in which the probabilities for
the different pairs of fission products are equal, i.e., whether
the channel functions differ principally in the phases, or both
the phases and amplitudes, of the components representing the
different fission products. A careful comparison of the abundances
of the different products, as furnished by the different resonances,
would give an answer.

    Let me stop here with the comparison of experiment and
theory. As I said, such comparisons are the lifeblood of all
physics but we all hope to learn about the contact between experiment and theory, as far as our subject is concerned, in the course
of our conference. Let me, instead, come to my last subject
which points, I am afraid, to a fundamental weakness of our
present statistical theory of the nucleus.

4. Hamiltonian or Collision Matrix?

    It was mentioned before that, at least above the binding

energy of any particle, the Hamiltonian's spectrum is continuous and
if we admit the interaction with the radiation field, this applies
in fact to the whole spectrum. One can ask, therefore, what the
energy spectrum of the compound nucleus really represents, what
is the Hamiltonian the properties of which we are trying to re-
capture with our ensembles. As far as I can see, there are
only two possible answers to this question.

The first answer is that we should not take the whole
question, and in fact the statistical theory, too seriously. It
applies in an intermediate energy region in which the level
structure is too complex to be interesting but not yet dense
enough for the resonances to overlap significantly. In this point
of view, the statistical theory is, by its very nature, an approximate theory.

The second answer appears to be that of R matrix theory. This
claims that one should use a Hamiltonian which is confined to a
finite region of configuration space (if one disregards the center
of mass coordinates), with boundary conditions on the wave function
which render the Hamiltonian self adjoint even if restricted to
that finite region. It will then have a point spectrum and the
statistical theory applies to that.

The difficulty with this interpretation is that it depends on
the "finite region" to which the R theorem is applied. If we change
that region, both the density of the characteristic values, and
also the statistics of their spacings, changes. If we increase
the region, the density of the characteristic values increases.
The distribution of the spacings, in terms of the average spacing,
remains approximately invariant, but only approximately. If one
goes to the limit of a very large internal region - this is the
name of the region to which one applies the R theorem - the
spacing becomes very nearly the average (very small) spacing
throughout and the fluctuations converge to zero. This is, perhaps, not surprising because, in this case, we look at the characteristic value distribution of the Hamiltonian as applied to
some very large, essentially empty, domain in which the nucleus
occupies a very small volume. There is a good deal to be added and
elaborated on this last statement but, on the whole, it seems to
me uncertain that this second point of view is much more attractive
than the first which simply admits that there is no precise formulation for the statistical theory.

The difficulties which we are having with the statistical
theory based on the Hamiltonian remind us that, when dealing with
the continuous spectrum - and we are, except for the normal state,

always in the continuous spectrum if we do not disregard the coupling with the electromagnetic field – the collision matrix provides the appropriate characterization of the situation. This
suggests that if we wish to extend the theory to regions of
higher energy, and if we wish to give it a more rigorous formulation, we should try to do this in terms of the collision matrix
rather than the Hamiltonian which we have to restrict, quite
artificially, so that it has a discrete spectrum. To formulate
the theory in terms of the collision matrix is, naturally,
relatively easy in the region of well separated resonances where
we can simply translate the present theory into the language of
the collision matrix using, perhaps, the Humblet-Rosenfeld
theory. The translation though, I fear, will not have the
simplicity of the present formulation, nor its natural character.
Also, I should admit <u>some</u> difficulties if we wish to be truly
precise and wish to take reactions such as $(\gamma, 2\gamma)$ and similar
processes into account. This may be, though, unnecessary. The
serious difficulty, at any rate, is to be expected in the higher
energy region where any new theory will have to stand on its own
feet. Thus a statistical theory in terms of the collision matrix,
if at all feasible, is a task for the future.

$$P = c \exp \left(\alpha \sum_i M_{ii} - \beta \sum_{ik} |M_{ik}|^2 \right)$$

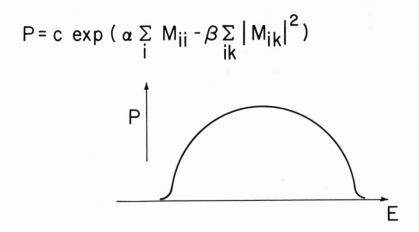

Fig. 1. Wishart ensemble.

DISCUSSION

KRISHNAIAH
     I wish to make a comment about the terminology "Wishart Dist-
ribution." Nuclear physicists call the distribution pxp matrix M
as Wishart Distribution where the elements of M are independently
and normally distributed with zero means and the variances of
diagonal elements are equal to 2 while the variances of the off
diagonal elements are equal to 1. The distribution of M is <u>not</u>
Wishart Distribution. The distribution of S=XX' is known as Wishart
Distribution where S is a pxn matrix whose columns are distributed
independently as multivariate normal with zero mean vector and covariance matrix Σ; here X' denotes the transpose of X. We note that
a random vector y is distributed as a multivariate normal with zero
mean vector and covariance matrix Σ if its distribution is given by:

$$f(y) = C \exp [-1/2y^1\Sigma^{-1}y]$$

where C is a normalizing constant and Σ^{-1} is the inverse of Σ. A
very brief summary of some of the literature on the distribution
problems connected with the eigenvalues of the matrix S is given
in the following paper: P.R. Krishnaiah and T.C. Chang. "On the
Exact Distributions of the Extreme Roots of the Wishart and Manova
Matrices." <u>Journal of Multivariate Analysis</u>, Vol.1, pp.108-117,1971.

MOLDAUER (Argonne)
     I just wanted to make the comment that one respect in which
all the matrix models differ from physics is that the matrix models
all deal with finite matrices, while physics deals with an infinite
matrix, and it seems to me that one must always, in interpreting
the results of the matrix model, be careful not to use those results
which specifically have to deal with the finiteness of the matrix.
I just wanted to make that clear.

WIGNER (Princeton)
     Dr. Moldauer is absolutely right. Except I am firmly convinced
that Dyson's Distribution could be made infinite, and so could be
this m†m matrix be made infinite. In that case, you start with an
infinite dimension, but nobody has done it right.

BLOCH (Saclay)
     I would like to mention that a way to derive distributions which
differ from (how shall we call it--Wigner Distributions if not Wishart)
in such a way that one obtains a level density which is not the semi-
circle law which was presented I think two years ago by Balian. I
don't have the exact reference now, but it was presented in the Kyoto
Conference on Statistical Mechanics. It was based on the use of
information theory. One states as a postulate that the distribution
should contain the minimum information--information being defined as
usual as ∫P log P where, it is consistent with some constraints.

It is very easy to show that if as constraints you take the values
of the averages of the matrix and its mean square, then you obtain
the Wigner Distribution. But, you could take different constraints.
In particular, you could take the level density as the constraint.
Then it is very easy to obtain the full distribution, and one can
see that it will also give rise to level repulsion as usual.

WIGNER
      Do I understand you right that this anticipates to a consider-
able extent the work which I attributed to Dyson?

BLOCH
      I think it has a very different starting point. It is based
on a rather arbitrary postulate that Dyson also used but it gives
a way to introduce a method for obtaining different level density
from just the semicircle law distribution.

WIGNER
      Is there any information available that the spacing distri-
bution is the same as calculated by Gaudin and Mehta?

BLOCH
      The spacing distribution itself is rather difficult to evaluate,
but the complete probability distribution P of $(E_1, E_2, \ldots E_n)$ is
very simple. It's very similar to the Wishart Distribution. It
involves the product of all the differences of eigenvalues two by
two multiplied by a function of E_1 times the same function of E_2
times the same function and so on, exactly as in the Wigner Distribution except that here it is not a Gausian function but an arbitrary
function which has to be determined by the level density.

WIGNER
      That is very much the same as Dyson's work, and I am sorry that
I did not mention Balian's work. I was not familiar with it. I am
still not familiar with it. Thank you very much.

FESHBACH   (M.I.T.)
      Just a comment about your very last paragraph of Professor
Wigner's talk. There is another procedure which was carried out
by a student of mine, Victor Newton. One takes a model of coupled
equations in the continuum--in other words, a scattering problem
determined by the solution of many Schroedinger equations coupled
together. In this model, one takes the coupling part of the
potential matrix as random and solves the resultant equation.
In this way, one simultaneously considers both the continuum and
the statistical hypothesis.

WIGNER
      That seems rather opposite to what French and Wong do who
put in many zeros. Thank you.

MEHTA   (Saclay)
    I just wanted to give the reference of Balian's work. It is
Nuovo Cimento B57, 183 (1968) which deals with the information
theory treatment of the random Hamiltonians.

WIGNER
    Thank you.

ROSENZWEIG   (SUNY/Albany and Argonne)
    First one comment,I would like to tell you, that during
the last few weeks Mehta analyzed more recent data than was
available to Bohigas and Flores to test the difference between
their model and the conventional theory. We hope that in
tomorrow's session, we will be able to show our results maybe
after or during Bohigas' talk. We find that the disagreement
between the experiment and the conventional theory is not great
at all. That was just a comment since you brought it up at this
time. The other things I would like to ask you: Are the ensembles
considered by French and Wong and by Bohigas and Flores--could
they conceivably be generated as a special case of Dyson's ensembles?

WIGNER
             I believe not, because in Dyson's ensembles,
you have a direct repulsion between any two levels, and I shouldn't
speak about it, but I think French should be here to answer these
questions.

ROSENZWEIG
    But Dyson is not here.

WIGNER
    But Dyson I can represent. In Dyson's ensemble, essentially
all the matrix elements are present and there are no zeros. It
is much too general for that.

ROSENZWEIG
    So the specific things that French and Wong find as consequences
could not be obtained.

WIGNER
    Not as consequences--as assumptions different from the
assumptions of Dyson. In Dyson's theory there is an electrostatic
repulsion between any two levels.

FRENCH (Rochester)
    Perhaps I should make something clear about the zeros in
matrices. It is indeed true that if one uses a picture or a
representation in space which is simple, a Hartree-Fock kind of
thing, of describing it in configurations. Then when you put in

two body interactions, you do get a fair number of zeros. I
don't think you get as many as has been suggested. But, that we
regard as not the real essence of the thing, that is done. The real
essence instead, is that constraints on all the matrix elements
are imposed by the condition that we have a two-body interaction.
Ane we have, for example, verified that if you take large matrices
and put them and think of them as being in block form and impose
only the zeros corresponding to the fact that you cannot move two
particles from here to there, that is much more likely to give
you a semicircular form than to have a Gaussian form. If now,
however, you impose a further condition that the matrix elements
have to satisfy a large number of constraints then the distribution becomes Gaussian. In other words, I don't think that one
should put too much stress on inserting zeros. That doesn't
really convert you from semicircular to Gaussian.

WIGNER
    But isn't it a necessary consequence of your theory that
there are those zeros. In fact, the higher up you go in energy
the more zeros you have because at the higher energy you have
many configurations which differ by more than two particles.

FRENCH
    Yes, indeed I feel it is a necessary consequence, but it is
not however, sufficient, I believe, to produce the Gaussian form--
necessary, but not sufficient.

WIGNER
    Thank you.

SURVEY OF RESONANCE REACTION THEORIES

H. FESHBACH

Massachusetts Institute of Technology

I.   Introduction

Resonances, will play a central role in the discussions of
this Conference. On the one hand we shall be concerned with the
extraction of the parameters describing a resonance, its energy,
its half width, etc., from experiment. On the other hand we shall
want to be able to interpret these results in terms of the structure
of the interacting nuclear systems--ultimately in terms of the wave
function of the system or (a lesser but physically as significant
a goal) for statistical parameters suitably related to that wave
function. Ancient as these problems (the description of a resonance,
the extraction of resonance parameters from experiment, their interpretation) are, their solution has remained the subject of a debate
which has been argued by some with fanatical fervor. One of my
objectives in today's talk is to take the heat out of this quarrel.
It is my thesis that each of the methods (although they may have had
some deficiencies in their original formulation) have been or can
be adapted by sufficiently clever men so as to become equivalently
valid. I will not therefore, be an advocate for any particular
formalism. I like my own [1,2,3,4] but all that means is that I am
more comfortable in using its constructs and concepts,not that it
is superior. The final expressions obtained for experimental observables are in fact very similar so that, as far as fitting the data
is concerned, the experimentalist or theorist can choose whichever
suits his fancy. In passing, it would however, be useful to have
programs which would translate between possible forms.
    Another part of the debate has involved the physical inter-
pretation of the parameters. It is my contention that there is
considerable danger of over interpretation and I would like briefly
to discuss this.

25

Still another debate has been concerned with statistical averaging. This will be discussed together with an about-to-be published work of Kerman, Kawai, and McVoy.

In all of these remarks I shall repeatedly refer to the methods used in analyzing data. As we shall see, the natural criteria employed automatically excludes a range of resonance parameter values and their averages which are a-priori possible. This simplifies the discussion since every singular example which can be conceived does not need to be discussed. On occasion I shall use the formalism my colleagues and I have developed. As I said earlier, this is because for me it provides a convenient framework. Others giving this talk would perhaps have used other starting points and formalisms convenient for them. But I am convinced that they would arrive at identical results.

II.   Simple Consequences of Unitarity

In looking at the history of resonance theory one is struck by the invariance with authors of the final form which is to be compared with the experiment. This might be referred to as "formalism invariance." If one looks at Bethe's pre-World War II description, the Wigner R-matrix, the Kapur-Peierls formalism, and those which have been advanced in the recent years, one finds certain common elements.

An important reason for the similarity lies in the requirements of unitarity. This will be discussed much more thoroughly by Professor Humblet later. We shall make only some simple comments here. Unitarity when applied to a single channel process requires that the S-matrix satisfy:

$$|S|^2 = 1 \quad \text{or} \quad S = e^{2i\delta}, \ \delta \text{ real} \qquad (1)$$

This condition is satisfied by the following form:

$$S = e^{2i\phi_P} \frac{1-iK}{1+iK} = S_P S_R \qquad (2)$$

where ϕ_P and S_P are smoothly varying functions of energy, and where K is a real function which contains the part of S, S_R, which fluctuates rapidly with energy. The words "smoothly" and "rapidly" are not quantitatively spelled out and so the split of eq. (2) is not completely defined. Let us come back to this important point later. From eq. (2) we can now form T; the transition matrix:

$$T = T_P + e^{2i\phi_P} \frac{K}{1+iK} \qquad T_P = -e^{i\phi_P} \sin\phi_P$$

$$\equiv T_P + S_P T_R \qquad\qquad T_R \equiv \frac{K}{1+iK} \qquad (3)$$

We see that the scattering amplitude consists additively of a
smoothly varying scattering amplitude T_p, and rapidly varying one.
The first of these is referred to as potential scattering, the
remaining term, the resonant, or fluctuation amplitude. Note that
the resonant contribution will be larger whenever the potential
scattering is larger, the effect being carried by the S_p factor.

If the experimental situation is such that it is possible to
decompose the scattering amplitude into a potential scattering term
and a single resonance, then:

$$K = \frac{1}{2} \frac{\gamma_0}{E-E_0} \qquad\qquad \gamma_0 > 0 \qquad (4)$$

and

$$T = T_p + e^{2i\phi_p} \left[\frac{\frac{1}{2}\gamma_0}{E-E_0 +\frac{i}{2}\gamma_0} \right] \qquad (5)$$

We could have obtained this form directly if we had required that
the scattering amplitude T consist of a smoothly varying amplitude,
a single resonance, and satisfy unitarity. Form (5) is then unique.
The parameter γ_0 varies slowly with the energy because of its
dependence on the potential scattering. This dependence, often
exhibited explicitly by means of penetrabilities, is most important
near thresholds or when the resonance is very broad.

Suppose there is more than one resonance of importance. This
is the so-called case of overlapping resonances [3]. Some time
ago, I derived a set of sum rules [1] which apply in this case.
It was A. Michaudon who pointed out that these follow from unitarity
requirements. Let us see how this works out. We write T_R in two
alternative forms

$$T_R = \frac{1}{2} \Sigma \frac{A_i}{E-\xi_i} \qquad (6)$$

the expansion in S-matrix "poles" ξ_i and A_i complex and

$$T_R = \frac{\frac{1}{2} \sum_i \frac{\gamma_i}{E-E_i}}{1 + \frac{i}{2} \sum_i \frac{\gamma_i}{E-E_i}} \qquad (7)$$

γ_i and E_i real. As in the case of the single resonance, these parameters are slowly energy dependent. The S-matrix expansion eq.(6)
does not manifestly satisfy unitarity. For n resonances there are
4n real constants in form (6), while in form (7) there are only 2n.

The relation between (6) and (7) will therefore give rise to 2n relations between the sets A_i and ξ_i. The fewer constants in eqs.(7) results from the required unitarity of S. Hence these 2n conditions are just consequences of unitarity. And in fact, after they are inserted in (6), one can verify that unitarity is satisfied. The conditions on A_i and ξ_i can be readily determined. The dependence of A_i, ξ_i on E does not affect the identities being obtained here. Note that as a consequence γ_i and E_i are energy dependent. Toward this end, expand both forms in a series in inverse powers of E:

From (6)

$$T_R = \frac{1}{2E} \Sigma A_i + \frac{1}{2E^2} \Sigma A_i \xi_i + \ldots$$

From (7)

$$T_R = \frac{1}{2E} \{ \Sigma\gamma_i + \frac{1}{E} [\Sigma\gamma_i E_i - \frac{i}{2} (\Sigma\gamma_i)^2] + \ldots \}$$

from which it follows that

$$\Sigma A_i = \Sigma\gamma_i \tag{8}$$

or

$$Im\Sigma A_i = 0$$
$$Re\Sigma A_i = \Sigma\gamma_i$$

and

$$\Sigma A_i \xi_i = \Sigma\gamma_i E_i - \frac{i}{2} (\Sigma\gamma_i)^2 \tag{9}$$

and so on

Another set of relations proceeds from the requirement that the poles of T_R occur at the same values of E for each form. Thus the roots of the equation

$$\Pi_j (E-E_j) + \frac{i}{2} \Sigma_i \gamma_i \Pi_{j\neq i} (E-E_j) = 0 \tag{10}$$

must be ξ_i. In (10) the coefficient of $-(E^{n-1})$ must be the sum of the roots. Hence

$$\Sigma_j (E_j - \frac{i}{2} \gamma_j) = \Sigma_i \xi_i \tag{11}$$

or

$$\Sigma E_j = Re \Sigma_i \xi_i$$
$$\frac{1}{2} \Sigma\gamma_j = - Im \Sigma \xi_i = \frac{1}{2} \Sigma A_i$$

and so on. Although the numerators and denominators of (6) are not related as in the Breit-Wigner formula, that relation does prevail on the average according to eq. (8) and eq.(11). Eq. (8),

(9), and (11) were obtained by dynamical consideration in references [1], [2], and [3]. The important point to be gained here [4], [5] is that the form of T is fixed by unitarity and does not depend upon the dynamic formalism or the theory behind the nuclear structure implied by the resonance. It is also important to realize that it is always possible to transform from the S-matrix to the K-matrix form in a number of ways. A particular method has been elaborated on by Adler and Adler [6]. These results have been used extensively by the Oak Ridge group (de Saussure, Perez, et. al.).

III. Potential Scattering

Although the form of (6) and (7) is fixed, there are still questions which must be decided before one can confront the experiment--namely, how many terms in the series for K should be taken-- and then there is the associated question, what to choose for T_p-- the potential scattering. If the early version of some of the formalisms is used, T_p is taken to be hard sphere scattering. It then follows that the series must be infinite. This can be seen most readily in two-body potential scattering. In that case, by Norman Levinson's theorem, the difference in the phase shift at zero energy and at infinite energy is given by $n_B \pi$ where n_B is the number of bound states. This condition is not satisfied by hard sphere scattering (e.g. $\delta_0 = -kR$) and can be compensated only by an infinite series of resonant-type terms.

But, and here another principle emerges, experimentally we are never concerned with infinite energies. The forms (6) and (7) are fitted to the experimental data over a finite energy range. The principle is an obvious one--namely, one must carefully examine the procedures actually employed in data analysis. Certain extreme possibilities, or criteria, which have a Cartesian a-priori rationality can in fact be discarded because they do not conform to the practicalities of the experimental situation or the manner in which the data is actually analyzed. We shall see several cases of this later. But to return to (6) and (7): it is clearly more appropriate to use a finite number of terms. One can add up the "tails" of the remaining terms and appropriately modify ϕ_p so it is no longer the hard sphere scattering phase shift. Or as occurs in some theories, ϕ_p is the actual solution of a two-body potential scattering problem; choosing the potential is then part of the fitting procedure. In that case, one has some guidance from the optical model potential. In any event, the ϕ_p one obtains is a function of the energy which satisfies Levinson's theorem. But then how many terms should be used in determining S_p or K? Here the criteria must surely be the minimum number which yields a χ^2 in comparison with experiment which is reasonable and, secondly, that there be a stability of γ_i and E_i against the subtraction or addition of a term in the series.

Let us discuss this in more detail. We require that S_p should vary slowly with energy while T_R should vary rapidly with energy. But this depends very much on how many terms are included in the series in (6) and (7) and the size of the energy interval ΔE over which the fit is made. If only a few terms are used or if the

energy interval is small, S_p may vary rapidly. The first of these,
too few terms, is obvious since if too few resonances are assumed,
the potential scattering will need to be rapidly energy dependent
in order to fit the data. If ΔE is small, the tails of the nearby
resonances will again cause S_p to vary rapidly. When ΔE and n are
too small, the fitting of the cross section will be unstable against
variations in ΔE and n.

How does one determine whether S_p is varying too rapidly? I
will give a result which is appropriate for the projection operator
formalism but I am certain that equivalent criteria can be developed
using other formalisms. In this method, one develops a potential
V which would be used to compute S_p. The parameters of the potential
which would have say the simple Wood–Saxon form and the resonance
parameters are then fitted to the data. The potential determined
in this way would be energy dependent. If, for example, a resonance
has been omitted in T_R then the potential V_E will have a pole term
or will violate the causality condition [1]

$$\frac{\partial \, \mathrm{Re} V_E}{\partial E} \leq 0$$

If on the other hand, it is only that ΔE is small and the tails of
nearby resonances make significant contributions, one would find
that V_E varies more rapidly with energy than typical optical model
potentials. If now ΔE is increased, the energy dependence of V_E
should be reduced to more reasonable values and stability of the
fit to the experimental data achieved.

One final remark on this subject: The penetrabilities are
employed to specify the major energy dependence of the widths Γ,
the width being generally written as the penetrability times a
reduced width. These penetrability factors are particularly
important near zero energy or if the resonance is unusually broad
so that they vary over the width of the resonance. It is important
to note that when the hard sphere phases are modified, the penetrabilities must also be modified. In the potential description
given above the wave function for that potential can be used to
obtain the new penetrabilities.

IV.   Doorway States and Intermediate Structure; Asymmetry

Let us now turn to intermediate structure and doorway states.
These can again be introduced in a formalism independent way. We
can rewrite eq.(2) for S as follows

$$S = S_p S_D S_r \quad ; \quad S_R = S_D S_r \tag{12}$$

where S_p is the S-matrix for potential scattering, S_D for the
resonant doorway state scattering and S_r describes the fine structure.
Equation (3) now takes on the form

$$T = T_p + S_p T_D + S_p S_D T_R$$

As this indicates, three sorts of scattering amplitudes contribute, the doorway amplitude being most observable within the "single" particle giant resonance", the fine structure being enhanced because of both doorway and single particle effects. Let us see in more detail how this develops. Let

$$
S_p = S_P \, S_D = e^{i\chi} \left[\frac{1 - iK_P}{1 + iK_P} \cdot \frac{1 - \dfrac{i}{2}\dfrac{\Gamma_D}{E - E_D}}{1 + \dfrac{i}{2}\dfrac{\Gamma_D}{E - E_D}} \right] \tag{13}
$$

where χ is some phase which is constant over the energy interval of interest and we have introduced the doorway resonance explicitly. Equation (13) may be rewritten as follows:

$$
S_p = e^{i\chi} \left[\frac{1 - i\left[K_P + \dfrac{1}{2}\dfrac{\Gamma'_D}{E - E'_D}\right]}{1 + i\left[K_P + \dfrac{1}{2}\dfrac{\Gamma'_D}{E - E'_D}\right]} \right] \tag{14}
$$

where

$$
\Gamma'_D = \Gamma_D \, (1 + K_P^2); \quad E'_D = E_D + \frac{K_P \, \Gamma_D}{2} \tag{15}
$$

We note the enhancement of Γ_D because of the presence of potential scattering.

To see the effect of S_p on the fine structure, we now evaluate $S_p S_r$ in the neighborhood of the K_r matrix pole at E_r taking S_r as

$$
S_r = \frac{1 - \dfrac{1}{2}\dfrac{\gamma_r}{E - E_r}}{1 + \dfrac{i}{2}\dfrac{\gamma_r}{E - E_r}} \tag{16}
$$

We shall not give a precise evaluation of S as that is much too involved algebraically. However, when E_r is at some distance from E'_D it is an easy matter to obtain an approximate solution:

(17)

$$S = e^{i\chi} \frac{1 - iK}{1 + iK} \; ; \quad K = \frac{1}{2} \frac{\gamma_r (1 + K_p^2)}{E - E_r - \frac{1}{2} \Gamma_r K_p}$$

where

(18)

$$K_p \equiv K_p (E_r) + \frac{1}{2} \frac{\Gamma_D'}{E_r - E_D'}$$

From (17) we see that the pole is shifted from E_r and that the width γ_r is enhanced:

(19)

$$\gamma_r' = \gamma_r (1 + K_p^2)$$

But K_p is a linear combination of potential scattering K_p and a doorway amplitude. The former is insensitive to the energy E_r over the domain Γ'_D but of course the doorway amplitude varies more rapidly, being negative for $E_r < E'_D$ and positive for $E_r > E'_D$. Hence the total K_p will vary across the doorway resonance width producing a modulation in γ_r' according to Eq. (19). The resultant dependence of γ_r' on E_r is not symmetric about the doorway state energy E'_D as a consequence of the interference between the potential and doorway state scattering. We see that this asymmetry is a general phenomenon not limited to the particular case of the isobar analog resonance [7]. This generality has been noted by Kerman and de Toledo-Piza [8], Iachello [9], McVoy [10], and Feshbach [4]. In the doorway state formalism of Feshbach, Kerman and Lemmer [11] it arises when it is possible for the compound nuclear resonance to be formed directly without the intermediate step of forming the doorway state. In another model [12], the doorway state amplitude gives the effect on a given compound resonance of the nearby levels while the potential scattering represents the effect of the distant levels. Regardless of the detailed physical significance of these results, the forms (12) with $S_P S_D$ given by Eq. (13) and S_r by Eq. (7) can be used to fit the experimental data in the one channel case. Again these results are formalism invariant.

One remark of interest. If a resonance E_r falls near E'_D it will be very much broadened so that one can expect in the final form (17) to find several considerably broader levels than might be otherwise expected.

V.   Physical Interpretation of RESONANCE PARAMETERS

We now turn to the more difficult discussion of the physical
significance of the poles of the S matrix or the K matrix and the
comparison between the two. As we have indicated these are related
quantities. In the case of a single channel the poles of the S
matrix correspond to the energies at which the solution of the
Schroedinger equation satisfies outgoing wave boundary conditions
while the poles of the K matrix occur at energies at which the
solution has a phase shift of $\pi/2$ or an odd multiple thereof.
One can debate which of these is more fundamental but not with me.
Rather let us consider (1) whether the position of the poles is
physically significant and (2) the very much related question of
the validity of the representations (6) and (7).
    We have already seen one example in which the poles are not
meaningful--when the potential phase shift is the hard sphere
phase shift. Then S_p does not satisfy Levinson's theorem although
S does. The error must be corrected by S_R and particularly by the
terms with large values of the resonance energies, and these of
course do not have any real significance.
    A second example is furnished by say a weak square well, as
weak as you please. Such a potential of course does not give
rise to any resonances. Nevertheless such a system possesses an
infinite number of S matrix poles. Clearly the physical significance of these poles is moot.
    One can also construct examples in which important poles are
omitted. To make the point here it is necessary to realize that
the K matrix can have complex poles, the imaginary part going to
zero as the maximum phase shift becomes $\pi/2$. If this phase
shift is slightly smaller than $\pi/2$, the imaginary part of the
pole energy is correspondingly small. For this case there would
be no term in representation (7) so that the K matrix representation in terms of real poles would omit a pole with physical
significance.
    This brings us to the question of the forms used to represent
S and K. We use the potential model once again as an example.
Reaction theory is presumed to apply here as well as to the many
body case. Since the many body case may have essential differences
these remarks remain cautionary in nature. The infinite number
and range of the values of the S matrix poles for the square well
is characteristic of "finite" wells such as the square and
gaussian wells. It, however, follows from very general quantum
mechanical principles that the actual potentials seen by an
incident projectile must decay exponentially. The singularity
structure of the S matrix, the poles etc., in the exponential case
differ markedly from that of the finite wells. This is illustrated
for the case of a superposition of exponential wells. All of the
poles and singularities are limited to the shaded region formed by
the intersection of the indicated hyperbolae (See Fig. 1). In
addition to the poles there is also a branch line. A sum over an

infinite number of poles clearly cannot describe this situation.
A more reasonable procedure would be to take a finite number of
poles into account explicitly, the branch line behavior being then
a property of the potential scattering term.

Suppose then for all these reasons we do use a finite number
of poles. Then the potential scattering is no longer hard core
scattering. The potential scattering will in fact depend upon how
many pole terms are employed in describing S_R. And as we mentioned earlier the widths and positions of these resonances will
depend upon the choice of the potential. It is clear that the
physical significance of each of the parameters describing all but
the narrow isolated resonances cannot be precisely given from a-
priori considerations. Additional criteria are needed. We have
already given one example--namely that the potential scattering
be that given by a reasonable one channel potential with parameters
which vary slowly with energy.

These ambiguities in the analysis of experimental data have
their complete counter parts in the reaction formalism. There are
also theoretical ambiguities. They are present as far as I know
in all formalisms. But in order to be even-handed in my discussion,
let me discuss the ambiguities in the projection operator formalism.
The fundamental equation is

$$[E = H_{PP} - H_{PQ} \frac{1}{E^+ - H_{QQ}} H_{QP}] \ P\Psi = 0 \tag{20}$$

where $P\Psi$ is asymptotically an open channel, or a set of open
channels, or all the open channels, or some mix of open channels
and important closed channels. The rest of the wave function is
$Q\Psi$. This series of possibilities indicates the freedom of
choice of P which is one of the ambiguities of this formalism.
It provides for a flexibility in tailoring the choice of P according to the physics of the problem--e.g., if there are some very
important closed channels one would like to include them in the
explicit   considerations of $P\Psi$.   There are other possibilities,
one of which, as I recorded in reference [2], gives rise to a form
of the R matrix theory. The choice of P however represents only
one ambiguity. Another lies in the choice of the potential describing the potential scattering. A number of examples will make
the point:

$$(1) \quad V_p = H_{PP} \tag{21}$$

$$(2) \quad V_p = H_{PP} + H_{Pq} \frac{1}{E^+ - H_{qq}} H_{qP} \tag{22}$$

where q selects out all the states whose resonance energies ,which
are directly related to the eigenvalues of H_{QQ}, fall outside the
energy interval ΔE under consideration. This has been the method
used in reference [1] – [4]. It is the method of choice when we
are dealing with isolated resonances.

$$(3) \quad V_p = H_{PP} + H_{PQ} \, \frac{1}{E^+ - \bar{H}_{QQ}} \, H_{QP} \qquad\qquad (23)$$

where $\bar{H}_{QQ}$ is some energy average of H_{QQ} in the energy interval,
ΔE. In this case, an average potential has been removed so that
only the fluctuations will be present in the remainder and in
addition the energy dependence for values at large separations
from ΔE is already included. Kerman, Kwai and McVoy use
$\bar{H}_{QQ} = H_{QQ} + iI/2$ where I is a constant so that V_p becomes identified
with the optical model potential. Choice (23) is useful in the
case of many levels. It is not appropriate for the single isolated
level. In other formalisms these possibilities (21), (22), and
(23) correspond to differing ways of taking distant levels into
account.

The consequence of these varying choices of V_p lies in the
corresponding variation in resonance energies and widths. Of
course the variation is not very significant when the widths are
small and the spacing large--that is in the case of isolated
resonances. The effects mentioned are significant when the
resonances overlap to some degree or the resonance is very broad.
But in that case by choosing V_p and ΔE one can effect the widths
as well as the spacing of the resonances.

This is not intended to be negative--but rather to point out
that in theoretical discussion one must bear in mind how the data
was analyzed and in particular how the potential scattering was
described. On the other hand the experimentalists should include
these important facts in their reports.

To summarize, a pole of the S or K matrix is of physical
significance providing direct information on the nuclear Hamiltonian
when the corresponding resonance is isolated and sufficiently narrow.
When the resonances are overlapping and/or the resonance is broad
then it is no longer possible to uniquely separate potential and
resonance scattering so that the individual parameters describing
the resonances are not individually significant. T_R will depend
upon the choice of S_P. A broad level which is in T_R could just as
well be included in S_P and vice versa but with a corresponding
change in the values of the position of the poles of the S_R matrix!
Of course the total amplitude T is not changed.

The time dependent description of this problem is instructive.
For a single narrow resonance the prompt amplitude which is generated by the potential scattering and the time delayed component
generated by the narrow resonance will be well separated in time.

However, as the width of the resonance increases or if it is
necessary to consider a number of resonances because they overlap,
the separation in time will be considerably reduced. As the
overlap between the prompt and delayed amplitude becomes significant the question of how much to assign to each will arise and
this is the problem to which we alluded in the preceding paragraphs
in another language.

VI. Averaging and the OPTICAL POTENTIAL; Transmission Factor

We now then turn to the optical model and the averaging
process about which there has been a great deal of controversy.
I shall not attempt to discuss that here. In the case of isolated
resonances there is no controversy. The problem starts when the
resonances overlap--and when the widths are relatively large.
But it is just here that the value of the widths, and value of
the spacings is not meaningful. One can for example drastically
change the nature of the distribution in widths by incorporating
the large ones into the potential scattering or regarding them as
an intermediate resonance. Again the choice of the potential
scattering amplitude is critical. A second point of interest
has to do with the choice of the averaging function. In principle it should be related to the one employed in the experiment
or in the analysis of the data. And indeed I constructed a set
of such functions [4] which go all the way from a Lorentzian
average (n = 1) to the Gaussian average. These are

$$\rho_{n,\delta}(E,E') = \left(\frac{\sqrt{n}\,\delta}{2}\right)^{2n-1} \frac{\Gamma(n)}{\Gamma(\frac{1}{2})\,\Gamma(n-\frac{1}{2})} \frac{1}{\left[(E-E')^2 + \frac{n\delta^2}{4}\right]^n} \quad (24)$$

The average will be sensitive to the power n and the parameter
δ if the number of important fluctuations of the amplitude included in the effective averaging interval changes sharply with
n and δ . This is sometimes called the "edge effect" appearing
most clearly for the box average. It would be very difficult to
obtain useful averages, (one would have to examine each experiment
with a great deal of care) if the edge effect were important.
However, when averages are measured, or computed, it is the
usual procedure to extend the range over which the average is
made until it is no longer sensitive to that range. In other
words the data presented is viewed as significant only when such
stability is exhibited. In any event, I would argue that this
is the appropriate procedure. In that case edge effects will not
be important and an average which is to a great extent independent
of the particular experimental set up can be obtained. We may,

therefore, choose the most convenient n, n = 1, the Lorentzian average.

This is the approach of Kerman, Kawai and McVoy [13] who employ

$$E^+ - H_{QQ} = E + \frac{iI}{2} - H_{QQ}$$

so that their optical potential reads

(25)

$$H_{opt} = H_{PP} + H_{PQ} \frac{1}{E + \frac{iI}{2} - H_{QQ}} H_{QP}$$

Here P projects on to the open channels. The full problem now reads

(26)

$$\left(E - H_{opt} - V_{PQ} \frac{1}{E^+ - H_{QQ}} V_{QP} \right) (P \Psi) = 0$$

where V_{PQ} is modified from the original H_{PQ}:

(27)

$$V_{PQ} \equiv H_{PQ} \sqrt{\frac{iI/2}{E - \bar{H}_{QQ}}}$$

Note that V_{PQ} will tend to zero as E increases corresponding to choosing an interval denoted in this case by I to average over. Solving (26) yields

(28)

$$P \Psi = \Psi_{opt} + \delta \Psi$$

where

(29)

$$(E - H_{opt}) \Psi_{opt} = 0$$

The nomenclature H_{opt} is justified by the result

(30)

$$\overline{P \Psi} = \Psi_{opt}$$

where the bar represents the energy average; in this case with a
Lorentzian weighting function. The choice (25) is the first important step. Its value is as indicated by (30) that it becomes
possible to treat fluctuations more consistently which we shall
now discuss briefly. From (26) we now may write the matrix
elements of S as follows

$$\text{(31)}$$

$$S_{cc'} = (S_{opt})_{cc'} - i \sum \frac{\gamma_{qc} \, \gamma_{qc'}}{E - E_q} = (S_{opt} + S_{f\ell})_{cc'}$$

with

$$\text{(32)}$$

$$\overline{S}_{cc'} = (S_{opt})_{cc'} \quad \text{and} \quad (\overline{S}_{f\ell})_{cc'} = 0$$

This is a pole expansion. The matrix elements γ_{qc} are

$$\text{(33)}$$

$$\gamma_{qc} = \sum_{c'} < q \mid V_{qc'} \mid \Psi^{opt}_{c'c} > = \sum_{c'} \int d\vec{r} \; V_{qc'}(\vec{r}) \; \Psi^{opt}_{c'c}(\vec{r})$$

$\Psi^{opt}_{c'c}$ is the coupled channel wave function, c represents the
incident channel, c' the outgoing one. The matrix element $V_{qc'}$
is the appropriate integral over all the target nucleon coordinates. It will be the random variable in the following.
The fluctuation problem familiarly arises in the calculation of
the cross-section

$$\text{(34)}$$

$$\overline{S^+ S} = S^+_{opt} S_{opt} + \overline{S^+_{f\ell} S_{f\ell}}$$

When Eq. (31) is inserted and the averages taken the following
quantity makes its appearance

$$X_{cc'} \equiv \overline{\gamma_{qc} \gamma^*_{qc'}} = \sum_{c_1 c_2} \iint d\vec{r}_1 d\vec{r}_2 \; \Psi_{c_1 c}(\vec{r}_1) \; V_{qc_1}(\vec{r}_1) V^*_{qc_2}(\vec{r}_2) \Psi^*_{c_2 c'}(\vec{r}_2)$$

where the superscript opt has been dropped. Under a random phase
assumption (the averages are made with respect to q)

$$X_{cc'} = \sum_{c_1} \iint d\vec{r}_1 d\vec{r}_2 \; \overline{\Psi_{c_1c} \, V_{qc_1}(\vec{r}_1) \, V^*_{qc_1}(\vec{r}_2)} \; \Psi^*_{c_1c'} \tag{35}$$

$$= \sum_{c_1} \int d\vec{r}_1 \; \Psi_{c_1c} \; | \, V_{qc_1}(\vec{r}_1) \, |^2 \; \Psi^*_{c_1c'}$$

Note that the random phase has been employed not only with respect to the indices c_1 and c_2 but also with respect to the coordinates r_1 and r_2:

$$\overline{V_{qc_1}(\vec{r}_1) \, V^*_{qc_2}(\vec{r}_2)} = \delta_{c_1c_2} \, \delta(\vec{r}_1 - \vec{r}_2) \, \overline{|V_{qc_1}(\vec{r}_1)|}^2 \tag{36}$$

Note $X_{cc'}$ is hermitian and

$$X_{cc} = \frac{\overline{\Gamma}}{qc} \tag{37}$$

With these assumptions it is easy to verify that the transmission factor $\mathcal{J}_{cc'}$ is

$$\mathcal{J}_{cc'} \equiv \overline{(S^+_{f\ell} S_{f\ell})}_{cc'} = (1 - S^+_{opt} S_{opt})_{cc'} \tag{38}$$

But from the explicit representation of $S_{f\ell}$ and assuming that the number of channels is sufficiently great so that fluctuations in the total width Γ can be neglected one obtains:

$$\mathcal{J}_{cc'} \simeq \frac{2\pi}{D\Gamma} \left[X \; \mathrm{Tr} \, X + X^2 \right]_{cc'} \tag{39}$$

In principle one can solve this equation for X in terms of and evaluate $\mathcal{J}$ using (38). A rough solution is valid when

$$(X^2)_{cc'} \ll X_{cc'} \; \mathrm{Tr} X \tag{40}$$

a condition which is satisfied when the number of channels is large.   Then it follows that

$$\text{Tr } X \simeq [\ \frac{\Gamma D}{2\pi}\ \text{Tr} \mathcal{J}\ ]^{1/2} \tag{41}$$

and

$$X \simeq \sqrt{\frac{\Gamma D}{2\pi}}\ \ \frac{\mathcal{J}}{\sqrt{\text{Tr } \mathcal{J}}} \tag{42}$$

A more exact result can be obtained by solving Eq. (35)

$$X = -\frac{1}{2}\ \text{Tr } X + \frac{1}{2}\ \left[(\text{Tr}X)^2 + 4\ \frac{\Gamma D}{2\pi}\ \right]^{\frac{1}{2}} \tag{43}$$

and taking the trace of both sides to obtain an equation for TrX:

$$(1 + \frac{N}{2})\ \text{Tr}X = \sum_{c} \left[(\text{Tr}X)^2 + 4\ \frac{\Gamma D}{2\pi}\ \mathcal{J}_{cc} \right]^{\frac{1}{2}} \tag{44}$$

An approximate solution is obtained if one assumes

$$\nu \mathcal{J}_{cc} \simeq \text{Tr } \mathcal{J}$$

where ν is some effective number of channels, assumed to be weakly dependent on c.   Then

$$(\text{Tr } X)^2 = \frac{N^2}{\nu\ (N+1)}\ \frac{\Gamma D}{2}\ \text{Tr } \mathcal{J} \tag{45}$$

replaces (41). X can then be obtained from Eq. (43).
     Returning to more general considerations the fluctuation cross-section $\sigma_{cc'}^{f\ell}$ $(\sim|S_{cc'} - \overline{S}_{cc'}|^2)$ is

for approximation (42)

$$\sigma_{cc'}^{fl} \sim \frac{\mathcal{J}_{cc}\mathcal{J}_{c'c'} + |\mathcal{J}_{cc'}|^2}{\sum_c \mathcal{J}_{cc}} \tag{46}$$

The first term, which for $c \neq c'$ dominates generally, gives the familiar Hauser formula. The second term is important for $c = c'$, i.e., for compound elastic scattering for which a factor of 2 makes it's appearance.

$$\sigma_{cc'}^{fl} \sim 2 \frac{|\mathcal{J}_{cc}|^2}{\sum \mathcal{J}_{cc}} \tag{47}$$

VII Thomas-Lane and Wigner Single Particle Giant Resonance

Let me conclude with a discussion of the Thomas, Lane and Wigner[14] giant resonance as it appears in the projection operator formalism. The problem here is to relate the single particle resonance to the fine structure which is revealed if a high resolution study is made. We write the effective Hamiltonian for the open channel as follows:

$$H_{eff} = (V_p - V_p') + V_p' + H_R \tag{48}$$

Here V_p is the potential scattering potential, V_p' is a potential adjusted to have a bound state in the appropriate energy range, while H_R is that part of the Hamiltonian which gives rise to the fine structure resonances. V_p' could be a simple factorable potential of the type appearing in H_R with the projection operator involving a suitable approximation of the single particle resonance wave function. If one neglects H_R a single particle resonance will occur near the bound state of V_p'. A complete analogy exists between this effective Hamiltonian and the one which applies in a more familiar decomposition of the Hamiltonian which leads to potential scattering, a doorway state, and resonant scattering. The role of the doorway state is now played by the single particle bound state, the potential scattering by V_p-V_p'. Since the doorway state situation has already been discussed, one can simply adapt those results. The single particle resonance acts then as an envelope to the fine resonance structure. And an additional result is a predicted asymmetry in the fine structure resonance widths, the asymmetry being with respect to the single particle resonance energy.

We thus have a coherent heirarchial picture emerging. Because of the single particle resonance, the widths of the fine structure resonances are enhanced. Because of the doorway state resonance there is a substructure to this enhancement, the enhancement being greater in the neighborhood of the doorway state energy. Upon averaging over the fine structure using an averaging width small compared with the doorway state width, the doorway state resonance broadens. Of course if this broadening is too

great it may no longer be visible. This in fact happens to the
single particle resonance, the imaginary part of the optical
model being simply so large that the single particle resonance
can no longer be observed in the scattering cross-section.

REFERENCES

[1] H. Feshbach, Ann. Phys. (N.Y.) 5 357 (1958).
[2] H. Feshbach, Ann. Phys. (N.Y.) 19 287 (1962).
[3] H. Feshbach, Ann. Phys. (N.Y.) 43 410 (1967).
[4] H. Feshbach, Latin American Summer School (1968).
[5] K. McVoy, Ann. Phys. (N.Y.) 54 552 (1969).
[6] D. B. Adler and F. T. Adler, Trans. Am. Nucl. Soc. 5,53(1962).
[7] D. Robson, Phys. Rev. 137B, 353 (1965).
[8] A. K. Kerman and A. de Toledo-Piza, Ann. Phys. (N.Y.) 43,363
       (1967); Ann. Phys. (N.Y.) 48 173, (1968).
[9] F. Iachello, Ann. Phys. (N.Y.) 52 16 (1969).
[10] K. McVoy, Ann. Phys. (N.Y.) 54 17 (1969).
[11] H. Feshbach, A. K. Kerman, R. H. Lemmer, Ann. Phys. (N.Y.)
       41 230 (1967).
[12] A. de Toledo-Piza, M.I.T. Thesis.
[13] Kerman, Kawai, McVoy, Private Communication.
[14] R.H. Thomas, A.M. Lane, and E.P. Wigner, Phys. Rev. 98 18
       (1953).

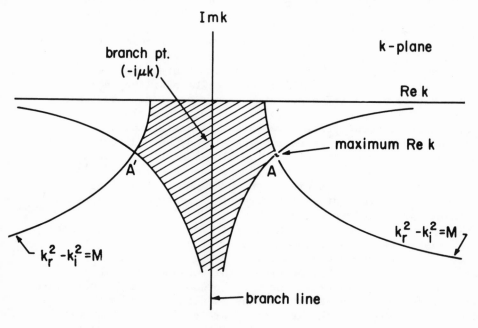

Fig. 1.

NEWSTEAD (Saclay)

You have written down an S matrix for the Doorway states, and I wonder if this implies that you can define an optical potential which will describe the Doorway behavior.

FESHBACH

Yes, that's in the paper with Kerman and Lemmer. In the absence of the fine structure resonance, the Doorway state and the ordinary incident channel--let's take the simple case, then form a set of coupled channels. The Doorway resonance is then a resonance in the resulting set of coupled Schroedinger equations.

NEWSON (Duke)

I am afraid I missed a point. You said in your earlier remarks that you expected a rapidly changing function for the resonance parts, a slowly changing function for the potential part, and later on, you said you couldn't tell the two apart.

FESHBACH

What I said was that making a split between a rapidly varying part and a slowly varying part was a feature which I couldn't really be quantitative about--I said that originally. Now, I explained that in the case where the resonances are over-lapping, or very broad, that the attempt to make that separation becomes ambiguous. You can make that separation, but you have to say what you're doing in order to be useful. If you just say, well here are the resonance parameters and you don't say what you did about the potential scattering, then it's meaningless. You have to take both pieces of information. After all, all that is really observed is the crossection, and that should not change. But the resonance parameters will change if you change the potential scattering. So tell me what the potential scattering is, as well as tell me what the resonance parameters are. In the case of isolated resonances, there are no problems.

NEWSON

You wouldn't expect to see a potential scattering curve that went this way, would you?

FESHBACH

That's exactly the criterion that I was describing. I mean, if the potential had that form it would violate causality.

WIGNER

I have two questions. First, I am not sure I understood your causality criterion. Can one formulate a causality criterion merely in terms of observable quantities?

FESHBACH

This is a causality criterion which I derived in a paper in 1958--the first one I wrote on projection operator reaction theory. It comes out of the following consideration. If you look at the effective potential in the presence of open channels, then there is a dispersive relation between the real and the imaginary part of the potential, which I don't think will surprise you. If you now insert that relation into the expression for the real part of the potential, you get a fixed sign of the energy derivative of the real part of the effective potential.

WIGNER

I was under the impression that if the potential extends sufficiently far, then all these rules are violated. Am I wrong?

FESHBACH

No, you're right. The statement, as usual, is for short-range potentials. Short range in the sense that they decay exponentially with distance.

WIGNER

My second question concerns the essential singularities of the collision matrix, those which manifest themselves as threshold phenomena, cusps, and similar anomalies. How do these essential singularities enter the expression for the collision matrix which you proposed?

FESHBACH

These effects are contained in the γ's which are energy dependent. This energy dependence produces the cusps etc. at threshold.

UNITARY PARAMETRIZATION OF NUCLEAR REACTIONS

J. Humblet, A. Lejeune and M.A. Nagarajan

Theoretical Nuclear Physics, University of

Liège, Sart Tilman, 4000 Liège 1, Belgium

1. INTRODUCTION

For low energy compound nuclear reactions, the determination of resonance energies, total, partial and
reduced widths, background,...from experimental data is
based on approximations of the relevant elements of a
parametrized collision matrix. Several types of parame-
trizations have been introduced for that purpose, all
having a priori advantages of their own. Nevertheless,
since only approximations of parametrized collision matrices can be of any use, it has long been recognized
that such approximations also have major shortcomings.
The main one is certainly that the approximations can
rarely be fully justified, with the result that elaborate
"theories" (involving e.g. a matrix inversion or a com-
plex-energy plane !) eventually give parametrized cross
sections whose phenomenological character is more obvious
than their true theoretical foundation.

In particular, recent experimental data have been
fitted to approximate collision matrices which obviously
are not unitary (1), while it proved difficult to evaluate objectively and quantitatively the lack of unitarity.
One has considered the possibility of reintroducing unitarity in the form of a set of relations to be a priori
satisfied by the parameters which enter into such approximate collision matrices (2). This is a rather hazardous
procedure to which one must certainly prefer an approximate collision matrix whose unitarity has not been lost

45

because of the necessary approximations (3). The R-matrix
approximations are of the latter type, at least at their
early stage. Approximate collision matrices rigorously
satisfying the unitarity relation can also be derived
from any few-level approximation of a modified K matrix,
the definition of which is such that its elements are meromorphic functions of the total energy E .

2. THE K-MATRIX PARAMETRIZATION

While the element $K_{c'c}$ of the conventional K matrix is defined as the ratio of the amplitudes of wave
functions whose radial Coulomb factors are G_ℓ, and F_ℓ
in channels c' and c respectively, here we define it
as the ratio of the amplitudes of $\overline{G}_\ell$, and $\overline{F}_\ell$, with

$$\overline{F}_\ell = F_\ell/(\epsilon_\ell k^{\ell+1}) \quad , \quad \overline{G}_\ell = \epsilon_\ell k^\ell (G_\ell - f F_\ell) \tag{1}$$

where

$$\epsilon_\ell = (2\ell+1)!! C_\ell = \left[(\ell^2+\eta^2)\dots(1^2+\eta^2)2\pi\eta/(\exp(2\pi\eta)-1)\right]^{\frac{1}{2}}/\ell! \tag{2}$$

f is chosen so as to make $\overline{G}_\ell$, like $\overline{F}_\ell$, an integral
function of $k^2 = E2M\hbar^{-2}$, namely, with $\ell_\psi = \Gamma'/\Gamma$,

$$f = \eta C_0^{-2}\left[\psi(1+i\eta) + \psi(1-i\eta) - 2\log\eta\right] \quad . \tag{3}$$

Under such conditions, the level expansion of $K_{c'c}$
simply is

$$K_{c'c} = \sum_{n=1}^{\infty} \left[\frac{g_{c'n}\, g_{cn}}{E - \mathcal{E}_n} + \frac{g^*_{c'n}\, g^*_{cn}}{E - \mathcal{E}^*_n}\right] \quad . \tag{4}$$

Defining diagonal matrices p , η according to

$$p_c = \epsilon_c k_c^{\ell+1/2} \quad , \quad \eta_c = p_c^2(1-if_c) \quad , \tag{5}$$

the relation between the K and S matrices reads

$$1-S = 2ipK(1+i\eta K)^{-1}p = 2ip(1+iK\eta)^{-1}Kp \quad ; \tag{6}$$

the formal analogy with the corresponding R-matrix relation is obvious (4). It can be used to extend to the present parametrization many of the results obtained in the
frame of R-matrix theory. The main differences are that,
on the one hand, here most of the g_{cn} , $\mathcal{E}_n$ are complex,
while on the other hand, the former quantities and K ,
p , η are uniquely defined, i.e. independent of channel radii and boundary condition constants. For a neutron
channel, f_c vanishes for any ℓ and in a diagonal element the threshold factor p_c^2 can be replaced by the

conventional penetration

$$P_c = k_c a_c / (F_\ell^2 + G_\ell^2) \qquad , \qquad (7)$$

K_{cc} and g_{cn}^2 being modified accordingly.

We proved that the constant "reduced widths" g_{cn}^2 satisfy sum rules, while in contrast with the R- and S-matrix residues, here $g_{cn}^2 \simeq 0$ when Re $\mathscr{E}_n$ lies well below the threshold of channel c . This is an important property, from a physical and practical point of view : a pole of K brings a negligible contribution to the matrix elements between channels which open far above the energy corresponding to the real part of that pole.

3. THE ONE-CHANNEL CASE

In the case of scattering of a neutral spinless particle by a central potential of finite range a typical distribution of the poles of K_ℓ is given on fig. 1. When $\mathscr{E}_{\ell n}$ is real, the cross section σ_ℓ has a maximum at $E = \mathscr{E}_{\ell n}$, while when $\mathscr{E}_{\ell n}$ is complex, one expects that σ_ℓ has a maximum or an inflection at $E = $ Re $\mathscr{E}_{\ell n} \equiv E_{\ell n}$. There does not exist, however, a one to one correspondence between the poles of the K matrix and those of the collision matrix. In order that a pole of the K-matrix corresponds to a <u>physical resonance</u>, it should give rise to a positive maximum of the phase derivative $d\delta_\ell/dk$. It can easily be seen that a pole $\mathscr{E}_{\ell n} = E_{\ell n} > 0$ corresponds to a physical resonance only when $g_{\ell n}^2 > 0$. If the residue is negative, the phase derivative is negative and the corresponding pole is by definition associated with an <u>echo resonance</u>. No such simple criterion exists in the case of the <u>complex</u> poles of K_ℓ : an <u>approximate</u> criterion is that the phase derivative will be positive at $E = E_{\ell n}$ when Re $g_{\ell n}^2 < 0$. The case of s-wave scattering studied, however, indicates that the above condition is sometimes too strong and that <u>all</u> the complex poles should be associated to physical resonances. The question now arises as to the importance of the various poles in terms of their contribution to the cross-section. Firstly, it has been pointed out earlier that the residues of poles occurring at energies far below the threshold are small. Hence, their contribution to σ_ℓ would be negligible. In the case of s-wave scattering, potentials of 5 MeV and 50 MeV with $\hbar^2/2Ma^2 = 1$ MeV were considered. In the former case, there exist no negative energy poles, but only one echo pole and complex poles. In fig. 2 we have compared the exact phase and phase de-

Fig. 1. A typical distribution of the poles of the K-
matrix in the complex energy plane.
Fig. 2. Comparison of the phase and phase derivative,
for s-wave scattering by a potential of depth 5 MeV, with
one echo pole approximation to K_o .

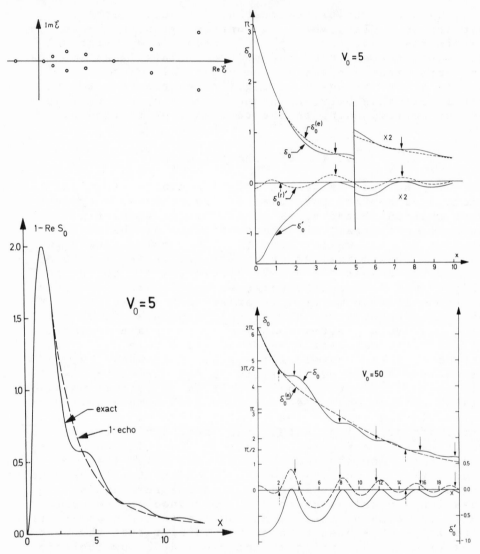

Fig. 3. Comparison of 1- Re S$_o$ with one echo pole approximation. The potential is the same as in fig. 2.
Fig. 4. Comparison of the phase and phase derivatives,
for s-wave scattering by a potential of depth 50 MeV,
with a two-echo-pole approximation to K_o .

rivative with the echo pole approximation to K_0 . In fig. 3, a similar comparison is shown for $1-\text{Re } S_0$. It is seen from the above figures that the dominant energy dependence of the cross section is determined by the echo pole, while the small fluctuations are due to the complex poles. In fig. 4, for $V_0 = 50$ MeV , we compare the exact phase and phase derivatives with the approximation corresponding to the two echo poles. The same features are observed again.

In order to study the role of the real poles of the K matrix corresponding to physical resonances (rather than the echoes only), we next considered a p-wave scattering for $V_0 = 9.33$ MeV . This potential is too weak to possess a bound state. The K matrix has two real poles at 0.19 and 5.61 MeV with positive and negative residues respectively. These poles thus correspond to a physical and an echo resonance. In fig. 5, the exact phase shift is compared with the two-pole approximation with two different penetrability factors $p_1 = x^3$ and $P_1 = x^3/(1+x^2)$. In fig. 6 a similar comparison is shown for the dimensionless number $1-\text{Re } S_1$ with the value obtained with the two real poles only. In both cases it is seen that at larger energies there are small fluctuations whose maxima correspond to the contributions of the complex poles of K_1 .

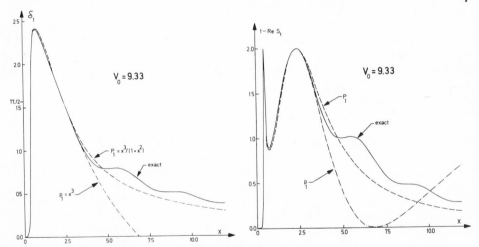

Fig. 5. Comparison of the exact phase shift and its two-pole approximations with different threshold factor p_1 and penetrability P_1 .

Fig. 6. Comparison of the dimensionless cross section $1-\text{Re } S_1$ with the two-pole approximations. The fluctuations around $x = 5.5$ and $x = 8.5$ are due to the complex poles. P_1 yields a better agreement than p_1 .

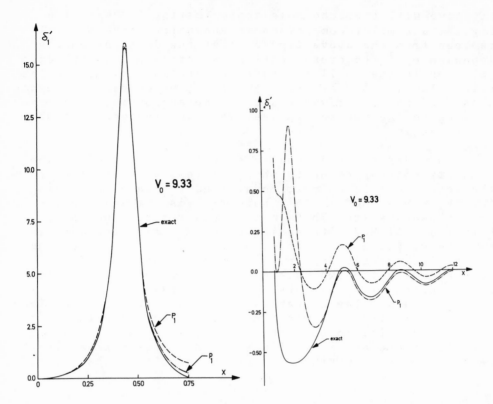

Fig. 7. Comparison of the derivatives of the total (exact) and resonance phases, δ_1' and $\delta_1' - \delta_e'$, where δ_e is the phase corresponding to the echo-pole term of K_1. They agree very well near the scattering resonance at $x = 0.44$. Fig. 8. The same at higher energies where the resonances correspond to the complex poles of K_1.

However, the magnitude of these maxima are here an order of magnitude smaller than the principal maximum due to the scattering resonance. Hence, the main features are extremely well described by the scattering resonance which dominates the low energy cross section and the echo resonance which yields the correct background cross section. These results are further amplified in the comparison of the derivatives of the total and resonance phases, as shown in figs. 7 and 8. The secondary maxima due to the complex poles occur at values of the dimensionless parameter $x = ka$, which are separated from each other approximately by π. These maxima are small compared with that of the scattering resonance, indicating that the dominant time delay arises from the real pole of K_1 and the complex poles account for small fluctuations of the time delay from the mean. Just as in the other cases, it is seen that here also the echo pole does not correspond to a time delay.

The study of potential scattering is by no means sufficient to make general statements about the behaviour of the poles of the K-matrix. The many channel situation has to be studied. But we have indications that the main character will not be drastically changed and that the possibility of being able to calculate the background cross section explicitly in terms of the echo resonances and to neglect its poles well below the threshold will remain a very strong motivation to the study of the K-matrix.

REFERENCES

(1) L. Kraus, I. Linck and D. Magnac-Valette, Nucl. Phys. A136 (1969) 301
A. Lejeune, Nucl. Phys. A141 (1970) 123
L. Kraus, Thesis, University of Strasbourg (1971) unpublished
(2) K.W. McVoy, Ann. Phys. (N.Y.) 54 (1969) 551
L. Rosenfeld, Acta Phys. Pol. A38 (1970) 603
(3) J. Humblet, Phys. Letters 32B (1970) 533 and Nucl. Phys. A151 (1970) 225
(4) A.M. Lane and R.G. Thomas, Revs. Mod. Phys. 30 (1958) 257.

SESSION II

ANALYSIS OF RESONANCE REACTION EXPERIMENTS

Chairman   - J. Humblet, Liege University
Scientific Secretary - S. Wynchank, Brooklyn College

HIGH RESOLUTION NEUTRON CROSS SECTIONS

M. S. Moore

University of California, Los Alamos Scientific

Laboratory, Los Alamos, New Mexico

ABSTRACT

Current techniques in the measurement and analysis of high resolution neutron cross sections are reviewed. In the energy region in which the resonance structure is resolved, only for fissile target nuclei is the present situation unsatisfactory. This is attributed to insufficient data, particularly with respect to resonance spins. Fission fragment anisotropies are found to be correlated both with fission widths of the resonances and with the fission mass distribution, and such measurements are expected to be fundamental to resonance analysis. In the unresolved resonance region, the fissionable targets still are a major problem because of intermediate structure in fission. While a systematic approach to the problem is still lacking, the use of a runs statistic (following a suggestion by G. D. James) seems to hold promise.

I. INTRODUCTION

Measurements of neutron cross sections can be categorized into three classes: (1) the resolved resonance region, (2) the unresolved resonance region, and (3) the region of smooth average cross sections. In determining the statistical properties of nuclei, it is the resolved resonance region which has been most useful. Here, one might reasonably hope that from a complete experiment (one in which all the quantities of interest are measured), he could obtain enough information on the distributions and possible correlations of the variables to form a basis for extrapolation. The unresolved (or incompletely resolved) resonance region contains a great deal of information, and forms a natural pool on which one can try to

float the model he has constructed from studying the resolved region. However, with a few exceptions, the unresolved region has
been largely ignored in data analysis.

It is perhaps worth considering at what point the resolved
region changes to the unresolved region. For the heaviest targets,
the level spacing ranges from $\lesssim$ 1 eV for odd A to ~ 10-20 eV for
even-even targets. For these materials, present measurements are
invariably made by neutron time of flight.

As shown in Fig. 1, the current state of the art in time-of-
flight measurements with white sources gives an energy resolution
comparable to the Doppler width for an uncooled sample up to a few
keV neutron energy. Thus, one might expect to see at least half
the structure present up to 1-2 keV for the odd mass targets, and
perhaps up to ~ 10 keV for the even-even targets. In either case,
the measurements should give a statistical sample of several hundred resonances for analysis. The resolved region can be extended
by a number of techniques: cooling the sample, extending the

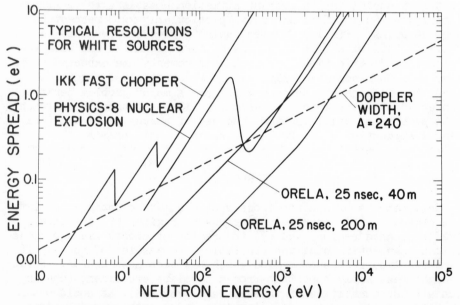

Fig. 1. Effective energy resolution of various neutron time-
of-flight spectrometers, as a function of neutron energy. Shown
as the dashed line is the Doppler spread for a typical heavy nucleus
(A = 240).

flight path, and running with shorter pulses. While all these have
been done, and done successfully, any measurement represents a
compromise between intensity and resolution. Practically, the useful statistical sample of resonances rarely exceeds a few hundred
for even-even targets, and is often less than 100 for the odd ones.
Furthermore, the information obtained is usually incomplete.

At this point, it may be of interest to define what is implied
by a "complete" measurement. What are the quantities needed?
First of all, one can list the parameters used in the single-level
Breit-Wigner formula: the resonance energy and partial widths for
each resonance, and if the spin of the target nucleus is non-zero
or if the resolved resonance region permits higher angular momenta
than zero, the spins and parities of each resonance. This simplest
example, then, requires a measurement of the total and/or all partial cross sections plus, in most cases, the angular distribution
of the scattered neutrons or the reaction rate of a polarized neutron beam incident on a polarized target. Next, one requires the
partial radiation widths, obtained from a measurement of the cap-
ture-gamma spectra and the coincidence spectra and angular correlations. If fission is present, one needs the partial fission widths,
obtained from measurements of fragment angular distributions when
the target nuclei are oriented. One might also ask for measurements
of other quantities peculiar to fission, such as the variation of
the fragment mass and kinetic energy distributions, the variation
of the number of neutrons emitted per fission, or the variation of
long-range alpha emission in fission. In most cases, the definitive
measurements have yet to be made.

What, then, can be done with the limited data presently available? Analysis techniques currently in use are not unique, but this
limitation appears to be due to insufficient data, particularly in
the case of fission resonances. Both the fission width distribution and the fission product mass distribution show a correlation
with the K-quantum number, which presumably characterizes the partial fission widths. Finally, there is the question of intermediate structure in fission, which may be important in the unresolved
resonance region. Here the use of a distribution-free runs statistic, as was first suggested by James,[1] seems to show considerable
promise.

II. TECHNIQUES FOR RESONANCE ANALYSIS

For measurements on non-fissile nuclei, the assumption is most
often made that the single-level Breit-Wigner formula applies, and
the resonance parameters are generally extracted from the data by
area analysis. The Atta-Harvey code,[2] with options for both shape
and area analysis, is perhaps the earliest of these which is still
used for the analysis of transmission (total cross section) data.

For the simultaneous area analysis of capture, transmission, scattering, and self-indication measurements, the code TACASI,[3] written by Fröhner at General Atomic (now Gulf Radiation Technology) is most widely used. The utility of the single-level approach to the fitting of cross-section data is greatly increased by the use of interactive graphics, as discussed, for example, by Marshall et al.[4]

When the resonances are not well isolated, one occasionally finds difficulties with the single-level formula. For s-wave resonances, the equation is usually coded as

$$\sigma_{n,n}^{J}(E) = \frac{\pi}{k^2}\, g_J\left(4k^2a^2 + \sum_\lambda \frac{\Gamma_{\lambda n}^2 + 4\Gamma_{\lambda n}ka(E_\lambda - E)}{(E_\lambda - E)^2 + \Gamma_\lambda^2/4}\right), \tag{1a}$$

$$\sigma_{n,x}^{J}(E) = \frac{\pi}{k^2}\, g_J \sum_\lambda \frac{\Gamma_{\lambda n}\Gamma_{\lambda x}}{(E_\lambda - E)^2 + \Gamma_\lambda^2/4}, \tag{1b}$$

where $\sigma_{n,n}^{J}(E)$ and $\sigma_{n,x}^{J}(E)$ are the cross sections at energy E for elastic scattering and for any partial cross section x for a resonance with spin J, respectively; k is the neutron wave number; a is the nuclear radius; g_J is a statistical weight factor; $\Gamma_{\lambda n}$ and $\Gamma_{\lambda x}$ are the partial widths for elastic scattering and the reaction s, respectively, for resonance λ; Γ_λ is the total width, and E_λ is the resonance energy. As it is usually coded, the formula gives a symmetric line shape for all partial cross sections except elastic scattering, and takes into account interference of resonance and potential scattering. If there is more than one resonance, the cross sections for each resonance are added, and then the potential scattering cross section is added in. If two resonances lie close together, the formula is not appropriate, but is still occasionally used. In this case, sometimes the superposition of negative interference terms exceeds the potential scattering, and the calculated cross section is negative. The simplest way to avoid the problem involves superposition of the resonance contributions in the collision matrix. This approach has become known as the multilevel Breit-Wigner equation (so designated by Gregson et al.[5]). Here the formula for elastic scattering is written

$$\sigma_{n,n}^{J}(E) = \frac{\pi}{k^2}\, g_J\, \left| e^{2ika} - 1 + \sum_\lambda i\Gamma_{\lambda n}/(E_\lambda - E - i\Gamma_\lambda/2)\right|^2, \tag{2a}$$

and, if interference effects are present in any of the partial cross sections, one can take them into account as

$$\sigma_{n,x}^{J}(E) = \frac{\pi}{k^2}\, g_J\, \left| \sum_\lambda i\sqrt{\Gamma_{\lambda n}}\,\sqrt{\Gamma_{\lambda x}}/(E_\lambda - E - i\Gamma_\lambda/2)\right|^2. \tag{2b}$$

This formulation is that of Feshbach, Porter, and Weiskopf;[6] it is expected to be valid in cases for which the width-to-spacing ratio is small.

For the analysis of cross sections of fissile nuclei, this approximation is also not adequate, and a more nearly complete multilevel approach is required. The formulations currently used are conveniently discussed as specializations of the theory of Wigner and Eisenbud,[7] in which the resonance contributions are additive in the R matrix. Here, following the notation of Lane and Thomas,[8] one writes, for any neutron cross section c,

$$\sigma_{nc}^{J}(E) \frac{\pi}{k^2} g_J |\delta_{nc} - U_{nc}|^2, \tag{3}$$

and the collision matrix U is given by the matrix equation

$$U = O^{-1}(1 - RL)^{-1} (1 - RL^*)O^*. \tag{4}$$

The diagonal matrices O and L describe the properties of the external region; their elements are known. The R-matrix is written as

$$R = \sum_\lambda \frac{\gamma_\lambda \times \gamma_\lambda}{E_\lambda - E} , \tag{5}$$

where the elements of the vectors γ_λ are the reduced widths, related to the observed widths $\Gamma_{\lambda c}$ through the known elements of matrices L and O.

The problem in using the general R-matrix formula lies in the matrix inversion in Eq. (4). Vogt,[9] following Wigner,[10] assumed an expansion of the form $(1 - RL)^{-1} = 1 + \sum_{\mu\nu} A_{\mu\nu}(\gamma_\mu \times \beta_\nu)$, where $\beta_\nu = L\gamma_\nu$. This expansion yields $(A^{-1})_{\mu\nu} = (E_\lambda - E)\delta_{\mu\lambda} - \sum_c \gamma_{\mu c}\beta_{2c}$. If the assumption is then made that all but a few levels can be neglected, the Wigner level matrix A can be inverted, and the problem is solved. Reich and Moore[11] retained the channel inversion for a few channels only by partitioning the matrix and eliminating the radiative capture channels by the above technique, assuming the random phase approximation for the off-diagonal elements. Here the expressions to be evaluated are again given by Eqs. (3) and (4), with the matrix R redefined as

$$R = \sum_\lambda \frac{\gamma_\lambda \times \gamma_\lambda}{E_\lambda - E - i\Gamma_{\gamma\lambda}/2} , \tag{5'}$$

where $\Gamma_{\gamma\lambda}$ is the radiative capture width for the λth resonance, and R is now a matrix of rank k, where k is the number of channels retained.

Both the above specializations are somewhat cumbersome, because a matrix inversion must be performed at each energy point. Adler and Adler[12] introduced parameters which are to a first approximation energy independent, and in which the matrix inversion is handled implicitly. Diagonalizing the inverse of the Wigner level matrix A^{-1} by the transformation $D^{-1} = S^{-1}A^{-1}S^{-1}$ leads to the expression

$$A_{\lambda\nu} = \sum_c S_{\lambda c}S_{\mu c}/(a_c - E), \qquad (6)$$

where the elements of S and a are complex. The collision matrix then has the form

$$U_{cc'} = e^{i(\varphi_c + \varphi_{c'})}\left[\delta_{cc'} + i\sum_k (a_k - E)^{-1}\sum_{\lambda\mu}\gamma_{\lambda c}\beta_{\mu c'}S_{\lambda k}S_{\mu k}\right], \qquad (7)$$

or, in the usual notation,

$$\sigma_{nc}(E) = \frac{C}{E}2(1 - \cos\omega)$$

$$+ \frac{C}{\sqrt{E}}\sum_k \frac{\nu_k(G_k^c \cos\omega + H_k^c \sin\omega)+(\mu_k-E)(H_k^c \cos\omega-G_k^c \sin\omega)}{(\mu_k - E)^2 + \nu_k^2}, \qquad (8)$$

where ω = 2ka for the total cross section and zero for any reaction cross section; and $C = 6.52 \times 10^5$ barns-eV. The Adler-Adler approach is a specialization for low energy neutrons of the Kapur-Peierls[13] parameterization; formally, insofar as the parameters can be considered energy independent, it is equivalent to the S-matrix formulation of Humblet and Rosenfeld[14] for low energy neutrons.

The Adler formula lends itself quite readily to Doppler broadening corrections and to least-squares extraction of the level parameters. The code CØDILLI, written by Adler and Adler,[15] is widely used. Least-squares R-matrix fitting is much less convenient; codes have been written by Auchampaugh[16] and by Derrien[17] for this purpose. Transformation of R-matrix parameters to Adler parameters is straightforward, for example by the code LEMA, written by Adler and Adler,[18] or by PØLLA, by de Saussure and Perez.[19] Harris[20] has shown that to first order in the perturbation transformation of R-matrix to Adler parameters, the μ_k and ν_k of Eq. (8) are the R-matrix resonance energy E_k and the total width $\Gamma_k/2$. It was shown by Adler and Adler[21] that the inverse transformation can also be done, under the restriction of a single R-matrix fission channel.

Auchampaugh[16] has carried out a computer study of the uniqueness of R-matrix parameters, using the two-fission-channel multilevel least-squares fitting code MULTI. The conclusions of this study are that certain of the parameters are unique: the resonance energy E_λ, and the partial widths $\Gamma_{\lambda n}$, $\Gamma_{\lambda\gamma}$, and $\bar\Gamma_{\lambda f} = \Gamma_{\lambda f1} + \Gamma_{\lambda f2}$. The phase relations between the fission vectors are not unique, leading to multiple solutions which describe the cross section nearly equally well. It is important to note, however, that this degeneracy is not inherent to the approach. It is due entirely to inadequate data: If the partial fission cross sections from each fission channel were provided, the fit would be unique. This is true not only for computer-generated data (as in Auchampaugh's study) but for actual data as well.

The R-matrix parameters obtained by Auchampaugh as multiple solutions in this study were converted to Adler parameters by the code PØLLA. Small differences were observed, particularly in the interference parameters H_k. The differences were small enough that it could reasonably be expected that iteration by CØDILLI, using these as initial parameters, would have converged to the same result.

The fitting of actual data is more difficult. De Saussure and Perez of ORNL, with Kolar of BCMN, Geel, have studied the problem of uniqueness by fitting the ORNL-RPI capture and fission cross sections of ^{235}U.[22] Fitting the data from 30 to 60 eV, with the least-squares R-matrix code MULTI, gave the results illustrated in Fig. 2. They then converted the R-matrix parameters to Adler parameters by means of PØLLA, and used these as input parameters for a further least-squares fit, varying only the G_k and H_k parameters. The improvement in the fit, shown in Fig. 3, is significant. The final Adler parameters obtained are listed in Table 1; shown for comparison in the table are the parameters de Saussure and Perez[23] obtained earlier with the same program, fitting the same data. The differences in the listed sets of resonance parameters are certainly at least in part due to the different fitting procedure (non-linear in one case, linear in the other). The fit to the data which is shown in Fig. 3 is fairly respectable, however -- comparable in quality to that which was reported at Helsinki.[23] Inspection of the parameters obtained in the two cases suggests two effects: (1) a change in the parameter ν is accompanied by a corresponding change in the symmetric terms G_c and G_f, such that the resonance area is preserved; and (2) a change in the resonance energy can be compensated, within the measurement error, by a change in the coefficients H_c and H_f.

It is perhaps not unreasonable to conclude that the present situation for fissile nuclides is untenable. The simplest analysis techniques are not adequate, and the data are not sufficient for the more complicated ones to yield consistent results. What is needed is a single definitive experiment -- a determination of

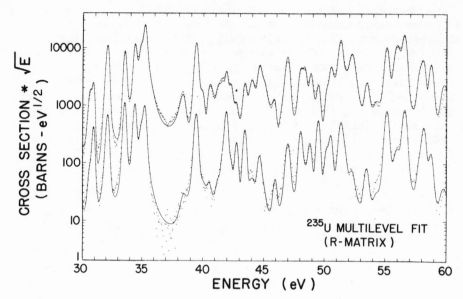

Fig. 2. R-matrix multilevel fit to the fission and capture cross sections of ^{235}U from 30 to 60 eV.

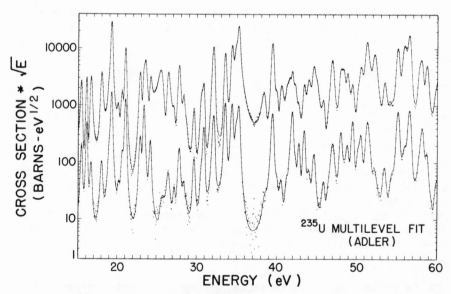

Fig. 3. Multilevel fit to the fission and capture cross sections of ^{235}U from 15-60 eV. These results were obtained by varying the Adler parameters G_k and H_k from Fig. 2. These results and those of Fig. 2 are the work of R. B. Perez, G. de Saussure, and W. Kolar, and were communicated for inclusion in the present paper.

Table 1. Comparison of multilevel parameters for ^{235}U from 30 to 48 eV

μ (eV)		ν (eV)		G_c (b-eV$^{3/2}$)		H_c (b-eV$^{3/2}$)		G_f (b-eV$^{3/2}$)		H_f (b-eV$^{3/2}$)	
(A)	(B)	(A)	(B)	(A)	(B)	(A)	(B)	(A)	(B)	(A)	(B)
30.59	30.60	0.09	0.09	1.41	9.64	-0.99	-1.66	12.53	19.00	-3.72	-2.84
30.91	30.89	0.07	0.03	44.76	35.40	10.44	-0.23	21.95	18.26	11.46	2.89
32.10	32.12	0.08	0.05	83.36	58.88	5.24	9.83	135.24	85.36	12.63	31.36
33.55	33.53	0.07	0.05	123.20	113.91	14.48	1.80	95.21	85.86	4.06	-7.37
34.39	34.37	0.09	0.06	142.82	114.97	20.02	7.21	139.89	112.54	-8.19	-16.35
34.85	34.92	0.14	0.10	51.16	60.66	-30.97	13.02	113.40	139.38	-18.75	51.14
35.21	35.19	0.11	0.07	133.79	116.64	10.35	-14.07	344.31	290.39	-2.53	-61.53
35.75	35.59	0.49	0.54	-13.41	-4.80	-14.72	4.79	22.87	57.41	-2.86	23.42
38.42	38.42	0.19	0.23	2.37	2.56	-1.73	-1.31	29.38	33.24	19.08	18.77
39.44	39.43	0.09	0.07	100.65	93.74	8.57	5.54	173.72	161.95	8.47	3.45
39.92	40.05	0.10	0.13	5.13	4.60	-2.73	5.29	19.61	22.25	-4.82	20.06
40.51	40.51	0.15	0.20	6.93	13.10	0.17	1.53	30.59	57.73	-15.07	-23.58
41.39	41.26	0.14	0.16	4.04	12.52	6.14	-0.19	28.45	38.14	15.81	-11.44
41.64	41.62	0.06	0.04	8.12	14.46	5.79	-2.11	22.59	23.62	9.43	2.20
41.91	41.90	0.08	0.03	116.62	79.97	8.35	-3.55	55.77	35.40	3.87	-6.88
42.16	42.26	0.13	0.09	6.34	12.11	-24.56	-2.16	25.90	24.23	-27.25	-4.72
42.70	42.74	0.06	0.03	19.46	20.45	-8.75	-0.98	1.98	4.46	0.91	0.91
43.41	43.41	0.04	0.04	37.32	38.97	-2.68	1.81	16.39	19.22	-3.96	0.32
44.04	44.00	0.04	0.21	3.71	16.38	2.05	3.45	16.03	60.93	9.45	2.07
44.75	44.62	0.15	0.09	28.04	19.58	18.97	-2.14	44.47	58.70	77.33	3.65
44.97	44.97	0.28	0.19	0.48	1.54	-18.58	0.58	115.80	52.26	-25.84	-6.67
45.79	45.92	0.17	0.07	4.62	4.09	-2.13	3.46	19.94	8.75	0.81	10.07
46.92	46.96	0.06	0.10	23.63	25.28	46.08	-1.92	76.90	90.98	122.36	16.53
48.06	48.00	0.12	0.05	63.65	41.00	15.86	-0.46	57.16	38.99	27.75	0.88
48.35	48.37	0.20	0.16	0.34	15.12	-31.20	-9.91	111.16	93.36	-31.57	0.79

(A) G. de Saussure, R. B. Perez, and H. Derrien, in "Nuclear Data for Reactors," Vol. 2, p. 757 (1970).

(B) G. de Saussure, R. B. Perez, and W. Kolar, private communication (1971).

resonance spins and K quantum numbers for each resonance. It is commonly assumed that the K-quantum number is characteristic of the fission channel; the measurement would remove the degeneracy of the R-matrix analysis, or, as an alternative, would permit the transformation from Kapur-Peierls to R-matrix parameters by the method outlined by Adler and Adler.[20] A measurement of the relative variation of the fission fragment anisotropy from resonance to resonance has been reported by Pattenden and Postma.[24] However, in order that these data be of real value to multilevel analysis, knowledge of the resonance spins is essential.

In this connection, it is of interest to note that such an experiment is presently underway. Keyworth et al at LASL, and Dabbs et al at ORNL plan a collaborative experiment to determine resonance spins of ^{235}U, ^{233}U, and ^{237}Np by measuring the fission neutron production rate when polarized neutrons from ORELA are incident on polarized target nuclei. Since the polarization of such a target nucleus also orients it effectively, they plan to determine K at the same time by measuring the fission neutron angular distribution.

It is of more than casual interest that the variation of K seems to be connected significantly to other parameters in the fission process: in particular, to the variation in the fission fragment mass distribution and to the fission widths. Recently, Cowan et al[25] reported measurements of R, the ratio of valley-to-peak mass yields, as a function of neutron energy for ^{235}U. One can plot the measured valley-to-peak ratio vs A_2, the second Legendre coefficient, as determined from the measurements of Pattenden and Postma.[24] The correlation, shown in Fig. 4, is striking. If one goes through the exercise of calculating the correlation coefficient, one finds a coefficient of 0.80 for 28 degrees of freedom, or a significance level, under the null hypothesis, of $< 10^{-5}$. There is also a marked correlation between A_2 and the resonance width (Adler parameter ν), as shown in Fig. 5. The resonance width reflects primarily the fission width variation. Here the correlation coefficient is 0.54 for 61 degrees of freedom, giving a significance level of $< 10^{-3}$. These results lead one to calculate the correlation between A_2 and a number of other quantities which have been observed to vary from resonance to resonance, such as $\bar{\nu}$, the number of fission neutrons emitted per fission (Weinstein et al,[26] Ryabov et al[27]), the total kinetic energy and ternary alpha yield (Melkonian and Mehta[28]), or the other partial widths Γ_n^0 and Γ_γ, as determined by Perez, de Saussure, and Kolar from their R-matrix multilevel fitting results mentioned above. The calculated correlation coefficients are given in Table 2. The correlation of the radiation width and A_2 also appears to be significant, but here the results may be questioned because the R-matrix fitting code was constrained to give radiation widths no larger than 0.05 eV.

The observed correlations between A_2 and the fission mass

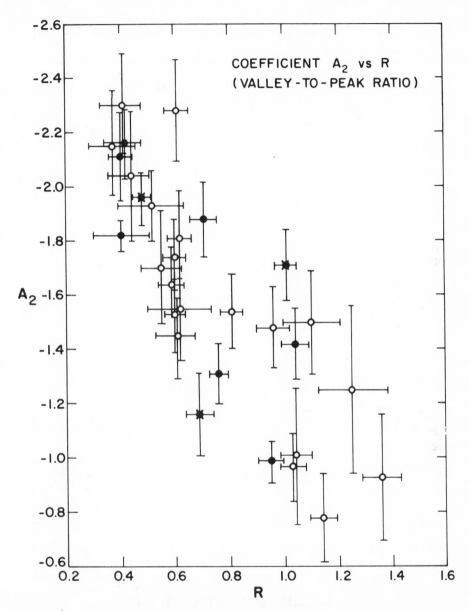

Fig. 4. The second Legendre coefficient A₂, as reported by Pattenden and Postma,[24] vs R, the valley-to-peak ratio in the fission mass distribution, as reported by Cowan et al,[25] for a number of resonances in (^{235}U + n). Resonances thought to be of spin 4 are designated by filled circles; resonances thought to be of spin 3 are designated by x's; the others are unknown.

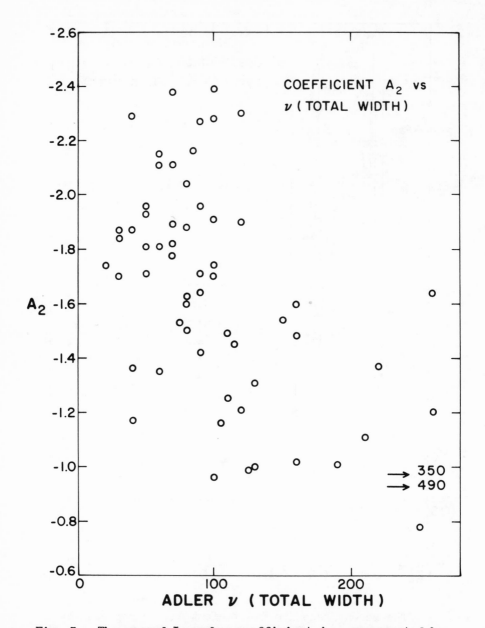

Fig. 5.   The second Legendre coefficient A_2, as reported by
Pattenden and Postma,[24] vs the resonance width parameter ν, as reported by de Saussure et al,[23] for (^{235}U + n).   The variation in
the width is primarily due to the fission component.

Table 2.   Correlation coefficients between A_2 as determined by Pattenden and Postma[24] and a number of other fission variables for ($^{235}U + n$).

Variable	Correlation coefficient	No. of degrees of freedom	Approximate significance level
Valley-to-peak ratio in mass distribution (Ref. 25)	0.80	28	10^{-8}
Total width (Adler parameter v) (Ref. 23)	0.54	59	$<10^{-5}$
No. of neutrons/fission - Weinstein et al (Ref. 26)	0.06	11	0.42
No. of neutrons/fission - Ryabov et al (Ref. 27)	0.08	22	0.35
Total kinetic energy (Ref. 28)	-0.16	28	0.19
Ternary alpha emission (Ref. 28)	0.27	28	0.07
$\Gamma_n^{\,0}$ (de Saussure et al, preliminary results, 1971)	-0.04	50	0.39
Γ_γ (de Saussure et al, preliminary results, 1971)	-0.42	50	0.001

distribution and the fission width seem to be just the reverse of those which might have been expected from the channel theory of Bohr,[29] in that the larger fission widths seem to be characteristic of higher K bands rather than lower ones. The higher valley yields (presumably arising from fission events with higher excitation at the scission point) are also characteristic of the higher K bands, rather than the lower ones. These results are quite different from those of ^{239}Pu fission, where large fission widths and high valley mass yields are characteristic of resonance spin 0^+, and narrow widths and low valley yields of spin 1^+.

III.   THE UNRESOLVED RESONANCE REGION AND THE RUNS STATISTIC

The practical interest in determining the statistical behavior of neutron cross sections in the resolved resonance region lies in the extension to the unresolved region. One hopes that the average parameters will provide an adequate description of such phenomena as the variation of the effective (self-shielded) capture cross section or the variation of fission neutron production with temperature. Before this can have any hope of success, one requires some way of assessing the importance of possible non-statistical effects, such as intermediate structure in fission.

James[1] has suggested the use of a distribution-free statistic based on a consideration of the number of runs in a sequence. James' analysis was carried out on the basis of a runs statistic whose properties were described by Wald and Wolfowitz,[30] which is a special case of the following general approach. Suppose that we consider a sequence of events, to be tested under the hypothesis that they are uncorrelated, i.e., that the probability of each event's occurring on either side of an assumed median value is a constant which is independent of all other events in the sequence. The simplest example is the tossing of a coin. Suppose we toss a coin n times, and obtain a sequence of heads and tails. In this sequence, the probability that heads will have occurred n times is given by the first term in the binomial expansion

$$\alpha^n + n\alpha^{n-1}(1-\alpha) + \frac{n(n-1)\alpha^{n-2}(1-\alpha)^2}{2!} + \ldots + (1-\alpha)^n \qquad (9)$$

where α is the probability of tossing a head, and $1-\alpha$ is the probability of tossing a tail. For a fair coin, $\alpha = (1-\alpha) = 1/2$. In this expansion, the probability that heads will occur $(n-1)$ times and tails once is given by the second term, the probability that heads will occur $(n-2)$ times and tails twice by the third term, etc. We next want to ask what is the number of runs in the sequence, where a run is defined as an unbroken subsequence of either heads or tails. For example, the probability of a single run is given by the sum of the first and last terms in the expansion (either all heads or all tails), or

$$P(1) = \alpha^n + (1-\alpha)^n. \qquad (10)$$

Two runs will occur only when each of the other terms consists of an unbroken subsequence of heads followed by an unbroken subsequence of tails, or the reverse; thus

$$P(2) = 2\left(\alpha^{n-1}(1-\alpha) + \frac{2\alpha^{n-2}(1-\alpha)^2}{2!} + \ldots + \alpha(1-\alpha)^{n-1}\right) \qquad (11)$$

The most useful results are obtained if the probabilities are equal, i.e., if $\alpha = (1-\alpha) = 1/2$. In this case, the general term is given by

$$P(k+1) = \frac{1}{2^{n-1}} \binom{n-1}{k} \qquad (12)$$

where $\binom{n-1}{k} = (n-1)!/(k!(n-k-1)!)$, and k ranges from 0 to n-1. The average value is easily shown to be

$$E(U) = \frac{1}{2^{n-1}} \sum_{k=0}^{n-1} \binom{n-1}{k} (k+1) = \frac{n+1}{2} \qquad (13)$$

and the variance is*

$$\sigma^2(U) = \frac{1}{2^{n-1}} \sum_{k=0}^{n-1} \binom{n-1}{k} \left[(k+1)^2 - E(U)^2 \right] = \frac{n-1}{4} \tag{14}$$

Wald and Wolfowitz[29] considered a single term in the binomial expansion (Eq. 9), e.g., that with precisely h heads and t tails, and investigated the probability distribution of runs for such a term. They showed that the mean and variance are given by

$$E(U) = \frac{2ht}{h+t} + 1, \tag{15}$$

and

$$\sigma^2(U) = \frac{2ht(2ht - h - t)}{(h + t)^2 (h + t - 1)} \tag{16}$$

This is very nearly the same, for terms near the center of the distribution (h ≈ t), as the perhaps more general Eqs. (5) and (6), which can be recast for comparison, as

$$E(U) = \frac{2ht}{h+t} + 1/2 \tag{13'}$$

and

$$\sigma^2(U) = \frac{ht(4ht - h - t)}{(h+t)^3}, \tag{14'}$$

where h = t = n/2. It should be noted that the Wald-Wolfowitz statistic is independent of the median, whereas the other assumes that the median value is known. In some cases, this may be an advantage; for example, it would seem to be a more appropriate statistic to use in testing limited ranges of data for possible structure, where the median value may be given accurately by data outside the region of interest.

The runs distribution rapidly tends to normality as h and t increase beyond about 5, and it is customary to evaluate significance levels by assuming that the expression $X = (|U-E(U)|-1/2)/\sigma(U)$ is normally distributed with zero mean and unit variance. (The term 1/2 is included to take account of the discrete nature of the distribution.) As shown in Fig. 6, there is some error introduced by using the normal approximation at high-significance levels, since the wings of the runs statistic are slightly lower than those of the normal curve. However, the error is in the direction that it does not mislead the user.

---

*A method of proving the identities implied in Eqs. 13-14 was provided by Miss Gail Rein, a student in mathematics at the University of New Mexico in Albuquerque. The method involves successive differentiations w.r.t. x of $(a+x)^n$, and evaluation of the expansion at a = x = 1.

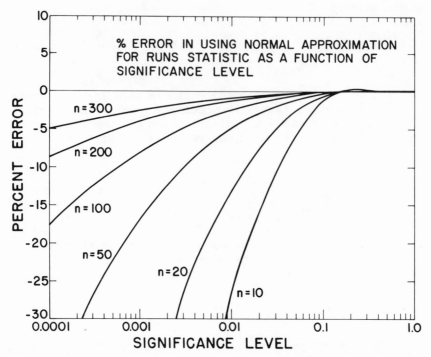

Fig. 6.   The percentage error to be expected as a function of
the significance level, for various sample sizes, in using the normal approximation for the runs statistic based on equal probability.
The error in using the normal approximation for the Wald-Wolfowitz
statistic is virtually identical to this, provided h ≈ t, as described in the text.

James[1] suggested the runs statistic as a sensitive test for
the presence of intermediate structure in fission, applying it to
the study of fission widths in ^{234}U, reduced neutron widths in
^{232}Th (no structure expected), and serial correlation coefficients
in ^{239}Pu and ^{235}U. He concluded that significant structure is
present only in ^{234}U and ^{239}Pu.

The fission widths of ^{244}Cm, reported by Moore and Keyworth,[31]
provide an example of the utility of this statistic.   The data are
shown in Fig. 7, and the results of the statistical test in Fig. 8.
Moore and Keyworth suspected that the structure in the fission
widths near 800 eV is non-statistical, and found that the significance level of testing the widths under the assumption of a chi-
squared distribution with 1 or 2 degrees of freedom was 0.02 and
0.01, respectively.   Using the runs statistic, the significance
level is found to be 1.7×10^{-3} (under the normal approximation).

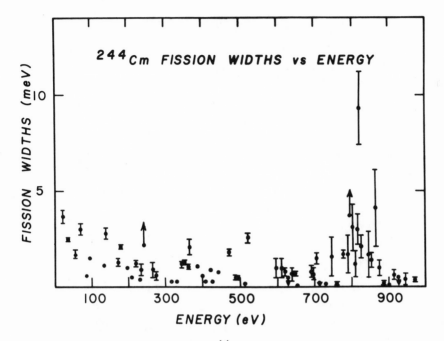

Fig. 7. Fission widths of (^{244}Cm + n), derived from an analysis
of fission and capture cross sections[31] under the assumption that
the capture width is 37 meV.

The reduced neutron widths, tested in the same way, show precisely
what is expected under the null hypothesis of random fluctuation.

Extension to the unresolved region is perhaps best done by
considering ratios of cross sections measured simultaneously, rather
than correlograms or serial correlation coefficients as was done by
James. The cross section of ^{235}U may serve as an example to show
why this is necessary. It is well known that the fission cross
section of ^{235}U shows strong fluctuations in the energy region of
10-100 keV or higher.[32-35] Such fluctuations could arise either
from intermediate structure in fission, or from fluctuations in the
average neutron widths and overlapping of resonances (probably p-
wave). The latter seems to predominate. Figure 9 shows a comparison of the total cross section of ^{235}U, measured by Böckhoff and
Dufrasne[36] and the fission cross section of ^{235}U, reported by Lemley
et al.[34] It is readily seen that the structure is, for the most
part, correlated.

This does not necessarily rule out the possibility of inter-

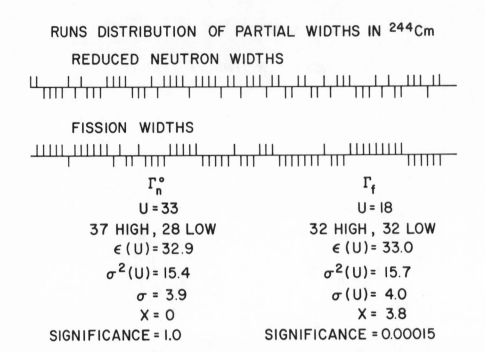

Fig. 8. Analysis of the reduced neutron widths in (^{244}Cm + n) using the Wald-Wolfowitz statistic. The fission-width distribution shows significant non-statistical effects.

mediate structure, however; it simply means that before tests for intermediate structure are likely to be meaningful, one must first eliminate the structure due to neutron width fluctuations. The most attractive technique is to test alpha, the capture-to-fission ratio, since the fluctuation in the radiative capture widths is expected to be small. As an example, ^{244}Cm from 1-5 keV will serve, as shown in Fig. 10. If the data are channelized into 80 energy bins of 100 eV (much larger than the resolution), one finds the number of runs to be 23, with $E(U) = 41$ and $\sigma(U) = 4.486$. Use of the normal approximation to obtain the significance gives a level of 6×10^{-3} under the null hypothesis, showing again strong evidence for intermediate structure.

Up to this point, the runs statistic has been used only to detect the existence of possible intermediate structure. It would also seem to be useful to have a test to tell which of the fluctuations in the cross section ratios are significant (to be interpreted as intermediate structure) and which are not. Physical information to be extracted with this test is the D_{II} spacing.

Such a test is easily devised. Suppose we again consider the

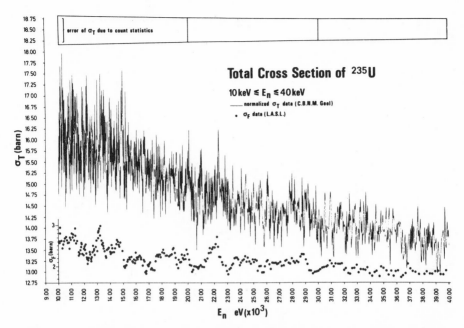

Fig. 9. Comparison of structure observed in the total cross section of (^{235}U + n), as measured by Böckhoff and Dufrasne,[36] and in the fission cross section of (^{235}U + n) as measured by Lemley et al.[34] This figure was supplied by K. H. Böckhoff prior to publication. It shows that the structure in (^{235}U + n) is most likely not intermediate structure in fission.

binomial expansion (Eq. 9), with $\alpha = 1 - \alpha = 1/2$, and we ask, "What is the probability of observing a run of length k, in a sequence of length n?" Except for the first and last terms of the expansion, there are two or more runs of various lengths in each possible sequence. The total number of such runs in all the 2^n possible sequences is $2^{n-1}(n+1)$. The probability of observing a subsequence of length n (all heads or all tails) is $2/(2^{n-1}(n+1))$, and the probability of observing a subsequence of length $k \neq n$ is given by

$$P(k) = 2^{-k}(1 + \frac{2-k}{n+1}) \tag{17}$$

As is obvious from the above equation, the probability of occurrence of a given length of run is almost independent of the length of sequence considered. Slightly over half the runs are expected to have length of 1, one-fourth have length 2, slightly less than 1/8 have length 3, etc. We might choose, as is customary, the 5% significance level as the criterion for non-statistical effects. We

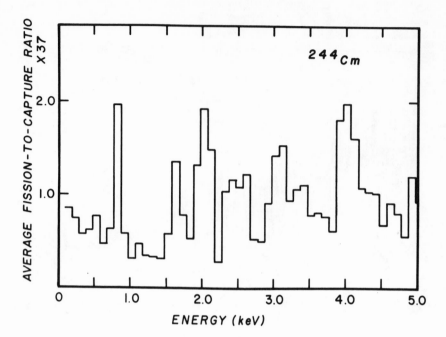

Fig. 10. Ratio of average fission to capture in (^{244}Cm + n) below 5 keV.[31] The ordinate gives the average fission width in meV under the assumption that the average capture width is 37 meV.

then should look for runs of length ≥ 5 (which actually lie at a significance level of ~ 3%).

As an example, let us consider again the case of the ^{244}Cm fission-to-capture ratio, with the data averaged in 20 eV bins. The median about which the structure is to be tested is evaluated in the region of 250-700 eV, and shows no significant non-statistical effects. The test shows evidence for seven regions below 5 keV for which the average fission width is enhanced, in agreement with the conclusions reached by Moore and Keyworth[31] from visual inspection (Fig. 10).

It is perhaps unnecessary to point out that this test will not pick out non-statistical structure whose natural width is less than five resolution widths, since the point spacing should be at least as large as the resolution width. However, in cases such as ^{244}Cm, in which the Class II strength seems to be spread over a fairly large number of Class I states, it should prove useful.

ACKNOWLEDGMENTS

I should like to take this opportunity to thank R. B. Perez and G. de Saussure of the Oak Ridge National Laboratory, and W. Kolar of the Central Bureau of Nuclear Measurements, Geel, Belgium, for undertaking the study of R-matrix and Adler parametric fitting, and for communicating their preliminary results for incorporation into this paper. Thanks are also due to Miss Gail Rein, at the University of New Mexico, for devising a simple method of proving the relationships implied in Eqs. 12-14. This research was performed under the auspices of the U. S. Atomic Energy Commission.

REFERENCES

1. G. D. James, UKAEA Report AERE-R6633 (1971), to be published.
2. S. E. Atta and J. A. Harvey, USAEC Report ORNL-3205 (1961).
3. F. H. Fröhner, USAEC Report GA-6906 (1966).
4. N. H. Marshall, O. D. Simpson, J. R. Smith, and J. W. Codding, in Neutron Cross Sections and Technology, Knoxville (1971).
5. K. Gregson, M. F. James, and D. S. Norton, UKAEA Report AEEW-M-517 (1965).
6. H. Feshbach, C. E. Porter, and V. F. Weisskopf, Phys. Rev. $\underline{96}$, 448 (1954).
7. E. P. Wigner and L. Eisenbud, Phys. Rev. $\underline{72}$, 29 (1947).
8. A. M. Lane and R. G. Thomas, Rev. Mod. Phys. $\underline{30}$, 257 (1958).
9. E. Vogt, Phys. Rev. $\underline{112}$, 203 (1958).
10. E. P. Wigner, Phys. Rev. $\underline{70}$, 606 (1946).
11. C. W. Reich and M. S. Moore, Phys. Rev. $\underline{111}$, 929 (1958).
12. D. B. Adler and F. T. Adler, Trans. Am. Nucl. Soc. $\underline{5}$, 53 (1962).
13. P. L. Kapur and R. Peierls, Proc. Roy. Soc. (London) $\underline{A166}$, 277 (1938).
14. J. Humblet and L. Rosenfeld, Nucl. Phys. $\underline{26}$, 529 (1961).
15. D. B. Adler and F. T. Adler, USAEC Report COO 1546-3 (1966).
16. G. F. Auchampaugh, Nucl. Phys. (to be published).
17. H. Derrien, J. Blons, and A. Michaudon, in Nuclear Data for Reactors, IAEA, Vienna (1970), Vol. I, p. 481.
18. D. B. Adler and F. T. Adler, USAEC Report COO-1546-4 (1967).
19. G. de Saussure and R. B. Perez, USAEC Report ORNL-TM-2599 (1969).
20. D. R. Harris, USAEC Report LA-4327 (1969).
21. D. B. Adler and F. T. Adler, in Nuclear Data for Reactors, IAEA, Vienna (1970), Vol. II, p. 777.
22. G. de Saussure. L. W. Weston, R. Gwin, R. W. Ingle, J. H. Todd, R. W. Hockenbury, R. R. Fullwood, and A. Lottin, in Nuclear Data for Reactors, IAEA, Vienna (1966), Vol. 2, p. 233; see also G. de Saussure et al, USAEC Report ORNL-TM-1804 (1967).
23. G. de Saussure, R. B. Perez, and H. Derrien, in Nuclear Data for Reactors, IAEA, Vienna (1970), Vol. 2, p. 757.
24. N. J. Pattenden and H. Postma, Nucl. Phys. $\underline{A167}$, 225 (1971).

25. G. A. Cowan, B. P. Bayhurst, R. J. Prestwood, J. S. Gilmore, and G. W. Knobeloch, Phys. Rev. C2, 615 (1970).

26. S. Weinstein, R. Reed, and R. C. Block, in Physics and Chemistry of Fission, IAEA, Vienna (1969), p. 477.

27. Yu. V. Ryabov, So Don Sik, N. Chikov, and N. Yaneva, JINR Preprint P3-5297 (1970).

28. E. Melkonian and G. K. Mehta, in Physics and Chemistry of Fission, IAEA, Vienna (1965), Vol. II, p. 355.

29. A. Bohr, in Proc. Intern. Conf. Peaceful Uses At. Energy, United Nations, New York (1956), Vol. 2, p. 151.

30. A. Wald and A. Wolfowitz, Ann. Math. Stat. 11, 151 (1940).

31. M. S. Moore and G. A. Keyworth, Phys. Rev. C3, 1656 (1971).

32. B. H. Patrick, M. G. Sowerby, and M. G. Schomberg, J. Nucl. Energy 24, 269 (1970).

33. C. D. Bowman, M. L. Stelts, and R. J. Baglan, in Nuclear Data for Reactors, IAEA, Vienna (1970), Vol. II, p. 65.

34. J. R. Lemley, G. A. Keyworth, and B. C. Diven, Nucl. Sci. Eng. 43, 281 (1971).

35. C. D. Bowman and G. S. Sidhu, in Neutron Cross Sections and Technology, Knoxville (1971).

36. K. H. Böckhoff and A. Dufrasne (to be published).

DISCUSSION

CHRIEN (Brookhaven National Laboratory)
     I have one comment and one question. First of all, the
comment. One way of determining spins of resonances in U235 as
used in other nuclei is to examine high energy gamma ray transitions. We now know the level scheme of U236 well enough to be able
to establish some 2+ levels and perhaps, even a 5+ level has been
identified. This technique has been used at B.N.L. to fix the
spins of something like six resonances now in U235 plus neutron
and we see all the assignments are pretty good ones. We're confid-
ent of them. The interesting thing that has come out of this work
is that the fission widths for the spin 3 resonances in U235
appear to be smaller than the average fission width for all
resonances. Now, the simple Bohr channel theory would predict
quite the converse of the prediction of about 2-1 ratio of spin

three to spin four resonances. Our sample size is small, but we
feel it is significant enough to tell us that the ratio is not
2-1. The question I have, and this result incidentally, will be
reported in the Tuscon meeting hopefully. The question I have has
to do with the Wald-Wolfowitz statistic. Is this a more efficient
way for testing for intermediate structure than say, plotting
serial correlation coefficients on a correlogram.

MOORE (Los Alamos Scientific Laboratory)
     Yes, well, I personally feel it is, simply because with a
correlogram you are always plagued by end effects that you don't
understand. I think Paris showed this very nicely in his paper at
the Vienna Conference on physics and chemistry of fission, and it
has made me doubt all correlogram results, at least those that
rely on a picket fence distribution of spacings. O.k.. I would
also like to comment on your earlier remark, in the sense that
there have been a number of ways to test spins for U235, all of them
indirect, and the results are so conflicting, or they have been
in the past, that the definitive measurements, I think, are
polarized neutrons and polarized nuclei. Those are incontrover-
tible.

NEWSON
     I should have been able to work out the answer to this in my
head--as we went along, but one of your slides you showed as
intermediate structure which seemed to give a peak about every
kilovolt. Did I read you right?

MOORE
     Of course, it was a very, very broad resolution data.

NEWSON
     Yes, but I mean roughly, how many resonances were there in the

kilovolt interval blocks?

MOORE

This is the unresolved region. We did not resolve them. The spacing in Curium 244 seems to be about 13 ev. We astimated there were about eight resonances in each one of the blocks that I showed, not the intermediate structure blocks, but in each of the bins.

NEWSON

So there'd be about eighty per kilovolt, right?

MOORE

Yes.

NEWSON

R. Block (R.P.I.) was very interested in those A_2 variations across a resonance. Could you comment on these quasi-resonances which were thought up by Eric Lynn several years ago? Is this a handle of telling a quasi-resonance from a resonance?

MOORE

Gee, I hadn't thought about that. I suppose it would be, yes.

NEWSTEAD

As I understand your proposal to search for intermediate structure in fissile nuclei, it is to look for such structure in the α measurements. Now, if you do this, how can you distinguish between nonstatistical effects in capture as opposed to intermediate structure in fission?

MOORE

I can't. But I'm assuming that those in capture are negligible compared to those in fission. That's the assumption that goes with it. Yes, I agree with you.

NEWSTEAD

I see. Then, I wonder if Dr. Lane would care to comment whether, in that energy region, one would expect this to be the case. Would non-statistical effects be negligible in such heavy nuclei?
                         (No response)
We will just have to wait for further experimental clarification.

MICHAUDON (Saclay)

You mentioned some test to detect intermediate structure in fission crossections. Have you applied those tests to well-known cases of intermediate structure in Np 237, Pu 240, etc? What's the significance level you obtained if you did carry out the answers?

MOORE

No, I haven't. My own feeling is that the test is useful
where the eye cannot tell whether there's intermediate structure.
I've never really done it, but James has. He applied it in U234
and one other. I've forgotten what it was--but it got very, very
significant results. It's very significant.

PATRICK (Harwell)

Just two comments: One, the Pu^{329} data in which intermediate
structure in the fission cross section we think we have seen.
This appears beautifully in α, the ratio of capture to fission.
Two, the Geel data on U235 in which they see structure in both the
total cross section and the fission cross section. I don't think
that's completely proved yet.

MOORE

No, I will agree it has not proved the absence of intermediate
structure.

PATRICK

No, we have to wait for the ratio.

MOORE

The resonances very often in the total cross section appear
much narrower than in the fission cross section and not always in
the right places.

PATRICK

This has to do also with the difference in resolution of the
two measurements.

MOORE

Right, and a slight energy shift that was uncovered in one
of the measurements that we're testing against.

HIGH–RESOLUTION NEUTRON RESONANCE SPECTROSCOPY IN NATURAL FLUORINE, ALUMINUM, CHLORINE AND POTASSIUM*

U.N. Singh and J.B. Garg, State University of New York

at Albany, Albany, N.Y., U.S.A., J. Rainwater, W.W.

Havens, Jr., S. Wynchank, M. Slagowitz and H. Liou, Columbia University, New York, N.Y., U.S.A.

In order to understand the features of nuclear reactions, we need to be able to relate the cross sections to the internal properties of the nucleus. The simple Breit Wigner formule[1] has been intensively used with the necessary modifications for Doppler broadening in the analysis of resonances observed in neutron interactions with heavy nuclei. Another important theory is due to Kapur and Peierls[2] which employs two sets of eigenstates and eigenvalues. The boundary conditions required for their definition require the logarithmic derivatives of the interior wave function at each channel entrance to be equal to the logarithmic derivative of the radial outgoing or ingoing wave function, respectively, in that channel. These two sets of eigenstates are mutually, but not separately, orthogonal, and the necessary expansions must be made in terms of both of them. The resulting cross section expressions are very familiar in form to those of S-matrix theory, but the parameters are all energy dependent. The S-matrix formulation of nuclear reaction theory by Humblet and Rosenfeld[3] has the advantage of a physically clear definition of resonance energy and rather simple expressions for the cross sections. Much of the latter advantage is lost, however, by the greater number of parameters and the complicated correlations that implicitly exist between them. Because of this, it seems that R-matrix theory[4], which has the great advantage of having unitarity built into it, is simpler in practice to handle in spite of its more complicated formulae. This theory has been used in the analysis of neutron resonance data in the medium weight nuclei[5]. The theory is based on an assumption that the many dimensional configuration space of the nucleons of the system could be divided into two parts. The external region containing pairs of nucleons interacting with each other weakly through long range forces and the interior region

81

where the nucleons interact strongly.

In the R-matrix theory the boundary conditions chosen to generate the eigenstates are real and energy independent and the corresponding value and derivative parameters are real and independent of energy.

When the neutron scattering is the predominant channel (which is the case in these nuclei under investigation) one can usually introduce the auxilliary matrix of smaller order so called the reduced R-matrix.[6,7,8] The expressions for various cross sections are:

$$\sigma_t(E) = \frac{2\Pi}{k^2} \sum_J g_J [1 - \text{Real } U_J(E)]$$

$$\sigma_n(E) = \frac{\Pi}{k^2} \sum_J g_J |1 - U_J(E)|^2$$

$$\sigma_r(E) = \frac{\Pi}{k^2} \sum_J g_J (1 - |U_J(E)|^2)$$

Where k is the neutron wave number and g_J is the spin statistical weight factor $= 2J+1/2(2I+1)$ where J is the spin of the compound state and I is the ground state spin of target nucleus. $U_J(E)$ is called the collision function and for neutrons of orbital angular momentum 1 is

$$U_J(E) = \exp(-2i \, \phi_\ell) = \frac{1-(S_\ell - iP_\ell) R_J(E)}{1-(S_\ell + iP_\ell) R_J(E)}$$

where ϕ_1 is the hard sphere scattering phase shift, S_1 is the energy shift factor, P_1 is the barrier penetration factor and $R_J(E)$ is the reduced R-function defined as

$$R_J(E) = \sum_\lambda \frac{\gamma^2_{\lambda Jn}}{E_{\lambda J} - E - (1/2)i\Gamma_{\lambda J\gamma}} \qquad (1)$$

Where the sum is over levels λ of angular momentum J with the eigenvalues $E_{\lambda J}$ and reduced neutron widths $\gamma^2_{\lambda Jn}$. The width $\Gamma_{\lambda J\gamma}$ is the total reaction width of level λ which is usually dominated by the capture width at low energy of neutrons. Neutron width ($\Gamma_{\lambda Jn}$) is related to reduced width ($\gamma_\lambda^2{}_{Jn}$) by

$$\Gamma_{\lambda Jn} = 2P_\ell \, \gamma^2_{\lambda Jn}$$

In applying this formalism to the analysis of data we recognize that the energy range covers only a limited number of resonances, and we may in general determine the detailed parameters of only these local resonances within the range. Despite this, the levels that are just outside the range of measurement may have a marked effect on the cross section within the range of measurement and the "giant resonance" effect of far away levels will add a constant real part to R_J which affects the potential scattering cross section. Thus we divide $R_J(E)$ into two components: R_J^{local} which is the sum in equation (1) over the local levels within the measured energy range, and R_J^{res}, the residual part of R_J from levels outside the range.

$$R_J^{res}(E) = A_J + B_J(E - E_{1/2})$$

where $E_{1/2}$ is meadian energy of observation and μ signifies the level outside the measured energy range.

$$A_J = R_J^{\infty} + \sum_{\mu} \frac{\gamma_{\mu Jn}^2}{E_{\mu J} - E_{1/2}}$$

$$B_J = \sum_{\mu} \frac{\gamma_{\mu Jn}^2}{(E_{\mu J} - E_{1/2})^2}$$

In order to determine the resonance parameter $E_{\lambda J}$, $\gamma_{\lambda Jn}^2$ and possibly $\Gamma_{\lambda J\gamma}$ we employ a numerical least square analysis. By fitting the known thermal data on capture cross section it is possible to estimate the radiative capture width of strong levels or the presence of negative energy resonance.

Strength Function: Knowledge of resonance parameters enables one to determine strength function. S_1, the strength function for neutron of orbital angular momentum 1 is defined as

$$S_\ell = \frac{1}{(2\ell+1)\Delta E} \sum_i g \Gamma_{ni}^{(\ell)}$$

ΔE is the measured energy region and $\Gamma_{ni}^{(\ell)} = \frac{\Gamma_{ni}}{\sqrt{E_{oi}}(ev)} \frac{1}{v(\ell)}$

$$v(\ell) = \frac{P(\ell)}{ka}$$

We have analysed the high resolution neutron total cross section data (up to about 200 keV) on natural F, Al, Cl and K. These measurements were performed at the Nevis Synchocyclotron Laboratory of Columbia University. A detailed description of this system is given elsewhere.[9] A flight path of 200 meters was used for all the measurements. The overall resolution achieved was about 0.5 keV at 100 keV for F and Al measurements and about 0.25 keV at 100 keV for chlorine and potassium measurements. Theoretical cross sections were corrected for the instrumental resolution width.

Samples of various thicknesses were used to obtain the reliable value of peak cross section. The cross section between resonances was obtained from the thickest sample, and in case this was saturated, from the next thickest sample. The intermediate cross sections were obtained from samples of medium thickness and the cross section on or near the peak was obtained from the thin samples. In general, the procedure was that whenever the observed

$n\sigma t$ becomes greater than unity, the cross section at that energy was obtained from the next thinner sample.

Results

1. Fluorine: Three samples of $1/n$ = 5.6, 27.2, 106.0 b/atom were used in the measurement. R-matrix fit to the data is shown in Figure 1. Resonance parameters, <D> and S_1 are given in Table 1. Our results agree very well with the work of Newson et al[10]. The agreement with the results of Hibdon[11] is not very good. Our results agree fairly well with the capture work of Block et al[12] and Nystrom et al[13]. We have used their results for Γ_γ.

2. Aluminum: Samples of natural Al (grade 2S) of thickness $1/n$ = 2.15, 6.47, 17.17, 51.51 b/atom were used. A typical R-matrix fit to data is shown in Figure 2. Resonance parameters, <D> and S_o and S_1 are given in Table 2.

According to Hibdon[14], the 35 keV resonance is 3^+ whereas we assign it as 2^+. By fitting the thermal capture cross section we have also determined the Γ_γ for this resonance as 8.30 eV.

3. Chlorine: Samples of natural Cl (Cl^{35} =75.53%, Cl^{37} =24.47%) of thickness $1/n$ = 5.05, 32.58, 130.9 b/atom were used. A typical R-matrix fit to the data is shown in figure 3. Resonance parameter for Cl^{35} and values of <D>, S_o, and S_1 are given in Table 3 and similar quantities for Cl^{37} are given in Table 4. The information obtained from this investigation is much more complete and new than that published in BNL 325[15]. A negative energy resonance belonging to Cl^{35} at 185 eV with Γ_γ=0.53 eV was assumed to fit the $\sigma_{n\gamma}$ at thermal energy.

4. Potassium: Samples of natural K (K^{39} =93.10%, K^{41} = 6.88%) of thickness $1/n$ = 2.45, 7.60, 25.86, 79.33 and 261.8 b/atom were used. A typical R-matrix fit to the data is shown in figure 4. Resonance parameters for K^{39}, <D>, S_o and S_1 are given in Table 5 and similar quantities for K^{41} are given in Table 6. A negative energy resonance at 1 keV belonging to K^{39} with Γ_γ = 1 eV was used to fit the $\sigma_{n\gamma}$ thermal energy. The information obtained from this investigation is a more or less a new piece of information.

Table 1

F^{19} (I^{Π}= $1/2^+$)

E_o(keV)	ℓ	J^{Π}	Γ_n(keV)
27.02 ± 0.07	1	2^-	0.37
49.10 ± 0.17	1	1^-	1.31
97.00 ± 0.45	1	1^-	14.31

<D> = 66.7 keV, S_1 = (14.31 ± 12.53) x 10^{-4}

TABLE 2

Al^{27} ($I^{\Pi} = 5/2^+$)

E_0(keV)	ℓ	J^{Π}	Γ_n(keV)
5.903±0.008	1	1^-	0.01
34.70±0.10	0	2^+	3.42
88.50±0.39	1	3^-	11.28
120.0±0.60	1	2^-	3.49
145.0±0.80	1	3^-	15.14
159.0±0.90	1	4^-	2.84
204.8±1.3	1	2^-	4.26
208.1±1.4	1	2^-	4.35

$\langle D \rangle$ = 28.1 keV, S_0=0.3397x10^{-4}
S_1 = (10.13±5.81) x 10^{-4}

TABLE 3

Cl^{35} ($I^{\Pi} = 3/2^+$)

E_0(keV)	ℓ	J^{Π}	Γ_n(eV)
-0.185	0	2^+	13.6
0.398±0.0005	1	0^-	0.3
4.249±0.002	1	0^-	1.7
14.797±0.015	0	1^+	45.5
16.350±0.017	1	0^-	26.4
17.131±0.018	1	1^-	29.9
25.680±0.031	1	0^-	185
26.630±0.033	1	1^-	150
44.190±0.066	1	0^-	110
55.430±0.090	1	0^-	307
57.800±0.096	1	1^-	147
101.160±0.211	1	0^-	725
103.480±0.218	1	1^-	468
113.471±0.248	1	2^-	582
127.700±0.294	1	0^-	669
133.798±0.314	0	1^+	814
136.550±0.323	1	1^-	688
139.250±0.333	1	0^-	1017
143.200±0.346	1	1^-	931
149.937±0.370	1	0^-	731
165.708±0.427	1	0^-	443
183.242±0.493	1	2^-	1503
188.390±0.513	1	1^-	824
190.769±0.522	1	0^-	940

$\langle D \rangle$=8.7 keV, S_0=(0.08±0.07)x10^{-4}
S_1=(1.65±0.55)x10^{-4}

TABLE 4

Cl^{37} ($I^{\Pi} = 3/2^+$)

E_0(keV)	ℓ	J^{Π}	Γ_n(eV)
1.101±0.0008	1	0^-	0.003
2.312±0.0014	1	0^-	0.012
8.317±0.007	1	3^-	73.1
25.24±0.03	0	1^+	138
27.36±0.034	1	0^-	506
27.727±0.035	1	0^-	65.8
27.820±0.035	1	0^-	75.2
29.172±0.037	1	0^-	35.0
33.9±0.046	1	0^-	315
34.33±0.047	1	0^-	57.6
46.600±0.071	0	2^+	350
51.570±0.082	1	0^-	184
54.920±0.089	0	2^+	166
56.670±0.093	1	0^-	190
62.043±0.106	1	0^-	330
62.726±0.107	1	2^-	275
66.850±0.117	1	1^-	185
68.260±0.121	1	2^-	450
90.323±0.180	0	2^+	304
93.120±0.188	1	2^-	807
141.0±0.339	1	2^-	739

$\langle D \rangle$= 9.5 keV, S_0=(.12±0.09)x10^{-4}
S_1=(2.87±1.06)x10^{-4}

TABLE 5 $K^{39}(I^{\Pi} = 3/2^{+})$			
E_0(keV)	ℓ	J^{Π}	Γ_n(eV)
-1.00	0	1^+	45.5
9.38±0.008	1	2^-	46.2
25.540±0.031	1	2^-	113
42.600±0.063	0	2^+	554
46.009±0.070	1	1^-	84.4
57.850±0.096	0	2^+	865
59.940±0.101	1	1^-	137
68.805±0.122	1	1^-	2070
87.700±0.172	1	1^-	358
96.750±0.198	0	2^+	329
99.268±0.205	1	0^-	746
101.173±0.211	1	0^-	684
102.239±0.214	1	0^-	451
104.500±0.221	1	0^-	544
108.600±0.234	0	2^+	1255
116.240±0.257	1	0^-	913
118.62±0.265	1	0^-	682
121.300±0.273	1	0^-	836
127.300±0.293	1	0^-	771
132.700±0.310	1	1^-	687
135.000±0.318	1	1^-	459
136.690±0.324	1	0^-	1017
144.400±0.350	1	0^-	1288
145.820±355	1	0^-	1474
147.736±0.362	1	0^-	1673
149.110±0.367	1	0^-	526
154.320±0.385	1	1^-	1591
162.500±0.415	1	1^-	879
178.100±0.473	1	1^-	890
193.520±0.533	1	3^-	166
199.500±0.556	0	1^+	199

<D>= 6.7 keV,
$S_0=(0.37\pm0.23)\times10^{-4}$
$S_1=(2.71\pm0.82)\times10^{-4}$

TABLE 6 $K^{41}(I^{\Pi} = 3/2^{+})$			
E_0(keV)	ℓ	J^{Π}	Γ_n(eV)
1.111±0.0008	1	1^-	0.59
2.029±0.0012	1	2^-	3.82
3.205±0.002	1	0^-	1.74
3.278±0.002	1	1^-	2.86
4.038±0.003	1	0^-	0.64
5.536±0.004	0	1^+	118
8.461±0.007	1	0^-	7.2
12.148±0.011	1	1^-	10.9
12.615±0.012	1	3^-	12.6
13.340±0.013	1	1^-	11.9
14.291±0.014	1	1^-	17.5
15.026±0.015	1	0^-	34
15.947±0.016	1	2^-	25.5
16.585±0.017	0	1^+	81.8
16.710±0.017	1	2^-	48
16.785±0.017	1	1^-	52
16.840±0.018	1	1^-	57.6
20.299±0.023	1	1^-	69.1
24.740±0.030	0	1^+	143
24.950±0.030	1	3^-	81
27.651±0.035	1	1^-	99
32.200±0.043	1	1^-	65
32.510±0.043	1	1^-	52
33.350±0.045	1	3^-	136
33.808±0.045	1	1^-	42
37.085±0.052	1	1^-	60
37.55±0.053	1	3^-	117
38.045±0.054	1	1^-	116
38.250±0.054	0	1^+	310
38.582±0.055	1	3^-	170
44.645±0.067	1	2^-	41.6
52.152±0.083	1	3^-	137
54.620±0.089	1	3^-	262
55.789±0.091	1	2^-	43.5
79.890±0.151	1	3^-	437
81.20±0.154	1	3^-	647
90.621±0.180	1	1^-	318
91.524±0.183	1	1^-	284
93.100±0.187	0	2^+	355

<D>= 5.13 keV
$S_0 = (0.12\pm0.06)\times10^{-4}$
$S_1 = (4.52\pm1.18)\times10^{-4}$

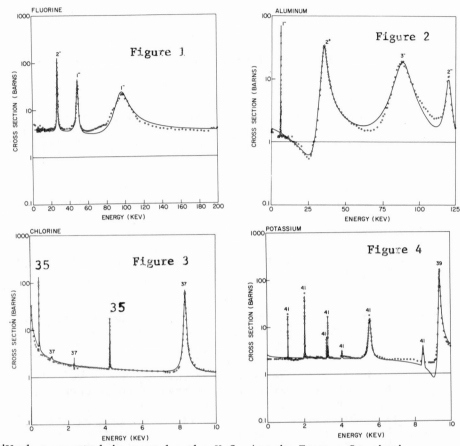

*Work supported in part by the U.S. Atomic Energy Commission.
1. G. Breit and E. Wigner, Phys. Rev. 49, 519 (1936).
2. P.L. Kapur and R.E. Peierls, Proc. Roy. Soc. (London), A 166, 277 (1938).
3. J. Humblet and L. Rosenfeld, Nucl. Phys. 26, 529 (1961).
4. E.P. Wigner and L. Eisenbud, Phys. Rev. 72, 29 (1947).
5. J.B. Garg, J. Rainwater and W.W. Havens, Jr., Phys. Rev. C 3, 2447 (1971).
6. T. Tiechmann and E. Wigner, Phys. Rev. 87, 123 (1952).
7. R.G. Thomas, Phys. Rev. 97, 224 (1955).
8. F.W.K. Firk et al, Proc. Phys. Soc., 82, 477 (1963).
9. J. Rainwater, et al, Rev. Sci. Instr. 35, 263 (1964).
10. H.W. Newson et al, Annals of Phys, 14, 365 (1961).
11. C.T. Hibdon, Phys. Rev. 133, B353 (1964).
12. R.C. Block et al, RPI Linear Acc. Proj. Rep. (Oct'67-Sept'68).
13. G. Nystrom et al (Private Communication).
14. C.T. Hibdon, Phys. Rev. 114, 179 (1959).
15. Neutron cross sections, BNL 325, Vol. I (1964).

DISCUSSION

NEWSON

Both potassium and chlorine are very difficult things to get pure. Could you tell me the precautions you took to be sure that you got pure compounds?

WYNCHANK (Columbia)

Perhaps I can help. The samples were analyzed by a company of analytical chemists in New York City. I can give you the name of the company and details of their spectroscopic analysis later, if you wish.

NEWSON

What sort of figures did they give for bromine in chlorine, rubidium in potassium?

WYNCHANK

Less than the order of one part in 10^4 I think, if my memory serves me correctly.

NEWSON

That was too small to account for any of the peaks that were shown on the slide. Thank you.

NEWSTEAD

I would like to ask a general question. I suppose we might say it's a philosophical question, and perhaps I could open it up to be answered by the speaker or one of the R matrix experts in the audience. It's this--I suppose R- matrix theory is a square-well theory, and so the penetrability one uses is the square-well penetrability. Yet, of course, we know physically that isn't the way the nucleus looks, and in the optical model calculations, one would use the penetrability corresponding, say to the Saxon-Woods shape. So, what penetrability should one really use in determining these reduced neutron widths and comparing them to the optical model calculations?

WIGNER

As far as the R matrix theory by itself is concerned, the situation is simple. One can use for the radius of the square-well, any radius large enough so that the nuclear interaction outside this radius can be neglected. Naturally, the penetrability factor will depend on the radius chosen, but the variation of the penetrability will be compensated by the variation of the R matrix. The exact position of the levels and the reduced widths will depend on the radius chosen and the resulting change of R will compensate for the changed penetrabilities. If this compensation is mathematically exact, the calculated collision matrix will be independent of the radius. This is called "Radius of the Internal Region," in R matrix theory.

The transition to the optical model is more complicated if one wants to make it directly and not via the collision matrix. Naturally, if one has the collision matrix, one is no longer interested in the optical potential so that only a direct transition is of interest. There is no single best prescription known to me to make the transition. The most attractive method was, I believe, proposed by E. Vogt in his Rev. Mod. Physics article. It is based on placing the surface of the square-well inside the nucleus, into the region where the Saxon-Woods potential is about half its maximum value. Vogt gave the modification of the R matrix theory which is necessary for such a position of the surface of the square-well.

May I ask a question which bothers me greatly? I hope I am not out of order. Dr. Moore had some data on the crossection of uranium and angular distributions of fission fragments at an energy of a tenth of an electron volt. He had an A_2 of minus one which means that he did not have pure S waves. Have I misunderstood this?

MOORE

I think I can clear that up, but perhaps Dr. Postma would like to talk about this. The angular distribution arises because one has aligned nuclei. These are S wave resonances. When the nuclei are aligned, then the angular distribution of the fission fragments can have an expansion.

NEWSTEAD

I just want to make one more comment. It's more an experimental question. What does one do about the errors on the strength function when one has so few levels? I don't suppose one can take the Porter-Thomas Distribution into account because then the error would be 150% if you have one level, for example. If so much of the strength is concentrated in that one region, it does seem that these few levels are more representative of the strength function than the usual error assignment would indicate. But then, how does one assign the error?

SINGH (S.U.N.Y./Albany)

Well, it's hard to assign the error on the strength function if you have, say, less than three levels. For example, we have only one s-wave level in aluminum and I did not assign any error in it, because error in the strength function is given by the square root of 2.3 divided by n where n is the number of resonances. If you have only one level, n=1, then the error is meaningless.

FESHBACH

I want to reply to a question Dr. Newstead raised in another form to Dr. Wigner's reply. If you remember, when Mr. Singh exhibited his comparison with the R-matrix theory, he had expressions coming from distant levels. That was in $R\infty$. Now you can get that

by the R—matrix theory, if you wish, or you can assign that to
potential scattering. Now, one way or the other, it doesn't
matter. I would say that there is one advantage of assigning it
to potential scattering, namely, you know something about the
optical potential. That's not the optical potential that is
giving rise to those terms, but you would think that the form of
the potential would be similar to the optical potential, and you
could use those parameters describing the potential to fit the data.
They're equivalent, but that's another way of doing it. I think
that's the question you were asking.

BOLLINGER (Argonne)
     I have two questions of detail. First of all, I noticed that
for the fluorine and I believe also for the aluminum, the fit is
rather poor by modern standards between resonances, and I wonder
if there's some easy explanation for that. Secondly, for the
fluorine resonance at about 90 kilovolts, there appeared to be
clear evidence for structure that you ignore, and I wonder again
if there is good reason for that.

SINGH
     Yes, in case of fluorine there is a level at about 97 kev.
Some people have reported fine structure with about two or three
levels there, but we could not fit it. We don't agree there are
other levels sitting there. We tried, but we could not improve
the fit. In the case of aluminum, we were not able to fit the
minimum of the 89 kev. resonance, and that may be due to the
experimental uncertainty in the data.

COMPARISON OF FESHBACH AND R-MATRIX REACTION THEORIES FOR

^{12}C(n,n)^{12}C *

B.A. ROBSON   and   W.J. VAN MEGEN

Research School of Physical Sciences

The Australian National University, Canberra, Australia

1. INTRODUCTION

Two main approaches to a microscopic description of nuclear reactions are Feshbach's unified theory 1) and the Wigner-Eisenbud R-matrix theory 2).   It is the main aim of the present work to compare the results of both these theories for a realistic problem, namely the elastic scattering of neutrons from ^{12}C below the inelastic threshold of 4.43 MeV, when the same model is employed to describe the nuclear structure.   For this energy region, the cross section is dominated by positive parity states and for simplicity the negative parity contributions are ignored.   Both a shell model and a collective model are used to describe the positive parity states of ^{13}C.

2. CALCULATIONS

2.1 Shell Model

Following Barker 3), the model adopted for the ^{13}C system is that of a 2s or 1d neutron weakly coupled to either the 0^+ ground state or 2^+ 4.43 MeV first excited state of ^{12}C, which are taken to belong to the configuration $(1s)^4(1p)^8$ and partition [444]. The basis functions for describing this system are taken to be eigenfunctions of the Hamiltonian

$$H_o = H(^{13}C) + h(sp) \qquad , \qquad (1)$$

where $H(^{12}C)$ defines the states of ^{12}C and h(sp) is a single

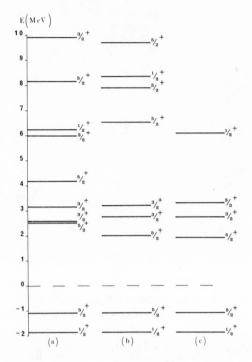

Fig.1. Energy levels of ^{13}C. Eigenvalues $(J \leqslant 5/2^{+})$ of (a) shell model Hamiltonian (b) particle-rotator model Hamiltonian compared with (c) experiment.

particle energy operator of the form

$$h(sp) = T(sp) + V(r) \quad , \tag{2}$$

$T(sp)$ being the kinetic energy operator and $V(r)$ a Woods-Saxon potential which for convenience is chosen such that both the 2s and 1d states are bound by a few MeV. In order to allow the bound states of ^{13}C to have the correct binding energies, a single particle residual interaction of the form

$$r(sp) = B\,\ell\,(\ell+1) + S\,\underset{\sim}{\ell}\cdot\underset{\sim}{s} + V_{o} \tag{3}$$

is assumed, where S is taken to be -2.03 MeV corresponding to a d3/2 - d5/2 level splitting of 5.08 MeV found for ^{17}O. The constants B and V_{o} are adjusted so as to fit the binding energies of the $\frac{1}{2}^{+}$ (-1.86 MeV) and $5/2^{+}$ (-1.09) bound states. The interaction between the core nucleons and the extra neutron is taken to be a simple zero range two body residual interaction of the form 4)

$$H_I = -497.5 \sum_i (0.865 + 0.135 \, \underline{\sigma} \cdot \underline{\sigma}_i) \, \delta(\underline{r} - \underline{r}_i) \text{ MeV} \qquad (4)$$

where $\underline{r}_i$ denotes the coordinate of a 1s or 1p core nucleon.

2.2 Particle-Rotator Model

In this collective model 5), the ^{13}C system is also considered to consist of a 2s or 1d neutron weakly coupled to the ground and 4.43 MeV states of ^{12}C which are assumed to be the I=0 and 2 states of a rigid rotator. The total Hamiltonian for the system

$$H = H(^{12}C) + h(sp) + r(sp) + H_I \qquad , \qquad (5)$$

where $H(^{12}C)$ defines states of an axially symmetric rotator and H_I is taken to be

$$H_I = -\beta f(r) \, Y_2^0(\theta', \varnothing') \qquad , \qquad (6)$$

where the angles are measured in the body system and $f(r)$ is a Woods-Saxon derivative form factor with radius and diffuseness parameters R_I and d_I.

2.3 Results

Fig.1 shows the level spectrum $(J \leqslant 5/2^+)$ for (a) the shell model Hamiltonian (b) the particle-rotator model Hamiltonian $(\beta = -16$ MeV·fm, $R_I = 4$ fm, $d_I = 0.65$ fm) compared with experiment (c).   For the energy region of interest the collective model gives a much better description of the positive parity states.   The discrepancies of the shell model spectrum, notably the small separation of the $5/2^+$ and $3/2^+$ states near 2.5 MeV and the $5/2^+$ state at 4.2 MeV, persist in the subsequent calculations of the total neutron cross section.

Figs.2(a) and 2(b) show the results for the Feshbach and R-matrix theories compared with experiment for the shell and particle-rotator models, respectively.   The Feshbach and R-matrix calculations essentially follow the corresponding works of Lovas 6) and Purcell 7) except that more realistic (Woods-Saxon) single particle radial functions have been used and all higher configurations (3s, 2d, etc.) have been omitted. The inclusion of higher configurations was found not to lead to any improvement in the fit.   For both models, the Feshbach theory predicts two sharp resonances superimposed upon a broad $3/2^+$ resonance and mainly $\frac{1}{2}^+$ background.   The broad $3/2^+$ resonance arises primarily from the $3/2^+$ level predicted to lie near 6 MeV (shell model) or 8 MeV

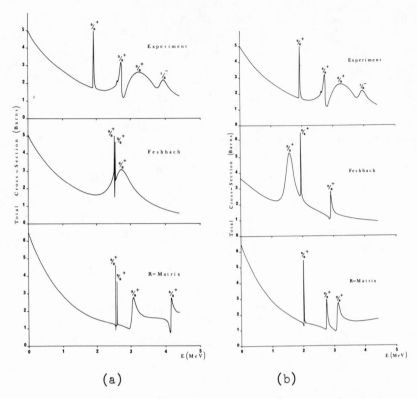

(a)   (b)

Fig.2.   Total elastic scattering cross section for Feshbach and
R-matrix theories using (a) shell model (b) particle-rotator model
compared with experiment.

(particle-rotator model).   The coupling of the open and closed
channels leads to a large level shift for this state. Consequently,

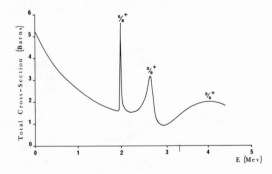

Fig.3.   Total elastic scattering cross section for modified
Feshbach theory.

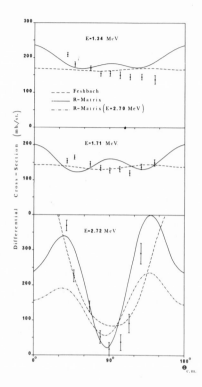

Fig.4. Differential cross section at 1.34, 1.71 and 2.72 MeV for modified Feshbach and R-matrix theories compared with experiment.

the resonance appears at a much lower energy and by interference destroys the single particle $3/2^+$ resonance which in the absence of coupling would occur near 4 MeV for our choice of scattering potential (similar to that used by Lovas). Thus contrary to the work of Lovas the overall agreement with the experimental total neutron cross section is not very satisfactory. By contrast, the R-matrix results (the channel radii are 6.7 fm (shell model) and 6.5 fm (particle-rotator model)) show that in this theory the level shifts arising from allowing the states to decay are very small so that in the collective model the theory gives a qualitative description of the data.

Fig.3 shows the prediction of the Feshbach theory for the particle-rotator model when only contributions from the levels of fig.1(b) in the energy region of interest are included. This rather artificial procedure (modified Feshbach theory) leads to substantially better agreement with experiment.

For the particle-rotator model both the R-matrix theory (fig.2(b)) and the modified Feshbach theory (fig.3) give a

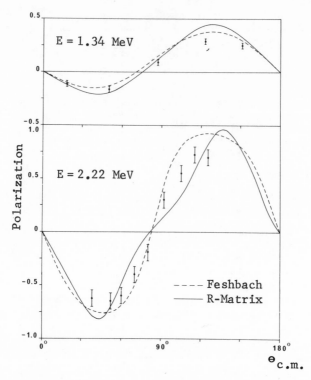

Fig.5. Polarization at 1.34 and 2.22 MeV for modified Feshbach
and R-matrix theories compared with experiment.

reasonable description of the total cross section. It is of
interest to compare the corresponding predictions for more
sensitive quantities such as the differential cross section and
polarization.   However, these quantities depend critically upon
small ℓ =1 and ℓ =2 components in the amplitude and while these
contributions may be well described in the Feshbach theory, this is
not the case in our R-matrix calculations where no account has
been taken of the distant levels.   For this reason ℓ =1 and ℓ =2
potential scattering terms were included in the R-matrix approach.
The parameters of this potential were taken to be the same as
those used to obtain the elastic scattering wave functions in the
Feshbach calculations.   Figs.4 and 5 show some theoretical
predictions for the modified theories compared with experiment.
The comparisons have been restricted to energies less than 3 MeV
since the method adopted for including distant levels in the
R-matrix theory leads to difficulties in the region of the $3/2^+$
single particle resonance at 3.2 MeV.

3. DISCUSSION

The foregoing calculations allow a comparison to be made between the Feshbach and R-matrix theories.

The Feshbach theory, by means of projection operators, separates the resonances into two classes whereas the R-matrix theory treats all resonances on an equal footing. In each case the nuclear structure information enters the calculation via model Hamiltonians and basis functions. It is very desirable that the reaction theory provide results which are consistent with this input data. This means that the resonance energies should lie close to the eigenvalues of the model Hamiltonians. While the R-matrix theory satisfies this 'small level shift' criterion, it has been found that for the Feshbach method this is not the case for some of the higher levels. Thus the weak coupling approximation inherent in the usual treatment of the Feshbach method is not generally valid for the nuclear problem. Substantially better agreement with experiment was obtained for the Feshbach theory when the contributions from the higher states were ignored.   The R-matrix theory contains the channel radii as parameters but provided these are not too small the results are not very sensitive to the actual values. The R-matrix theory is much faster to calculate, especially when many bombarding energies are required, since most of the calculation is energy independent. In the present work the Feshbach calculations were roughly 100 times slower than the corresponding R-matrix computations.

The simple shell model adopted for the positive parity states of ^{13}C appears to be inadequate whereas the particle-rotator model allows a good description of these states for the energy region of interest. Using the latter model, the R-matrix theory leads to a fairly satisfactory description of the resonances observed in the total cross section below 4.43 MeV. In addition, both the modified Feshbach theory and the R-matrix theory including potential scattering terms as discussed above give a satisfactory description of some differential cross section and polarization data below 3 MeV.

* The paper was presented by Dr. K. J. LeCouteur.
1) H. Feshbach, Ann.Phys. 5 (1958) 357, 19 (1962) 287.
2) W. Tobocman and M.A. Nagarajan, Phys.Rev. 138(1965) B1351.
3) F.C. Barker, Nucl.Phys. 28 (1961) 96.
4) G.E. Brown, L. Castillejo and J.A. Evans, Nucl.Phys. 22 (1961) 1.
5) D. Kurath and R.D. Lawson, Nucl.Phys. 23 (1961) 5.
6) I. Lovas, Nucl.Phys. 81 (1966) 353.
7) J.E. Purcell, Phys.Rev. 185 (1969) 1279.

HIGH RESOLUTION PROTON RESONANCE REACTIONS

E. G. Bilpuch

Duke University and Triangle Universities Nuclear Laboratory

Durham, North Carolina

For the past several years we have been measuring proton excitation functions for all available even-even isotopes in the mass region $40 \leq A \leq 64$. These excitation functions extend over energy ranges greater than 1 MeV and on the average were taken in energy steps of about 200 eV. One can quickly calculate that for the 18 isotopes studied or being studied this represents a tremendous amount of data. My graduate students, past and present, who performed these experiments are as follows:

^{40}Ar	Dr. G. A. Keyworth[1]
^{40,42,44,48}Ca	Mr. W. M. Wilson (in progress)[2,3]
^{46,48,50}Ti	Dr. N. H. Prochnow[4]
^{50,52,54}Cr	Dr. J. D. Moses[5]
^{54,56,58}Fe	Dr. D. P. Lindstrom[6]
^{58,60,62,64}Ni	Dr. J. C. Browne[7,8]

In addition we have studied the (p,p') and (p,n)[9] reactions for the above isotopes whenever possible. We are now engaged in high resolution (p,γ) measurements on the analogue fine structure states in ^{54}Cr and ^{58}Fe and Mr. W. C. Peters is responsible for these measurements.[10] The high resolution techniques used in these experiments were developed in collaboration with Professor H. W. Newson of Duke University. In recent years I've enjoyed the close collaboration of Professor G. E. Mitchell of North Carolina State University and The Triangle Universities Nuclear Laboratory.

In the 18 isotopes studied we have resolved the fine structure of over 40 analogue states. These analogue states are commonly called "micro-giant resonances". The analogue state (the $T_>$ state) mixes with the normal states of the compound nucleus (the $T_<$ states). The number of fine structure levels that one observes is a function of the level density of the compound states (with the same J^π as the analogue state) and of the spreading widths of the analogues. In these data we have observed analogue states as single levels in isotopes with large spacings and we have observed up to 20 fine structure states in isotopes with spacing (for a particular J^π) of the order of 5 keV. The kinds of information one can obtain from the fine structure of analogue states include:

1. Spectroscopic factors of the parent state.
2. Spins of the parent states.
3. Coulomb displacement energies of the isobaric pair.
4. Shape of the fine structure distribution.
5. Spreading widths of the analogue state.
6. Local density measurements due to enhancement of $T_<$ states by the analogue state.

Away from the analogue state, or if one looks at a specie of levels with spins different from the analogue state, we can obtain statistical information such as:

1. Level density changes with energy for states of a particular J^π.
2. The effect of the density changes on the spacing distribution.
3. The effect of the density and strength function changes on the width distribution.
4. Average strength functions for various J^π levels and changes in the local strength function as a function of energy.
5. Correlation of the partial widths of the elastic channel with the (p,p') channel, the (p,n) channel, and the (p,γ) channels.

The subject of correlations between partial widths of different channels is rather important; our results on this topic will be discussed by Professor Mitchell in Session 4 of this conference.

The quality and the scope of our data are best illustrated in Figure 1. Here we see the ^{48}Ti(p,p) cross section over an energy range of about 1.4 MeV. These data are taken at 4 or 5 angles for the elastic channel and 3 or 4 angles for the inelastic channel. Over this energy range our resolution is constant and equal to about 300 eV. The target thicknesses are from one to 2 μgm/cm^2 in all cases and counting statistics are 1 percent, i.e., $\sim$10,000 counts per point. In the lower two curves we have expanded the scale to illustrate the details in the excitation function. In these data there are 5 species of compound nuclear levels; $1/2^+$, $1/2^-$, $3/2^-$, $3/2^+$ and

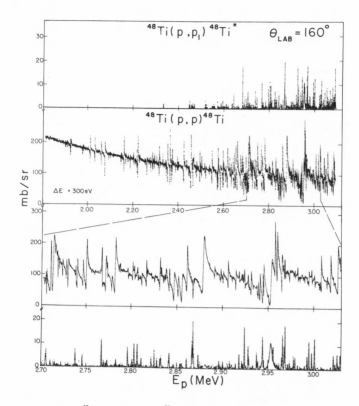

Fig. 1. The ^{48}Ti(p,p) and ^{48}Ti(p,p$_1$) differential cross section at 160° over the entire energy range. Lines are fits to the data.

5/2$^+$ states. These are the normal compound nuclear levels and comprise the background states (T$_<$ states) for the analogue states. Sometimes these states are not observable because of their narrow widths and I shall show examples of these types of data. When the T$_<$ states are observed, as in this case, then the analogue states are not obvious by inspection. In these data the analogues of the 17th (1/2$^-$) and the 18th (3/2$^-$) excited states of ^{49}Ti are at E$_p$ = 2.85 and 2.95 MeV, respectively. This will be clearly demonstrated when the results of the analysis of these data are shown later in this paper. In this figure the lines are fits to the data in both the elastic and inelastic channels. The fit is obtained with a multi-level multi-channel code[11] based on R-matrix theory. The expressions upon which this code is based are given in the detailed treatment by Lane and Thomas.[12] The R-matrix theory is very general in that nuclear reactions may be treated without introducing a detailed nuclear model. In our case we use the R-matrix theory to parametize the observed resonances in terms of their total angular momentum, energy, parity and their total and partial widths.

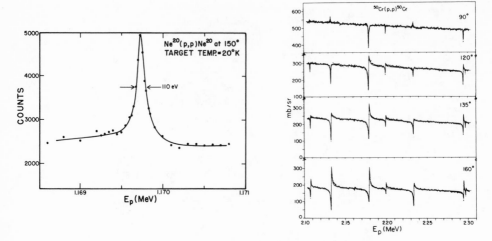

Fig. 2. Narrow resonance illustrating beam resolution.
Fig. 3. Typical fits for solid target data.

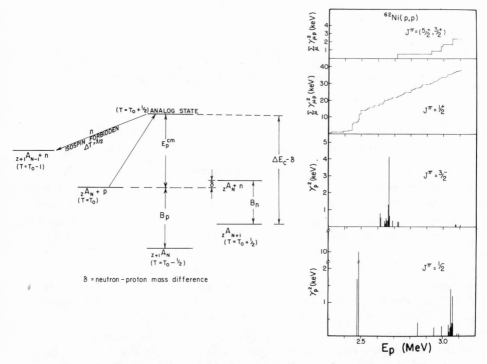

Fig. 4. Energy relationship between the analogue state and
the parent state.
Fig. 5. Summary of reduced widths for ^{62}Ni(p,p).

Figure 2 is shown to demonstrate our overall resolution (110 eV) which includes the incoherent beam spread, Doppler broadening in the target, the target thickness and the natural width of the resonance. These data were taken with our windowless cryogenic target chamber[13] which was voltage modulated to cancel the time dependent energy fluctuations of the proton beam.[14] The temperature of the neon gas was 20° K, the angle of observation was 150° and the energy steps over the resonance were 25 eV. At these temperatures and energy the target Doppler is about 40 eV and the target thickness is even less. This resonance is the 5th excited state of ^{21}Na and a single level fit shows it to have a laboratory width of only 8 eV with $J^\pi = 5/2^+$. Therefore, in our measurements with gas targets most of our energy spread is in the proton beam from our RF ion source. This is approximately 100 eV.

For our data on solid targets the best resolution that we have been able to obtain is 300 eV for 1 MeV $< E_p <$ 3 MeV. The fact that we seem to be limited to a resolution of 300 eV when we have a beam spread of only 100 eV apparently is related to the target Doppler. Measurements of very narrow levels indicate broadening only when our targets have exceeded 3 μgm/cm^2. Since our targets are on ∼ 10 μgm/cm^2 carbon foils, there is an energy loss in the carbon backing. If we assume that the heat conduction is small along the carbon foil, then the target can only lose heat by radiation. With 5 to 10 μA on target, the temperature increase in the target leads to a Doppler broadening which is consistent with our observed resolution. We are presently investigating this phenomenon in greater detail.

Figure 3 shows the ^{50}Cr(p,p) excitation function at four angles in the laboratory. The solid line through the data is a fit to the data using a Gaussian resolution function of 300 eV FWHM. The fit contains $1/2^+$ (indicated by dips at 90°) $1/2^-$ and $3/2^-$ levels. For example the level at 2.20 MeV has $J^\pi = 1/2^+$ and $\Gamma_p = 60$ eV. In these data only the elastic channel is open and therefore we can assume that $\Gamma_p/\Gamma = 1$. The capture channel is small and can be neglected.

In analogue state experiments the location of the analogue state can be determined very simply. As shown in Figure 4, the proton center-of-mass energy ($E_p{}^{cm}$) required to reach the analogue of the ground state of the parent nucleus ($_Z A_N + 1$) is

$$E_p{}^{cm} = \Delta E_c - B_n$$

where ΔE_c is the Coulomb energy of the additional proton in the compound nucleus (daughter) and B_n is the binding energy of the neutron in the parent nucleus. Experimentally, it is found that the spacing between analogue states in the daughter nucleus reproduces to a remarkable accuracy the spacing of the low-lying states in the parent nucleus.

In general we use the above expression to determine which parent states can be observed in our energy interval and we use the spectroscopic factors to estimate the strength of the analogue. For the target nucleus (^{62}Ni) we expect to observe the analogues of the ground state (1/2$^-$), the 2nd excited state (3/2$^-$) and the 3rd excited state (1/2$^-$) of ^{63}Ni. We therefore expect to find enhancement of the T$_<$ states in ^{63}Cu at E_p = 2.48, 2.65 and 3.04 MeV. As shown in Figure 5 the analysis of these data show this dramatically. In this figure proton energy is plotted on the abscissa and reduced widths (γ_p^2) are plotted on the ordinate. In the compound nucleus we are at an excitation of approximately 9 MeV and the average level width for the 1/2$^-$ and 3/2$^-$ compound states are such that they are not observable with our resolution. The analogue state enhances the compound states (T$_<$ states) and they become observable in the vicinity of the analogue state. Because of this enhancement we can therefore obtain a measurement of the level density for a particular J$^\pi$ which is not otherwise obtainable. For the 1/2$^+$ states we have plotted $\Sigma\gamma_p^2$ as a function of E_p. The slope of this histogram is the strength function:

$$S.F. = \frac{\Sigma\gamma_p^2}{\Delta E_p} = \frac{\langle\gamma_p^2\rangle}{\langle D\rangle}$$

Above 2.5 MeV the slope is constant and the value of the strength function is only slightly greater than the black nucleus value. In our units ($\langle\gamma_p^2\rangle/\langle D\rangle$) the black nucleus value is approximately 0.04. Blatt and Weisskopf[15] give the relation $\Gamma_n = 2kR\gamma_p^2$. Using this relation the strength functions in terms of $\langle\gamma_p^2\rangle/\langle D\rangle$ should be divided by ~400 to obtain the strength function in terms of $\langle\Gamma_n^\circ\rangle/\langle D\rangle$. If we divide the excitation function into intervals of 100 keV and sum the reduced widths in each interval we obtain a local strength function. In this isotope such a curve would produce an anomaly in the energy region 2.4 MeV to 2.5 MeV.

In Figure 6 we show the elastic proton data for two of the iron isotopes (^{54}Fe where Z = 26, N = 28 and ^{58}Fe where Z = 26, N = 32). These data illustrate the difference in level densities (greater than an order of magnitude) that can occur in two isotopes. The three upper curves for the target nucleus ^{54}Fe show isolated resonances that have ℓ = 0, 1, 2. When the data is taken at 4 angles the ℓ-values can be assigned by inspection. In the lower curve, the data on ^{58}Fe show the same ℓ-values but since the spacing is much smaller, the average reduced widths are also much smaller if the strength functions of each J$^\pi$ remain essentially constant. In ^{54}Fe the resonances at 2.243 (3/2$^-$) and 2.751 (1/2$^-$) represent analogue states that appear as single levels. The spacing in this nucleus is so large that the analogue state is not close enough to a T$_<$ level to mix. In contrast, the region around 2.98 MeV in ^{58}Fe shows many fine structure levels. In this region we expect to see the 3rd excited state (3/2$^-$) of ^{59}Fe. The fact that we do observe this level is shown in Figure 7 which shows the analysis of the 3/2$^-$ levels.

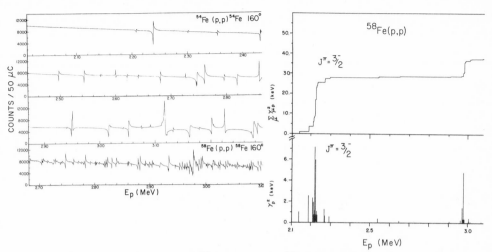

Fig. 6.   Comparison of ^{54}Fe(p,p) and ^{58}Fe(p,p) data.
Fig. 7.   Summary of reduced widths of 3/2$^-$ states in ^{58}Fe(p,p).

The two steps in the reduced width plot represent the analogues of the ground state and 3rd excited state of ^{59}Fe.   This represents a beautiful example of the fragmentation of the analogue state.   For ^{54}Fe we see more 1/2$^-$ and 3/2$^-$ states than we do in the region between the analogue states in ^{58}Fe due to a larger spacing and a lower value of A.   As the spacing increases the strength must appear in fewer levels.   In addition as A decreases the 2p giant resonance is nearer and thus the p-wave strength function is larger.

In Figure 8 we show the data for ^{52}Cr.   The spacing for 1/2$^+$ states is somewhat larger than the spacing of the 1/2$^+$ states in ^{58}Fe (a factor of about 4).   In these data the 8th (3/2$^-$) and the 11th (1/2$^-$) excited states of ^{53}Cr should be strong enough to be observed.   These states occur at 2.77 MeV and 3.14 MeV, respectively.   It is not very obvious that the normal compound states have been enhanced, because the enhancement is occuring in the midst of T$_<$ states which have average reduced widths large enough to be seen with our resolution.   That these two analogue states are present is illustrated by the results of the analysis of these data which are shown in Figure 9.   The 3/2$^-$ state is clearly indicated and the 1/2$^-$ state is less pronounced. From this work the spectroscopic factors for these two levels are 0.19 and 0.044, respectively.   The smallness of the spectroscopic factor for the 1/2$^-$ state is the reason this analogue state does not stand out in the reduced width plot.   The plot of the 1/2$^+$ reduced widths is very interesting.   A plot of the local strength function would show an anomaly around 2 MeV and also a strong increase near 3 MeV.   It is also interesting that the 5/2$^+$ strength is greater than 3/2$^+$ strength.   We infer that mass 52 is closer to the 2d$_{5/2}$ giant resonance than to the 2d$_{3/2}$ giant resonance.   Again a plot of the local

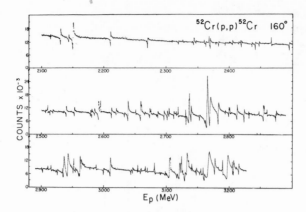

Fig. 8. ^{52}Cr(p,p) data at 160°.

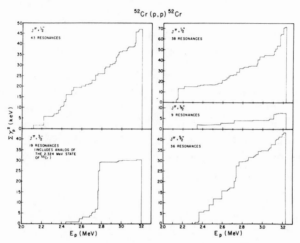

Fig. 9. Summary of reduced widths for ^{52}Cr(p,p).

strength function for the $5/2^+$ levels would show an anomaly around 2.7 MeV. In all of our data we observe these anomalies frequently for J^{π}'s other than that of analogue states. Because of their frequency we do not believe that these anomalies are statistical fluctuations and we tentatively consider them to be "doorway states".[16]

In Figure 10 we show the analysis of our ^{54}Cr data. These data are similar to the ^{62}Ni data except that we see no $\ell = 2$ resonances. The analogue states are those of the ground state $(3/2^-)$, the first excited state $(1/2^-)$ and the third excited state $(3/2^-)$ of ^{55}Cr. Except in the vicinity of the analogue states, the negative parity states have very small widths and cannot be observed. Above 2.5 MeV we begin to observe a few $1/2^-$ levels but these states have laboratory widths of the order of 20 to 30 eV. Because the spacing is larger and we are closer to the $2p_{1/2}$ giant resonance, we should expect to see the normal 1/2⁻ levels before the 3/2⁻ levels.

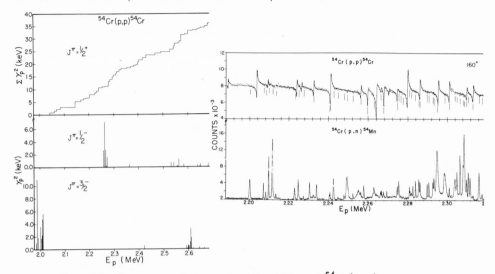

Fig. 10. Summary of the reduced widths of ^{54}Cr(p,p).
Fig. 11. Comparison of the (p,p) and (p,n) channels for the target ^{54}Cr.

    In this particular nucleus the (p,n) channel opens at about 2.2 MeV. Figure 11 shows the elastic data in the upper curve and the (p,n) yield in the lower curve. It is immediately obvious that there are more observable levels in the neutron channel than in the elastic channel. All of the proton resonances shown have $J^\pi = 1/2^+$ or $1/2^-$. Since the ground state of ^{54}Mn is 3^+, the proton resonances decay by neutron emission with $\ell' = 2$ or 3. In this case the proton penetration is about 500 times greater than the neutron penetration at 20 keV above the neutron threshold. Therefore the neutron channel is effectively closed. The situation is different for $5/2^+$ and $3/2^-$ resonances which decay with $\ell' = 0$ and $\ell' = 1$ neutrons, respectively. For the $5/2^+$ levels neutrons can decay with $\ell' = 0$ and the neutron penetration factor is larger than that for the proton. This, coupled with the fact that the s-wave neutrons are near the $3s_{1/2}$ giant resonance, makes Γ_p/Γ so small that no $\ell = 2$ resonances are observed in the elastic channel.

    In Figure 12 we show the excitation functions for the ^{48}Ti(p,p) and ^{48}Ti(p,p') reactions for the proton energy range of 2.5 to 3.10 MeV. The lines through the elastic and inelastic data are fits. The partial widths Γ_p and Γ_p' are determined for each resonance, subject to the condition $\Gamma_p + \Gamma_p' = \Gamma$. The analogues of the 17th and 18th excited states of ^{49}Ti are located near $E_p = 2.85$ and 2.95 MeV, respectively. The two analogues are not readily apparent in the cross section data but can be seen easily after the analysis, as shown in Figure 13. In this figure (on the left hand scale) we show the

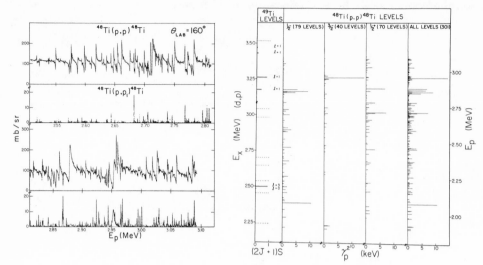

Fig. 12. Comparison of the (p,p) and (p,p₁) channels for the
target ⁴⁸Ti. Lines are fits to the data.

Fig. 13. Summary of reduced widths for ⁴⁸Ti(p,p) and comparison with the (d,p) results.

(d,p) data of Wilhjelm et al.[17] for excitation energies of 2.25 to 3.50 MeV
showing the 8th to the 21st excited states. On the abscissa we have plotted
the spectroscopic factor times $2J + 1$. The horizontal dotted lines indicate
non-stripping states and the solid lines stripping states with their assigned ℓ-
value. On the right we have plotted the reduced widths of all the levels
analyzed. In the three center plots we have separated the $1/2^+$, $1/2^-$ and
$3/2^-$ levels to show their distribution with energy. The fragmentation of the
17th ($1/2^-$) and the 18th ($3/2^-$) excited states of ⁴⁹Ti can be seen in the
figure. The large $1/2^-$ state at E_p = 2.1 MeV is not an analogue state. Due
to penetrability considerations the $\ell = 3$ stripping state at E_x = 2.50 MeV
is not seen in our data.

In Figure 14 we show a comparison of the ⁴⁶Ti and ⁴⁸Ti data for both
elastic and inelastic channels. In each case the lines are fits to the data
and one can see by inspection that the level density is quite different in the
two isotopes. In the case of ⁴⁶Ti we expect the level spacing to be larger
than that in ⁴⁸Ti. As I shall show later, we are approaching the $2p_{1/2}$ giant resonance. These two effects combine to give a larger average reduced
width for the negative parity states. This effect is clearly shown in these
data since we observe $1/2^-$ and $3/2^-$ levels lower in energy in the ⁴⁶Ti(p,p)
reaction than in the ⁴⁸Ti(p,p) reaction.

Figure 15 shows on an expanded scale the ⁴⁶Ti data at four elastic
angles and one inelastic angle. Again the lines are fits to the data. Com-

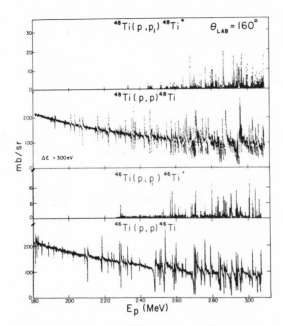

Fig. 14. Comparison of the ^{48}Ti(p,p) and ^{46}Ti(p,p) reactions. Lines are fits to the data.

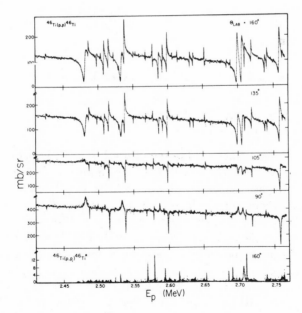

Fig. 15. Fits to the ^{46}Ti(p,p) reaction at 4 angles and the ^{46}Ti(p,p$_1$) reaction at 160°.

parison of this figure with Figure 12 indicates that for ^{46}Ti the spacing is larger
and the levels on the average have bigger widths. The small resonance at
$E_p \cong 2.55$ MeV is an $\ell = 3$ resonance with a laboratory width of only 1 eV,
and represents the smallest level that we have analyzed. In Figure 16 we
show the 160° ^{46}Ti elastic data from $E_p = 2.825$ to 2.96 MeV and the
^{46}Ti(p,p$_1$) differential cross section at 160°, 135° and 105°. This figure
illustrates the anisotropic angular distribution for the inelastic data of the 3/2$^-$
resonances. The four largest inelastic resonances in the bottom curve from
2.850 to 2.925 MeV are 3/2$^-$ and clearly have an anisotropic angular distribution. The comparison is further illustrated by the two p-wave resonances near
2.950 MeV. One is very well defined and has an anisotropic angular distribution (3/2$^-$ with a small width); the other is very broad and has an isotropic
angular distribution (1/2$^-$ with a large width).

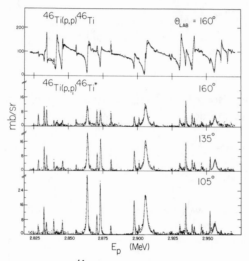

Fig. 16. Fits to the ^{46}Ti(p,p$_1$) reaction at 3 angles and the
^{46}Ti(p,p) reaction at 160°.

Figure 17 summarizes all the levels used in the ^{46}Ti(p,p) fit. As in
Figure 13 we show a comparison with the (d,p) data.[18] Except for the first
$\ell = 1$ resonance in the (d,p) work shown, all the $\ell = 1$ resonances have
$J^\pi = 3/2^-$ and, because of the level density and spreading width, they do
not have well-developed fine structure. Near $E_p = 2.50$ MeV the 1/2$^-$ plot
shows two very strong 1/2$^-$ levels (reduced widths of 42 keV and 31 keV).
If these levels are analogue states, they would have spectroscopic factors
close to one and it should be impossible to miss them in a (d,p) experiment.
If they are normal $T_<$ states, then they have about 10 percent of the Wigner
limit $(\hbar^2/2ma^2)$. As will be shown later the 1/2$^-$ strength function shows a
peak in this region due to the presence of these levels. In addition the density of 1/2$^-$ levels decreases in the vicinity of these two levels. The two

levels in question might be doorway states which are not fragmented.

Because of the number of levels present in the Ti isotopes, the data on these nuclei lend themselves to a statistical analysis[19]. In ^{49}V(^{48}Ti + p) we observed 70 $1/2^+$ levels over an interval of 1 MeV. In Figure 18a we

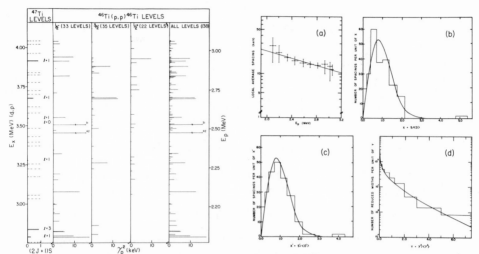

Fig. 17. Summary of reduced widths for ^{46}Ti(p,p) and comparison with the (d,p) results.
Fig. 18. (a) Local level spacing of the $1/2^+$ states in ^{49}V.
(b) Spacing distribution for the $1/2^+$ states. (c) Corrected spacing distribution. (d) Width distribution for the $1/2^+$ states.

show a plot of the local level spacing as a function of the proton energy. These data show an exponential dependence of the form $<D> = D_o e^{-E_p/T}$. The slope of the curve yields the nuclear temperature. In this case it has a value of 1.5 MeV, in agreement with values for this mass region.[20] In Figure 18b the histogram is the experimental spacing distribution normalized to a Wigner[21] distribution. The fit is reasonable but a χ^2 test gives only a 28% probability that the histogram corresponds to a Wigner distribution. However the Wigner distribution assumes a constant density and our data show a factor of two change in density. If we correct the data to constant density and replot our spacing histogram we obtain a better fit, as shown in Figure 18c. Here a χ^2 test increases the probability to 80%. However, if the density of levels increases, then we should see an effect in the average level widths. In Figure 18d we show the Porter-Thomas[22] distribution normalized to our experimental widths. The fit is better than one would expect. In Figure 19 we see why the width distribution fits so well. As shown in the three top figures, as the local level spacing decreases by a factor of two the local strength function increases by a factor of two. In terms of the local average reduced width, the two effects cancel each other, giving

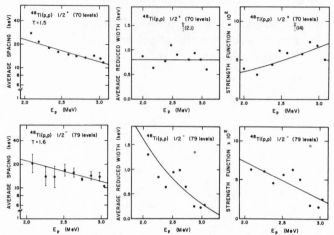

Fig. 19. Local spacings, local average reduced widths and local strength function for the $1/2^+$ and $1/2^-$ states in ^{49}V.

a constant reduced width over our energy interval.

As a check on this phenomenon we consider the $1/2^-$ states. Again we plot the local average spacing which shows a factor of two change and a nuclear temperature of 1.6 MeV in agreement with the $1/2^+$ states. In this case the local strength function also decreases by a factor of two and the two effects on the local reduced width are in the same direction. In the center bottom plot of Figure 19 we see that the local average reduced width does indeed decrease by a factor of four over our energy interval. The high point in the reduced width plot (open circle at 2.85 MeV) is due to the $1/2^-$ analogue state and has been omitted in these data.

In Figure 20 (left hand side) we have plotted the values of the local strength function per 200 keV for the $1/2^+$ states in the Ti isotopes. In each isotope we see trends as illustrated by the lines drawn through the data. The point for ^{51}V is a single point since we have only measured the excitation function over an energy range of 140 keV. These data suggest that if we translate the ^{47}V data down by an MeV and the ^{51}V point up by an MeV we would obtain a continuous curve as is shown in the right hand side of this figure. It now appears that we are observing a large fraction of the $3s_{1/2}$ giant resonance.[23] Qualitatively Figure 21 shows why we can shift the data as we have done. The real potential in the optical model contains a symmetry term $[24(N-Z)/A]$ which deepens the well for the heavier isotopes as indicated in this figure. The position of the giant resonance shifts as the well deepens. In this figure we have schematically represented the Coulomb barrier and the cross-hatched region is our effective laboratory energy. Qualitatively, then, as we move to heavier isotopes, we should

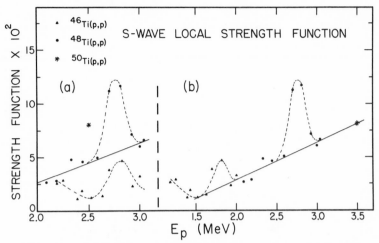

Fig. 20. (a) Local strength functions for the $1/2^+$ states in ^{47}V, ^{49}V, and ^{51}V. (b) Shifted local strength functions as discussed in the text.

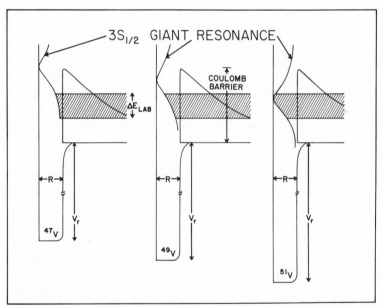

Fig. 21. Qualitative effect of the symmetry potential on the $3s_{1/2}$ giant resonance.

in effect move up on the giant resonance. An optical model calculation shows that this trend is correct but as yet we have not tried to fit the data by varying parameters in the optical model.

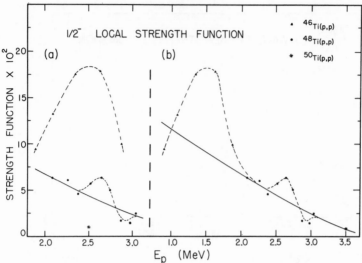

Fig. 22. (a) Local strength functions for the 1/2⁻ states in ⁴⁷V, ⁴⁹V and ⁵¹V. (b) Shifted local strength functions as discussed in the text.

In Figure 22 we show a similar plot for the $1/2^-$ resonances for the Ti isotopes. Again we have plotted the $1/2^-$ local strength function for each isotope and connected the points of each isotope by a line. In these plots we have removed the strength of the analogue states. If we shift each isotope by the same amount as we did for the $1/2^+$ states, then we obtain the curve shown on the right hand side of this figure. The lines through the data are simply to emphasize the trend. Again we see a large fraction of the giant resonance, but this time it is the $2p_{1/2}$ resonance. The single point for ⁵¹V is only an estimate of the $1/2^-$ strength since our limited energy range (140 keV) contains the analogue of the first excited state of ⁴⁹Ti which has $J^\pi = 1/2^-$. The big anomaly in the ⁴⁷V data is produced by the two large $1/2^-$ levels discussed previously. At this point it is tempting to say that we are looking at a "doorway state". We can shed more light on these data by plotting the density and the local reduced widths of the $1/2^-$ states for the ⁴⁶Ti(p,p) reaction (Fig. 23). The left hand plot shows a hole in the density near the two anomalous $1/2^-$ states. The solid line is the calculated density from Gilbert and Cameron.[20] We obtain the expected density at the low end of our energy interval and also at the high end of our interval. In the right hand side of this Figure we have plotted the local reduced width which shows the effect of the two big $1/2^-$ states which occur precisely where the spacing is greatest.

In order to summarize our data in strength functions we have calculated an average strength function for each isotope and for a particular

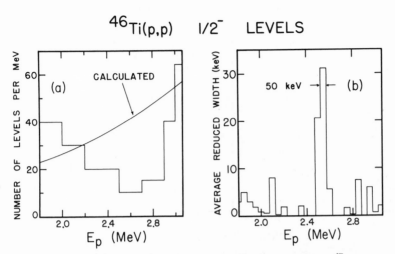

Fig. 23. (a) Local level density of 1/2⁻ states in ⁴⁷V.
(b) Local reduced widths of the 1/2⁻ states in ⁴⁷V.

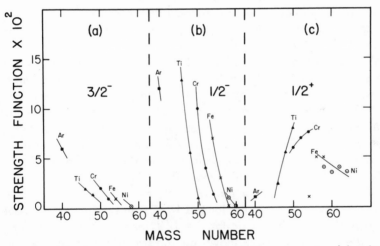

Fig. 24. Summary of proton strength functions for (a) 3/2⁻
states, (b) 1/2⁻ states and (c) 1/2⁺ states.

specie of levels (1/2⁺, 1/2⁻ and 3/2⁻).[24] These data are shown in Figure 24. The lines are merely to indicate trends--different symbols represent different elements. The data clearly show the $2p_{1/2}$ giant resonance. The small values of the 3/2⁻ strength function indicate that, as expected, the $2p_{3/2}$ giant resonance peaks below the $2p_{1/2}$ giant resonance (at lower values of A). The 1/2⁺ strength functions show an anomaly around the mass 50 region but the data seems to scatter. This is presumably the $3s_{1/2}$ giant

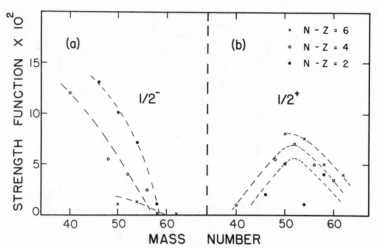

Fig. 25. (N-Z) effect on the proton strength functions for
(a) $1/2^-$ states and (b) $1/2^+$ states.

resonance. (The low point at mass 54 is for ^{54}Fe and is only based on two
resonances and probably should be discounted.) We plan to do some of
these measurements at higher energies to improve our strength function values. The strength function values shown in this figure are tentative; some
of the low values are estimates. In the last figure (Figure 25) we have re-
plotted the strength function data for the $1/2^+$ and $1/2^-$ states as a function
of N - Z in order to show the effects of the neutron excess. Except for
the $1/2^+$, N - Z = 2 plot the data show a remarkable symmetry. (The low
point in the N - Z = 2 for $1/2^+$ states is due to ^{54}Fe and should be discounted as noted above.) The plots for the $1/2^-$ states also show definite
trends which are indicated by the solid lines. The data for the $3/2^-$
states are too sparse for a plot of this type. These data show that the absolute value of the N - Z plots for the $1/2^+$ states increases as N - Z increases, but this behavior is reversed for the $1/2^-$ states. An optical model
calculation shows this same behavior for the $1/2^-$ and $1/2^+$ strength functions. Most of the data that I have shown on the statistical properties are
unpublished since we have only very recently been in a position to bring
our results together.

Up to now the statistical properties of highly excited nuclei have
been almost exclusively studied via neutron resonance reactions and very
little information has been obtained from proton resonance reactions. As
I have tried to show the high resolution proton resonance reactions can provide statistical information over greater energy ranges and for several spins
and parities in a single isotope. For this mass region $40 \leq A \leq 64$ these
data should add greater scope to the existing neutron data. Unfortunately,

this quality of data cannot be extended throughout the Periodic Table with our present experimental techniques. At higher A values, the higher Coulomb barrier and the increase in level density would increase the complexity of the excitation function beyond our ability to analyze it. Except for the lighter isotopes of Zn and perhaps a few isotopes around the mass 90 region (N = 50) such excitation functions would require an improvement in our resolution by a factor of ~3. At the present time we do not see how to accomplish this with solid targets. However, for some of the isotopes discussed in this paper which have a large level spacing, we now have plans underway to increase our measurements of the excitation functions to about 5 MeV. In addition to the s-, p-, d- and f-wave resonances already observed we should see g- and h-wave resonances in our increased energy range.

Throughout this series of experiments we have been encouraged by several nuclear theorists through discussions and correspondence. My colleagues and I therefore greatfully acknowledge the help of Dr. A. M. Lane of Harwell, Professor D. Robson of Florida State University, Professor W. Greiner of the University of Frankfurt, Professors L. C. Biedenharn and R. Y. Cusson of Duke University and Professor H. Feshbach of Massachusetts Institute of Technology.

References

1. G. A. Keyworth, G. C. Kyker, Jr., E. G. Bilpuch and H. W. Newson, Nuclear Physics 89, 560 (1966).

2. J. C. Browne, G. A. Keyworth, P. Wilhjelm, D. P. Lindstrom, J. D. Moses, H. W. Newson, and E. G. Bilpuch, Physics Letters 28B, 26 (1968).

3. P. Wilhjelm, G. A. Keyworth, J. C. Browne, W. P. Beres, M. Divadeenam, H. W. Newson and E. G. Bilpuch, Phys. Rev. 177, 1553 (1969).

4. N. H. Prochnow, H. W. Newson, E. G. Bilpuch and G. E. Mitchell. To be published.

5. J. D. Moses, H. W. Newson, E. G. Bilpuch and G. E. Mitchell. Submitted for publication (Nuclear Physics).

6. D. P. Lindstrom, H. W. Newson, E. G. Bilpuch and G. E. Mitchell, Nuclear Physics A168, 37 (1971).

7. J. C. Browne, H. W. Newson, E. G. Bilpuch and G. E. Mitchell, Nuclear Physics A153, 481 (1970).

8. J. C. Browne, D. P. Lindstrom, J. D. Moses, H. W. Newson, E. G. Bilpuch and G. E. Mitchell, Nuclear Isospin, p. 573 (Academic

Press Inc., New York, 1969).

9.   J. D. Moses, J. C. Browne, H. W. Newson, E. G. Bilpuch and G. E. Mitchell, Nuclear Physics A168, 406 (1971).

10.   G. E. Mitchell, E. G. Bilpuch, J. D. Moses, W. C. Peters and N. H. Prochnow, Session 4, This Conference.

11.   D. L. Sellin, "Excited States in ^{19}F", Ph.D. Dissertation (Duke University, 1969).

12.   A. M. Lane and R. G. Thomas, Revs. Mod. Phys. 30, 257 (1958).

13.   P. B. Parks, P. M. Beard, E. G. Bilpuch and H. W. Newson, Rev. of Sci. Instr. 35, 549 (1964).

14.   P. B. Parks, H. W. Newson and R. M. Williamson, Rev. Sci. Instr. 29, 834 (1958).

15.   J. M. Blatt and V. F. Weisskopf, Theoretical Nuclear Physics (John Wiley and Sons, Inc., New York, 1952).

16.   B. Block and H. Feshbach, Annals of Physics 23, 47 (1963).

17.   P. Wilhjelm, O. Hansen, J. R. Comfort, C. K. Bockelman and P. D. Barnes, Phys. Rev. 166, 1121 (1968).

18.   J. Rapaport, A. Sperduto and W. W. Buechner, Phys. Rev. 143, 808 (1966).

19.   E. G. Bilpuch, N. H. Prochnow, R. Y. Cusson, H. W. Newson and G. E. Mitchell, Physics Letters 35B, 303 (1971).

20.   A. Gilbert and A. G. W. Cameron, Can. J. Phys. 43, 1446 (1965).

21.   E. P. Wigner, Proc. Inter. Conf. Neutron Interactions of Nuclei, p. 49 (Columbia University, TID-7547, 1957).

22.   C. E. Porter and R. G. Thomas, Phys. Rev. 104, 483 (1956).

23.   E. G. Bilpuch, N. H. Prochnow and G. E. Mitchell. To be published.

24.   E. G. Bilpuch, J. D. Moses and G. E. Mitchell. To be published.

DISCUSSION

NEWSTEAD

As I understand you, you said for the 3/2's, as opposed to the 1/2, you have a different behavior with the symmetry potential. In one case you have an increase in the strength function one way, and in the other case it goes the opposite way. Is that correct?

BILPUCH (Duke)

I'm afraid you misunderstood. I said there is a difference between the 1/2+ and the 1/2-. The 3/2 data is so sparse that we can't say anything about it except that it peaks below the 2P 1/2 resonance. Of course, everybody knows that. We don't know where yet. But the N -Z curves show a flip as we go from one spin state to the other. An optical model gives you the same kind of thing.

NEWSTEAD

If you put the symmetry potential into it, you mean?

BILPUCH

I didn't put the symmetry potential into these N-Z curves because they are plots. I am trying to take out the symmetry potential by plotting it as a function of N-Z. Therefore, what I'm left with is just our dependence on the width. That's all we try to do.

NEWSTEAD

O.K., because the point I was trying to make was that the 1/2- and the 1/2+ peak at different mass numbers. The effect of the symmetry potential with mass number is very different whether you're in a strength function maximum or minimum. In the one case it would tend to increase the strength function, and in the other case it would tend to decrease it with mass number, because you're decreasing or increasing the width of the size resonance as you change the symmetry potential.

BILPUCH

Well, the 1/2- peak is in a completely different place than the 1/2+ as a function of A. The neutron peaks around 30 or 35. We don't know where the proton peaks are yet. Maybe it peaks in the same place. The data seem to indicate that it's peaking about 40, but that's a guess. The 3S 1/2 peaks around 54, I'd say. It isn't as big as the neutron 3S 1/2, and we say presumably it is the 3S 1/2 giant resonance. There's some work by Schiffer which says it should peak around mass 70. Maybe it's a two-hump thing. I don't know. We have to expand our energy range, and for some of the lighter isotopes that have S-wave spacings of 100 kev., we can certainly go up 2 Mev. in energy. In fact, we can go to a tandem if we must. We can go up to 10 Mev., but the difficulty increases

as you go up in energy. You start picking up more and more species
of ℓ-values.For example, if you went and picked up G and H-waves,
we would have eleven species of level on an even target and a minimum of something like 20 in a spin target. So you have to start
thinking about how you're going to separate all these things. What
we have done here is to separate spins.

HAUSMAN (Ohio)
        For, say, an ℓ=2 resonance, the shapes of the $j=\ell\pm1/2$ reso-
nances are very similar. How do you make your spin assignments in
that case?

BILPUCH
        You can sometimes look at the inelastic scattering because
usually the first excited state is 2+, so you have an inelastic
scattering that goes in as a d-wave and comes out as s and d-waves.
That helps sometimes. The other thing is that the interference dip
on the high energy side is kind of broad. You can get some idea
down to maybe 30 ev. You can separate them. After that you can't,
and in our data, if you notice, we have put parentheses around
anything under 30 ev. because we don't know which it is unless
we have the inelastic channel. Sometimes, with the inelastic
channel and the anisotropy, then we know what the spin has to be.
Also, I've done an experiment with Kyker some time ago on (p,p'γ)
on some of these isotopes, on chromium, titanium and iron; and
we've looked at the same levels there. We were seeing 5/2+ states
pretty cleanly. We agree very much with that analysis.

VONACH (Munchen)
        I noticed you observed just as many 1/2- levels as 3/2- levels,
whereas theoretically you might expect that the levels might be
about two times as frequent. Can you comment on this?

BILPUCH
        There are twice as many 3/2- states, and you're further away
from the giant resonance, so that the 3/2- are enhanced less than
the 1/2- states. This, added to the fact that there are twice as
many 3/2- levels, means that less strength is divided between more
levels and therefore they're not observable with our resolution.
I think that's the answer.

NUCLEAR SPECTROSCOPY UTILIZING AN (ALPHA, NUCLEON) RESONANCE REACTION

E. Sheldon, W.A. Schier, G.P. Couchell, B.K. Barnes,
D. Donati, J.J. Egan, P. Harihar, S.C. Mathur and
A. Mittler

Dept. of Physics, Lowell Tech. Inst., Lowell, Mass.

Resonance reactions offer that vital element of simplicity required in nuclear spectroscopy, especially when they involve entrance and exit channels of channel spin 0 or 1/2, since then the formation and decay of any one of several possible intermediate states J^{π} in a compound nucleus can take place only via a unique pair of incoming and outgoing partial waves l, l' associated with angular distributions of highly individualistic shape and magnitude. The potentialities have already been intensively, and fruitfully, exploited [1] in studies of the $^{13}C(\alpha,n_0)^{16}O$ reaction to arrive at spin and parity assignments to high-lying states of ^{17}O occupying the region of level overlap where coherent interference between a pair of levels of unlike parity manifests itself in disruption of the 90° symmetry of the ensuing angular distribution.

In this talk I shall deal with our commensurate investigation of the $^{31}P(\alpha,p_0)^{34}S$ reaction (Q = +0.631 MeV) at incident energies from 3.25 to 5.50 MeV, populating levels in the compound nucleus ^{35}Cl at excitation energies between 9.9 and 11.9 MeV, several facets of which have proved to be among the most interesting and rewarding thus far. In particular, it has provided intimations of systematic statistical trends in resonance cross section data that immensely facilitated our quest for spin-parity assignments to overlapping twin CN states.

The highly-structured character of the ground-state proton excitation function as measured in 5-keV steps is depicted in Figs. 1 and 2. It displays 48 peaks, at each of which we measured an on-resonance angular distribution. The count rates were converted to absolute differential cross sections preparatory to subjecting the distributions to shape and magnitude analysis

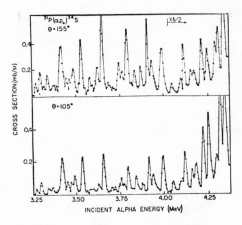
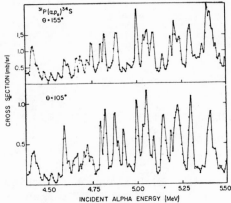

Fig. 1 Excitation functions.   Fig. 2 Excitation functions.

designed to furnish spin and parity assignments simultaneously for
either single, pure levels or twin interfering levels in the ^{35}Cl
compound nucleus. Our expectation of twin-level involvement was
not only repeatedly borne out in practice but was also substantiated
by estimates of the mean level separation as deduced from the high-
energy Gilbert-Cameron level-density formula; the mean separation
of $1/2^{\pm}$ levels ranged from 20 keV at the lowest energy to 7 keV
at the highest (cf. our incident energy spread was less than 15 keV).

     From Wigner-Eisenbud reaction theory applied to angular
distribution calculations by Blatt and Biedenharn and specialized
to resonance reactions by Lustig [2] the cross section formula is
expressible as a Legendre series in function of the emission angle
θ, wherein the weighting coefficients are dependent upon the energies, momenta and parities featured in the transition sequence.
Aside from familiar Racah factors to take account of vector momentum coupling through products of $\overline{Z}$-coefficients there are quantitatively unknown interaction amplitudes that comprise elements of
the unitary scattering matrix and that evince a complicated dependence upon energy, momentum and parity. Even when reduced to
partial widths they remain intangible parameters that preclude
numerical computation, except in the statistical limit of a quasi-
continuum of multiply overlapping CN states when they go over to
the readily calculable Hauser-Feshbach penetrabilities

$$\tau = T_{lj}(E_\alpha)T_{l'j'}(E_p)/\Sigma T_{l''j''}(E)$$

that are furnished by optical-model codes. In simplified form
the cross section formula in the case of overlapping twin CN
levels $J_1^{\pi_1}/J_2^{\pi_2}$ that coherently interfere at resonance ($\sin^2 \delta_R \to 1$,
with δ_R the resonant phase shift) can be written [3]

$$\sigma(\theta) = \tfrac{1}{8}\lambda_\alpha^2 \{\Sigma_\nu [\overline{Z}(1_1 1_1 J_1 J_1 ; \nu \tfrac{1}{12}) \overline{Z}(1_1' 1_1' J_1 J_1 ; \nu \tfrac{1}{12})(\Gamma_{\alpha_1}\Gamma_{p1}/\tfrac{1}{4}\Gamma_1^2)]P_{\nu_1}(\cos\,\theta)$$

$$+ [\overline{Z}(1_2 1_2 J_2 J_2 ; \nu_2 \tfrac{1}{2}) \overline{Z}(1_2' 1_2' J_2 J_2 ; \nu_2 \tfrac{1}{2})(\Gamma_{\alpha_2}\Gamma_{p2}/\tfrac{1}{4}\Gamma_2^2)]P_{\nu_2}(\cos\,\theta)$$

$$\pm 2[\overline{Z}(1_1 1_2 J_1 J_2 ; \nu_{122}\tfrac{1}{2}) \overline{Z}(1_1' 1_2' J_1 J_2 ; \nu_{122}\tfrac{1}{2})(\sqrt{\Gamma_{\alpha_1}\Gamma_{\alpha_2}\Gamma_{p1}\Gamma_{p2}}/\tfrac{1}{4}\Gamma_1\Gamma_2)$$

$$\times \cos(\phi_{\alpha_1}+\phi_{\alpha_2}-\phi_{p_1}-\phi_{p_2})]P_{\nu_{12}}(\cos\,\theta)\}$$

where the ϕ's are unknown potential phase shifts for the α- and
p-channels. Supposing there to be a simple numerical relationship
between the interaction-amplitude (partial-width) factors and the
corresponding penetrability terms τ of the form

$$\Gamma_{\alpha_1}\Gamma_{p1}/\tfrac{1}{4}\Gamma_1^2 \to N_1\tau_1 \text{ and } \Gamma_{\alpha_2}\Gamma_{p2}/\tfrac{1}{4}\Gamma_2^2 \to N_2\tau_2$$

$$\text{whence } \sqrt{\Gamma_{\alpha_1}\Gamma_{\alpha_2}\Gamma_{p1}\Gamma_{p2}}/\tfrac{1}{4}\Gamma_1\Gamma_2 \to \sqrt{N_1 N_2 \tau_1 \tau_2}$$

and gathering the phase constants of the mixed interference term
into a single unknown variable parameter δ, i.e., setting

$$\pm 2\cos(\phi_{\alpha_1}+\phi_{\alpha_2}-\phi_{p_1}-\phi_{p_2}) \to 2\cos\,\delta$$

we arrive at a formula suitable for quantitative application to the
experimental data once the values of the normalization factors N_1
and N_2 have been empirically determined. The phase variable $\cos\,\delta$
remains the only adjustable parameter in the analysis. In the case
of a pure single-level resonance, only the first term in the formula remains and the Legendre expansion runs only over even orders
ν, with $\nu \leq 2J$. This imposes a shape condition on the angular
distribution, yielding the characteristic forms for the various
J-values depicted in Fig. 3. By first selecting the symmetric
distributions and testing their shapes we could identify those
corresponding to single-level excitation
and determine the spin J. Complementing
this with magnitude analysis, we could not
only confirm the J-assignment but, also,
deduce the parity π, thanks to the very
pronounced parity selectivity exhibited
by the excitation function for any given
J-value. This welcome feature arose from
the drastic differences in barrier penetrability for the various 1, 1' combinations coupled to the state J^π. The
action of this "parity filter" is to
promote order-of-magnitude differences
between cross sections for the same spin
J but opposite parity π, as illustrated
in Fig. 4. The empirically selected

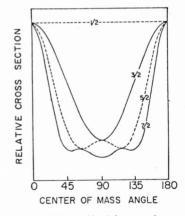

Fig. 3 Distribution shapes.

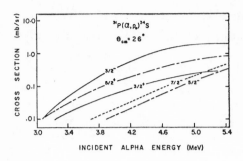
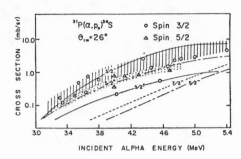

Fig. 4 Theoretical dσ/dΩ(26°).   Fig. 5 Experimental dσ/dΩ(26°).

single-level data in Fig. 5, far from being broadly dispersed about
the mean, likewise grouped themselves systematically in bands whose
energy dependence paralleled that of statistical theory and, moreover, could be brought into conjunction with the theoretical curves
on boosting the latter by a factor of four throughout. The need
for an enhancement factor for resonances of the same J^π is expected,
since the angular distributions are taken on resonances whereas
Hauser-Feshbach penetrabilities parameterize the average cross
section. On finding that the statistical cross section is related
to the mean on-resonance cross section by the same factor for all J^π
($N_1 = N_2 = \ldots = 4$) we could embark on the systematic analysis of all 48
on-resonance angular distributions, pure and mixed alike.

Of the 24 pure resonances, we identified instances of $1/2^{(\pm)}$,
$3/2^+$, $3/2^-$ and $5/2^+$ levels and present examples of fits in Figs. 6-9.
It should be noted that only in the J=1/2 case of Fig. 8 does an
ambiguity arise due to the weak parity selectivity (i.e. unlike the
other J^π's the parity selectivity is at most 1/3 for $\frac{1}{2}^+/\frac{1}{2}^-$ over the
energy range considered). The convention adopted in these and the
remaining figures is to display the theoretical cross section as a
solid curve on an absolute scale and as a dashed curve when redrawn
arbitrarily normalized for shape comparison.

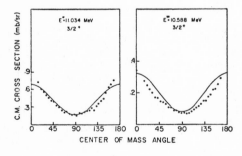
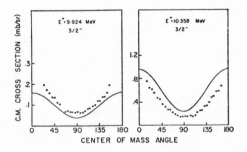

Fig. 6 Pure $3/2^+$ distributions.   Fig. 7 Pure $3/2^-$ distributions.

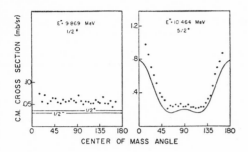
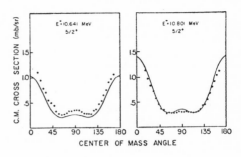

Fig. 8   Pure $1/2^{(\pm)}$ and $5/2^+$ cases.   Fig. 9   Pure $5/2^+$ cases.

Of the remaining 24 impure resonances, we have currently made
conclusive spin-parity assignments to 12 pairs of twin overlapping
levels and tentative assignments in cases where the analysis is still
incomplete. It appears there will be essentially no case where
a satisfactory fit can not be achieved and although all pairs of
levels of either parity up to spin 13/2 are allowed to mix, only
a few ambiguous cases may be forthcoming. Examples of on-resonance
angular distributions analysed with mixed J^π were chosen in Fig. 10-
15 to demonstrate a variety of observed shapes. Measured distributions displaying similar structure are paired in each figure.
When $\pi_1 = \pi_2$ the index ν_{12} is even and the interference term changes
the structure of the distribution symmetrically whereas when $\pi_1 \neq \pi_2$
positive or negative values of $\cos \delta$ cause forward or backward
peaking, respectively, becoming increasingly pronounced as $\cos \delta$
increases in magnitude.

The results of the analysis are listed in Table I with each
resonance energy tabulated in terms of both an incident alpha energy
and an excitation energy in the ^{35}Cl compound nucleus. The spin-
parity values listed are unambiguous assignments except for the $1/2^{(\pm)}$
case discussed earlier and the $5/2^{(\pm)}$ level at $E_\alpha = 4.477$ MeV. Tentative J^π assignments for 12 mixed resonances have been excluded from
this tabulation pending completion of the data analysis.

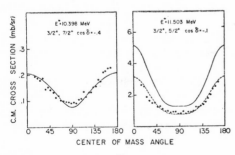
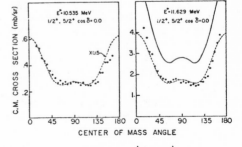

Fig. 10   Mixed J^π distributions.   Fig. 11   Mixed $1/2^+$, $5/2^+$ cases.

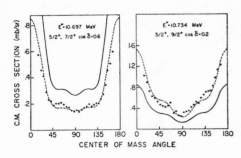

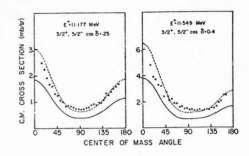

Fig. 12   Mixed J^π cases.   Fig. 13   Mixed $3/2^+, 5/2^-$ cases.

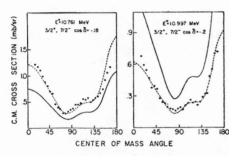

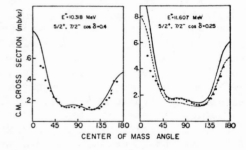

Fig. 14   Mixed $3/2^+, 7/2^-$ cases.   Fig. 15   Mixed $5/2^+, 7/2^-$ cases.

A continuing study of channel-spin 1/2 reactions is in progress at Lowell Technological Institute. The $^{19}F(\alpha,p_o)^{22}Ne$ reaction has yielded over 20 resonances in the ^{23}Na compound system at excitation energies between 12.5 and 14.8 MeV and analysis of the measured on-resonance angular distributions is now underway. We have also completed the experimental setup to study the $^{29}Si(\alpha,n_o)^{32}S$ reaction for the elucidation of level structure in ^{33}S at similar excitation energies.

The systematic statistical trend of partial widths evinced in the resonance studies of the $^{31}P(\alpha,p_o)^{34}S$ and $^{13}C(\alpha,n_o)^{16}O$ reactions has facilitated an analysis based on a statistical formalism that in most cases allowed unambiguous spin-parity assignments for single and twin-interfering levels in the compound nucleus. We are encouraged by the quantity of spectroscopic information obtained thus far and furthermore by the ease with which the experimental and theoretical phases of this program have proceeded. The success of the statistical theory as applied to this work serves as a strong incentive for exploring its usefulness as a spectroscopic tool in other (alpha, nucleon) reactions as well as its wider application to other nuclear resonance phenomena.

TABLE I: NUCLEAR SPIN AND PARITY ASSIGNMENTS TO STATES IN ^{35}Cl

E_α(lab) [MeV]	$E^*(^{35}Cl)$ [MeV]	Pure J^π	E_α(lab) [MeV]	$E^*(^{35}Cl)$ [MeV]	$J_1^{\pi_1}, J_2^{\pi_2}$	$\cos \delta$
3.242	9.869	$1/2^{(\pm)}$	3.425	10.030		
3.305	9.924	$3/2^-$	3.580	10.168		
3.335	9.951	$3/2^-$	3.750	10.318	$5/2^+, 7/2^-$	0.40
3.490	10.088	$5/2^+$	3.840	10.398	$3/2^+, 7/2^+$	-0.40
3.540	10.132	$3/2^-$	3.995	10.535	$1/2^+, 5/2^+$	0.00
3.635	10.216	$3/2^-$	4.155	10.677		
3.655	10.234	$3/2^-$	4.178	10.697	$5/2^+, 7/2^+$	0.60
3.795	10.358	$3/2^-$	4.220	10.734	$3/2^+, 9/2^+$	0.20
3.877	10.431	$5/2^+$	4.250	10.761	$3/2^+, 7/2^-$	-0.18
3.915	10.464	$5/2^+$	4.296	10.802		
3.938	10.485	$5/2^+$	4.313	10.817		
4.055	10.588	$3/2^+$	4.418	10.910		
4.114	10.641	$5/2^+$	4.517	10.997	$3/2^+, 7/2^-$	-0.20
4.295	10.801	$5/2^+$	4.590	11.062		
4.343	10.843	$3/2^-$	4.613	11.082		
4.477	10.962	$5/2^{(\pm)}$	4.720	11.177	$3/2^+, 5/2^-$	0.25
4.558	11.034	$3/2^+$	4.795	11.244		
4.658	11.122	$3/2^-$	4.873	11.313		
4.680	11.142	$5/2^+$	4.890	11.328		
4.740	11.195	$3/2^-$	5.050	11.469		
4.815	11.261	$3/2^-$	5.088	11.503	$3/2^+, 5/2^+$	0.10
4.995	11.421	$3/2^-$	5.140	11.549	$3/2^+, 5/2^-$	0.40
5.392	11.772	$3/2^-$	5.205	11.607	$5/2^+, 7/2^-$	0.25
5.403	11.782	$3/2^-$	5.230	11.629	$1/2^+, 5/2^+$	0.00

The authors wish to express their gratitude to Professor L. E. Beghian for his support and encouragement throughout this program. We are also deeply indebted to Messrs. J. Correia, P. Ketchian, R. LeClaire, J. Mason, F. Prevo, P. Quinn, and L. Sanin for their valuable assistance in the measurement phase of this experiment.

[1] A.D. Robb, W.A. Schier and E. Sheldon, Nucl. Phys. A147, 423 (1970), and earlier references therin.

[2] H. Lustig, Phys. Rev. 117, 1317 (1960), and earlier references therein.

[3] R.B. Walton, J.D. Clement and F. Boreli, Phys. Rev. 107, 1065 (1957), and earlier references therein.

SESSION III

STATISTICS OF RESONANCE PARAMETERS
(PART I)

Chairman   - N. Rosenzweig, State University of New York
                          at Albany and Argonne National Laboratory
Scientific Secretary - N. Cue, State University of New York at
                          Albany

Fluctuations in Charged Particle Cross Sections*

W. R. Gibbs

University of California
Los Alamos Scientific Laboratory
Los Alamos, New Mexico 87544

In 1960 Ericson (1) pointed out that when the excitation energy of the nucleus is high enough that the average width of the compound nucleus levels greatly exceeds their average spacing the cross section as a function of energy should exhibit random fluctuations. In the limit of $\Gamma/D \gg 1$ the coherence width of the fluctuations depends only on the average lifetime of the compound nucleus. The mathematical picture that he presented at that time remains valid in its essentials today. Let us spend a few minutes reviewing this picture.

The scattering amplitude can be represented by an expression of the type

$$f(E) = f_D + \sum \frac{a_\lambda}{E - E_\lambda + \frac{i\Gamma_\lambda}{2}} \qquad (1$$

where f_D is a constant and is to represent the direct, or short-time component of the amplitude while the sum is to represent the compound nucleus contribution. The assumption is made that $\Gamma_\lambda/D \gg 1$ for all λ so that the terms in the sum are strongly overlapping. It is also assumed that the Γ_λ are independent of λ. This is a good assumption if $\Gamma/D \gg 1$ because there will be many channels contributing to Γ and it will tend to have very small fluctuations from level to level. If $\Gamma/D \approx 2$ or 3 a correction should be made to the formalism to include the effects of fluctuations in the width. Another modification to this assumption is with regard to angular momentum. Clearly Γ must be a function of the angular momentum of the compound level. In practice one can use an average Γ with little error. If we now assume some simple statistical properties for the a_λ

$$\langle a_\lambda a^*_{\lambda'} \rangle = \text{const } \delta_{\lambda\lambda'} \; ; \quad \langle a^2_\lambda \rangle = 0 \tag{2}$$

we have a well defined statistical model. The first and basic result to be derived from this model is

$$R(\varepsilon) \equiv \frac{\langle \sigma(E)\sigma(E + \varepsilon) \rangle - \langle \sigma(E) \rangle^2}{\langle \sigma(E) \rangle^2}$$

$$= (1 - y^2)/(1 + \varepsilon^2/\Gamma^2) \tag{3}$$

where y is the fraction of the average cross section due to direct reaction.

A careful application of this simple picture leads to predictions for:

a) $R(\varepsilon)$ as a function of ε
b) $R(0)$ as a function of y
c) $R(0)$ as a function of the spins of the particles
d)   the size of the fluctuations in $R(\varepsilon)$ for large ε
e)   correlations between different angles
f)   correlations between different final states
g)   the number of peaks in a given energy interval as a
      function of Γ
h)   probability distributions of the cross section
i)   dependence of the apparent values of $R(0)$ and Γ on
      the resolution used.

Much of the initial experimental work was done to check these predictions and indeed they were verified in almost every instance.[2] It was realized very early that y did not enter into the measurements in a sensitive manner and so an accurate determination of y would be difficult. This is due to the way in which y enters the expression for $R(0)$ and the fact that the number of independent points which is contained in ΔE is $\Delta E/\pi\Gamma$ and not $\Delta E/\Gamma$ as might be naively expected. In spite of these difficulties several direct reaction cross sections were measured and comparison with theory gave indication that they were, at least approximately, correct. [3-5]

The investigation of fluctuations in elastic cross sections has several differences from the corresponding investigation in the case of reaction cross sections. Unitarity requires that a unitary theory be used (although the autocorrelation function is sufficiently general that an estimate of the width using this method would be substantially correct). In principle a phase shift analysis may be done to separate the various J-values before the analysis is done. In practice this is difficult and has been done in only one case.[6] Calculations using unitary models have led to interesting, and somewhat controversial, results.[7] Research is continuing in this area.(8)

The measurements of Γ by means of fluctuations have provided us with a large body of data to show how the width varies with atomic number, excitation energy, and bombarding particle. Figure 1 illustrates how the measured width is related to these last two

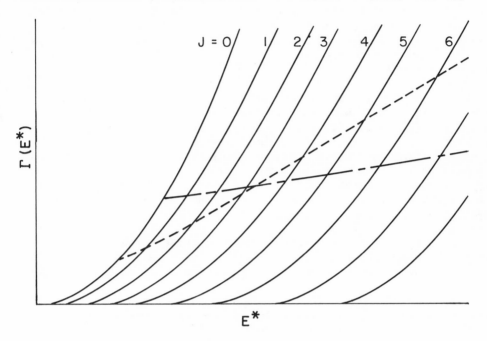

Figure 1

The coherence width, Γ, as a function of excitation energy. The dotted curve represents a light particle trajectory contrasted with the heavy ion trajectory represented by the dot-dash curve.

variables. An approximate expression relating the observed width to the J dependent widths is

$$\Gamma(E*)^{-1} = \sum X_J(E) \Gamma_J^{-1}(E*) \tag{4}$$

where $X_J(E)$ is the probability of the reaction proceeding through a level with spin J, E* is the excitation energy and E is the bombarding energy. As the excitation energy is increased the Γ_J all increase in an exponential fashion as shown. In general the widths decrease with increasing J, as shown, but this is not absolutely necessary and in some cases they may increase with J (especially if alpha-decay is the dominant contributor to the total width). The $X_J(E)$ functions depend on the reaction used to study the width and peak around a few J-values for a fixed E. At energies where the peak of the distribution falls on an integer J-value we may think of the Γ which results from Eq. 4 as being the Γ_J associated with that value of J. At the threshold for the reaction only J = 0 is involved and thus at that energy $\Gamma = \Gamma_0$. As we increase the bombarding energy (and hence the excitation energy) the peak of the X_J distribution moves through J = 1, 2, 3, etc. If we put a dot on each Γ_J curve at the appropriate E* we may trace out the resultant $\Gamma(E*)$.

Since heavy ion reactions cause the angular momentum to increase more rapidly with energy than light particle reactions, the apparent $\Gamma(E*)$ curve for heavy ions is flatter than for light particles. Note that if one keeps the excitation energy and the projectiles fixed but goes to various final states the higher excited states involve less energy for the final particle and hence a lower angular momentum weighting for the reaction. Thus in general one would expect the apparent width to increase with increasing final excitation energy. Of course if the Γ_J increase with J these trends will be reversed.

By performing several experiments of these types one may hope to learn more about the structure of the Γ_J curves, which, in turn, will give us information about the level densities at high energies as a function of J.

To date only one attempt to present such a coherent study has been reported.(9) Its interpretation has been somewhat of a puzzle so far and an understanding of these results will, no doubt, give a valuable clue to a deeper understanding of level densities.

By using properties of fluctuations verified earlier one may extend the measurement of widths to regions where the experimental resolution is not adequate to do a standard analysis. This may be done by investigating the variation of the autocorrelation with experimental resolution. The equation

$$\tilde{R}(0) = R(0)/(\rho/\pi\Gamma + 1) \qquad (5$$

expresses this variation to a good approximation for a rectangular resolution function. This equation may be used in two ways. For $\rho/\pi\Gamma \sim 1$ the resolution may be varied and the resulting curve fitted with the parameters R(0) and Γ chosen to obtain the fit.(3) If $\rho/\pi\Gamma \gg 1$ the differences in curves will not be sufficient and R(0) must be estimated independently. This can often be done in a reasonably reliable manner.(10)

Using the methods described above total widths for a large number of nuclei have been measured at high excitation energies. These data are allowing a study of the level densities of nuclei at high energies. At low energy it was found that the level density depends markedly on the pairing, presence of shells and deformation. Whether or not these effects persist at higher energy is an interesting question which may be answered by these studies.

REFERENCES

*Work performed under the auspices of the U. S. Atomic Energy Commission.

(1)   T. Ericson, Phys. Rev. Letters 5, 430 (1960).
(2)   See M. G. Braga Marcazzan and L. Milazzo Colli, Progress in Nuclear Physics 11, 145 for a complete review of this area.
(3)   V. Bobyr, M. Corti, M. G. Marcazzan, L. Milazzo Colli and M. Milazzo, Energia Nucleare 13, 312 (1966).
(4)   G. Dearnaley, W. R. Gibbs, R. B. Leachman and P. C. Rogers, Phys. Rev. 139B, 1170 (1965).
(5)   R. B. Leachman, p. 773, Proceedings of the International Nuclear Physics Conference, Gatlinburg, Tenn. (1966), Academic Press, New York.
(6)   P. P. Singh, B. A. Watson, J. J. Kroepfl, and T. P. Marvin, Phys. Rev. Letters 18, 31 (1966).
(7)   P. A. Moldauer, Phys. Rev. Letters 18, 249 (1967); 19, 1047 (1967); Phys. Rev. 157, 907 (1967); 171, 1164 (1968); W. R. Gibbs, Phys. Rev. 181, 1414 (1969).
(8)   A. K. Kerman and K. McVoy (private communication).
(9)   M. L. Halbert, F. E. Durham and A. Van Der Woude, Phys. Rev. 162, 899 (1967).
(10)   P. Fessenden, W. R. Gibbs, and R. B. Leachman, Phys. Rev. Letters 15, 796 (1965), Phys. Rev. C3, 807 (1971).

DISCUSSION

PEREZ   (Oak Ridge)
    Those peaks that you showed in the auto-correlation function
are typical of end effects. Could you comment on them?

GIBBS   (Los Alamos)
    The   data finite sample size, you mean? Yes, they're end
effects in a sense but they're due to the whole range. If you look
at the expressions that one gets for R(ϵ) but do it for a finite
sample size, you can indeed calculate the size of the fluctuation
as a function of the interval size.

COHEN   (Pittsburgh)
    Could you give a reference for those curves that come down and
go up again? (Illustrated on the blackboard).

GIBBS
    You can go back to Ghoshal.

COHEN
    I've measured a lot of curves, and I've never seen anything
like that. I've seen things like that, but it's due to something
else. I'm just wondering if there's a confusion here.

GIBBS
    Well, there are a lot of curves which look like that and which
are supposed to be due to the mechanisms I said. Fessenden and
Leachman measured gross excitation functions which in fact looked
very much like what I've drawn on the board. They perhaps took
only two or maybe three, not four as I've drawn, but they took them
as a function of angle. In fact, the data behaved exactly as one
thought they should. You have doubts?

COHEN
    It looks very much like something else than what we measured.

ROSENZWEIG
    Did I understand you correctly that the ratio of resolution to
$\pi\Gamma$ could be 20 or 30 and it would still work?

GIBBS
    Of course the $\pi\Gamma$ is helping you. The limit to what you can
measure is in this quantity R(o). Your counting statistics tend
to wipe you out. You can in practice, measure R(o) down to about
.03. Experimentally you have a lot of difficulties below that.
That's about the limit. If this is so indeed the ratio is about 30.

NEWSON
    I can't resist remarking that in order to make this analysis,

if I understand you correctly you have to put a value of ρ in there. ρ is the energy resolution. My experience is that people are not even honest when it comes to estimating the resolution.

GIBBS

Well, what was done in these particular measurements was to, in fact, increase the resolution. They didn't use the minimum resolution. They used a thicker target than they actually had to in order to know the resolution well.

NEWSON

How do you measure the thickness of the target?

GIBBS

I didn't do the experiment, and I can't say how it was done exactly. The error you see, is only proportional to the error of the resolution. These widths you see quoted have errors which are 20% if you're very lucky, and 90% if you're not so lucky. So the error and your knowledge of the resolution is probably small potatoes.

NEWSON

In other words, you think the widths that are important in your average widths are the ones you will resolve if you don't even try.

GIBBS

Excuse me?

NEWSON

Well, you say that the resolution doesn't matter much. That means that your result depends mostly on the larger widths.

GIBBS

No. You're saying that the error in the width you're measuring is proportional to the error in the resolution.

NEWSON

That is often larger factors than I care to quote. My serious remark is this is a good idea, and I think it's the only way to do certain things, but you have really got to measure the resolution and know what it is.

GIBBS

This was hopefully done in the experiment.

NEWSTEAD

I suppose one expects the direct reaction to be generally strongly forward peaked. Why is it that you have quite a bit of direct reaction at higher energy in the backward direction. What causes that?

GIBBS
    I extended these curves probably a little too high in energy.
I should have cut them off sooner.

VONACH
    I just wanted to ask if there is any physical picture for this
increase of Γ with J . As far as I know, unless you have a quite
strange energy dependence for the moment of inertia you will always
get that Γ decreases with J. Normally the spin cutoff, sigma,
increases with energy. Unless sigma decreases with energy, I cannot
quite see how Γ can increase with J.

GIBBS
    Of course, one does put in what you might regard as a strange
moment of inertia. That is, one assumes the moment of inertia is
changing with excitation energy, in the final reactions when one
gets this effect. Generally, one can't get it without that. It's
coming from the α particles which have a tendency towards high
angular momentum anyway.

VONACH
    I see, then it's o.k.. Is there any physical reason for
assuming this?

GIBBS
    It goes to rigid at high energy.

VONACH
    Then it has to increase, not to decrease.

GIBBS
    Increase, yes--that's what I said.

WIGNER
    One of the curves representing the auto-correlation function
was the theoretical one. I am wondering what the origin of the
other curve, with the many dots, was. I presume it was experimental
and perhaps we could learn about the particular reaction it refers to.

GIBBS
    It was supposed to be a typical one like experimentalists show
a typical spectrum. That is, it's probably the best one I could
pick up. It was an experimental one, and I've forgotten the exact
reaction. I think it was $A^{27}(p,\alpha)$, an experiment done by
Leachman's group at Los Alamos some time ago.

COHERENCE WIDTHS IN HIGHLY EXCITED COMPOUND NUCLEI*

K. A. Eberhard and A. Richter

Sektion Physik der Universität München, Germany, and
Institut für Experimentalphysik der Ruhr-Universität
Bochum, Germany, and Max-Planck-Institut für
Kernphysik, Heidelberg, Germany

INTRODUCTION

The existence of cross section fluctuations in nuclear reactions involving highly excited compound nuclei
is now well established {1} since its theoretical prediction {2}. Next to the mean square deviation of the individually fluctuating cross sections from the average cross
section, the important quantity desired experimentally
is the mean width (FWHM) of the fine structure. This mean
width is closely related to the mean level or mean coherence width Γ of the highly excited compound nucleus levels. Besides the fact that there is no other experimental access to this quantity in the region of overlapping
levels (continuum region) than via cross section fluctuation measurements, its knowledge together with the mean
level spacing D in the continuum is rather fundamental.

A statistical model with various approximations has
been used to calculate the mean level widths as a function of energy and angular momentum of compound nuclei
excited in the continuum region. The results are compared
with all experimentally determined coherence widths
available between mass numbers A = 16 and A = 116. Furthermore, for some nuclei the ratio of the mean level
width to the mean level spacing, which is useful in
Hauser-Feshbach calculations, is also calculated and
compared with experiment.

---

*The paper was presented by A. Richter.

MODEL FOR $\Gamma^{J\pi}(E,A)$

The model described in this section in brief is that of Ref.{3}. Based on the well known relation between the average total level width $\Gamma^{J\pi}(E)$ of levels with angular momentum J and parity π located at an excitation energy E in the compound nucleus, the average level spacing $D^{J\pi}(E)$ and the transmission coefficients $T_{c''}$ for decay into various possible channels

$$2\pi\Gamma^{J\pi}(E)/D^{J\pi}(E) = \Sigma_{c''}T_{c''}^{J\pi} \qquad (1)$$

the width $\Gamma^{J\pi}(E)$ is explicitly {2}

$$\Gamma^{J\pi}(E)=\frac{D^{J\pi}(E)}{2\pi}\sum_{\nu}\sum_{\ell=0}^{\infty}\int_{0}^{\varepsilon_{\nu}^{max}}d\varepsilon_{\nu}T_{\ell}(\varepsilon_{\nu})\sum_{s=|J-\ell|}^{J+\ell}\sum_{j=|s-i_{\nu}|}^{s+i_{\nu}}\rho_{\nu}(E_{\nu},j). \qquad (2)$$

Here the index ν stands for the different kinds of particles which the compound nucleus emits. These have the kinetic energy ε_{ν} and their spin is denoted by i_{ν}. The level densities of the corresponding residual nuclei depending on the excitation energy E_{ν} and on the spin j of the level populated are written as $\rho_{\nu}(E_{\nu},j)$. Transmission coefficients T_{ℓ} depending only on the orbital angular momentum ℓ and on the energy of the emitted particles are used. For actual calculation the width Γ and the spacing D are assumed to be independent of parity π. This is supported by the fact that both parities might occur as often in the compound nucleus as in the residual nuclei.

Three basic assumptions {3} are made for an analytic evaluation of the right hand side of Eq.(2): (i) the transmission coefficients are treated in a "sharp cut-off" model, (ii) the level densities $\rho(E,J)$ of the compound and residual nuclei are factored into an energy and angular momentum dependent part, and Fermi-gas expressions are used for both quantities, and (iii) the sums over channel spins and angular momenta which occur in Eq.(2) are replaced by integrals with the help of Euler's summation rule. With these assumptions and some additional approximations for cases of practical interest (Ref.{3}) Eq.(2) simplifies to

$$\Gamma^{J}(E)=\sum_{\nu}\Gamma_{o}^{\nu}(E)\exp\{-J(J+1)(\frac{1}{2\sigma_{\nu}^{2}(1+\omega_{\nu})} - \frac{1}{2\sigma_{C}^{2}})\} \qquad (3)$$

Here the exponential term describes the angular momentum dependence of the mean level width, which is determined mainly by the spin cut-off parameters σ_ν and σ_C of the residual and compound nuclei, respectively. (The correction term ω_ν to σ_ν^2 is small for neutron and proton decay of the compound nucleus but is important in case of heavy particle emission {3}.) The energy dependence of the coherence width is contained in $\Gamma_o^\nu(E)$ which is given by

$$\Gamma_o^\nu(E)=(\pi\hbar^2)^{-1}(2i_\nu+1)A_\nu R_\nu^2\theta^2\frac{\sigma_C^2}{\nu\sigma_\nu^2}\frac{\rho_\nu(E-B_\nu-C_\nu)}{\rho_C(E)} \qquad (4)$$

and hence is essentially determined by the ratio of the level densities of the residual nuclei, ρ_ν, taken at the excitation energy ,E, of the compound nucleus minus the binding energy, B_ν, of the emitted particle ν and its respective Coulomb barrier height, C_ν, and the level density ρ_C of the compound nucleus. The quantity R_ν is the interaction radius between the emitted particle ν of mass A_ν and the residual nucleus of mass $A_{C-\nu}$ and is given by $R_\nu = r_{\nu o}(A_\nu^{1/3}+A_{C-\nu}^{1/3})$ with $r_{\nu o}$ being the usual interaction radius ($r_{\nu o}\approx1.4f$). Finally θ_ν is the thermodynamic temperature {3}.

CALCULATION OF COHERENCE WIDTHS AND COMPARISON WITH EXPERIMENT

In order to calculate the coherence widths over wide ranges of compound nuclei of mass A and of the variables E and J, only average parameter values for the spin cut-off parameters (rigid body moment of inertia value) and for the level density parameter{4} a (a = A/7.9 MeV^{-1}) for calculating Γ from Eqs.(3) and (4) have been used. To account for the fact that the emission of charged particles from the compound nucleus sets in already when energy ε_ν of the particle reaches about 70 % of the Coulomb barrier (as can be seen from the behaviour of the transmission coefficients) the interaction radius R_ν was adjusted properly for calculating the height, C_ν, of the Coulomb barrier. Since effective excitation energies (true excitation energy E minus pairing energy) enter always the level density expressions, pairing energies {5} were taken into account.

As a first test of the model for $\Gamma^{J\pi}(E)$, experimentally determined coherence widths have been compared with the predictions of Eqs.(3) and (4). However, since the experimental width is presumably a weighted average over

individually excited widths of levels with different
angular momenta in the compound nucleus the theoretically
predicted coherence widths have also been weighted. Here
the relation {6}

$$\Gamma_{EFF} = \sum_J \sigma_J / \sum_J (\sigma_J / \Gamma^J) \qquad (5)$$

was used. The cross sections integrated over angle,
σ_J, for the formation of the compound nucleus were calculated with the help of optical model transmission
coefficients.

The results of the calculation are compared with
all experimentally determined coherence widths available
between mass numbers A = 16 and A = 116. As seen in Fig.1,
where the comparison for an excitation energy of about
20 MeV in the compound nucleus is made, the agreement
achieved with experiment in general is good. Without
adjusting parameters from nucleus to nucleus the experimental values which go over five orders of magnitude
are reproduced for the most part within a factor of two.

Similarly, the dependence of the coherence width on
the excitation energy is also described satisfactorily
by Eqs.(3) and (4), and is demonstrated as an example
for the compound nuclei ^{28}Si, ^{31}P, ^{32}S, ^{27}Aℓ, ^{57}Co and
^{90}Zr in Figs. 2-5, respectively.

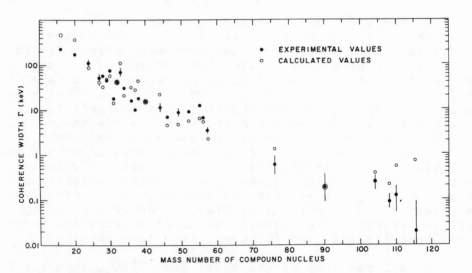

Fig.1   Experimental and calculated values of the
         coherence width Γ as a function of the mass
         number of the compound nucleus.

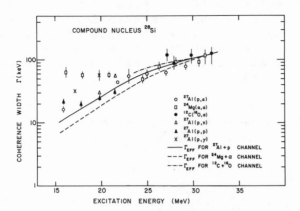

Fig.2   Comparison between experimental and calculated
         coherence widths at different excitation energies
         for the compound nucleus ^{28}Si reached by the
         reactions indicated in the figure.

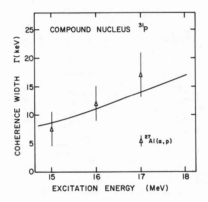

Fig.3   Same as Fig.2 but for ^{31}P.

         Finally, the quantity Γ/D_o as determined experiment-
ally by Huizenga et al {7} for the compound nuclei ^{56}Fe
and ^{60}Ni is compared with calculated values Γ_{EFF}/D_o as
a function of excitation energy. The level density, $1/D_o$,
was obtained from the Fermi-gas model using the Lang-
LeCouteur formula (see,e.g., {2}) and using the same
parameters which entered the calculation of Γ from
Eqs.(3) and (4) in Fig.6. The good agreement between the
experimental and theoretical values seems to indicate
that the ratio Γ/D_o can be estimated with good

confidence with the help of the present model. This is
important for Hauser-Feshbach calculations of energy-
averaged cross sections, since the knowledge of this
ratio in principle replaces the often tedious computation
of the capture cross sections of all open compound-
nucleus decay channels.

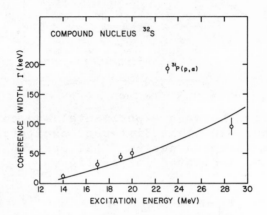

Fig.4   Same as Fig.2 but for ^{32}S.

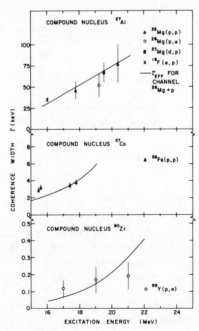

Fig.5   Same as Fig.2 but for large (^{27}Aℓ), intermediate
(^{57}Co) and small widths (^{90}Zr).

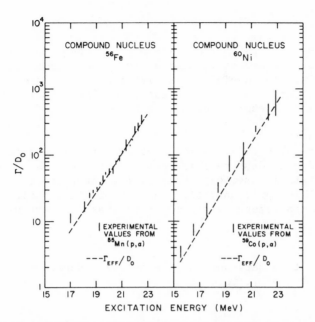

Fig.6   Experimental and theoretical ratio of the mean
         level width to the mean level spacing for the
         compound nuclei ^{56}Fe and ^{60}Ni.

     A more complete account of the work presented here
briefly, which will also include references to the
experimental material used for comparison, will be
given elsewhere {8}.

REFERENCES

{1}. T. Ericson and T. Mayer-Kuckuk, Ann. Rev. Nucl. Sci.
     16, 183 (1966)
{2}. T. Ericson, Advan. Phys. 9, 425 (1960)
{3}. K.A. Eberhard, P. von Brentano, M. Böhning, and
     R.O. Stephen, Nucl. Phys. A.125, 673 (1969)
{4}. E. Gadioli and L. Zetta, Phys. Rev. 167, 1016 (1968)
{5}. A. Gilbert and E.G.W. Cameron, Can. J. Phys. 43,
     1446 (1965)
{6}. P. Fessenden, W.R. Gibbs, and R.B.Leachman,
     Phys. Rev. Letters 15, 796 (1965)
{7}. J.R. Huizenga, H.K. Vonach, A.A. Katsanos, A.J. Gor-
     ski, and C.J. Stephan, Phys. Rev. 182, 1149 (1969)
{8}. K.A. Eberhard and A. Richter, (to be published)

DISCUSSION

VONACH

In the cases of light nuclei, the energy dependence of Γ became slower with higher excitation energy. You had a curve this way, but in the heavier nuclei, like yittrium, it was a kind of parabolic curve. Have you any qualitative explanation for this change of behavior of energy dependence of Γ ?

RICHTER (Heidelberg)

Well, it probably reflects the fact that the moment of inertia becomes energy dependent, and this energy dependence is different for heavy nuclei than for light nuclei. I think there is a contributed paper in this conference (3.2) where for light nuclei at very high excitation energies the behavior that it flattens off may be explained.

VONACH

I cannot quite understand why the energy dependence can get steeper in the case of yttrium or so.

RICHTER

The excitation energy in the case of yttrium is not quite as high as it is in the case of light nuclei, so the calculated coherence width may flatten off at a high excitation energy. We have not looked at that in particular.

VONACH

You think it will flatten off again?

RICHTER

Yes. As I remarked, we have to look at the behavior of the moment of inertia. At present we have not adjusted anything in the calculations. We have just taken the rigid body moment of inertia to obtain the spin cut off parameters which, as has been pointed out many times, may be be the best approximation to use.

SHELDON (Lowell)

The prescription you used for your transmission coefficient cutoff is rather a drastic one. Have you tried carrying through the calculations with actually computed transmission coefficients, and do you discern any difference in your results in that case?

RICHTER

Yes. This is a good point since the transition going from the reflective to the absorbtive region is a smooth function. However, it turns out that Γ does not depend very much on the particular form of the transmission coefficients used. We have compared our calculations with exact calculations. [See e.g. Vonach et.al.,

Nucl. Phys. A122, 465 (1968)] Namely, taking exact formula (2) of our paper with transmission coefficients from the optical model one gets in the case of S^{32} agreement within 5% compared to the calculations where one takes the transmission coefficients in the drastic smooth cutoff assumption. There is another case where this was tested (Fe^{56}) where the agreement is better than 30% not varying any parameters.

CUE (S.U.N.Y./Albany)

In extracting Γ in the mass 90 region one has to be careful because one sees a lot of isobaric analogue resonances in (p,α) reactions.

RICHTER

I'm aware of your work. I do not know exactly, however, if the proton energy in the experiment by Leachman and collaborators correspond to the region where you have seen isobaric analog resonances.

CUE

For the specific case of 93Nb target which we investigated, we start seeing resonances from proton energies of about 5 Mev.

GILAT (SNRC-Israel)

How sensitive are these calculated results to your assumption of the parameters? To put in in an opposite way, can you really draw the conclusion that the parameters you used are more or less representative of physical level densities?

RICHTER

Yes. It has been shown in the literature, again, mainly by Huizenga and his collaborators, that one can draw firm conclusions about level densities in nuclei with the help of fluctuation theory and arguments like the ones presented here.

ROSENZWEIG

Am I correct that you are asking whether a = A/7.9 (or something like it) represents nuclear level densities realistically? Is that what you are asking?

GILAT

What I'm really asking is--to what extent are these reasonably good fits sensitive at all to the assumptions, whether it's the parameter a, the rigid body moment of inertia, the transmission coefficients and so on.

RICHTER

There are really only two parameters in the calculation, actually. That's the spin cutoff parameter and the level density parameter, and since the coherence widths depend exponentially on both, you can, within reasonable assumptions which you get from

other experimental information, vary those two parameters in order
to obtain agreement with the experiment. I think what you might
do at least with this analysis is putting fairly tight bounds on
those parameters.

WIGNER
    Could you explain why the average Γ does not depend on the
angle at which the experiment was undertaken? Since the average
involves averaging over several angular momenta J, one would
expect that at small angles, the Γ of the larger J are most
relevant. At larger angles, near 90°, the Γ of lower J are most
relevant. Why do the quantities discussed not depend on the
angle of the reaction or scattering?

RICHTER
    You are correct in pointing out that the coherence width
should also be dependent on scattering angle. This angular dependence could be taken into account by making the weighting
factors σ_J in the relation (5) of the contributed paper,

$$\sum_J \sigma_J \ / \ \sum_J (\sigma_J \ /\Gamma)$$

dependent on angle. However, in most cases no dependence of Γ
on angle was found experimentally. Furthermore, the theoretical
dependence of the coherence width of the compound states on J is
found to be weak also for two reasons: 1) the dependence of
on J is weak for the low values of J which are relevant in light
ion reactions and 2) for a given reaction only a few values of J
are important.
    Of course, for heavy ion reactions the above statement has
to be modified. The simple model discussed here is applicable
only if $J \sim \sqrt{2\sigma^2 \sqrt{aE}}$.

DISTRIBUTION OF NEUTRON AND FISSION WIDTHS

A. MICHAUDON

Département de Physique Nucléaire

C.E.N. Saclay, 91 - Gif-sur-Yvette, France

1. INTRODUCTION

The rate of decay of a compound nucleus state λ by neutron emission or by fission is usually expressed in terms of its partial widths Γ_n or Γ_f, respectively. Neutron emission can occur only above the neutron threshold, i.e. for excitation energies greater than the neutron separation energy S_n in the compound nucleus. Typical values of S_n are between 6 and 8 MeV for most nuclei. There is not such a clear-cut threshold for fission since its rate is governed by the dynamics of penetration of the fission barrier. Fission can then occur for energies below the top of the barrier, about 5 to 6 MeV in heavy nuclei in the actinide region. But the fission probability decreases very rapidly below the top of the barrier for decreasing energy and becomes negligible compared to the other modes of decay (γ-ray emission, for example).

We therefore have to consider neutron emission and fission for states excited high enough in energy so that their great complexity and great number justify a study based on statistical assumptions. The distributions of neutron and fission widths for these states can be best studied experimentally by measuring these widths individually for a great number of levels. This can be achieved by slow neutron spectroscopy, using the excellent resolution of time-of-flight methods. In this manner, many neutron resonances can be measured which correspond to compound nucleus states excited just above the energy S_n. The neutron widths Γ_n (or more exactly the products $g\Gamma_n$ of the width Γ_n by the statistical factor g) can easily be obtained from an area analysis of the resonances as they appear in the total cross sections. For nuclei in the actinide region, these resonances can also have an appreciable fission width which can be obtained from an analysis of the measured total and fission cross sections.

Most of these low energy resonances are induced by "s" wave neutrons and have spin and parity $J^{\pi} = 1/2^+$ for even-even target nuclei. For other target nuclei, having spin and parity I^{π}, the "s" wave resonances can have $J^{\pi} = (I + 1/2)^{\pi}$ or $J^{\pi} = (I - 1/2)^{\pi}$. The "p" wave contribution is generally weak except at relatively high energy and/or for nuclei having a large "p" wave strength function S_1 (i.e. in the neighbourhood of A = 100 and A = 220). Therefore "p" wave resonances will be neglected unless specifically mentioned.

The theoretical framework for the study of the distribution of reaction widths of compound nucleus states has been provided by Porter and Thomas in their well known 1956 paper [1]. The basic assumption in the calculations is that the compound nucleus states are so numerous and so complex that the values of the nuclear matrix elements have a Gaussian distribution centered on the value zero. For the decay of a compound nucleus state λ by process α, the partial width $\Gamma_{\alpha\lambda}^i$ for the channel i is proportional to the square of the small and real transition amplitude from the initial state λ to the final state.

Under the assumptions recalled above, the distribution of the reduced widths x normalized to their average over many states λ, for one single channel i, is simply given by the so-called Porter-Thomas distribution

$$P(x,1) = \frac{1}{\sqrt{2\pi}} \frac{1}{\sqrt{x}} \exp(-\frac{x}{2}) \cdot \tag{1}$$

When several exit channels, in number ν, contribute to the whole process α, then the corresponding width $\Gamma_{\alpha\lambda}$ is simply the sum of the relevant partial widths $\Gamma_{\alpha\lambda}^i$

$$\Gamma_{\alpha\lambda} = \sum_{i=1}^{i=\nu} \Gamma_{\alpha\lambda}^i \cdot \tag{2}$$

The distribution of the reduced widths x for ν exit channels can be simply derived from the Porter-Thomas distribution (1) if the relevant partial widths $\Gamma_{\alpha\lambda}^i$ are uncorrelated and have equal values $<\Gamma_{\alpha\lambda}^i>$ averaged over many states λ. One thus obtains, by folding expression (2) with itself (ν - 1) times, the well-known chi-squared distribution $P(x,\nu)$ with ν degrees of freedom

$$P(x,\nu) = \frac{\nu/2}{\Gamma(\nu/2)} (\frac{\nu x}{2})^{\nu/2 - 1} \exp(-\frac{\nu x}{2}) \tag{3}$$

where $\Gamma(\nu/2)$ is the value of the gamma function for $\nu/2$.

Plots of $P(x,\nu)$ versus x, for various values of ν, are given in Fig. 1 which shows narrower distributions for increasing values of ν. This effect is consistent with the calculated variance of

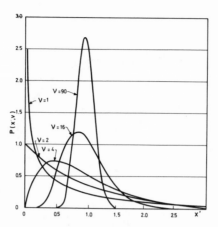

Fig. 1 - *The chi-squared distribution P(x,ν) given by formula (3) in the text, is plotted as a function of x for various values of ν (after Porter and Thomas [1]).*

distribution (3), which is equal to :

$$\text{Var } (x) = \frac{2}{\nu} . \tag{4}$$

This clearly shows the effect of the number ν of exit channels on the fluctuations of the widths.

This "black box" approach to the study of the distribution of neutron resonance widths will be used as a starting point for the examination of the large amount of data on neutron and fission widths. We shall show also that the distributions of these widths cannot all be explained in terms of this simple theory and that more sophisticated models must be developed to explain the data, especially the energy dependence of the widths. In this respect, this paper somewhat overlaps with Mahaux's paper [2] since some aspects of intermediate structure need to be considered for a full understanding of the behaviour of neutron and fission widths.

2. DISTRIBUTION OF THE NEUTRON WIDTHS

For low energy neutron resonances, decay by neutron emission has only one single exit channel which is available for energy consideration, the entrance channel, since inelastic neutron scattering is not allowed. Therefore, the reduced neutron widths Γ_n^o should obey the Porter-Thomas distribution given in (1). The reduced neutron widths $\Gamma_n^o = \Gamma_n E_n^{-1/2}$, where E_n is the neutron energy, are considered rather than Γ_n in order to remove the well-known $E_n^{1/2}$ average smooth energy dependence of Γ_n.

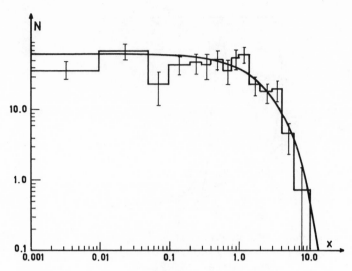

Fig. 2 – *Distribution of the reduced neutron widths* $x = \Gamma_n^o/\langle\Gamma_n^o\rangle$ *for the slow neutron resonances observed in even–even target nuclei. N is the number of resonances, multiplied by $\sqrt{x}$, having values of x in 0.1 intervals. The smooth solid curve is the Porter–Thomas distribution (after Garrison* [3]*).*

The observed neutron widths actually show extensive fluctuations. A detailed comparison of the data with the Porter–Thomas distribution has been carried out by Garrison [3], first for even–even nuclei because all their "s" wave resonances belong to the same family ($J^\pi = 1/2^+$). Such a comparison is illustrated in Fig. 2 where the distribution of reduced neutron widths for a set of even–even nuclei is plotted together with the Porter–Thomas law. Good agreement is observed. More quantitatively, for an ensemble of even–even and odd–A nuclei, the number of degrees of freedom for the neutron width distribution, as determined by the likelihood method, is equal to $\nu = 1.04 \pm 0.06$.

In some cases, however, apparent disagreement can be seen between the experimental and the expected distributions, especially for small neutron widths. Careful examination of the data, with the help of computed artificial cross sections, generated by Monte–Carlo techniques, shows that the disagreement is not significant. A dearth of small neutron widths usually occurs for odd nuclei. It is interpreted (for example in ^{235}U [4]) as being due to missed levels which are in relatively larger proportion for these nuclei because the level overlap is greater, as a consequence of the absence of repulsion between levels having different J^π values. This effect is also accompanied by a lack of small level spacings [4]. In other cases, the experimental distribution shows, on the contrary, an excess of

small neutron widths which is currently interpreted as being due to
"p" wave levels. As recalled in the introduction, their number is
relatively important at high energy and/or for mass numbers around
A = 100 and A = 200. Since they are weak, they also appear best in
high accuracy measurements. This renders difficult an accurate verification of the Porter-Thomas distribution despite the steady and
impressive increase in the quantity and the quality of the data. The
likelihood method has also been applied to reduced neutron width distributions truncated at small neutron width values to remove the
spurious effect of the missed "s" wave levels and/or the excess of
"p" wave levels. The values of ν thus obtained are also consistent
with $\nu = 1$, but with larger errors since chi-squared distributions
are most sensitive to the value of ν for small widths. Therefore,
one can conclude that, taking into account the spurious effects
discussed above, the observed reduced neutron width distributions
obey the Porter-Thomas law ($\nu = 1$).

Another approach to the study of the neutron widths is the
shell model description of the formation of compound nucleus states
by successive two-body interactions [5,6]. In the first step, the
nuclear system is simply represented by the target nucleus A in its
ground state coupled to the incoming neutron. This is the so-called
one-particle configuration. In the second step, the incident neutron interacts, through a two-body force, with *one* of the nucleons
in the target nucleus. This nucleon is raised to an empty orbit
while the incident neutron is caught in another empty orbit ; a 2
particle-1 hole (or in short 2p-1h) configuration is thus obtained.
During the following steps of the interaction, 3p-2h, 4p-3h and more
complicated configurations are excited. During the whole process of
excitation of compound nucleus states, an intermediate state having
a simple configuration (2p-1h for example), may have a relatively
long life-time and play a dominant role in the reaction mechanism.
Such states, called "doorway states" by the M.I.T. School (see for
example [7]), can enhance the neutron widths of the resonances when
they have the same energy, spin and parity. This effect of the doorway states should be seen in the total neutron cross section σ_T
when suitably averaged by folding it into a theoretical weighting
function or into the experimental resolution. An intermediate structure may appear in the averaged cross section provided that the
width Γ_G of the weighting function is intermediate between the mean
spacing <D> of the resonances and the width $\Gamma_{s.p.}$ of the single-
particle resonance (usually a few MeV wide) ($<D> \ll \Gamma_G \ll \Gamma_{s.p.}$).
If the doorway states are too broad because they strongly couple to
more complicated excited states, their damping width becomes larger
than their spacing and the intermediate structure washes out.

Many cross sections σ_T have been treated and examined in this
manner. Fluctuations are actually observed in such averaged cross
sections but no convincing evidence seems to have been provided as

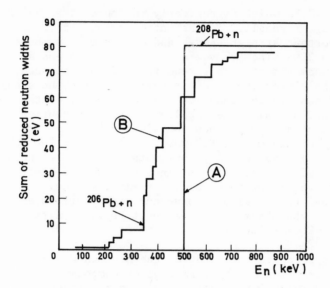

Fig. 3 - *The cumulative sum of the reduced neutron widths of "s"*
wave resonances is plotted for ^{206}Pb (curve B) and ^{208}Pb (curve A)
as a function of neutron energy. For ^{208}Pb there is only one big "s"
wave resonance at 500 keV (after Farrell et al. [9]).

yet which could prove that these fluctuations are actually due to
doorway states in the entrance channel. Careful examination of σ_T
for 18 elements from Mg to Bi, as carried out by Carlson and
Barschall [8], shows that, when fluctuations are observed in the
averaged σ_T, they can be accounted for by those of the widths and
spacings of the resonances.

Nevertheless, a possible doorway state seems to have been observed in the system ^{206}Pb + n [9], near closed shells where the
damping width is expected to be the lowest. In the cross section σ_T
for the doubly magic nucleus ^{208}Pb, one of the few "s" wave resonances is at E_n = 500 keV, with Γ_n = 58 keV, and is assumed to be a
2p-1h state. For ^{206}Pb, the "s" wave resonances are much denser than
for ^{208}Pb, but the strongest of them seem clustered around 500 keV,
the energy of the 2p-1h ^{208}Pb resonance already mentioned. This is
illustrated by the histograms A and B of the cumulative sums of the
reduced neutron widths plotted in Fig. 3 for ^{208}Pb and ^{206}Pb. The
sum of the ^{206}Pb widths, integrated over the whole 500 keV cluster,
is very close to the width of the 500 keV ^{208}Pb resonance. These
results are interpreted as being due to a doorway state in ^{207}Pb of
a similar configuration to that of the 2p-1h ^{209}Pb state responsible
for the 500 keV ^{208}Pb resonance. The sum rule seems to apply here
for the widths of ^{206}Pb and ^{208}Pb so that the strength of the 2p-1h
state appears to be split among the more complicated states of ^{207}Pb.
For both isotopes ^{207}Pb and ^{209}Pb, the 2p-1h configuration is be-

lieved to be due to the interaction of the incident neutron with one proton in the 82-proton shell. This results in a 2 particle (1 neutron-1 proton) - 1 hole (proton) state. For ^{207}Pb, the two $p_{1/2}$ neutron holes in the 126-shell of the target nucleus ^{206}Pb do not seem to play an important role since the 2p-1h configuration occurs at about the same neutron energy for both lead isotopes. Damping of these 2p-1h excitations strongly depends on the density of other more complicated excitations. The damping width is small for ^{209}Pb : in fact, it cannot be deduced from the measurements. The damping width is larger for ^{207}Pb, as expected, since other more complicated excitations have a greater level density. This damping width can be estimated from histogram B (Fig. 3) corresponding to ^{206}Pb + n, which shows that, in addition to fluctuations, the reduced neutron widths are distributed along a Lorentzian line from which the damping width is found to be equal to about 150 keV, in qualitative agreement with theoretical estimations.

Neutron widths for a few further cases are also discussed in the next section on fission widths.

3. DISTRIBUTION OF THE FISSION WIDTHS

3-1. Concept of Fission Exit Channel

As recalled in the introduction, fluctuations of the resonance widths Γ_α for a given process α strongly depend on the number of exit channels for that process. In the case of fission, an exit channel can be defined as the splitting of the fissioning nucleus into a given pair of fragments in well-defined quantum states ; then the number of these fission exit channels is very large (about 10^{10} for low energy fission) and the fission widths should be practically constant from resonance to resonance ; this is in contradiction to what is observed in the experimental data since the fission widths fluctuate strongly, nearly as much as the neutron widths.

Bohr and Wheeler [10] introduced the concept of fission exit channels in terms of fission saddle-point configurations which are energetically available. The number N of such "open" fission exit channels is given by the following relation :

$$N = 2\pi \frac{<\Gamma_f>}{<D>} \tag{5}$$

where $<\Gamma_f>$ and $<D>$ are respectively the average fission width and spacing of the resonances. This expression simply comes from the fact that the ensemble of nucleons finds itself in all allowed configurations (including the N fission saddle-point configurations) once every period $\tau \sim 2\pi\hbar/<D>$. The life-time τ_f for fission is thus $\tau_f \sim 2\pi\hbar/N<D>$ from which the fission width and relation (5) are then deduced.

A. Bohr, in his so-called channel theory of fission [11], also emphasized the importance of saddle-point configurations for fission properties. For excitation energies slightly above threshold, the fissioning nucleus is "cold" when it passes over the saddle-point since most of the excitation energy is then in the form of potential energy of deformation. Therefore, only a few quantum states (called "transition states") are energetically available and they are expected to have a spectrum similar to that of the observed low-energy excitations of the ground state. In this situation, there are a few transition states i, in number ν and having the same spin and parity J^π, which significantly contribute to the fission decay of compound nucleus states having spin and parity J^π. These transition states play the role of fission exit channels i and, for each of them, there is a specific barrier having penetrability P_i. One can now give a more appropriate interpretation of the number N of fission exit channels defined in (5). Rather than considering the number N of available fission saddle-point configurations, one can define an effective number of fission exit channels N_{eff}, for given spin and parity J^π, which is the sum of the penetrabilities P_i for all the ν channels i having the same spin and parity J^π. Thus :

$$N_{eff} = 2\pi \frac{<\Gamma_f>^{J^\pi}}{<D>^{J^\pi}} = \sum_i P_i . \qquad (6)$$

This channel theory of Bohr has greatly stimulated the study of many aspects of fission at low energy. We shall consider in the next subsections only those aspects related to the fission widths.

3-2. Distribution of the Fission Widths for Fissile Nuclei

We consider first the simplified situation where there is only one saddle-point, i.e. where the fission barrier presents one single hump, as obtained in liquid drop model (LDM) calculations. The channel theory of Bohr seems to apply to fission induced by resonance neutrons in fissile nuclei since the excitation energy, slightly greater than S_n, is not much above the fission threshold. Moreover, for common fissile nuclei, such as ^{235}U, ^{239}Pu, etc..., the available transition states are of collective character, hence in still more limited number, at most a few for given spin and parity J^π.

The partial fission width Γ_f^i, for one channel i, is expected to have Porter-Thomas fluctuations of the type (1). If several fission channels contribute to the total fission width Γ_f (the only fission width which can be obtained experimentally), its fluctuations are then expected to obey the more general chi-squared distribution of the type (3), provided that the average partial fission widths $<\Gamma_f^i>$ are equal. This happens when the transition states i are at about the same energy since the penetrabilities P_i are then all equal, or when the exit channels are all fully open ($P_i = 1$ for all i). Otherwise more complicated distributions must be used.

A quantitative comparison of theoretical and experimental fission width distributions requires the determination of spin and fission width of many resonances and a knowledge of the spectrum of transition states. A great amount of data has been accumulated on this subject these last few years (see for example [12,13,14]). But, while the fission width is known for a rather large number of resonances, the spin has been measured for only a few of them, except for ^{239}Pu because its spin is low (I = 1/2) whereas the spin of the other common fissile nuclei is equal to or greater than I = 5/2 [15]. Information on the spectrum of transition states is also very limited and is usually known from (d,pf) or (t,pf) reactions (see for example [16]).

Given these conditions, the distribution of the fission widths can be best studied for ^{239}Pu, not only because the parameters Γ_f and J have been determined for a large number of resonances below 660 eV [17,18] but also because the transition states have different energies and properties for the two possible spins and parities $J^{\pi} = 0^+$ and $J^{\pi} = 1^+$. There is at least one fully open channel $(P_i = 1)$ for $J^{\pi} = 0^+$, that of the ground state at saddle-point since it lies about 1.6 MeV below S_n according to (d,pf) data. There are no 0^+ simple collective excitations since the K = 0^+ quadrupole vibration is excluded from the scheme of transition states because it is the main fission mode. In contrast, there are no fully open $J^{\pi} = 1^+$ channels since there are no 1^+ simple collective excitations. Only in the region of 2 quasi-particle excitations, about 1 MeV above S_n, can 1^+ transition states be found [16].

The fission widths of ^{239}Pu resonances are therefore expected not only to present wide fluctuations but also to have quite different average values $<\Gamma_f>_{0^+}$ and $<\Gamma_f>_{1^+}$ for $J^{\pi} = 0^+$ and $J^{\pi} = 1^+$ respectively. This is actually observed in the experimental results as can be seen in Fig. 4. The integral distribution A is plotted for more than 200 fission widths determined from a single level fit to the total and fission cross sections below 660 eV. Obviously, this histogram cannot be interpreted with one single chi-squared law. A break is observed around $\sqrt{x} = 0.3$ and two distributions having quite different average widths are needed to obtain a good fit. Spin assignments for more than 100 resonances by both direct and indirect methods seem to provide convincing evidence that these two families of resonances actually correspond to the two spin states $J^{\pi} = 0^+$ and $J^{\pi} = 1^+$ [14,15,18]. A least-squares fit to the data, using two chi-squared distributions, gives the following results.

Family of narrow resonances ($J^{\pi} = 1^+$) :

$$<\Gamma_f>_{1^+} = 0.039 \text{ eV} ; \nu = 1 ; (N_{eff})_{1^+} = 2\pi \frac{<\Gamma_f>_{1^+}}{<D>_{1^+}} = 0.07 . \quad (7)$$

Family of wide resonances ($J^{\pi} = 0^+$) :

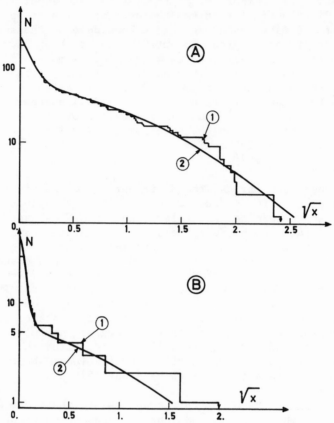

Fig. 4 – *The integral fission width distribution* ① *for the* ^{239}Pu
slow neutron resonances is plotted as a function of $\sqrt{x} = \sqrt{\Gamma_f/\langle\Gamma_f\rangle}$
*for two energy ranges : 0 to 660 eV (histogram A) and 550 eV to
660 eV (histogram B). These data are fitted with two chi-squared
distributions* ② *having the following parameters:*
A $\langle\Gamma_f\rangle = 35.5$ meV, $\nu = 1$ *B* $\langle\Gamma_f\rangle = 7$ meV, $\nu = 1$
   $\langle\Gamma_f\rangle = 2.27$ eV, $\nu = 1.4$ $\langle\Gamma_f\rangle = 1.8$ eV, $\nu = 1$
*(A constant fission width of 3.5 meV has been subtracted to
remove the calculated contribution of the (n,γf) process) (after
Blons et al.* [17]*).*

$$\langle\Gamma_f\rangle_{0^+} = 2.27 \text{ eV} ; \nu = 1.4 ; (N_{eff})_{0^+} = 2\pi \frac{\langle\Gamma_f\rangle_{0^+}}{\langle D\rangle_{0^+}} = 1.48 . \quad (8)$$

The values of ν and N_{eff} given in (8) are not very accurate because of the relatively small number of 0^+ resonances, as a consequence of the $(2J + 1)$ law and also because of level interference
which is rather pronounced for the wide 0^+ resonances. In fact, a
multilevel analysis, more appropriate to fit the cross sections,
gives $(N_{eff})_{0^+} \simeq 1$ [19]. This is in good agreement with the channel

theory of Bohr which predicts at least one fully open fission exit channel for 0^+ resonances. The values of ν and N_{eff} given in (7) show that fission for the 1^+ resonances proceeds through one partially closed exit channel and is therefore below threshold (at threshold $P_i = 1/2$). These values of ν and N_{eff} are also in good agreement with the channel theory of Bohr provided that the 1^+ transition state is located about 200 keV, not 1 MeV, above S_n. This led Griffin to postulate that this 1^+ transition state could be a collective excitation resulting from the coupling of the two octupole vibrational states $K = 0^-$ and $K = 1^-$ [16].

In the spectrum of transition states, more complicated collective excitations than the well-known 1-phonon states seem to be necessary to explain the behaviour of the fission widths. Rather sophisticated spectra of transition states have been proposed by various authors, especially by Lynn [20], in order to fit fission data.

It may be worth noticing that no ^{239}Pu resonances are observed with fission widths smaller than about 3 meV. This lower limit corresponds very well to the constant width $\Gamma_{\gamma f}$ for the many-exit-channel $(n,\gamma f)$ process which is usually included in the overall (n,f) reaction [24].

The fluctuations and average values of the fission widths of 0^+ and 1^+ ^{239}Pu resonances, as measured over the relatively broad energy range of 660 eV, seem to be in agreement with the channel theory of Bohr. The fission widths of the other common fissile nuclei seem to have distributions consistent with one chi-squared family. Two families can appear in the distributions only if they have quite different average widths. In this respect the ^{239}Pu case is rather unique. With a spectrum of transition states properly chosen, the fission widths of these other fissile nuclei also seem to be in agreement with the theory of Bohr [12,20] with perhaps some inconsistencies for ^{235}U for which, for instance, the measured value of N_{eff} seems too low [14]. One must point out, in this connection, that some systematic error may be introduced in the measured value of $<\Gamma_f>$ since, in a given energy range, the resonances cannot all be analysed and those which are missed have probably, on the average, a larger fission width. This effect leads to measured values of N_{eff} which are probably underestimated.

The contribution of the various transition states to the fission width has been studied by measuring the anisotropy of the fission fragments emitted in the fission induced by non-polarized neutrons in aligned nuclei [21,22]. A paper is presented on this subject at this Conference for ^{233}U, ^{235}U and ^{237}Np [23]. It may be worth noticing that the puzzling absence of the contribution of the $K = 0^-$ octupole band in the fission of ^{235}U observed in the earlier results [21b,22a] seems to have disappeared [22b,23] by renormalization of the

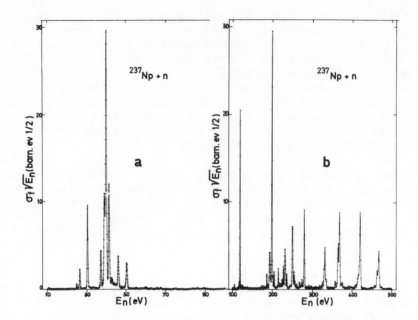

Fig. 5 – *The Saclay subthreshold fission cross section of* ^{237}Np,
multiplied by $\sqrt{E_n}$, *is plotted as a function of neutron energy* E_n
*between 10 eV and 500 eV. (Note the different energy scales for
plots a and b). This cross section clearly exhibits an intermediate
structure effect. Other fission clusters (not plotted here) also
appear in the cross section above 500 eV (after Michaudon* [28a]*).*

experimental data and by a more thorough analysis of the results.
Reasonable agreement with the theory of Bohr seems now to be obtained here also.

This kind of agreement between experimental and theoretical
fission widths is perhaps misleading since other aspects of the
fission cross sections such as intermediate structure have not been
considered in the comparison between data and theory. This aspect
will be studied in the following subsections.

3-3. Intermediate Structure in Subthreshold Fission Cross Sections

The study of the fission widths presented in subsection 3-2
implicitly assumes that their properties are independent of the excitation energy, provided that the energy range under consideration
is relatively narrow, as is the case for neutron resonance spectroscopy. This assumption can be strongly erroneous for fission below
threshold as was shown for fission induced by resonance neutrons,
first in ^{237}Np [25] and later in other non fissile nuclei : ^{240}Pu [26],
^{234}U, ^{238}Pu, ^{242}Pu, etc... [27,28].

The subthreshold fission cross sections σ_f of these nuclei clearly exhibit an intermediate structure effect. This is illustrated in Fig. 5 where the Saclay cross section $\sigma_f\sqrt{E_n}$ of ^{237}Np is plotted as a function of neutron energy E_n. Rather than being distributed in a uniform manner as a function of E_n, the observed fission resonances are clustered around definite energies. In the first cluster at 40 eV, overall experimental resolution is sufficiently good for the various resonances in this cluster to be resolved. At higher energy, especially above 150 eV, individual resonances in the clusters cannot be resolved because the resolution rapidly deteriorates for increasing neutron energy. The clusters then appear as big peaks in the fission cross section, with a spacing of about 50 eV, that is roughly two orders of magnitude larger than for the resonances, as observed in the total cross section. Between the clusters, fission components are so small that they can hardly be detected.

This short-range intermediate structure in σ_f does not seem to be due to "doorway" states in the entrance channel, for the total cross section does not present any anomaly around the energies of the fission clusters. This is confirmed by an analysis of the resonances, both in σ_T and σ_f, from which the widths Γ_n and Γ_f are obtained up to the energy limit of 150 eV. The cumulative sums of these widths Γ_n and Γ_f are plotted in Fig. 6 as a function of E_n. Both histograms reflect the usual fluctuations of the widths discussed previously but in addition, the fission histogram has a staircase behaviour with steep rises at the energies of the clusters,

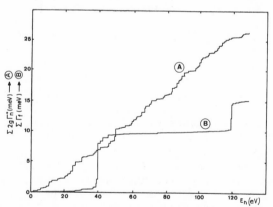

Fig. 6 - *The cumulative sums of the reduced neutron and fission widths, for the ^{237}Np resonances which could be analysed at low energy, are plotted as a function of neutron energy. Histograms A and B correspond to reduced neutron widths and fission widths respectively. This picture demonstrates that the intermediate structure is not due to the neutron entrance channel but to the fission exit channels (after Fubini et al. [25c]).*

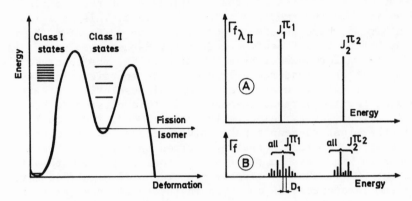

Fig. 7 – *Mechanism of intermediate structure in subthreshold fission cross sections* [29)]. *Clusters appear in the fission cross section when energy, spin and parity of a class-II state match those of the class-I resonances (at most two Jπ values are possible). The fission widths are drawn at the energy of the respective levels for class-II states (diagram A) and for the observed resonances (diagram B).*

in contrast to the neutron histogram, which has a smoother energy dependence, even across the clusters. This gives evidence that the intermediate structure does not come from the neutron entrance channel but rather is due to the coupling of the compound nucleus states to the fission exit channels, which is more intense at some discrete energies, those of the fission clusters [25b)].

Interpretation of this phenomenon was proposed, after the ^{240}Pu results confirmed those of ^{237}Np, by Lynn [27)] and Weigmann [30)] in terms of the double-humped fission barrier obtained by Strutinsky [31)] for nuclei of this category, having neutron number N in the vicinity of 146. The mechanism for this type of intermediate structure is illustrated in Fig. 7. The double-humped fission barrier provides the possibility for the existence of two types of compound states at energies below the top of the barrier and having different deformations. The class I states λ_I (in the terminology used by Lynn) are those which are usually considered with a deformation comparable to the permanent deformation of the nucleus in its ground state ; they correspond to the first well of the double-humped fission barrier. Class II states λ_{II} are more strongly deformed and correspond to the second well of the barrier. For increasing energy above the top of the barrier, distinction between these two types of states becomes hardly possible. The properties of class II states have been studied in great detail and compared to those of class I states [27)]. Essentially, i) they have larger fission widths $\Gamma_{f\lambda_{II}}$ because only the outer barrier needs to be penetrated for a fissioning state in the second well, ii) their neutron widths are smaller because there

is practically no overlap between the wavefunctions of class II
states and that of an outgoing neutron wave coupled to the target
nucleus in its ground state, and iii) they are more widely spaced
because,for the same total nuclear energy, the available excitation
energy is smaller in the second well.

With a few exceptions, the observed neutron resonances are
practically all of the class I type. For resonances with energy,
spin and parity matching those of a class-II state, fission is
greatly enhanced because of the admixture of this class-II state
which has a much larger fission width. Therefore, the class-II compound nucleus states act as doorway states in the fission exit
channels. But, in contrast to the doorway states in the entrance
channel discussed in section II, they have complicated configurations, as complex as those of the class I states. In that sense,
this type of intermediate structure is a noise (class-I states)
modulated by another noise (class-II states). It can exist because
the inner barrier (between the two wells) prevents too strong a
mixing between the two categories of states and also because the
outer barrier is high enough to keep the fission width of the class
II states smaller than their spacing.

There is at present no direct evidence that the proposed mechanism is correct, by showing for instance that class-II states have
a large quadrupole moment. But the mechanism predicts that all the
large fission resonances in the same cluster should have the same
spin and parity J^{π}, those of the class-II state responsible for the
existence of the cluster. This seems to be verified by the still
very limited data, for the group at 40 eV in ^{237}Np [23,30], the only
odd nucleus for which such a verification is possible at present.

3-4. Distribution of the Fission Widths in a Fission Cluster

In a fission cross section presenting an intermediate structure effect, the fission width of the resonances in a cluster is
mainly determined by the width $\Gamma_{f\lambda_{II}}$ of the doorway state and the
coupling matrix element H" between class-I and class-II states.
This subject has been studied by several authors, especially by
Lynn whose approach we shall follow from now on [27].

Basically, three different types of coupling between class-I
and class-II states can occur.

a) Narrow class-II state. Very weak coupling. This is the case
for class-II states having energy sufficiently below the tops of
the inner and outer barrier so that both their damping and fission
widths are small compared to their spacing. Hence, in a given fission cluster, there are a few resonances and their fission widths
are given by the following relations obtained from perturbation

theory :

$$\Gamma_{f\lambda'} = \Gamma_{f\lambda_{II}} \times \frac{|H''|^2}{(E_{\lambda_I} - E_{\lambda_{II}})^2} \tag{9}$$

$$\Gamma_{f\lambda''} = \Gamma_{f\lambda_{II}} \times \left\{ 1 - \sum_{\lambda_I} \frac{|H''|^2}{(E_{\lambda_I} - E_{\lambda_{II}})^2} \right\} \tag{10}$$

In these two expressions : i) indexes λ_I and λ_{II} refer to un-perturbed class-I and class-II states respectively ; ii) λ' and λ'' refer to perturbed class-I and class-II states respectively ; iii) H'' is a matrix element representing the coupling between the class-II and the class-I states under consideration and iv) E_{λ_I} and $E_{\lambda_{II}}$ are the energies of unperturbed class-I and class-II states. Because of the small value of H'', one of the fission resonances in the cluster (labelled λ'') is mainly class-II and has a large fission width but a small neutron width. This seems to be the case for the first cluster in ^{242}Pu where the following resonance parameters are obtained [33] :

$$E = 767 \text{ eV} ; \Gamma_n = 89 \text{ meV} ; \Gamma_f = 22 \text{ meV (mainly class-II)} \tag{11}$$

$$E = 799 \text{ eV} ; \Gamma_n = 465 \text{ meV} ; \Gamma_f = 2.5 \text{ meV (mainly class-I)} \tag{12}$$

b) Broad class-II state. Weak coupling. In this type of weak coupling, the inner barrier is still rather high, but the outer barrier is rather low compared to the energy of the class-II state so that it is broadened by fission decay and its width is mainly the fission width $\Gamma_{f\lambda_{II}}$. The class-II state is now sufficiently broad to encompass many resonances which have fission widths given by the following relation, also obtained by perturbation theory :

$$\Gamma_{f\lambda'} = \Gamma_{f\lambda_{II}} \frac{|H''|^2}{(E_{\lambda_{II}} - E_{\lambda_I})^2 + \frac{1}{4}(\Gamma_{f\lambda_{II}})^2} \tag{13}$$

This expression shows that the fission widths must be distrib-uted, *on the average*, along a Lorentzian line, but with fluctuations of the Porter-Thomas type reflecting those of $|H''|^2$.

Whereas the width of the cluster is essentially the fission width $\Gamma_{f\lambda_{II}}$ of the doorway class-II state, its damping width $\Gamma_\downarrow$ is given by the sum rule :

$$\sum_{\lambda'} \Gamma_{f\lambda'} = \Gamma_\downarrow \tag{14}$$

The neutron width of the class-II state is very small so that this broad state can only appear as a nearly undetectable residual background cross section beneath the narrow fission resonances.

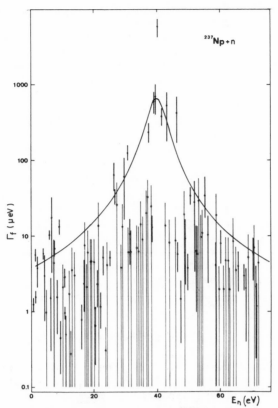

Fig. 8 – *The fission widths for the* ^{237}*Np resonances in the first fission cluster at 40 eV are plotted as a function of the resonance energy. For resonances having no detectable fission component, an upper limit is set for* Γ_f*. The solid curve is a Lorentzian line drawn through the greatest fission widths.*

The ^{237}Np intermediate structure seems to fall in this coupling category [25d,34]. The fission widths in the first cluster at 40 eV are plotted in Fig. 8 as a function of resonance energy. The largest widths seem actually to be distributed, *on the average,* along a Lorentzian line, with Porter-Thomas fluctuations. The other resonances have small fission widths, distributed in a uniform manner as a function of neutron energy, and indeed so small that they can hardly be detected. This behaviour of the fission widths can be simply interpreted in terms of the intermediate structure mechanism proposed in subsection 3-3. For ^{237}Np, "s" wave resonances can have spin and parity $J^\pi = 2^+$ or 3^+ but, in a cluster, such as the one at 40 eV, only those resonances having the J^π value of the doorway class-II state must be enhanced in fission. This explains the results plotted in Fig. 8, provided that the largest fission resonances all have the same J^π value.

c) <u>Class-II states broadened by moderate coupling to class-I</u>
<u>states</u>. This type of coupling is the reverse of the preceding one,
in the sense that now the inner barrier is lower than the outer one.
The fission width of the doorway state is then negligible compared
to its width $\Gamma_\downarrow$ for damping into the class-I states. Let us remark
that the inner barrier, though lower than for the previous coupling
cases, is nevertheless opaque enough to prevent too strong a
damping which would wash out the intermediate structure. In this
type of coupling, the calculated fission widths are given by the
following relation :

$$\Gamma_{f\lambda} = \Gamma_{f\lambda_{II}} \frac{|H''|^2}{(E_{\lambda_I} - E_{\lambda_{II}})^2 + \frac{1}{4}(\Gamma_\downarrow)^2} \qquad (15)$$

where the notation corresponds to that of expression (9). Again, in
this type of coupling, the fission widths must be distributed *on*
the average, along a Lorentzian line with Porter-Thomas fluctuations
due to those of $|H''|^2$. Whereas the width of the cluster is now the
damping width $\Gamma_\downarrow$ of the doorway state, the fission width of this
state is given by the following sum rule :

$$\sum_\lambda \Gamma_{f\lambda} = \Gamma_{f\lambda_{II}} . \qquad (16)$$

This moderate coupling situation probably applies to ^{234}U for
which the fission widths of the resonances in the first cluster at
700 eV are plotted in Fig. 9. This distribution of the widths seems
on the average, to obey the distribution (15), valid for all the
resonances since they are all expected to have the same spin and
parity ($J^\pi = 1/2^+$) in contrast to those of ^{237}Np, as discussed
earlier. Fluctuations of these ^{234}U fission widths about the mean
Lorentzian line are consistent with a chi-squared distribution
having a value of ν close to unity [35].

3-5. Intermediate Structure in the Fission Cross Section of Fissile Nuclei

For fissile nuclei, the fission barrier can also present two
humps but with heights below S_n. This fission barrier corresponds
to the deformation energy of the nucleus in its ground state. But,
as discussed in subsection 3-1, there is in fact one fission barrier
for each of the fission exit channels. For some of these channels,
fission can therefore occur below threshold and consequently, the
corresponding cross section may present an intermediate structure
effect.

This effect is hard to detect in general cases especially because of the wide fluctuations of both the neutron and fission
widths [28]. But, again, the situation seems favorable for ^{239}Pu.
This is because for spin 1^+ resonances, most of the fission contri-

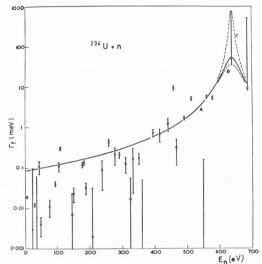

Fig. 9 – *The fission widths for the* ^{234}U *resonances in the first fission cluster at 700 eV are plotted as a function of the resonance energy. Curves A, B and C are Lorentzian lines. Curve A goes through the measured fission width at 638.4 eV whereas curves B and C go through points at one standard deviation from this datum (after James* [35]*).*

bution comes from one exit channel which is partially closed, as was shown in subsection 3-2. The 1^+ contribution to the fission cross section may thus present an effect of intermediate structure, which can be checked by the determination of the spins of the resonances over a wide enough energy range. At present, spins have been measured up to 660 eV [18] and the fission width $<\Gamma_f>_{1^+}$, averaged over resonances in 110 eV intervals, is plotted in Fig. 10 as a function of the mean energy of the interval. The energy dependence of $<\Gamma_f>_{1^+}$ shows rather extensive variations which are incompatible with a conventional statistical model. For example, the value of $<\Gamma_f>_{1^+}$ between 550 eV and 660 eV is very small, as confirms the comparison of the fission width distributions for the intervals 550 eV – 660 eV and 0 – 660 eV (Fig. 4). These energy variations of $<\Gamma_f>_{1^+}$ are interpreted as being due to the coupling of the 1^+ resonances to broad 1^+ class-II doorway states ; one of them may have an energy of about 400 eV, where $<\Gamma_f>_{1^+}$ takes the highest value.

Other cases of intermediate structure probably exist for other fissile nuclei, such as ^{235}U ; but their detection, not possible at present by a detailed analysis of the microscopic cross sections, would require the use of autocorrelation techniques which need to be developed.

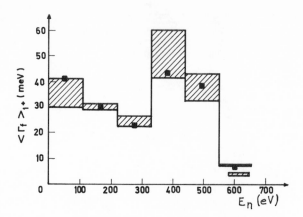

Fig. 10 – *The average fission width* $<\Gamma_f>_{1^+}$ *for the* 1^+ ^{239}Pu
*resonances situated in an energy interval of 110 eV, is plotted
as a function of the mean energy of this interval. Hatched areas
are between higher and lower limits of the values of* $<\Gamma_f>_{1^+}$ *as
determined from spin measurements. The dark squares (■) represent
the value of* $<\Gamma_f>_{1^+}$ *as obtained from a fit to the fission width
distribution corresponding to the 110 eV interval under
consideration. A constant value of 3.5 meV has been subtracted to
remove the calculated contribution of the* $(n,\gamma f)$ *process (after
Trochon et al. 18)).*

4. CONCLUSION

The relatively high excitation energy of slow neutron resonances in the heavy nuclei makes it natural to begin the study of
their neutron and fission widths using statistical assumptions.

This "black box" approach does not incorporate any knowledge
about the structure of the nucleus but, nevertheless, despite its
limited and simplified assumptions, proves remarkably successful in
explaining most of the neutron width distributions over a wide
range of nuclei. The shell model approach stimulated much investigation of the possible influence of doorway states in the neutron
entrance channel on the neutron widths. But detection of this intermediate structure effect, if present in the neutron total cross
section, is greatly hampered by the wide fluctuations of the neutron
widths themselves. In one possible case, for the system ^{206}Pb + n,
the behaviour of the neutron widths does seem to be caused by a
2p-1h excitation in the entrance channel. Other more sophisticated
approaches, based on the shell model but including the residual interaction treated in a statistical manner, are used for the study
of level spacing distributions [36]. Such a treatment of rather

highly excited states has not yet been applied for the distribution of neutron widths. The properties of the average neutron
width, their variation as a function of mass number A or spin J
for example, and explanation of the relevant data in terms of the
optical model or doorway states, have not been considered here
since this subject seems to fit in better with other papers presented at this Conference [37].

The study of the distribution of the fission widths in terms
of simple statistical models and of the fission exit channels proposed by A. Bohr also proved very successful for quite a long time.
But the energy dependence of the fission widths, as observed for
subthreshold fission, appeared to be in violent disagreement with
such simple conventional models. Under the theoretical framework
of Strutinsky, this intermediate structure in subthreshold fission
cross sections, together with other experimental results such as
the discovery of fission isomers, has played an important role towards a new and more thorough understanding of the fission process.
These new aspects of fission could not be presented here. Let us be
content to remark that one of these aspects, the existence of
strongly deformed states in the second well of a double-humped fission barrier, permits an explanation of the available data on the
fission widths. Conversely too, a knowledge of the fission widths
contributes to an insight into the properties of these strongly
deformed states which are also the subject of study by means of
other nuclear reactions.

REFERENCES

1) C.E. PORTER and R.G. THOMAS, Phys. Rev., 1956, 104, 483.
2) C. MAHAUX, "Intermediate Structure", Invited paper (This
   Conference).
3) J.D. GARRISON, "A Statistical Analysis of Neutron Resonance
   Parameters", Symposium on Statistical Properties of Atomic and
   Nuclear Spectra, Stony Brook, 1963 ; Ann. Phys., 1964, 30, 269.
4) A. MICHAUDON, H. DERRIEN, P. RIBON and M. SANCHE, Phys. Letters,
   1963, 7, 211 ; Nucl. Phys., 1965, 69, 545.
5) V.F. WEISSKOPF, Phys. Today, 1961, 14, 18.
6) C. BLOCH, "Many-Body Description of Nuclear Structure and Reactions", (Academic Press, 1966), p. 394.
7) H. FESHBACH, A.K. KERMAN and R.H. LEMMER, Ann. Phys., 1967, 41,
   230.
8) A.D. CARLSON and H.H. BARSCHALL, Phys. Rev., 1967, 158, 1142.
9) J.A. FARRELL, G.C. KYKER Jr., E.G. BILPUCH and H.W. NEWSON,
   Phys. Letters, 1965, 17, 286.
10) N. BOHR and J.A. WHEELER, Phys. Rev., 1939, 56, 426.
11) A. BOHR, Proc. Int. Conf. on Peaceful Uses of Atomic Energy,
    1956, 2, 163.

12) M.S. MOORE, "Low Energy Neutron Spectroscopy", (Measurements on Fissile Nuclides), Report IN-1222, (Jan. 1969).

13) J.E. LYNN, "The Theory of Neutron Resonance Reactions", (Clarendon Press, Oxford, 1968).

14) A. MICHAUDON, "Second Conference on Neutron Cross Sections", (Washington, March 1968), NBS Special Publication 229, Vol. $\underline{I}$, p. 427 ; J. Phys., 1968, C1, Suppl. n° 1, $\underline{29}$, 51.

15) A. MICHAUDON, "Nuclear Data for Reactors", (IAEA, Vienna, 1967) Vol. $\underline{II}$, p. 161.

16) J.J. GRIFFIN, "Physics and Chemistry of Fission", (IAEA, Vienna, 1965), Vol. $\underline{I}$, p. 11.

17) J. BLONS, H. DERRIEN and A. MICHAUDON, " Nuclear Data for Reactors", (IAEA, Vienna, 1970), Vol. $\underline{I}$, p. 513.

18) J. TROCHON, H. DERRIEN, B. LUCAS and A. MICHAUDON, "Nuclear Data for Reactors", (IAEA, Vienna, 1970), Vol. $\underline{I}$, p. 495.

19) H. DERRIEN, J. BLONS and A. MICHAUDON, "Nuclear Data for Reactors", (IAEA, Vienna, 1970), Vol. $\underline{I}$, p. 481.

20) J.E. LYNN, "Nuclear Data for Reactors", (IAEA, Vienna, 1967), Vol. $\underline{II}$, p. 89.

21) a) J.W.T. DABBS, F.J. WALTER and G.W. PARKER,"Physics and Chemistry of Fission",(IAEA, Vienna, 1965), Vol. $\underline{I}$, p. 39.
b) J.W.T. DABBS, C. EGGERMANN, B. CAUVIN, A. MICHAUDON and M. SANCHE, BAPS, Series II, 1968, $\underline{13}$, Nb 11, 1407 ;"Physics and Chemistry of Fission",(IAEA, Vienna, 1969), p. 321.

22) a) N.J. PATTENDEN,"Physics and Chemistry of Fission",(IAEA, Vienna, 1969), p. 330.
b) N.J. PATTENDEN and H. POSTMA, Nucl. Phys., 1971, $\underline{167}$, 225.

23) H. POSTMA, N.J. PATTENDEN and R. KUIKEN, Contributed paper (This Conference).

24) J.E. LYNN, Phys. Letters, 1965, $\underline{18}$, 31.

25) D. PAYA, H. DERRIEN, A. FUBINI, A. MICHAUDON and P. RIBON,
a) "Nuclear Data for Reactors", (IAEA, Vienna, 1967), Vol. $\underline{II}$, p. 128 and Vol. $\underline{III}$.
b) J. Phys., 1968, $\underline{29}$, Suppl. n° 1, 159.
c) A. FUBINI, J. BLONS, A. MICHAUDON and D. PAYA, Phys. Rev. Letters, 1968, $\underline{20}$, n° 24, 1373.
d) D. PAYA, J. BLONS, H. DERRIEN and A. MICHAUDON,"Physics and Chemistry of Fission",(IAEA, Vienna, 1969), p. 307.

26) E. MIGNECO and J.P. THEOBALD, Nucl. Phys., 1968, $\underline{A112}$, 603.

27) J.E. LYNN, "Nuclear Structure", (IAEA, Vienna, 1968), p. 463 ; "Physics and Chemistry of Fission",(IAEA, Vienna, 1969), p. 249.

28) A. MICHAUDON,
a) "Nuclear Structure", (IAEA, Vienna, 1968), p. 483.
b) Lectures given at the 14th International Meeting of Physicists, Duilovo, Yugoslavia, (Sept. 1969).
c) "Advances in Nuclear Physics", (Plenum Press), (under press).

29) S. BJØRNHOLM and V.M. STRUTINSKY, Nucl. Phys., 1969, $\underline{A136}$, 1.

30) H. WEIGMANN, Z. Physik, 1968, $\underline{214}$, 7.

31) V.M. STRUTINSKY, Nucl. Phys., 1967, $\underline{A95}$, 420 ; Nucl. Phys., 1968, $\underline{A122}$, 1.

32) F. POORTMANS, H. CEULEMANS, J. THEOBALD and E. MIGNECO, Paper
    VI.11, "Third Conference on Neutron Cross Sections," Knoxville,
    (March 1971).
33) C.D. JAMES, Nucl. Phys., 1969, A123, 24.
34) H. WEIGMANN, G. ROHR and J. WINTER, Phys. Letters, 1969, 30B,
    624.
35) G.D. JAMES and E.R. RAE, Nucl. Phys., 1969, A139, 471.
36) Session 3, This Conference.
37) Session 5, This Conference.

DISCUSSION

CHRIEN (B.N.L.)

Quite recently there was a report in Physical Review Letters from Los Alamos National Lab. concerning the distribution of neutron widths in Th^{232}. They find, in fact, a $\nu=2$ distribution. Would you care to comment on that?

MICHAUDON

In fact, I think that this deviation from the Porter-Thomas distribution comes from two sorts of experimental data--the neutron widths at low energy and the average capture cross section at high energy. At low energy where the individual resonances can be analyzed the evidence seems to come from the fact that there is a big gap between neutron widths which are supposed to belong to the s-wave resonances for the larger widths and the p-wave resonances for the smaller widths. This was observed in the Saclay data several years ago, also. The simple explanation which consists of saying that the small neutron widths belong to p-wave levels is not obvious, because some of these widths belong to s-wave levels. According to a discussion I had with Mike Moore yesterday, it seems that there is 1% probability that the data agree with the Porter-Thomas distribution. Therefore, I don't think it's possible to conclude for the moment, and we will need more experimental evidence to firmly establish that such a fundamental law (the Porter-Thomas one) is violated in the case of the "s" wave neutron widths.

DISCUSSION

POSTMA (Groningen University)
      During the past three years we have carried out experiments
concerning the angular distribution of fission fragments from
aligned ^{233}U, ^{235}U and ^{237}Np nuclei after capture of slow neutrons.
Neutrons were produced with the 45 MeV Harwell linear electron
accelerator using one of the standard Harwell time-of-flight
systems with a flight path of 10 m. In most of these experiments
the energy range covered for the neutrons was 0.4 to about 2000 eV.
Resonances could be well resolved below about 80 eV. Details of
the experimental set-up can be found in the paper, Nuclear Physics
A 167, 225 (1971). The angular distribution of fission fragments
from aligned nuclei is given by the expression:

$$W(\Theta) = 1 + A_2 f_2(I) \; P_2(\cos\Theta) \tag{1}$$

neglecting higher-order terms. $f_2(I)$ is the usual alignment parameter. Due to the symmetric emission of fission fragments, it
suffices to use aligned target nuclei and unpolarized neutrons.
The anisotropy parameter A_2 depends very strongly on the projection
K of the compound nuclear spin J along the nuclear symmetry axis
when the nuclear system is near the scission point; that is:

$$A_2(J;K) = \frac{15}{4} \frac{I}{I+1} \left\{ \frac{3K^2}{J(J+1)} - 1 \right\} \tag{2}$$

if only one channel with quantum numbers (J^{π};K) is open. If several channels (c) with different K-values are open the effective A_2-
value is given by:

$$A_2(\text{res.}) = \sum_c \Gamma_f^c \, A_2(J;K) \; / \sum_c \Gamma_f^c, \tag{3}$$

where Γ_f^c are partial fission widths for a resonance. The relative
openess of the fission channels is related to the fission widths
averaged over a large number of resonances.

      The first experiments were carried out with aligned ^{235}U.
In this case J^{π} is 3^- or 4^-. A_2-values could be obtained for 55
reasonably well separated resonances. Each datum was represented
in a pseudo-histogram by a gaussian curve with its maximum value
at the measured A_2-value and a width twice the experimental error,
not including an overall systematic error. The experimental
pseudo-histogram was found to be rather narrow with a maximum near
$A_2 = -2$. Numerical calculations to simulate the experimental distribution were carried out assuming channels with different K-values
being open and using a large number of resonances. To obtain a
theoretical A_2-value the Γ_f^c-values were selected randomly from
Porter-Thomas distributions. In this way A_2-values were calculated
for many resonances. Randomly chosen errors were assigned to each

A_2-value. In this way theoretical pseudo-histograms were obtained. A reasonable agreement could be obtained if it was assumed that the channels with K = 0 and K = 1 are fully open and that channels with K = 2 are only partially open. As can be seen in slide 1, the maximum values of the presented theoretical distributions occur at slightly lower A_2-values ($\simeq$ -2.3), but this difference with respect to the experimental pseudo-histogram is insignificant to our conclusion.

The results for ^{233}U with $J\pi = 2^+$ and 3^+ were similar. The experimental histogram is shown in slide 2; again the lowest channels with K = 0 and 1 are fully open. However, it seems that the K = 2 channel is more open in this case. Three examples of theoretical distributions given in slide 3 show how sensitive these curves are to the choice of the channels.

In the case of ^{237}Np ($J^\pi = 2^+$ and 3^+) we are dealing with subthreshold fission. The first group of resonances between 20 and 55 eV were well resolved. This group of resonances and the measured A_2-values are shown in slide 4. It follows that K = 3 or 2 if J = 3 or K = 2 if J = 2. The resonances at 26.6, 30.4 and 50.4 eV have been assigned as J = 2 by Poortmans, et al. Thus, the first group of subthreshold resonances is consistent with channel $(2^+;2)$, assuming that they have the same spin. The higher-energy groups observed below 1500 eV agree with channels $(2^+;2)$ and $(3^+;2)$.

The work reported here was done in collaboration with Pattenden and Kuiken.

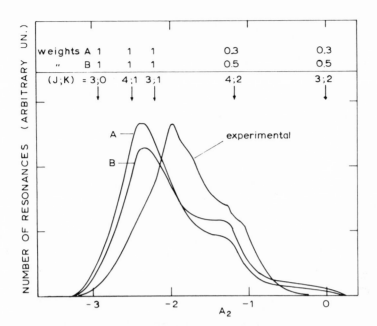

fig. 1. Experimental pseudo-histogram and two simulated distribution curves A and B for ^{235}U. A_2-values for various channels are indicated by arrows. Relative weights are given in the top of the figure.

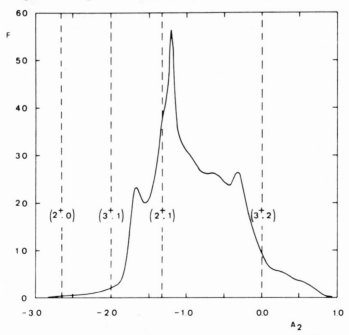

fig. 2. Experimental pseudo-histogram for ^{233}U.

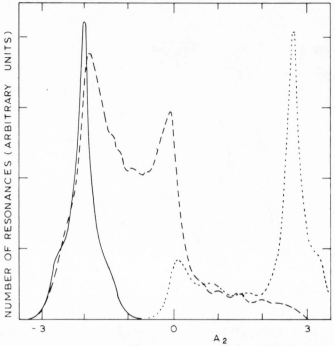

fig. 3.   Three simulated distribution curves for ^{233}U
assuming the following relative weights: solid curve
$W(2^+;0)=1$, $W(2^+;1)=W(3^+;1)=0.5$ others zero, dotted curve
$W(2^+;2)=W(3^+;2)=W(3^+;3)=1$ others zero, and dashed curve $W(2^+;0)=$
$W(2^+;1)=W(2^+;2)=W(3^+;1)=W(3^+;2)=0.5$ others zero.

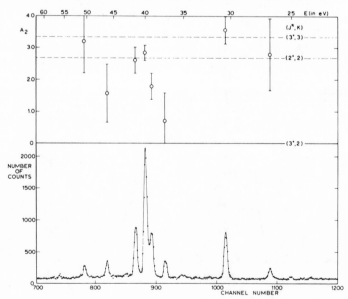

fig. 4.   Resonances and A_2-values of the first group of ^{237}Np.

ROSENWEIG (S.U.N.Y/Albany)

I would like to ask a question, Dr. Michaudon. In analyzing
these neutron widths, did you study whether or not there is a
correlation between the widths belonging to successive resonances?
According to the simplest statistical model, there ought to be
essentially no correlation, and in reading one of the papers from
the Columbia group, they seem to find a correlation in the E [166]
data, which otherwise follow the statistical model very well. Have
you examined your data from this point of view.

MICHAUDON

The Columbia data you mention were not available before the
conference. The other data examined so far, especially the extensive study carried out by Garrison, do not show any correlation
between the neutron widths belonging to different resonances.
I would like to say a few words about the slides projected by
Postma. First, I didn't quote these results in the oral presentation of my paper because I knew Postma would comment on them, but
they are mentioned in the written version of the paper. Secondly,
the Neptunium 237 results seem to provide the evidence that all the
large fission resonances in the 40 eV cluster have the same (J,K)
value, therefore in agreement with the proposed mechanism for
intermediate structure. The third comment I want to make is that
there is some puzzling disagreement about uranium 235, between the
previous measurements, as carried out at Saclay and Harwell, and
the new Harwell results. The previous results show absence of the
$K = 0$ contribution in the fission resonances. Now, according to
the new Harwell results, this apparent absence of $K = 0$ seems to
be removed by renormalization of the data and also by considering
the contributions of the various fission exit channels. Nevertheless, according to the results you presented here, if you take into
account two fully opened channels ($K = 0$ and $K = 1$), you certainly
come out with a fission width which is much larger than the observed
fission width of the resonances, in disagreement with the Channel
Theory of Bohr.

LEVEL SPACINGS, CORRELATION FUNCTIONS AND STATISTICS[*]

M. L. Mehta

C.E.N. de Saclay, 91 Gif-sur-Yvette, France

1. General remarks

Books on scientific subjects can be divided into two categories: Ones written on fast developing subjects and becoming even faster out of date; and those written in the old-fashioned way on subjects which are expected soon to die. At the time of publication in 1967, I considered the book "Random Matrices" to be of the second category. Fortunately my expectation turned out to be wrong. The random matrices are not only still living but are healthy and growing as can be witnessed by the number of papers published and by a part of this conference itself.

There have been excellent reviews of the subject by C. Bloch, E. P. Wigner and J. E. Lynn, a specialized presentation by N. Rosenzweig and a pedagogical discussion by A. Bohr and B. R. Mottelson. I will choose the arbitrary date 1965, the year of publication of C. E. Porter's book containing a detailed review article and reprints till then of various important papers in the field.

2. Clean analytical results

I see people complaining that "unfortunately limited success has been achieved in purely analytical exploration of the full implications of this (the Wishart) distribution." (See below equation (1).) Let me add to this in a low voice that the situation is not so unfortunate by presenting first the clean analytical successes. This is one way of justifying one's own existence. One starts with the joint probability density for the eigenvalues $x_1, \ldots, x_N$ of random matrices

---

[*] Partly under auspices of U. S. Atomic Energy Commission. This review was prepared while the author enjoyed the hospitality of the Argonne National Laboratory during the summer of 1971.

$$P_{N\beta}(x_1, \ldots, x_N) = C_\beta \prod_{j<k} |x_j - x_k|^\beta \cdot e^{-\sum_i^N x_j^2} \qquad \text{-Gaussian,} \quad (1a)$$

$$= C_\beta' \prod_{j<k} |e^{ixj} - e^{1xk}|^\beta \qquad \text{-Circular} \quad (1b)$$

The parameter β taking the value 1, 2 or 4 according as the random matrices are taken from the orthogonal, unitary or symplectic ensemble. Prior to 1965 one knew the function $E_\beta(0,s)$, the probability that a given interval of length s is empty of eigenvalues. The interval s is measured, as usual, in units of the mean spacing. One could then calculate by two differentiations the function $p_\beta(0,s)$, the probability density for a spacing of length S. One also knew the two level correlation function $R_{2\beta}(x_1,x_2)$, i.e., the probability of having one level at x_1 and another at x_2 and ignoring the positions of the other levels. One could then calculate the two-level cluster function $T_{2\beta}(x_1,x_2)$.

Now one knows the function $E_\beta(n,s)$, the probability that a given interval of length s contains exactly n eigenvalues. [Mehta and des Cloizeaux, 1970]. They are given by the formulae

$$E_1(0, s) = \prod_{\rho=0}^{\infty} (1 - \lambda_{2\rho}), \qquad (2a)$$

$$E_1(2r, s) = E_1(0, s) \cdot \sum g_e(j_1, \ldots, j_r)\{1 - \sum_{i=1}^{r} b_{j_i}\}, \qquad (2b)$$

$$E_1(2r - 1, s) = E_1(0, s) \cdot \sum g_e(j_1, \ldots, j_r) \cdot \sum_{i=1}^{r} b_{j_i}, \qquad (2c)$$

$$E_2(0, s) = \prod_{\rho=0}^{\infty} (1 - \lambda\rho), \qquad (3a)$$

$$E_2(n, s) = E_2(0. s) \cdot \sum g(j_1, \ldots, j_n), \qquad (3b)$$

$$E_4(n, \tfrac{1}{2}s) = E_1(0, s) \cdot \sum g_e(j_1, \ldots, j_n)$$

$$\cdot \left\{ 1 + \tfrac{1}{2} \sum_{i \neq (j)} \frac{\lambda_{2i} b_i}{1 - \lambda_{2i}} - \tfrac{1}{2} \left(\sum_{k=1}^{n} b_{j_k} \right) \left(1 + \sum_{i \neq (j)} \frac{\lambda_{2i} b_i}{1 - \lambda_{2i}} \right) \right\}$$

where   (4)

$$g_e(j_1, \ldots, j_n) = \frac{\lambda_{2j_1}}{1 - \lambda_{2j_1}} \cdots \frac{\lambda_{2j_n}}{1 - \lambda_{2j_n}}, \qquad (5a)$$

$$g(j_1, \ldots, j_n) = \frac{\lambda_{j_1}}{1 - \lambda_{j_1}} \cdots \frac{\lambda_{j_n}}{1 - \lambda_{j_n}}, \qquad (5b)$$

$$b_i = f_{2i}(1) \int_{-1}^{1} f_{2i}(x)\,dx \Big/ \int_{-1}^{1} f_{2i}^2(x)\,dx, \quad (6)$$

and the summations over (j) are on all integers with
$0 \leq j_1 < j_2 < \ldots$. The $f_i(x)$ are prolate spheroidal angle
functions and λ_i are their eigenvalues:

$$\lambda_i f_i(x) = \int_{-1}^{1} K(x - y) f_i(y)\,dy, \qquad (7)$$

with

$$K(x) = (\pi x)^{-1} \sin(\tfrac{\pi}{2} sx). \qquad (8)$$

One knows even the probability of having n levels at
$x_1, x_2, \ldots, x_n$ all lying within the interval s and no other
level within that interval.

One also knows now the n-level correlation function
$R_{n\beta}(x_1, \ldots, x_n)$ or equivalently the n-level cluster function
$T_{n\beta}(x_1, \ldots, x_n)$. [Dyson 1970, Mehta 1970a].

$$R_{n\beta}(x_1, \ldots, x_n) = \left\{ \det[\sigma_\beta(r_{ij})]_{i,\,j=1,\,2\ldots,\,n} \right\}^{\alpha} \qquad (9)$$

where

$$\sigma_1(r) = \begin{bmatrix} s(r) & D(r) \\ \\ J(r) & s(r) \end{bmatrix}, \qquad \sigma_4(r) = \begin{bmatrix} s(r) & D(r) \\ \\ I(r) & s(r) \end{bmatrix}, \qquad \text{(10a,b)}$$

$$\sigma_2(r) = s(r), \qquad\qquad\qquad\qquad\qquad\qquad\qquad\qquad \text{(10c)}$$

$$r_{ij} = \left| x_i - x_j \right|, \qquad\qquad\qquad\qquad\qquad\qquad\qquad \text{(11)}$$

and

$\alpha = 1$ for $\beta = 2$, while $\alpha = \frac{1}{2}$ for $\beta = 1$ or 4.

The determinant in equation (9) above is an $n \times n$ if $\beta = 2$ and is $2n \times 2n$ if $\beta = 1$ or 4.

In purely mathematical terms we have found a method to evaluate the multiple integral

$$\int \cdots \int \prod_{1 \leqslant j < k \leqslant N} \left| x_j - x_k \right|^\beta \prod_{j=1}^{n} g(x_j) dx_{n+1} \cdots dx_n \qquad \text{(12)}$$

for $\beta = 1$, 2 or 4, $g(x)$ any well-behaved function and $1 \leqslant n \leqslant N$, where the integrations are carried out within any limits (a,b) with or without the restriction that the variables $x_{n+1}, \ldots, x_n$ remain outside a certain interval containing the variables $x_1, \ldots, x_n$.

As a by-product one sees that in the limit $N \to \infty$ the n-level correlation functions for the Gaussian ensembles and those for the circular ensembles are identical for any finite n. In other words, for an infinitely long eigenvalue sequence all the statistical properties are the same whether the ensembles are circular or Gaussian. It is necessary to emphasize this point because some people seem not to be aware of it.

A result of pure academic interest is that

$$E_2(0, S) \exp(\pi^2 s^2/8) = \lim \int \cdots \int P(X)P(Y) \prod_{j=1}^{N} \left(\frac{x_j + t}{y_j + t} \right)^{\frac{1}{2}} dx_j dy_j, \qquad \text{(13a)}$$

with

$$P(X) \equiv P(x_1, \ldots, x_N) = \prod_{1 \leqslant j < k \leqslant N} (x_j - x_k)^2 \prod_{j=1}^{N} e^{-x_j} \qquad \text{(13b)}$$

$$Nt = \pi^2 s^2/16 \qquad\qquad\qquad\qquad\qquad\qquad\qquad \text{(13c)}$$

tends to zero as $s \to \infty$, while thermodynamic arguments indicated that this quantity remains finite. [Dyson, J. Math. Phys. 3, 152 (1962) equation (92)]. Thus thermodynamics gives correctly the dominant term, the second order term is not reliable and the third order term is definitely wrong.

It will be a pity if I do not mention a very elegant proof found by I. J. Good for the combinatorial theorem of Dyson. The earlier proof of K. G. Wilson and J. Gunsen are lengthy and tedious.

3. Various statistics

The next question is of finding suitable statistics for the purpose of comparing experimental results with theory. A "statistic" is a quantity which can be calculated from an observed sequence of levels alone without other information, and whose average value and variance are known from the theoretical model. A suitable statistic is one which is sensitive for the property to be compared or distinguished and is insensitive for other details. For example the Δ statistics of Dyson-Mehta are defined by

$$\Delta_a = \underset{A, B}{\text{Min}} \left\{ \frac{1}{2L} \int_{-L}^{L} \{N(E) - AE - B\}^2 dE \right\} \tag{14}$$

where $N(E)$ is the number of levels lying between $-L$ and E, and α is a parameter indicating how A and B are chosen. If $B = N(0), a = 1$, if $B = N(-L) + AL$, $\alpha = 2$ and if there is no restriction on A and B, $\alpha = 3$. The average values and the variances of the Δ are well known. [Dyson-Mehta, J. Math. Phys. 4, 702 (1963)]. They measure the long range order in the level-series. On the other hand their Q-statistic

$$Q = - \sum_{i<j} f(E_i, E_j) \ell n \left| (E_i - E_j)/R \right| + n \sum_{j} U(E_j)$$

$$- U_0 n^2 + \tfrac{1}{2} n \ell \, n(\pi R n/L) \tag{15}$$

with

$$f(x, y) = 1, \quad \text{for } |x - y| < R, \ |x| < L, \ |y| < L,$$

$$= 0, \quad \text{otherwise,}$$

$$U(x) = -R/L, \qquad |x| < L - R,$$

$$= -\frac{1}{2L}\{R + (L - |x|)(1 - \ell n[(L - |x|)/R])\},$$

$$L - R < |x| < L,$$

$$= 0, \qquad |x| > L,$$

$$U_0 = -R/L + \frac{1}{8}(R/L)^2,$$

$$n = 2L/D,$$

measures the short range order extending up to a distance R, R < L. The average value and the variance of Q are

$$\langle Q \rangle = n\{1 - \tfrac{1}{2}(\gamma + \ell n2) - \frac{1}{\pi^2 M}\}$$

$$\approx n\{0.365 - \frac{1}{\pi^2 M}\},$$

$$\langle Q^2 \rangle - \langle Q \rangle^2 = n\{\frac{3}{2} - \frac{\pi^2}{8} + \frac{\theta}{\pi^2 M}\}$$

$$\approx n\{0.266 + \frac{\theta}{\pi^2 M}\},$$

where n = 2L/D, M = R/D, D = mean spacing and θ is a coefficient of order unity. The statistic Q has however not been much used.

The Λ- and Λ^*-statistics of Monahan and Rosenzweig [1970] are designed to distinguish between a sequence of uncorrelated spacings each verifying separately the spacing distribution for orthogonal ensemble and a sequence of spacings between the eigenvalues of matrices of orthogonal ensemble itself. Their expressions appear in the accompanying contributed paper and therefore will not be repeated here. The average values and variances for these statistics are known only approximately from a Monte-Carlo calculation, and appear in the form of curves in the contributed paper.

The F-statistic of Dyson defined by

$$F_i = \sum_{j \neq i} f(x_{ji}), \qquad x_{ji} = (E_j - E_i)/L, \tag{17}$$

$$f(x) = \tfrac{1}{2} \ell n \left\{ \frac{1 + (1 - x^2)^{1/2}}{1 - (1 - x^2)^{1/2}} \right\}, \qquad \text{if } |x| < 1, \tag{18}$$

$$= 0, \qquad \text{if } |x| > 1.$$

is constructed to further purify an almost pure series. For example, in a series of neutron capture resonances by a zero spin target nucleus where most of the levels are believed to be s-wave, the F-statistic should enable, in principle, to detect the presence of a few p-wave levels or the absence of a few missed s-wave levels. The average value of F_i is

$$\langle F_i \rangle = n - \ell n(8n) - \gamma + 2 \approx n - \ell n\, n - 0.656, \tag{19a}$$

while its variance is

$$\langle F_i{}^2 \rangle - \langle F_i \rangle^2 \approx \ell n\, n. \tag{19b}$$

Thus F_i is expected to remain almost a constant for levels of a simple set, to rise for an extra outside level and to drop for a missing level. Such an analysis has recently been used to "correct" the level-series of various isotopes of Erbium and we will hear more about it from experimenters of the Columbia University.

Table 1 summarizes the situation as to which nuclear spectra have been analyzed with respect to what statistic and how good or bad is the agreement.

4. Partial violation of symmetries

Next I will speak about those problems which have been attacked more than once, but with very little success. One of them is how to treat the partial violation of symmetry, for example the time-reversal invariance. The spectrum of a system looks different according to whether it is or is not invariant under T, the operation of time reversal. If the Hamiltonian of the system is strictly invariant under T then it should belong to the orthogonal ensemble. On the other hand if the part of the Hamiltonian odd under T is equal in magnitude to the part even under T, then the Hamiltonian should belong to the unitary ensemble. If the entire Hamiltonian is odd under T, then it should belong to the ensemble of anti-symmetric matrices with pure imaginary elements. All these three extreme cases can be treated analytically. [For the last one see Mehta and Rosenzweig 1968].

Nuclear spectra have overall statistical properties in conformity with the provisions of the orthogonal ensemble. So

Table 1

Nucleus Energy range No. of Levels	Statistic	Experiment	Theory	Reference
U238 0-1 kev 57	Δ_1	1.778	0.84 ± 0.58	a
	Δ_2	1.299	0.53 ± 0.15	a
	Δ_3	1.278	0.40 ± 0.11	a
	Q	14.62	18.2 ± 3.9	a
Ta181 0-334 ev 68	Δ_1	3.411	1.53 ± 1.24	a
	Δ_2	1.443	0.93 ± 0.31	a
	Δ_3	1.437	0.87 ± 0.22	a
	Q	74.1	51.4 ± 8.6	a
Th232 1-4 kev 154	Δ_1	3.265	1.04 ± 0.72	a
	Δ_2	8.717	0.66 ± 0.18	a
	Δ_3	3.123	0.50 ± 0.11	a
	Q	61.17	52.4 ± 6.4	a
Er166 0-4.2 kev 109	Δ_3	0.455	0.468 ± 0.11	b
	$\mathrm{cov}(s_j, s_{j+1})$	-0.22	-0.271 ± ?	b
	$\Lambda(108)$	0.15	0.16 ± (0.10)	c
	$\Lambda^*(10)$	0.17	0.19 ± (0.04)	c

Table 1 (continued)

Nucleus Energy range No.of Levels	Statistic	Experiment	Theory	Reference
Er168	Δ_3	0.287	0.389 ± 0.11	b
0-4.9 kev	$\mathrm{cov}(s_j, s_{j+1})$	-0.29	-0.271	b
73-25+2	$\Lambda(48)$	0.12	0.16 ± (0.10)	c
	$\Lambda^*(10)$	0.20	0.19 ± (0.06)	c
Er170	Δ_3	0.359	0.341 ± 0.11	b
0-4	$\mathrm{cov}(s_j, s_{j+1})$	-0.093	-0.271	b
(62-31)	$\Lambda(30)$	0.06	0.17 ± (0.10)	c
	$\Lambda^*(10)$	0.17	0.19 ± (0.08)	c

[a] F. J. Dyson and M. L. Mehta, J. Math. Phys. 4, 701 (1963).

[b] Papers by the Columbia group. See references at the end.

[c] J. E. Monahan and N. Rosenzweig, (Phys. Rev.) Preprint.

T-invariance is quite good. But what about a small T-violating
part in the nuclear forces? Can it be detected by simply looking
hard at the statistics of nuclear spectra? Some prominent people
say yes, [Wigner 1967, p. 21], and they are probably right. The
question then is to study ensembles deviating slightly from the
orthogonal one. Real symmetric matrices with a small anti-
symmetric pure imaginary part added to them provide a valid model.
Quite a few papers exist on this subject either treating the small
imaginary part as a perturbation [McDonald and Favro, 1968,
McDonald, 1969, Mehta, 1970b], or effecting Monte-Carlo calculations on a computer [Mehta, Monahan and Rosenzweig 1968]. No
estimate being available for neglected terms the perturbation
theory is not at all reliable. This distrust is only strengthened
by the experience with 2×2 matrices where exact results are
known. Monte-Carlo calculations are not fine enough to detect
small differences. One may find an excuse by saying that the
experimental data are not clean enough either for that kind of
analysis. But the trouble is that experimenters are capable
people. If you tell them what they should look for and measure,
I am sure, they will sooner or later do it. What is lacking is
a theory.

We will learn much about the effect of a small mixture of
T-non-invariant part, if we could solve the partial differential
equations arising in Dyson's Brownian motion model. Despite some
serious attempts [Rosenzweig 1965] unfortunately very little is
known analytically about how the system evolves.

5. Random matrices with a given level-density

To get exponentially increasing level-densities we use one
type of theory, such as the Fermi-gas model, while to get finer
details of the spectra, such as spacings, we use another type.
Gaussian ensembles are derived by simplifying assumptions like
the statistical independence of the various matrix elements. This
has been objected to and some authors tried to repair this by
rederiving them by the principle of maximum randomness with given
constraints. [Balian 1968 a, b, Bloch 1968]. By this process
one can even construct ensembles of matrices which will have a
level-density given in advance, and which also has level-repulsion.
However, this line of enquiry has not been pursued.

6. Two-body random forces

Calculations for low-lying nuclear levels have always been
carried out with two-body forces. On the other extreme, in the
theory of random matrices we assume complete randomness implying
complicated many-body forces. To see what kind of differences
might be there in the two extremes, people recently undertook
Monte-Carlo calculations with random two-body forces. [French
and Wong, 1970, 1971, Bohigas and Flores, 1971]. The eigenvalue
density is no longer a semi-circle but it is a Gaussian. The
spacing distribution seems not to be different. However higher

order spacings, i.e. distribution of the interval between kth
neighbors seem different from that of the orthogonal ensemble
for k > 3 or 4. We will hear more about this from Dr. Bohigas.

7. Small metallic particles

Though this is a nuclear physics conference, let me draw
your attention to small metallic particles. When metallic
particles get smaller, $\sim 10^{-6}$ cm, the discreteness of the
electronic energy levels will show up, and their electric and
thermal properties will deviate from those of the bulk metal.
Now these properties depend on just how the electronic energy
levels are distributed. According to whether these electronic
levels are random without correlations or whether they follow
the same laws as the eigenvalues of a random matrix, the
properties such as the electronic specific heat or the magnetic
susceptibility, will be very different. I will content myself
by referring to the excellent review article on the subject by
R. Kubo.

8. Conclusion

In conclusion we may say that in spite of some nice
analytical results, much remains to be done. We urgently need a
method or methods to treat
   (i) a small admixture of time-reversal non-invariant part
in the nuclear Hamiltonian matrices;
   (ii) a small mixture of levels not belonging to the series
proper, and
   (iii) to see how the transition, if any, takes place when
nuclear interactions change nature from two-body to many-body
forces.
   In addition it will be illuminating to know how far the
simple spacing law does or does not depend on the details of the
interaction.

References

1. Reviews

   C. Bloch, Nuclear Physics, Les Houches, 1968, Gordon and
   Breach, p. 351.

   A. Bohr and B. R. Mottelson, Nuclear Structure, W. A.
   Benjamin, 1969, app. 2C.

   R. Kubo, in "Polarization, Matière et Rayonnement", Press.
   Univ. de France, Paris, 1970.

   J. E. Lynn, The Theory of Neutron Resonance Reactions,
   Clarendon Press, Oxford, 1968, chapter 5.

   M. L. Mehta, Random Matrices, Academic Press, New York,
   1967.

   C. E. Porter, Statistical Theories of Spectra: Fluctuations,
   Academic Press, New York, 1965.

N. Rosenzweig, Contemp. Phys., Trieste Symp. 1968, vol. II,
Int. At. Energ. Agency, Vienna, 1969, p. 381.

E. P. Wigner, SIAM Review $\underline{9}$, 1 (1967).

2. Experimental Papers

ANL group - (L. M. Bollinger, G. E. Thomas), Phys. Rev. $\underline{171}$,
1293 (1968), Contributions to this Conference.

BNL group - (M. R. Bhat, R. E. Chrien, S. F. Mughabghab)
Phys. Rev. $\underline{162}$, 1125 (1967), Contributions to this Conference.

Columbia group - (H. Camarda, J. S. Desjardins, J. B. Garg,
G. Hacken, W. W. Havens, Jr., H. I. Liou, J. S. Peterson,
F. Rahn, J. Rainwater, J. L. Rosen, M. Slagowitz, S. Wynchank).

Neutron Resonance Spectroscopy VI, Phys. Rev. $\underline{166}$, 1234 (1968),
Neutron Resonance Spectroscopy VII and VIII (to be published),
Contributions to this Conference.

3. Theoretical Papers

R. Balian, Nuovo Cimento, $\underline{B57}$, 183 (1968);
J. Phys. Soc. Japan, 26 Suppl, 30 (1968).

C. Bloch, J. Phys. Soc. Japan, 26 Suppl. 57 (1968).

O. Bohigas and J. Flores, Physics Letters, $\underline{34B}$, 261 (1971);
Contributions to this conference.

F. J. Dyson, Comm. Math. Phys. $\underline{19}$, 235 (1970);
Diagnostics of errors in a series of nuclear levels (to come).

L. D. Favro and J. F. MacDonald, Phys. Rev. Letters $\underline{19}$, 1254
(1967); J. Math. Phys. $\underline{9}$, 1429 (1968).

J. B. French and S. S. M. Wong, Phys. Letters $\underline{33B}$, 449 (1970);
$\underline{35B}$,5 (1971).
A. Gervois, Phys. Letters $\underline{26B}$, 413 (1968); Nuovo Cimento
$\underline{B69}$, 181 (1970).

I. J. Good, J. Math. Phys. $\underline{11}$, 1884 (1970).

J. F. MacDonald, J. Math. Phys. $\underline{10}$, 1191 (1969).

J. F. MacDonald and L. D. Favro, J. Math. Phys. $\underline{9}$, 1114 (1968);
J. Math. Phys. $\underline{11}$, 3103 (1970).

M. L. Mehta, Comm. Math. Phys. $\underline{20}$, 245 (1970a); Nuovo Cimento
$\underline{B65}$, 107 (1970b); On spacing distributions for large spacings
(unpublished report D. Ph.T.-71-12, Saclay).

M. L. Mehta and J. des Cloizeaux, The probabilities for
several consecutive eigenvalues of a random matrix (to be
published in the Indian J. Math.).

M. L. Mehta, J. E. Monahan and N. Rosenzweig, Nucl. Phys.
$\underline{A109}$, 437 (1968).

M. L. Mehta and N. Rosenzweig, Nucl. Phys. A109, 449 (1968).

P. A. Mello and M. Moshinsky, Contribution to this conference.

J. E. Monahan and N. Rosenzweig, Phys. Rev. C1, 1714 (1970); Contribution to this conference.

N. Rosenzweig, Nuovo Cimento 38, 1047 (1965).

W. H. Olson and V. R. R. Uppuluri, Contribution to this conference.

S. I. Sukhoruchkin, Contributions to this conference.

N. Ullah, J. Math. Phys. 8, 1095 (1967), 10, 2099 (1969); Contribution to this conference.

DISCUSSION

UPPULURI (Oak Ridge National Laboratory)

What are some of the properties possessed by the several statistics discussed by you? Are there any guidelines for proposing such criteria? You showed us a slide about the Poisson Distribution model. What is the parameter you have chosen for that?

MEHTA

For the Poisson, there is no parameter. I have drawn the curves for the probability that an interval of length chosen at random will contain exactly n points, when these n points are thrown at random with a given density.

UPPULURI

But don't you need to choose some average value of the Poisson distribution?

MEHTA

No. I measure the distance of the interval in terms of the mean distance between points. So there is no parameter. O.K? The mean distance is the parameter. I choose it to be one.

UPPULURI

Would you like to comment on the second question about the choice of one of the several statistics?

ROSENZWEIG

I'm afraid we didn't understand the first part of that question too well. Are you asking why one uses one or the other statistic? One would like a statistic that has as small a variance as possible. That would give the sharpest possible test. It seems that the Δ_3 statistic of Mehta and Dyson is the best in this respect because its figure of merit which is just the standard deviation divided by the mean is really quite small.

KRISHNAIAH (Aerospace Research Laboratory)

Would it be physically meaningful to assume that the mean values of the matrix elements in the random matrix are unequal? In the conventional theory it is assumed that the means of all the elements in the random matrix are equal to zero. Is it meaningful physically to assume that they need not be zero and they may be unequal?

MEHTA

For the diagonal elements, one can simply shift the scale. For the others, I was using the theorem that if the elements are independent random variables, and if matrices are also invariant

under unitary transformations, then according to Porter and
Rosenzweig, that the only thing possible is to have mean values of
zero for the off-diagonal matrix elements.

WIGNER
      It seems to me that if one looks for a mathematically simple
and precisely defined ensemble of hermitian or real symmetric
matrices, it is difficult to escape the postulate of orthogonal
invariance. The reason is that one cannot define a coordinate
system in Hilbert space without introducing a great deal of
arbitrariness, excepting the coordinate system spanned by the
characteristic vectors of the Hamiltonian. One does not want to
employ this coordinate system because, clearly, the Hamiltonian
is diagonal therein so that one cannot well define a reasonable
emsemble. If one disregards this coordinate system, all others
appear to be equivalent and if the ensemble is the same in all of
them, it is orthogonal invariant. I have skipped over the time
inversion invariance--this is what restricts the invariance to
real orthogonal, rather than to all unitary transformations.

      Drs. French and Wong do not assume invariance with respect to
orthogonal transformations. They use a restricted Hilbert space
and use therein a coordinate system in which the state vectors of
the axes correspond to independent particle states. This is a
possible choice--whether the independent particle picture is
reasonable in the region in which we want to use the statistical
model is, however, not clear. Feshbach has given arguments
against it, but the conclusions French and Wong, and also Bohigas
and Flores, arrived at remain interesting.

      When I started work on this subject, I also had a special
coordinate system in mind. It was the coordinate system in which
the operator of the kinetic energy was diagonal and all state
vectors were restricted to a definite finite spatial domain, the
"internal region" of R matrix theory. The potential energy would
provide the off-diagonal elements and these would decrease with
increasing difference between the energies of the two diagonal
elements, one of which is in the row, the other in the column of
the off-diagonal element in question. A high energy diagonal
element represents a state with a rapidly fluctuating wave function
and a low diagonal element represents a state with a low kinetic
energy, that is a relatively smooth wave function. The matrix
element connecting two such states will be small. This led me to
the concept of "bordered matrices," i.e., matrices in which the
off-diagonal elements vanish if the difference between the energies
of row and column diagonal elements is large. However, I never
succeeded to solve the resulting mathematical problem and this was
one of the reasons I abandoned the ensemble of bordered matrices.
The other reason was that its definition contained several rather
arbitrary parameters. In spite of this, physically, the model may

not be unreasonable.

ROSENZWEIG

Professor Wigner, if one wished to go beyond describing the
statistical properties of a few energy levels in a small interval
energy, that is, if one wished to build a "global" model valid for
large regions of energy, would it not be necessary to give up
orthogonal invariance?

WIGNER

Not necessarily. In my talk yesterday I gave some examples
of matrix ensembles which are invariant and would give a more
realistic description of the variation of the level density with
increasing energy.

FESHBACH

I just wanted to make a comment, also, on this subject. I
think the condition that the random matrix be invariant against
various transformations is a very nice mathematical condition, but
I'm not sure of any connection with physics. I think that's the
point that we should address ourselves to. In fact, I would take
another, let's say an opposing, point of view. Namely, that there
is a prejudice in Hilbert space. In other words, if I think of
the potential matrix which describes, let's say, a coupled channel
reaction, then I would say that I know that some of these are
attractive, some of them are repulsive. There is some information
that I have from the rest of the physics. I think this has to be
put in before we can say that we're testing the nuclear Hamiltonian.
For example, I would like to suggest the following thing: In the
coupled channel calculation one could say, O.K., I have some feeling as to the average value of the various coupling coefficients,
wherever it comes from. If I can do it from a Hamiltonian, so much
the better. And then I would like to add on to this a potential
matrix which is random with average value zero. That's what I
would propose as a possible model.

SOME PROPERTIES OF LEVEL SPACING DISTRIBUTIONS

O. Bohigas[+] and J. Flores[*]

[+] Institut de Physique Nucléaire, 91-Orsay, France
[*] Instituto de Física, UNAM, México

Recently new ensembles of random matrices have been introduced and treated numerically [1, 2] . They take into account the correlations between the elements of the N×N hamiltonian matrix induced by the Pauli principle and the two-body character of the interaction, when dealing with a system $(a, b, \ldots)\,^n_J$ of n identical fermions distributed among a set a, b, ... of degenerate single particle orbits and of total angular momentum J. They will be called two-body random hamiltonian ensembles (TBRE). The independent random variables are not, as in the case of the Gaussian orthogonal ensemble (GOE), the $N(N+1)/2$ elements of an N×N real symmetric hamiltonian matrix, but the m two-body matrix elements defining the two-body interaction $(m \ll N(N+1)/2)$. By standard shell model techniques, the N×N hamiltonian matrix corresponding to n particles and total angular momentum J is calculated in terms of the m independent two-body matrix elements, which are selected at random from a normal distribution centered at the origin ; the ratio, between the standard deviations (widths) of the distributions of non diagonal to diagonal two-body matrix elements is $1/\sqrt{2}$. After diagonalizing the N×N matrix a sufficiently large number of times in order to have good statistics, ensemble averages can be obtained for the different quantities in which we are interested.

[+] Division de Physique Théorique. Laboratoire associé au CNRS.

[*] Work supported by the Comisión Nacional de Energía Nuclear, México .

In a previous work [1, 2] it was found that the level density of TBRE is nearly normal (Gaussian) in contrast to the well known Wigner's semi-circle law that holds for GOE in the limit of large N [3] .

Let us turn our attention to local properties of the spectrum. Some local properties (e. g. level spacing distributions) derived in the framework of the statistical theory of energy levels of Wigner have often been compared to the experimental data. It is the purpose of this note, i) to go further, by means of Monte Carlo calculations, in the derivation of properties of GOE, ii) to compare the k-th neighbour spacing distributions p (k ;x) of GOE and TBRE, where x is the relative spacing S measured in units of the mean local spacing D and iii) to try to decide, once again, whether the experimental data are consistent with the theoretical predictions on spacing distributions.

To perform the Monte Carlo calculations we diagonalize the matrices of the ensemble and deduce from each matrix only one spacing, e. g. between the two eigenvalues closer to the center if we are interested in the center region (region of maximum constant density). We can double the statistics by taking advantage of the fact that $S_i^{(k)}$ and $S_{N-k-i}^{(k)}$ are statistically equivalent ; here $S_i^{(k)} = E_{i+k+1} - E_i$ and E_i is the i-th energy level.

Analytical expressions in the center region for GOE valid in the limit of large N have been derived by Mehta and Gaudin [4, 5] for k=0, by Kahn [6] for k=1 and recently Mehta and Des Cloizeaux have given the general expressions of p(k ; x) in terms of spheroïdal functions [7] . But numerical values can be given only for k⩽3 since the existing tables of these functions cover only a limited range of values of the arguments. However, we have obtained by means of Monte Carlo calculations estimations of the spacing distributions for k>3 . In fig. 1 we show the histograms resulting from a 688-member ensemble obtained by diagonalization of 344 matrices of dimensionality N=61. On the figure are also drawn the theoretical spacing distributions for k⩽3 . As can be seen, the Monte Carlo estimations of spacing distributions are accurate and the results valid in the limit of large N are attained.

As can be seen from fig. 1, the distributions p(k ; x) are asymmetric for k=0 and become more and more symmetric when k increases (this can be seen more precisely if one calculates the skewness γ_1 of the distributions ; one can then see that they

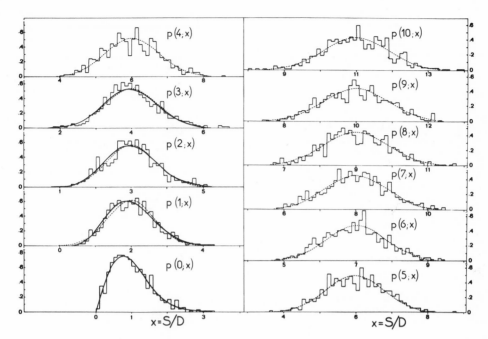

Fig. 1   Spacing distributions for the Gaussian Orthogonal Ensemble

decrease rapidly with increasing k). We have also drawn on fig. 1 normal distributions centered at k+1 and of width σ equal to the width of the corresponding distribution. The results show that this Gaussian approximation (valid for $k\sim2$) becomes better and better with greater k. Furthermore, it is not impossible that for large k σ_k tends to a limit (for σ_k, k=8, 9 and 10 we obtain respectively, .89 $\pm$ .02, .89 $\pm$ 02, .90 $\pm$ .02).

Let us turn our attention to the properties of TBRE. In figs. 2 and 3 we show results for some spacing distributions for TBRE. The histograms in fig. 2 correspond to a 950-member ensemble obtained by diagonalization of 475 matrices of dimensionality 44 corresponding to the case $(5/2, 9/2)^{n=5}_{J=9/2}$.

For comparison we have drawn on the fig. 3 the theoretical distributions corresponding to GOE for $k\leqslant3$ and for $k>3$ the GOE estimate given by the Gaussian approximation ; the Poisson distribution $(x^k/k\,!)\exp(-x)$ for a set of independent random levels is also shown. One can see that for k=0, GOE and TBRE results are in agreement [8, 9]. However, definite departures of TBRE results

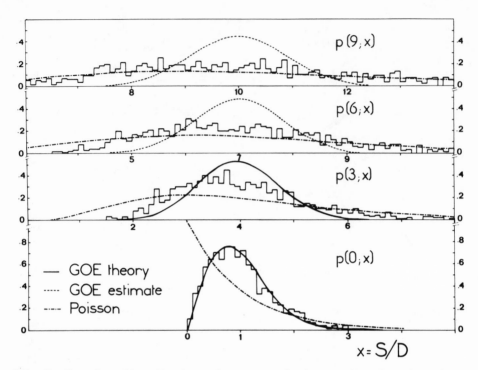

Fig. 2. Spacing distributions for a two-body random hamiltonian ensemble.

from GOE exist for k=3 [9], departures that increase with k. The effect of the correlations included in TBRE is to broaden the GOE distributions and results are intermediate between the GOE distributions and the Poisson distribution.

It is worthwhile to note that we do not know whether our spacing results do or do not depend strongly on the chosen configuration space and on the number of particles. At the present stage, we know how to examine this problem only by means of example, but not theoretically. In fig. 3 two further examples of TBRE are reported ; they correspond to a 1000 (2000)-member ensemble obtained by diagonalization of 500 (1000) matrices of dimensionality N=33 (21) corresponding to the case $(1/2, \ 3/2, \ 5/2)^{n=6}_{J=2}$ $\left((1/2, \ 3/2, \ 5/2)^{n=4}_{J=2}\right)$ respectively. The values of the standard deviation σ , which are a rough measure of the resulting distributions (k=0, 1, ... 9) corresponding to the different cases considered are shown in fig. 3 and curves are drawn to guide the eye. Values corresponding to GOE are obtained as explained in the preceeding

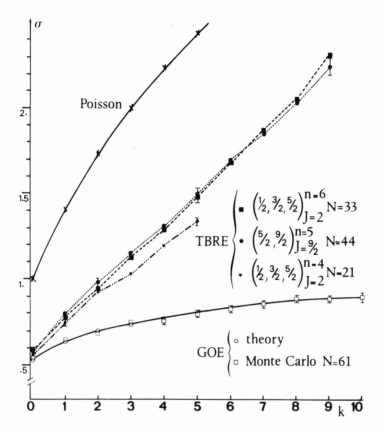

Fig. 3. Standard deviations σ of spacing distributions for TBRE.

paragraph ; for $k \leqslant 3$, as can be seen from the figure, the values
of σ agree with the theoretical ones valid in the limit of large N.
We have also drawn for comparison the values $(k+1)^{1/2}$ corresponding to the Poisson distribution. Inspection of fig. 3 shows
that we have not found any strong dependence on either the chosen
configuration space nor on the number of particles. This does not
mean, of course, that the question has been settled, but, insofar
as we know of no counterexample, we can hope that the TBRE
predictions are not strongly model dependent.

Let us now compare the different theoretical predictions with
the experimental data. There exist extensive experimental results
on the spacing distributions of levels observed in neutron resonance
spectroscopy. However, the fact that in order to test theoretical
models of level spacing distributions one must have a large number
of consecutive levels with same spin and parity limits drastically

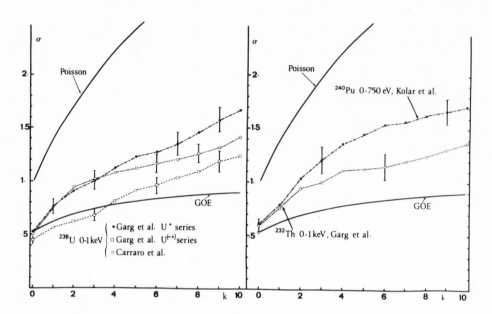

Fig. 4. Standard deviations σ of spacing distributions derived from experimental data.

the data that can be analyzed and has often made the comparison between the theoretical predictions of GOE and the experimental results very difficult. Up to now, the spacing distributions p (k ;x) have been compared to the theoretical predictions only for the nearest-and next-nearest-neighbour spacing distributions (k=0 and 1) and it has been generally concluded that the data support the GOE predictions. Our purpose here is to go further in this comparison : as explained above, on the one hand one has accurate estimates of p(k ;x) for GOE up to k~10 and on the other hand the predictions of the TBRE are very different from the GOE ones for k $\gtrsim$ 3.

The results of the analysis coming from the following data are drawn in fig. 4 : ^{238}U data [10, 11] , ^{232}Th data [10] and ^{240}Pu data [12]. As is well known the quality and the completeness of the data are higher for lower kinetic energy of the captured neutron. So, we have included in our analysis only the levels lying within the energy range 0-1 keV in all cases except for ^{240}Pu, where we take the range 0-750 eV . Two different analyses coming from the data of Garg et al. [10] on ^{238}U are reported : one includes all the levels up to 1 keV (U^{*} series in fig. 4); the other ($U^{(-*)}$ series in the figure) is obtained from the preceeding one when starred levels are excluded[10](levels that are either p-wave or uncertain). The error bars on the figure correspond to statistical errors arising from the finite number of points in the sample.

One can see in fig. 4 that all the distributions show a clear tendency, for $k \geqslant 1$, to be broader than the GOE predictions (except perhaps for the $U^{(-*)}$ series). Even if the quality of the data does not allow to deduce precise values of σ_k (see the different results obtained on ^{238}U), we think that there are strong indications that for $k \geqslant 1$ the predictions of the GOE are in disagreement with the experimental results. One can also see that, on the contrary, for $k=0$, all the data are compatible with the GOE prediction (practically the Wigner surmise). This explains why it is currently admitted that the data are in agreement with the GOE predictions.

We have tried to pursue this analysis extending the energy range considered, that is, calculating for example the widths of the spacing distributions in the range 1-2 keV and so on. The results should in principle be independent of the energy range considered. We have found that, except for the ^{232}Th data [10], which are stable in the range 0-3 keV , all the others are not (even if the mean spacing D is constant). We should point out, however, that the resulting values for the standard deviations never decrease to the GOE values.

We thus think that spacing distributions coming from experimental data are broader than the ones predicted by GOE (for $k \geqslant 1$); this effect (broadening of GOE spacing distributions) is qualitatively present in the examples of TBRE we have investigated.

We thank Drs M. L. Mehta and A. Michaudon for encouragement, comments and discussions.

[1] J. B. French and S. S. M. Wong, Phys. Lett. 33B(1970)449 ;
[2] O. Bohigas and J. Flores, Phys. Lett. 34B(1971)261 ;[3] E.P.Wigner, Ann. of Math. 67(1958)325 ;[4] M. L. Mehta,Nucl. Phys. 18(1960)395;
[5] M. Gaudin, Nucl. Phys. 25(1961)447 ; [6] P.B. Kahn, Nucl. Phys.41 (1963)159 ;[7] M. L. Mehta and J. des Cloizeaux, Saclay preprint DPh-T/70-74 ; [8] J. B. French and S. S. M. Wong,Phys. Lett. 35B(1971) 5 ; [9] O. Bohigas and J. Flores, Phys. Lett. 35B (1971)383 ;[10] J.B. Garg, J. Rainwater, J. S. Peterson and W.W.Havens Jr,Phys.Rev. 134 (1964)B985 ;[11] G. Carraro and W.Kolar, Conference on Nuclear Data for Reactors, Helsinki, IAEA 1970 ;[12] W. Kolar and K. H. Böckhoff, Journ. of Nucl. Ener. 22 (1968)299.

DISCUSSION

WONG

We have analyzed the eigenvalues obtained from Gaussian orthogonal and two-body ensembles in terms of Δ_1, Δ_2, Δ_3 and their variances as described by Dr. Mehta. The dimensionality of our matrices is 50 and there are 100 members in each of the ensembles. The question of finite matrix size is certainly present; nevertheless, it is possible to extract some information if one proceeds in the following manner.

Among the 50 levels obtained from matrix diagonalization, we take the central domain of a given range, 2L, and calculate the Δ's with respect to different sizes of the domain for each of the two ensembles. For large region sizes, say 40 levels, the two sets of Δ's are quite different from each other and from the analytical values of Dr. Mehta. As we reduce the size of the region taken, the two sets of Δ values approach each and also to the analytic values. Unfortunately, the Δ's become meaningless if there are too few levels. However, for the central eight levels or less the two sets of Δ values become equal to each other and to Dr. Mehta's values within the statistical errors. It, therefore, seems that the Δ statistics cannot distinguish the two types of ensembles.

Perhaps other kinds of staircase statistics should be designed which would give some differences between the two types of ensemble. We have a few ideas but no results at this moment.

NEWSTEAD (Saclay)

I just want to ask a brief question. What is the conclusion from all this? Can we say whether the nuclear forces are two-body or many-body?

FLORES

We attempted to address ourselves to that question when we found that our two-body random matrix ensembles gave values for the widths of the kth order spacing distributions which were substantially larger than the values predicted by the Gaussian orthogonal ensemble. The experimental values (based on relatively old published data) were intermediate between the two curves provided by the two theoretical models. We even carried out an interpolation which yielded the conclusion that the many-body force would be of rank three. But this conclusion is not to be taken seriously because more recent experimental data are in much closer agreement with the results of the Gaussian orthogonal ensemble. Furthermore, in our truncated shell-model space we are not dealing with the matrix elements of the nuclear forces but rather with an effective interaction which could have many-body components.

FESHBACH

In the last sentence, you said what I wanted to say--namely,
that you're not testing whether or not there are many-body forces
in the fundamental interaction, but rather, at this point, anyway,
whether in cutting your space, your Hilbert space in making your
calculations, if you wanted to do that exactly, the way to do that
is to introduce beside the two-body forces, the many-body forces
which come, not because of the fundamental nucleon interactions,
but rather because you've cut your Hilbert space. In other words,
if you started off with two-body forces in the infinite-dimensional
Hilbert space and wanted an accurate representation in a restricted
finite-dimensional vector space, the process of truncation would
give rise to effective many-body forces.

HACKEN (Columbia U.)

I'd like to address myself to Dr. Wong. His results don't
seem surprising, in the sense that as he goes near the center of
the semi-circle--Wigner's semi-circle--he will get better agreement. I think the next speaker, Dr. Camarda, will clear that up.
Did Dr. Wong compensate for the Wigner semi-circle density by
unfolding it before he did this? Or did he not do that?

WONG

The only interest we have is to compare the difference
between the semi-circular type of level density to the Gaussian.
I don't think the question was quite relevant in this point.

ROSENZWEIG

I would like to show a slide for the even Erbium isotopes
which illustrates Dr. Flores' statement that the recent Columbia
data (reported at this conference) seem to be in closer agreement
with the results based on Wigner's Gaussian orthogonal ensemble
(GOE) than with the results based on the two-body random matrix
ensembles (TBRE) studied to date. The analysis leading to the
slide was carried at Argonne National Lab. during the last few
weeks in collaboration with Dr. M. L. Mehta.

We find that what is believed to be the cleanest set of
s-wave levels yet measured (according to the authors) the resonances observed in ^{166}Er, gives experimental values for W(k), the
widths of the kth order spacing distribution--indicated by the
solid circles on the slide--which are in remarkably close agreement
with the values predicted by the GOE. The results for ^{170}Er, of
the three isotopes, based on the smallest and probably least
reliable set, deviate the most from the GOE curve in the direction
of the TBRE results. Similar results were obtained for the even
tungsten isotopes. The newly revised set of levels for ^{238}U and
^{232}Th give results much closer to the GOE prediction than the older
1964 Columbia data analyzed by Bohigas and Flores.

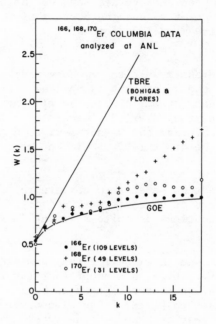

Slide shown by Dr. Rosenzweig.

RECENT EXPERIMENTAL NEUTRON RESONANCE SPECTROSCOPY
RESULTS AS A TEST OF STATISTICAL THEORIES OF SHORT
AND LONG RANGE ORDER FOR LEVEL SPACINGS *

H. Camarda, H.I. Liou, F. Rahn, G. Hacken,
M. Slagowitz, W.W. Havens,Jr., J. Rainwater
Columbia University, New York, N.Y.
and S. Wynchank, Brooklyn College, N.Y,N.Y.

During 1968 and 1970, we obtained large amounts
of high quality neutron resonance spectroscopy data
using the Columbia University Nevis Synchrocyclotron.
This report emphasizes our experimental results for
even-even nuclei having $150 < A < 190$. In the past,
attempts to make detailed comparison of experimental
resonance energies for nuclei for such "best test"
cases as Th^{232} or U^{238} with theory gave poor fits for
those tests which assumed that a single s level population only was present, and were sensitive to the inclusion of a partial extra p level population. For
$150 < A < 190$, the s level strength function S_0 is
sufficiently greater than the p level strength function
S_1 that all p levels tend to be weaker than all but a
very small fraction of the s levels, providing a better
separation of the two populations. Our results for
Er^{166}, Er^{168}, W^{182}, W^{184}, Sm^{152}, and Yb^{172} were of
particularly good quality, and give good agreement with
the following statistical tests. Except for Er^{168},
they seem to have only s levels, and for Er^{168} the p
levels can be cleanly separated out on the basis of
their strength.
    1. The Porter-Thomas (P-T) theory (single channel)
is believed to apply for the distribution of reduced
neutron widths, Γ_n^0, if intermediate structure fluctuations of $<\Gamma_n^0>$ (a priori) are not present over the
energy interval considered. The data should agree
well with the P-T distribution for the Γ_n^0 values if
a complete s population only is involved. Missing
weak s levels, and/or the partial inclusion of a p

level population should, with proper normalization, have the observed Γ_n^0 distribution in good agreement with P-T theory <u>except</u> for the lowest Γ_n^0 histogram box, where the discrepancy tells how many p levels were included (minus missed weak s levels). All of the above isotopes, except Er^{168}, gave a good fit of the observed Γ_n^0 values to the P-T distribution, including the lowest Γ_n^0 histogram box. For Er^{168}, the test showed how many p levels (weak levels) to remove.. These were all the weakest levels which a Bayes Theorem analysis indicated were most apt to be p levels. A good fit for all Γ_n^0 was obtained after this correction for Er^{168}. Figure 1 shows the $(\Gamma_n^0)^{1/2}$ histogram for Er^{166}, which was the experimental test isotope for which the largest clean level sample was found.

2. The single population nearest neighbor level spacing distribution is believed to be of the simple form surmised by Wigner, to within experimental precision. All of the above isotopes gave good fits to the Wigner shape, including Er^{168} after p level removal. Figure 2 gives the Er^{166} adjacent level spacing histogram.

3. The Dyson-Mehta Δ statistic for the mean square deviation of the staircase plot of the observed number of levels N(E) vs E is predicted to have $\langle\Delta\rangle$ increase with the number of levels only as $\ln N$, where $N \sim 20,000$ is needed before $\langle\Delta\rangle = 1.00$. Earlier Δ_{exp} values were all $\gg \langle\Delta\rangle_{D-M}$ due to the partial inclusion of spurious or p levels, and/or missed weak s levels. In all of the present cases, as shown in Table I, we obtain good agreement with the D-M value of $\langle\Delta\rangle$ within the theoretical S.D. of 0.11. This agreement was also found for the low energy ends of the data for the odd A isotopes W^{183}, Er^{167}, Yb^{171}, Yb^{173}, where the two randomly mixed s level populations for $J = (I\pm 1/2)$ are present.

4. The short range order expected from Wigner's random matrix theory leads to the prediction for the covariance of adjacent level spacings, $Cov(S_j, S_{j+1}) = -0.27$, according to Mehta. Table I shows that the experimental results are in good agreement with this prediction.

5. After seeing our new results, Prof. Freeman Dyson suggested that we apply his recently developed "F statistic" test to all of our observed level spacing distributions. His test, as we used it, evaluates a parameter F_j at each level, j, which is dependent on the positions of levels on either side of level j over an energy interval which contains ~ 10 levels on the average. The test is sensitive to ordering and was developed as a "test" to show the presence of missed

s levels, or included spurious or p levels. The above
isotopes, including Er^{168} after our p level subtraction,
gave good fits to the expected behavior for F_j values
for "clean single population data".

6. In addition, we obtained the unexpected result
for our best quality data, that for Er^{166} to 3 keV
(79 levels), that there seemed to be a definite negative correlation between adjacent level Γ_n^O values,
$Cov(\Gamma_{nj}^O, \Gamma_{n\ j+1}^O) = (-0.28\pm0.09)$. Also, other short
and long range tests of these Er^{166} Γ_n^O values showed
less fluctuation of the $<\Gamma_n^O>$ values over regions having $\geqslant 10$ levels than that expected for an uncorrelated
sequence of Γ_n^O values. The observed set of Γ_n^O values
agreed with the Porter-Thomas theory.

Since the data should be compared with some reasonable alternate theory, and since the Wigner distribution for adjacent level spacings is known to represent the data well, we also compared our results with
the behavior of levels obeying the adjacent level
spacing Wigner distribution, but otherwise uncorrelated
(U.W.). From Monte Carlo techniques, we found for this
U.W. model $\bar{\Delta} = n/(55-210/n)$, with S.D. of $\Delta = n/86$ for
$n \geqslant 15$. This calculation demonstrated the usefulness
of being able to construct probability distributions
for such predicted parameters as Δ. As a result, Monte
Carlo calculations were also performed using Wigner's
statistical model for the level position distribution
wherein sets of real symmetric matrices, having Gaussian randomly distributed elements, were generated and
diagonalized in accord with Wigner's formulation of the
problem. To convert the resulting Wigner "semi-circular" distribution to a more physical constant density
eigenvalue distribution (over a limited energy interval), we use the following method. First, we remove
~ 2 eigenvalues from each end of each set to minimize
end effects. Then the transformation $P_W(\epsilon)d\epsilon = P(y)dy$
was used, where $P_W(\epsilon)$ is Wigner's semicircular law and
$P(y)$ is set equal to a constant. The resulting set of
constant density eigenvalues was then used to evaluate
the mean, and the distribution of values for various
parameters such as the D-M Δ statistic and $Cov(S_j, S_{j+1})$.

When the values of Δ and $Var(\Delta)$ were calculated,
we obtained the expected, but previously undemonstrated
result that they were in agreement to within statistical fluctuations, with the D-M results using Dyson's
circular ensemble. This is shown in Table II. In
addition, it was found that $Cov(S_j, S_{j+1}) \approx -0.27$ for
matrices of large dimensions ($\geqslant 50 \times 50$). The eigen-

value behavior was also found to agree well with Dyson's
F statistic, giving further confirmation of the equi-
valence of Dyson's circular ensemble and the random
matrix model.

After discussions of these results with Professor
Dyson, he suggested that we try to use his Brownian
Motion Model to obtain, more rapidly, large sets of
results equivalent to those from diagonalizing large
dimensional matrices. With some help from Professor
Dyson, the approach was successfully pursued and gave
results in agreement with those obtained for the random
matrix diagonalization. With these results, a nearest
neighbor spacing distribution having ~ 97,000 spacings
(900 sets of 109 levels) was constructed. In addition
to the other tests, this sample was large enough to
show the small deviations from the simple Wigner distribution predicted by Mehta and Gaudin from a more
exact analysis.

To give a best distinction between the consistency of the experimental results with the "correlated"
and uncorrelated models, the quantity $[\Delta+Cov(S_j,S_{j+1})]$
was found to provide a more sensitive test than either
Δ or $Cov(S_j,S_{j+1})$ alone. The comparison of theory with
experiments, was made by determining the probability,
$P_<$, of finding $\leq$ the experimental value of $[\Delta+Cov(S_j,$
$S_{j+1})]$ for the correlated, $P_<^C$, and uncorrelated,
$P_<^{UC}$, models.

Figure 3 shows the N(E) vs E plot for Er^{166} to
4200 eV (109 levels). Figure 4 gives the cumulative
probability distribution of $[\Delta+Cov(S_j,S_{j+1})]$ for the
two models, and shows the experimental result. For
Er^{166} $[\Delta+Cov(S_j,S_{j+1})]$ = 0.235, giving $P_<^C$ = 0.590 and
$P_<^{UC}$ = 0.0004. This is dramatic experimental evidence
for long range correlations in the level position distribution. Table I lists the results for all of the
nuclei under consideration.

We also re-examined the experimental evidence for
Th^{232} and U^{238}, making partial use of our much better
new 1970 data to treat weak levels only. The idea was
to see if "selected populations" could be chosen which
fit all of the above statistical single population
tests. We must include all levels which are not very
weak since these stronger levels must be s levels.
These levels make up ~ 85% of the final population
choice, leaving only ~ 15% of the final population
choice to be taken from the excessive observed number

of weak levels which include many p levels.   The chal-
lenge was to see if such an s level selection could be
made from the observed weak level population which
would agree with all of the tests.   This would seem to
be a priori unlikely unless the s level population actually obeyed the above statistical theories.   We were,
in fact, able to find such sets for both Th232 and U238.

Other tests of our experimental results against
theory have been made by Rosenzweig and Monahan and are
reported in this conference.   We have submitted papers
on our experimental results for Er, and for these
theoretical studies, for publication in the Physical
Review.

Other results from the analysis of our 1968 and
1970 data include the following.   Er166: $10^4 S_O$=1.70
(N=112), <D>=37.6 eV for s levels,<Γ_γ>=96 meV (N=10),
$10^4 S_1$<0.75 meV.   Er167: $10^4 S_O$=1.89 (N=179), <D>=3.9 eV
for s levels, <Γ_γ> = 91 meV (N=52).   Er168: $10^4 S_O$=1.50
(N=105), <D>=93.6 eV for s levels, <Γ_γ>=87 meV (N=4),
$10^4 S_1$=(0.70±0.20).   Er170: $10^4 S_O$=1.54 (N=94), <D>=
149 eV for s levels, $10^4 S_1$=(0.80±0.25).   W182: $10^4 S_O$=
(2.42±0.30), <D>=66.4 eV for s levels.   W184: $10^4 S_O$=
(2.37±0.32), <D>= 89 eV for s levels.   Sm152: $10^4 S_O$=
2.72 (N=90), <Γ_γ>=65 meV (N=9).   Sm154: $10^4 S_O$=1.90
(N=33), <Γ_γ>=79 meV (N=3). Th232: $10^4 S_O$=0.84 to 5 keV,
<D>=16.8 eV for s levels, <Γ_γ>=(21.2±0.9) meV (N=84).
U238: $10^4 S_O$=(1.02±0.10) to ~ 4.3 keV, <Γ_γ>=22.9 meV
(N=41).   Yb172: $10^4 S_O$=1.68 (N=100), <D>=62.7 eV for s
levels, <Γ_γ>=75 meV (N=3).   Yb174: $10^4 S_O$=1.56 (N=78),
<D>=164 eV for s levels.   Yb176: $10^4 S_O$=2.29 (N=70),
<D>=185 eV for s levels.

Research supported in part by the U.S. Atomic
Energy Commission.

Table I   Summary of Experimental Results for Even-Even
           Isotopes and Comparison with Theory

	Er^{166}	Er^{168}	W^{182}	W^{184}	Sm^{152}	Yb^{172}
N	109	50	41	30	70	55
E_{max}	4200	4700	2607	2621	3665	3900
Δ_{exp}	0.455	0.287	0.259	0.446	0.400	0.412
Δ_{th}^{C}	0.468	0.389	0.369	0.338	0.420	0.399
Δ_{th}^{UC}	2.052	0.985	0.822	0.625	1.346	1.073
$Cov(S_j,S_{j+1})$	-0.22	-0.29	-0.37	-0.28	-0.26	-0.24
	±0.08	±0.14	±0.15	±0.18	±0.11	±0.13
$P_<^{C}$	0.590	0.180	0.103	0.705		0.610
$P_<^{UC}$	0.0004	0.0035	0.002	0.159	0.004	0.017

N = number of levels
E_{max} = upper limit of energy interval (in eV) con-
         taining the N levels

Table II

Predicted Behavior of $<\Delta>$ and $\sqrt{Var(\Delta)}$ as a Function of the Number of Levels for
Wigner's Random Matrix Model and Dyson's Circular Ensemble. It is seen that for
these tests the predictions of the two models are indistinguishable.

Monte Carlo Random Matrix Calculations					Theoretical Predictions of the Circular Ensemble	
No. of Matrices Diagonalized	Matrix Dimension	Eigenvalues Used	$\bar{\Delta}$	$\sqrt{Var(\Delta)}$	$<\Delta>$	$\sqrt{Var(\Delta)}$
78	21	13	0.257±0.011	0.098	0.253	0.11
78	31	23	0.324±0.012	0.106	0.311	0.11
78	41	33	0.364±0.014	0.124	0.347	0.11
78	50	42	0.372±0.011	0.098	0.372	0.11
62	81	77	0.424±0.012	0.094	0.433	0.11
† 900	120	109	0.470±0.003	0.093	0.468	0.11

Tabulated below are the average values of $Cov(S_i,S_{i+1})$ calculated using the average
sample spacing $\bar{s}$ and the true average spacing $<s>$. The difference between the
values of $\overline{Cov(S_i,S_{i+1})}$ obtained using $\bar{s}$ and $<s>$ (for the same number of levels) is
much less than the difference (1/n) found in the uncorrelated case.

No. of Matrices Diagonalized	Matrix Dimension	Eigenvalues Used	$\overline{Cov(S_i,S_{i+1})}_{\bar{s}}$	$\overline{Cov(S_i,S_{i+1})}_{<s>}$
78	21	15	-0.250±0.026	-0.236±0.026
78	31	25	-0.241±0.021	-0.233±0.021
78	41	35	-0.269±0.018	-0.266±0.018
78	50	44	-0.274±0.016	-0.272±0.016
62	81	77	-0.256±0.012	-0.256±0.012
† 900	120	109	-0.277±0.003	-0.276±0.003

† Calculated using Dyson's Brownian motion model.

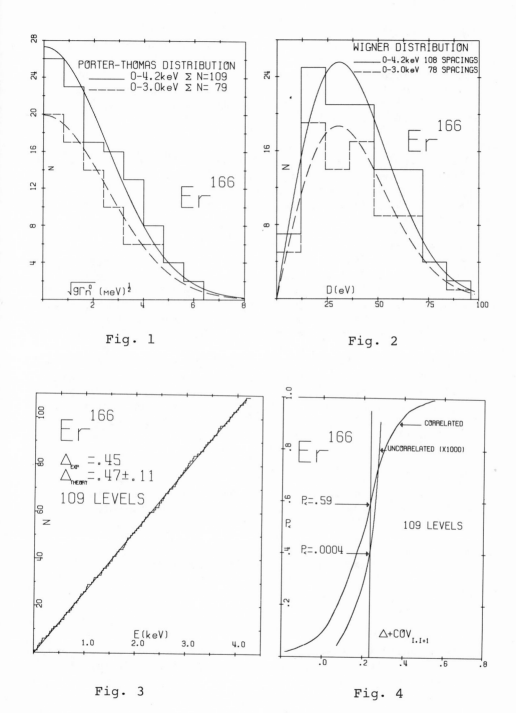

Fig. 1

Fig. 2

Fig. 3

Fig. 4

DISCUSSION

MICHAUDON (Saclay)
The correlation coefficient between the reduced neutron widths for Er^{166} is rather puzzling. Isn't it possible that the approach presented by Flores for the level spacing distribution and taking into account the two-body random interaction may be applied also for the study of the reduced neutron widths, and could it show a correlation of that kind?

CAMARDA
Right. In fact, I want to ask him if he had extracted the distribution of reduced neutron widths, too, sort of the same way you do in random matrix theory and see if there are in fact correlations. He should get correlations between the adjacent neutron widths.

NEWSTEAD (Saclay)
You stated that the Er isotopes are a particularly good region to look for these effects, because the p-wave strength functions of these isotopes are considerably smaller than the s-wave strength functions. But, in fact, I don't think that's true. In this mass region you're near the rotational splitting of the 4P size resonance, and that gives a broad bump in the region of the Er isotopes which predicts p-wave strength functions of the order of about 2×10^{-4}. This is quite comparable with what you might expect for the s-wave strength functions of the Er isotopes, which, after all, are in the minimum of the rotationally split 4s peak. In fact, in a contribution to this conference, we show that the p-wave strength function of ^{165}Ho is 1.63, and the s-wave strength function is 1.66, quite comparable.

CAMARDA
All right. I do have a comment about that. We also determined the p-wave strength function for Er^{168} and found it to be 0.6 and that the strength functions for the Er isotopes are about 1.7 or 1.8.

NEWSTEAD
The s-wave strength functions are 1.7 and 1.8?

CAMARDA
That's right.

NEWSTEAD
This is quite interesting then. If you have such a low p-wave strength function for Er^{168}. That's rather unusual. It is, of course, possible.

CAMARDA

I don't know if it's unusual. How much data is there on this
subject?

NEWSTEAD

Well, it's unusual compared to the optical model predictions
and compared to the measurement in holmium.

CAMARDA

I don't really trust those optical model predictions that
much for the p-wave strength function, personally.

DEVIATIONS FROM THE STATISTICAL DESCRIPTION OF NEUTRON LEVEL SPACING DISTRIBUTIONS AND STABILIZING EFFECTS OF NUCLEAR SHELLS IN POSITIONS OF NUCLEAR EXCITED STATES

S. I. Sukhoruchkin

Institute of Theoretical and Experimental Physics, Moscow

It is well known that the statistical model gives reliable description of properties of nuclear states at high excitation, say, in heavy nuclei. Statistical model is based on the complexity of nuclear forces, that results in the great number of nuclear models at present. Combined consideration of the existing experimental data on the level positions (E*) in many nuclei (in the broad scope of nuclei) shows rather complex distributions. But only analysis of experimental data on excitation levels can give the answer to the question if the energies are random or they are not fully independent (in different nuclei and shells). In principle it may be the case that the comparison of the spectra of different nuclei (including highly excited states) will show systematic deviations from statistical distributions, for instance, from Wigner distribution, in case of the neutron levels of a given nucleus. This paper is a review of studies of deviations of the level spacings in complex nuclei from statistical distributions carried out to date. The major part of the paper (two first sections) deals with analysis of 1)hyperfine structure - of distinguishing effect of certain spacings in neutron resonances of heavy nuclei (energy of the order of tens of ev). 2)fine structure - nonrandom distinguishing effect of certain energy intervals in the position of the neutron and low-lying levels in light and non-deformed nuclei (energy of the order of tens of keV). In the third section a description of the phenomenological model will be given. It makes possible intercomparison of these correlations and predicts some features of these deviations from statistical distributions. In the last two sections (one may consider them as a additional supplement) the correlations of the order of tens of MeV, reported in the literature, are discussed. They concern some aspects of the strong interactions; correlations in the binding energies of light nuclei and mass relations for

215

elementary particles. Considerations of these items was taken
under assumption that systematic behavior of different structural
effects, in case of their existence in reality, may be caused by
some effects with the energy scale of tens and of hundreds of MeV,
that is characteristic of strong interactions of nucleons.

1. HYPERFINE STRUCTURE OF THE ENERGY SPLITTINGS

In some papers (1,2,3,4) attention was drawn to the "structure"
in the spacing between neutron levels (D_{ij}) for some N-odd (compound)
heavy nuclei. N-odd nuclei targets were often selected as they were
monoisotopic. Besides they are of greater interest for studies of
effects that are being discussed in this paper. If single-particle
or to be more precise few-particle effects play any role in the
complex spectra of levels in heavy nuclei they will first of all
manifest itself in N-odd compound nuclei. These systems are thought
as N-even excited core plus neutron. If the structure in D_{ij} is due
to anomalously strong statistical fluctuation then, these data,
taken together for the large number of nuclei, will give smooth
distributions. But it doesn't happen and distinguishing effect in
D_{ij} near 5.5 eV maintains even in cases of levels with relatively
broad neutron widths which as one may expect, correspond to larger
contribution of single particle configurations. Distinguishing
effect of the interval (5.5 eV) has also been confirmed by the
analysis of the summary data on positions (E_O) of neutron resonances,
e.g. distances from levels to the binding energies of neutron(2,4,5).
The intervals in the neutron spacing level distributions are 5-6 eV
11 eV, 16 eV (4) and the intervals in combined distributions of
positions - in addition - 22 eV, 27 eV (4), 65 eV, 71 eV, and 93 eV
(4,5). It's worth mentioning that by the primary analysis of the
combined distributions of D_{ij} and E_O it was noticed that all these
intervals are rational to the period 5.5 eV (2). Further detailed
analysis of more accurate data revealed possible hyperfine structure
with the interval E_{ctc} = 1.35 eV that is one fourth period of 5.5 eV.
The only way to check such correlations is an independent analysis,
utilizing new data. The recent paper by K. Ideno and M. Ohkubo(7)
satisfies these requirements. They discovered the same 5.5 eV period
in the positions of neutron resonances of Sb^{123}. Besides these new
results allowed to undertake an additional analysis of: 1) data of
three even-even nuclei (Th^{232}(8), U^{238} (9), Pu^{240} (10)) that have
been studied to a larger extent because of practical importance in
reactor design and 2)data for a large number of nondeformed nuclei
(11-17). Combined distribution of the level spacings for three
isotopes, mentioned above, is shown in Fig 1. The arrows mark the
interval near 11 eV (5.5 eV). In the region under investigation
(0-100 eV) there is sequence of maxima which coincide with doubled
value of this interval (11 eV) or four-folded value of period of
5.5 eV.

Analysis of the values of the distinguished intervals for neutron
levels in individual isotopes was carried out in the following manner.
Distributions similar to shown in Fig 1, but for 20 independent sets

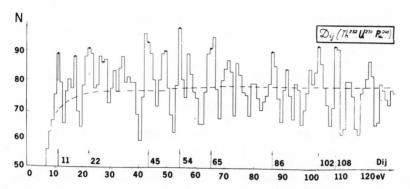

FIG. 1. The distribution of spacings between the neutron levels of
three even-even isotopes (the interval of averaging is 2 ev, the
step of idiohistogram is 1 ev). The arrows indicate the intervals
equal to even number times ∼5, 5 ev (see text).

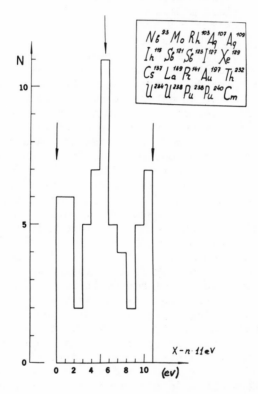

FIG. 2. The distribution of residual values after subtracting even
number of intervals 11 ev from values of spacings indicated between
neutron levels of many various nondeformed and even-even target
nuclei.

of data (for 20 isotopes or elements represented in the upper part
of Fig 2) were calculated. In each of these distributions (after
correction on recoil effect) the positions of three most prominent
maxima, e.g. values of the most frequently occuring level spacings
were looked for (the distribution in Fig 1 these maxima were at
X = 42, 54 and 66 eV but this summed distribution was not taken
into account and each of the three isotopes was analyzed separately.)
From such 60 values of "X" several numbers of 11 eV - intervals was
subtracted and the distribution of the "remainders" (shown in Fig 2)
gives the evidence for the "distinguishing" of the interval 5.5 eV,
marked by the arrows. For this analysis practically all non-
deformed and even-even nuclei were considered, each of which has
more than 40 neutron levels. That's why it is worth checking this
effect, utilizing independent experimental data.

    Now we want to draw attention to the fact that the same inter-
vals between the levels tend to manifest itself in positions of real
levels relative to the binding energy, e.g. to the mass difference
of nuclei (2,4,7). This correlation, if true, can't be understood
within existing nuclear models and may lead to a change in the present
understanding of the nuclear properties.

2.   FINE STRUCTURE OF NUCLEAR LEVELS

    An existence of fine structure (the intervals of the order of
1 KeV) of neutron and proton resonances in light nuclei has been
discussed in (6). The data on spacings (D_{ij}) available by 1968
year allowed the author to determine the parameter E_{tc} = 1.16 KeV
(arrows in Fig. 3). The similar structure has been investigated in
dE_g - distribution of mutual differencies of all precisely known
gamma-ray energies of the radiative transitions after thermal
neutron capture (from the compilation (18) the dE_g distribution for
nondeformed nuclei (upper curve on Fig. 4) shows only two expected
intervals and the major input in the structure of dE_g distribution
is brought by the data on Rh[103]. The discrete character of E_g-
distribution of this isotope have been reported by O. Sumbaev
et.al. (19). It's the reason why this information shouldn't be
considered as an independent. So the proof of fine structure is
still needed. The closeness of any summary distribution (over
many isotopes) depends on many factors including differences
between single particle states and splittings arising by the
residual nucleon interactions. As it has been noticed in (20),
both the energies of single particle states and small level splittings
are not fully independent. Even the highly excited states conserve
systematic correlations. In other words the above mentioned fine
structure seems to be the part of the general effect - systematic
distinguishing of nuclear level intervals of the order of tens and
hundreds of keV(2,4,21).This phenomenon has been discovered for the
first time by the analysis of the low-lying states in many nuclei
(5,21). Distinguished interval about 30 KeV was followed D_J=1+
change in momentum (J) and was often seen as a splitting of ground
states (because of residual interactions) of such odd-odd nuclei

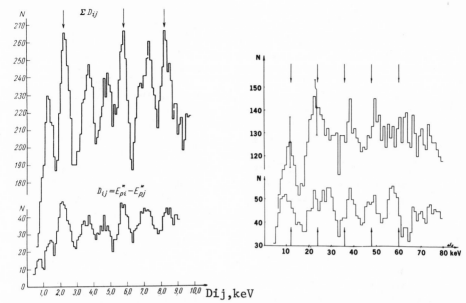

FIG. 3.   The idiohistogram of the integral distribution of Dij-
spacings between neutron and proton levels of light nuclei
(Fig. 1b and Fig. 2 of ref. [6]).   The averaging interval is
±0.25 kev, the arrows indicate the period ε_{Tc}=1,16 kev.

FIG. 4.   The idiohistogram of the distribution of the respective
differences $\Delta E\gamma$ of energies for radiative transitions.   Above the
integral distribution and below that for the isotope Rh^{104} are
given.   The averaging interval is ±0.25 kev, the arrows indicate
the period ε_{Tc}.

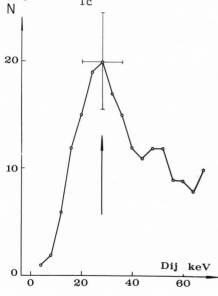

FIG. 5.   The idiohistogram
of the distribution of the
respective spacings for excited
levels of the nucleus K^{40}, the
averaging interval is ±8 kev.

as Al^{28}, K^{40}, etc. On Fig. 5 one may see the evidence of distinguishing of the 30 keV interval for a great number of highly **excited states** of K^{40} (22). This nucleus may be considered as a hole in the shell (Z=N=20) and a nucleon. The intervals of the order of tens of keV reflect themselves at high excitations(E*) as a superposition of small splittings on single-particle levels (or levels of the average field) and it permits to discover the stabilizing effect in single-particle levels of complex nuclei. Intervals of the order of 30 keV, 60 keV and 120 keV have been observed as groupings of energy excitations till E* up to 1 MeV (2), up to 3 MeV (4) and up to 5-6 MeV (20). An additional analysis of the excited state spacings for the nuclei having shells approximate to the closed ones was carried out and the dE* distribution for $0^{16,17}$, N^{16}, $F^{16,17}$, Ca^{40}, K^{40}, Sc^{41} and Ca^{41} (with $E \leqslant$ 5 MeV) $Tl^{206,208}$, $Pb^{206,209}$ and $Bi^{206,209}$ is seen in Fig. 6. The maxima are observed in just the same place (20) as in the distribution of excitation of low-lying states for all nuclei (arrows on Fig. 6). So we come to the conclusion that consideration of the level positions for a broad scope of nuclei (even in the scale of several keV and tens of keV) doesn't lead to statistical averaging. One must consider collective shell phenomena (possibly in the scale more than hundreds of keV) as possible explanation.

Non-random distribution of positions of neutron levels of light nuclei should be taken into account by the analysis of neutron strength function (5,23). Also one should take into consideration the non-random distribution of radiative transitions when constructing the level schemes (by analysis of doublets). It's worth mentioning that the consideration of the summary data on the gamma-ray transition energies in non-deformed nuclei (18, all E_g) has confirmed expected grouping of the order of tens and hundreds of keV though according to the statistical model we might expect random E_g distribution (over many nuclei).

Now then we shall discuss three correlations, that have not had satisfactory explanation within the present models but each of them may appear to be responsible for the observed deviations from statistical description. We give them as illustration of how difficult it is within existing models to depart from statistical theory in the description of grouping effect of highly excited states.

3. STABILIZING EFFECT OF NUCLEAR SHELLS

Now assume that fine and hyperfine structure are indications of the presence of the stabilizing effects in the scale of several MeV. The term "stabilizing effect of shells" was first introduced in the work of Canadian Group (24) to describe suprisingly accurate equality of binding energies of neutron in two isotopes of lead. The lead excited spectra have also coincident energy intervals which result in maximum in distribution on Fig. 6 by D_{ij}=150 keV. The author have already (4) payed attention to the coincidence of the exactly known value (25) of splitting of low-lying states (with J = 5$^-$) of double magic nucleus Pb^{208} (dE = 511.0 keV) with the splitting of low-lying levels of nearest single-particle nucleus Pb^{209} (dE = 512 keV, (26)). The first more accurate value was taken as parameter Δ = 511.0 keV because close magnitude was found in Bi$^{206-210}$ levels. In summed distribution of position of low-lying excited states (for broad scope of nuclei--see Figs. 1-3 in paper (20)) there is also a maximum at 510 keV.

An additional correlation was found in the levels of the lead isotopes; close excitation energies group around rational numbers of Δ -parameter which has been introduced by analysis of the low-lying states. This is remarkably seen in Fig. 7 at E*=8Δ = 4088 keV; Nuclei Si^{29} (Fig. 7), O^{16} (20) and some other isotopes (Pb^{196,200,202}) also show the distinguishing effect of E$_{ss}$=2 Δ interval (for instance the first state of negative parity in O^{16} (20) and some other isotopes (Pb^{196,200,202}) also show the distinguishing effect of E$_{ss}$=2Δ interval (for instance the first state of negative parity in O^{16} has E = 6131 keV while 6 E$_{ss}$=6132 keV (20) and binding energy of neutron in C^{14} = 8176 keV (27) is equal 8 E$_{ss}$ etc.). Consideration of the summary data (up to 1968 year, with the exception of rotational bands of deformed nuclei) has shown that non-random grouping trend of E* can be described by super-position of intervals E$_{ss}$=1022 keV and some small intervals like 30 keV interval (see above). The interval of about 1 MeV was deduced in (20) by the study of positions of single-particle states of light nuclei and binding energies of "peripheral" neutrons.

If the energy characteristics of light and closed shell nuclei have some features of regularity, formally described by parameters like E$_{ss}$, and heavier nuclei conserve this effect in some degree one can suppose then, that fine structure (with interval E$_{fs}$ about 1 keV and some others) originates in regularity of the shell intervals. Then it follows that "hyper-fine structure" arises because of splitting of real shell intervals of fine structure and the hyper-fine structure is determined by the interval of the order of 1 eV = 1keV (1 keV/1 MeV). If it is a result of existence of fine structure (as the splitting of the next order of magnitude of the dimensionless parameter about 1 keV/1 MeV) it possibly won't require any additional model except the presence of intervals like E$_{ss}$ and E$_{fs}$. The phenomenological model described above may be used (6) for independent determination of E$_{ss}$ taking E$_{fs}$ and E$_{hfs}$ from analysis of experimental data. The resulting value E$_{ss}$=1003 ±40 keV is not in contradiction with the value E$_{ss}$=2Δ =1022keV found in independent studies. Another proof for the model introduced could be the presence of relations in fine structure intervals analogous to those in hyperfine structure (distinguishing effect of second, fourth, 13-th and 17-th interval etc.).

4. CORRELATIONS IN BINDING ENERGIES OF LIGHT NUCLEI

If shell effects of the order of 1 MeV are really important in complex nuclei originating stabilizing effects and some fine structures then they ought to manifest themselves in the energy

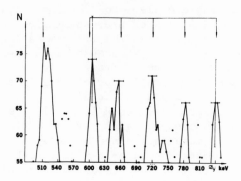

FIG. 6.   The idiohistogram of the distribution of respective spacings
between the excited levels of some extreme single-particle nuclei
(see the text).

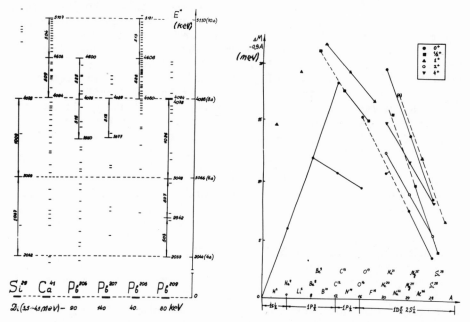

FIG. 7.   The excited levels of four lead isotopes, Si^{29} and Ca^{41}.
The arrows indicate the respective spacings between the levels near
the interval Δ=511 kev.

FIG. 8.   The relative values of the total binding energies (ΔM) of
light nuclei which differ by ΔZ=ΔN=2 with spins 0+, 1/2+, 1+, 2+, 4+.
On the horizontal scale of atomic weight A the regions of the first
closed shells are shown as well as the symbols of nuclei which states
are considered [32].

characteristics of the lightest closed-shell nuclei like C^{12}, O^{16}
and Ca^{40}. Everling (28) has already reported for these and the
nearest nuclei similarity in changing the total binding energies, if
the nuclei had alpha-cluster difference (one should consider ground
and excited states at the same time). Following the results of this
work it is possible to plot values of the total binding energies for
most of the light nuclei (see Fig. 8). One can see that number of
cases of lines with equal slopes really correspond to equal changes
of the total binding energies for a pair of nuclei with dZ=dN=2.
Now the intervals in Fig. 8 and some others reported in the work
(28) for Z=20-28 taken together allow to consider absolute values
of changes of the binding energy dM. The idiohistogram of dM in
Fig. 9 demonstrates grouping (the arrow upwards shows the interval
$E_{ss}=2\Delta$).
     The total binding energy of light nuclei is a sum of binding
energies of nucleons and clusters (and in case dZ=dN=2 as well).
That is why the distribution of differences of total binding energy
and the rational number of $E_{ss}=2\Delta$ cannot be random (Fig. 10).
Besides the maximum possibly conserves at zero after subtracting the
rational of 8.2 (8 per 2Δ =E_{ss}) interval-see right part of Fig. 10
(distributions are given for 10 mostly known closed-shell nuclei and
some others as well). The binding energies of the neutrons of closed-
shell nuclei, being the difference between correlated values, will
contain the tendency to preserve the same correlations (with the
interval about 1 MeV, see for example Fig. 5 in paper (20)). But
three shells (Z=N=6,8,20) provide too little experimental data to
make an estimation of possible random origin of the correlation
(modern theories give accuracy in calculation of binding energy not
better then 1 MeV). On the other side should the correlation be
true, the intervals of the order of 1 MeV could be considered as a
reflection of stabilizing effects of a greater order of magnitude.
In other words they are due to far correlations in the shell energies.
This analysis then shows that the magnitude of correlations is in
such a scale that is characteristic of the nucleon interaction, and
therefore we consider one more suggestion concerning the possible
origin of stabilizing shell effect.

5. THE PARTICLE MASS SPECTRUM ANALYSIS
     There are different semi-empirical relations between masses
of the particles (except well known relations of the unitary symmetry).
Some papers discussed the interval in the vicinity of pion (29-32).
One can see grouping near pion mass in Fig. 11 (32), while the
masses of lambda and omega hyperons with a good accuracy are
rational numbers of charged pion (29), Figure 12 illustrates this.
There are some other intervals on Fig. 12 which have been discussed
earlier in literature: k=235 MeV introduced by Takabayasi (29)
(seen in Fig. 12 as arrows); introduced by Wick (31) interval =
391 MeV =m_w=2 (double arrows), relating the masses of the particles
from different multiplets: the interval in the vicinity of muon
mass M (twisted arrow): the interval which is a difference of the

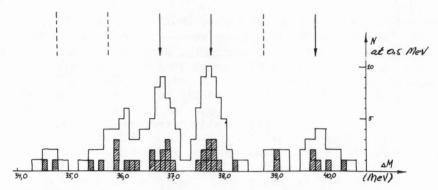

FIG. 9. The common distribution of the relative values of the total binding energies of light nuclei and those of $1f_{7/2}$ shell which differ by the α-particle. The dark squares correspond to the values $\Delta M^A - \Delta M^{A-4}$ itself, the distribution corresponds to the idiohistogram with the averaging interval ±0.25 MeV. The arrows above indicate the period ε_{oc}=1.02 MeV [32].

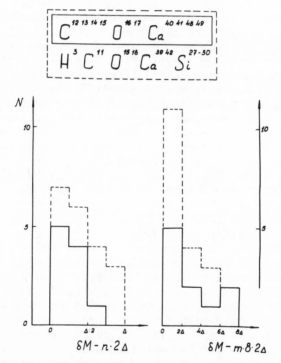

FIG. 10. The distribution of residual values after subtracting even numbers of intervals 2Δ=1022 kev (left) and 8.2Δ (right) from total binding energies of 10 and 20 (dotted line) light nuclei with the nearly closed shell.

nucleon and K-meson masses (Fig. 12, dashed curve); the difference
of the η^o and Π meson masses (crossed curve) the past two have
been introduced by Takabayasi and Sternheimer (31,33). For convenience X-axis is given as a third of Wick's interval and Y-axis
as a rational number of this interval. Under such representation
equal intervals look parallel to each other. Some authors including
Nambu (34) have pointed to the relation between the charged pion and
electron (mc^2=511.0 KeV) masses, namely mc^2 is equal to pion mass
divided by α /2 where " α " is a constant in **electrodynamics.**
All these relations should be considered as semi-empirical that will
possibly be interpreted within fundamental theory of strong interaction in the future. We try to find quantitive relation between
all these parameters. The study have been performed of 1) differences between masses of the particles (Fig. 13 from (32)).
2) splittings inside unitary multiplets and their mutual differences
(Fig. 14), all masses from (35), differences are listed in upper
part of Fig. 14. The structure was found with an interval about
8 MeV, that coincides with the double charge splitting of pion
(m $\Pi^{\pm}$ - m$_{\Pi o}$-m c^2), shown by arrows on Fig. 13 and 14. As in the
papers, quoted above the pion and muon masses appeared, the interesting question arises if this is occasional coincidence that
they are represented as 13-th and 17-th interval of the charge
splitting and this interval or charge splitting itself is equal to
8 mc^2 with an accuracy of several keV (4,21,32). Only more fundamental theory could give an answer to the question and Nambu-
relation (see above) would arise from the rational relations, manifesting also in the values of masses(for illustration, say pion mass
=17.16 mc^2+mc^2 etc). We are mainly interested how in such a semi-
empirical models (of mass structure) some fine effects discussed in
this paper might appear. First of all let us recall that distinguishing nuclear intervals Δ and 780 keV turned to be very close
to charge splittings of lepton (mc^2) and nucleon. Taking into
account splitting of pion (8 mc^2=4E_{ss}), those three correlations
seems to be systematic (and very astonishing). Now if one takes
another well known mass splitting namely mass of muon (m μ) then
the difference of mμ + mc^2 from the 13th interval of 16 mc^2 will
account dm = 120 keV or small value dm/m near 0.00115 (of the scale
of polarization correction of electrodynamics $\alpha/2\Pi$).

Let us for the moment assume that some dimensionless parameters
of that type might be real. From that hypothesis and from mass
intervals (starting with about 8 MeV and others marked in Fig. 12)
one will get the number of intervals of the scale of tens and hundreds keV (with the series of rational relations between them).In other
words the introduction of mass structure and dimensionless parameter would possibly permit to explain distinguishing effect of the
order of tens and hundreds of **keV.** From this assumption (and above
mentioned parameter) one should expect the distinguishing effect on
certain intervals, say for example on 4E_{fs} = 4.7 **keV;** 9.4 **keV;**
18.8 **keV** and on 13th and 17th intervals of these periods (besides,

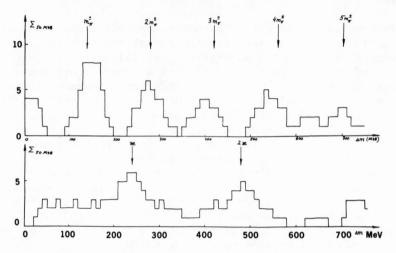

FIG. 11. The idiohistogram of the distribution of mass differences for barionic isosinglets [32]. The arrows indicate the interval equal to the charged pion mass. The averaging interval of the idiohistogram is ±25 MeV.

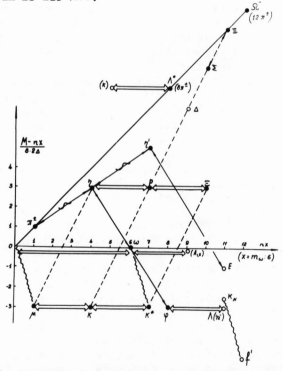

FIG. 12. Semiempirical relations of particle masses indicated by Takabaiasi and Onuki (arrows), Sternheimer (curved arrows), Wick (double arrows), Takabaiasi (crossed arrows). For convenience the even number $16 \times 8 \times 2\Delta \approx m_\omega : 6$ was subtracted from masses.

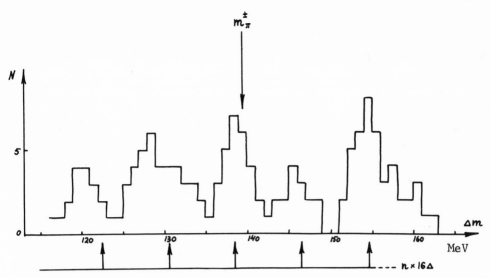

FIG. 13. The idiohistogram of the distribution of mass differences for isosinglets and single charged members of multiplets known with the error less than 8 mev. The averaging interval of the idiohistogram is ±2.5 MeV.

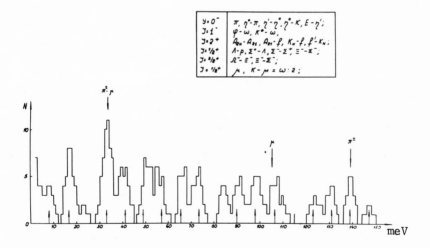

FIG. 14. The idiohistogram of mass shifts in unitary multiplets and their respective differences [35]. The averaging interval is ±2.5 MeV. The arrows above indicate the mass values of μ and π mesons, and those below indicate the period 8x2Δ.

similar relations are expected from the model discussed in third
section of the paper, where fine and hyperfine structure are
supposed to be correlated).

CONCLUSION

        Neutron physics as one of the most precise branch of nuclear
spectroscopy seems to touch with some non-statistical effects. We
have in mind hyperfine structure of the type discussed in some work
(for example from the Columbia University) and distinguishing effect
of position of neutron levels, recently discussed in paper by Ideno
and Ohkubo. But they arise some very difficult problems and this
paper should illustrate how analysis of small-scale correlations
(at high excitation energies) lead to finding of possible semi-
empirical correlations of the larger scale where we have many
unsettled problems of nuclear models and theory of strong interaction. So the statistical approach that doesn't demand the solution
of these fundamental problems will be for the long period of time
very useful for analysis of experimental data of neutron physics.

        I wish to thank Dr. M. Vlasoff for translating the
manuscript.

REFERENCES

1. J.B. Garg et al. Phys. Rev. 134, 985 (1964).
    W.W. Havens,Jr., Progress in Fast Neutron Physics, Univ.
    Chicago Press (1963), p.215.
2. S.I. Sukhoruchkin, Correlations In The Positions Of Single
    Particle Levels of Complex Nuclei. Program of 17th Meetings
    on Nuclear Spectroscopy. Charkov, 25-1-1967. "Nauka, p. 176.
3. G.B. Muradyan, et.al. Soviet Nuclear Physics, 8, 852 (1968).
4. S.I. Sukhoruchkin, Proc. Second Conf. Neutron Cross Section
    and Technology, Washington, 1968, 2, p.923.
5. S.I. Sukhoruchkin, Proc. Nuclear Data for Reactors, IAEA,
    Vienna, 1967, 2, p. 159.
6. S.I. Sukhoruchkin, Soviet Nuclear Physics, 10, 496 (1969).
7. K. Ideno, M. Ohkubo, Journ. Phys. Soc. Japan, 30, 3 (1971),
    p. 620.
8. P. Ribon, Dissertation, CEA-N-1149 (1969); see also (9).
9. J.B. Garg, et.al., Phys. Rev. 134 (1964), B 985; 136 (1964),
    B 174; 137 (1965), B 547; 166 (1968), 1234; 166 (1968),1246.
10. V. Kolar and K.H. Bockhoff, Journ. Nucl. Energy, 22 (1968),299,
    G. Carraro and W. Kolar, Nuclear Data for Reactors, IAEA,1970
    Vol. 1, P. 403.
11. G.V. Muradyan et. al., Soviet Nuclear Physics, 8 (1969),495;
    Ref. 5 report CN-23/107.
12. H. Weigmann, et.al., Third Conf. Neutron Cross Section and
    Technology, 1971, Report 6-19b.
13. H. Camarda, et.al., BNL-50276 (1970), p. 41.
14. J. Julien, Dissertation, CEA-R-3385 (1968), R. Alves,
    Dissertation, CEA-R-3602 (1969).
15. G.D. James and G.G. Slaughter, Nucl. Phys. A 139 (1969), 471.
16. T.E. Young and M.G. Silbert, NCSAC-31 (ANL-1970)p. 70.

17. G.A. Keyworth and M.S. Moore, BNL-50276 (T-603), 1970,p.136.
18. G.A. Bartholomew et.al., Nuclear Data 3-5 (1967-1969),
     Academic Press.
19. O.I. Sumbaev et.al., Program of 13th Meetings on Nuclear
     Spectroscopy, Kiev, 1963, p.43.
20. S.I. Sukhoruchkin, Soviet Journal Nuclear Physics, 10, 3,
     p. 496 (1969).
21. S.I. Sukhoruchkin, Proc. Meetings on Neutron Interaction with
     Nuclei, 9-12 June 1964, Preprint of JINI-1845(Dubna)p.39 and 44.
22. P.M. Endt and Van der Leun, Nucl. Phys. 105 (1967), 1; Nuclear
     Data, 3, p.447 (1967), O. Skoppstedt, Proc. Symp. Capture Gamma
     Ray Spectr. 1969, Studsvik.
23. S.I. Sukhoruchkin, Atomic Energy, vol. 29, nom. 3,p.187(1970).
24. W. Mc Latchie, et.al., Can. Journ. Phys. 42, 926 (1964).
25. A. Hedgren, Phys. Rev. 82, 138 (1951); see also (21).
26. J.R. Erskine and W.W. Buechner, BAPS 7, 4, p. 360 (1962).
27. J.H.E. Mattauch, et.al. Nucl. Phys. 67, (1965), 32.
28. F. Everling, Nucl. Phys. 40 (1963), 670; 47 (1963), 561.
29. T. Takabayasi, Progr. Theor. Phys. 29, 3, 472, (1963);
     T. Takabayasi and Y. Onnuki, Progr. Theor. Phys. 30, 272 (1963).
30. R.M. Sternheimer, Phys. Rev. 170, 1267 (1968).
31. R.M. Sternheimer, Phys. Rev. 136, 1364 (1964).
32. S.I. Sukhoruchkin, Program of 21th Meetings on Nuclear Spectro-
     scopy Moscow, 1971, part two, p.320.
33. T. Takabayasi, Progr. Theor. Phys. 32, 863 (1964).
34. Y. Nambu, Progr. Theor. Phys. 7, 595 (1952).
35. N. Barash-Schmidt, et.al., Particles Properties, Jan. 1970,
     UCRL-8030.

SESSION IV

STATISTICS OF RESONANCE PARAMETERS
(PART II)

Chairman   - R. C. Block, Rensselaer Polytechnic
                           Institute
Scientific Secretary - R. Hockenbury, Rensselaer Polytechnic
                           Institute

THE MEASUREMENT OF RADIATIVE CAPTURE WIDTHS[†]

R. E. Chrien

Brookhaven National Laboratory, Upton, New York

The history of resonance capture γ-ray spectroscopy dates back about 15 years when the first experiments with linac neutron sources were reported. Landon and Rae[1] used the observation of a high energy γ-ray to determine the spin of the 34 eV resonance of Hg-199 and Draper and his co-workers at Yale[2] looked for systematic variations in the low energy γ-ray production from resonances. Although thermal neutron capture resonance spectroscopy had been an active field for some 10 years previous to the first resonance measurements, the inherently low flux associated with the resonance region inhibited development in this direction for many years. The development of potent neutron sources, such as the high flux reactors and high current linacs, in conjunction with the advent of high resolution semi-conductor diode detectors, made neutron resonance γ-ray spectroscopy come alive.

Resonance capture γ-ray spectroscopy is rich in information relative to nuclear levels which are connected by the radiative transition. It complements the thermal neutron work in a crucial way, since the wide number of resonance states examined, either individually or in an averaged spectrum manner, can overcome the limitations of the Porter-Thomas distribution, and populate final states in a more uniform manner than is possible in the thermal work. Furthermore the importance of $\ell = 1$ or even $\ell = 2$ neutrons in resonance capture allows the population of a wider variety of states. The use of normal elements in resonance capture is often sufficient to isolate states masked by isotopic mixtures in the thermal capture work. The emphasis of the present discussion is on the bearing of resonance neutron capture spectroscopy on the

[†]Work supported by the U.S. Atomic Energy Commission

properties of the highly excited states near the excitation corresponding to the neutron binding energy. These states--with
typical widths about a millionth of the single particle widths,
for heavy nuclides--are expected to follow random decay patterns in accord with the statistical concept of the compound
nucleus. The γ-ray spectra from these states form a rich mine
of observations from which we may deduce many properties of the
resonances, such as their partial radiative widths, and their
spins and parities. The statistical properties of these
widths are of interest in testing the hypothesis that the
amplitudes describing radiative decay are random variables
with zero correlations among themselves and unrelated to specific
nuclear properties of either initial or final states.

One consequence of the statistical decay assumption is that
the partial radiative widths, which are proportional to the
squared amplitudes, follow a Porter-Thomas distribution, i.e.
a χ-squared distribution with $\nu = 1$.

$$P(x) = \{2^{\nu/2}\Gamma(\nu/2)\}^{-1}x^{(\nu/2-1)}e^{-\frac{1}{2}x} \quad x = \Gamma/<\Gamma>$$

Width distribution functions are generally analyzed in terms of
the chi-square class of functions. The most definitive test
of the Porter-Thomas distribution of radiative widths to date has
been reported by Wasson et al.[3] for (U-238 + neutron) where a set
of 15 γ-rays was studied for 23 S-wave resonances, in a joint BNL-
ORNL experiment at the Oak Ridge Electron Linear Accelerator.
Historically, the uranium case is interesting since attempts have
been made to study the width distribution with increasing neutron
and γ-ray energy resolutions since 1959. During this period
detector resolution has improved a factor of 70--from 7% with NaI
to about 0.1% with Ge(Li) detectors. This improved resolution has
been coupled with powerful new neutron sources such as ORELA. The
result has been a continuing decrease in the reported discrepancy
with the Porter-Thomas distribution. The ν values obtained for
the 15 transitions (10 E-1's and 5 M-1's) are shown in fig. 1 ,
where the error bars designate 10 to 90% confidence limits.
Eleven out of 15 γ rays appear to be consistent with $\nu = 1$.
Although we would expect that about 3 of the γ-rays fall outside
these limits for a set of 15, all 4 deviants are on the high
side, and the mean of the distribution of all γ-rays is clearly
higher than one.

An idea of the experimental difficulty is obtained from
fig. 2, where the radiation near 4067 keV is shown for the 37 and
67 eV resonances. No less than 3 peaks are in evidence (4061,
4067, and 4071) within a 10 keV span. Even with the excellent
resolution of 4.4 keV, there is difficulty in separating γ-ray
components. Deleting this complex group of γ-rays, one finds for
the remaining 12 sets over 23 resonances:

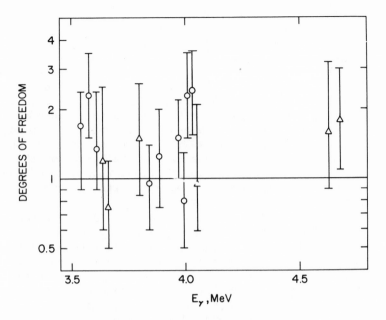

fig. 1. v-values for U-238

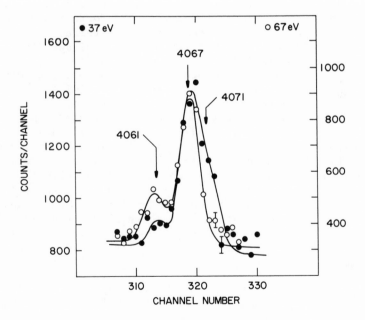

fig. 2. Complex structure near 4 MeV, U-238 (n,γ)

$$v = 1.20 \begin{array}{c} +0.20 \\ -0.15 \end{array} \quad \text{(10 to 90\% limits)}$$

This measurement represents the largest set of radiative widths ever tested and is indicative of presently available precision. Although statistical counting errors are included in the above error limits, one cannot exclude the presence of systematic errors which could cause the indicated 20% discrepancy in v-value. One example of such an error is the evaluation of background under a resonance peak, which amounts to as much as 10% at higher neutron energies. One concludes that there is no convincing evidence of inconsistency with the Porter-Thomas distribution. Other reported deviations can be challenged on the basis of insufficient sample size, insufficient knowledge of the final states, or systematic errors associated with background contamination of the resonance spectrum.

A second assumption of statistical radiative decay of the highly excited resonance states is the lack of any dependence of the transition probability on the nuclear structure of the final or initial states. Such a dependence would lead to width correlations, in violation of the underlying assumptions of the compound nucleus concept. Several reaction mechanisms have been suggested which lead to width correlations. All such mechanisms imply a simple structural component present in the highly excited states. Among such mechanisms are direct capture, channel capture or valence transitions, semi-direct capture, and decay of doorway states. The presence of such correlations can be perfectly consistent with the Porter-Thomas distribution; as for example in the case of the perfect correlation between neutron and radiative widths, envisaged for valence neutron transitions.

The problem of the significance of observed correlations is an important one when dealing with samples of limited population size, as is often the case in resonance capture. Figure 3 illustrates a Monte Carlo calculation for the expected distribution of ρ for samples of 5 pairs of random variables, drawn from uncorrelated Porter-Thomas distributions. For large samples the distribution is approximately Gaussian about zero, but for small samples the distribution function is quite asymmetric, showing a most probable value less than zero, and a long positive tail. The existence of the latter feature means that large positive sample correlation coefficients do not decisively demonstrate a functional dependence. The problem is especially severe in dealing with reported (n,γ), (d,p) correlations in thermal capture, since the number of final states is likely to be limited.

We would expect to usefully search for width correlations in regions where reduced widths are exceptionally large--that is near strength function peaks, and for final states of high spectroscopic factors. Hence we have searched near the 4s, 3s, and 3p giant resonances, and looked for enhanced transitions to low-lying $\ell = 0$, 1, and 2 states with large spectroscopic factors.

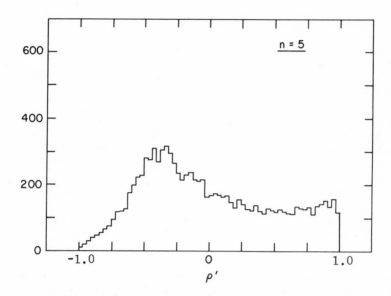

fig. 3.   Sample correlation coefficients, N = 5

At first much attention was focused near the 4s giant resonance, where the level spacing is such as to provide many resonances at low neutron energies. Significant correlations between
resonance neutron and radiative widths were reported by the BNL
chopper group for the following cases:[4,5,6]

Tm-169 $<\rho>_{J=1}$ = + 0.274 (8 res/15 final states)

Dy-163 $<\rho>_{J=3}$ = +0.35 (17 res/5 final states)

Yb-173 $<\rho>_{J=2}$ = +0.43 (4 res/11 final states)

In the rare earth region, the neutron widths are relatively
small, about 10^{-6} single particle size, and the correlations are
correspondingly small and difficult to establish. Very recently
the BNL group has initiated a collaborative effort with ORNL to
extend these width correlation studies. Using ORELA, it is possible to extend the useful energy range to two or three times
that available with the chopper. The first experiment, carried
out with the help of G. G. Slaughter of ORNL was an extension of
the thulium width correlation studies.

Figure 4 shows a time-of-flight spectrum, based on events detected in a 30cc detector above 145 keV of γ-ray energy deposited. Approximately 70 resonances in a sample of 180 gms of TmO_2 were recorded, of which 23 have been analyzed to date, up to an energy of 260 eV. The spectra were recorded with an over-all energy resolution of 0.1% . Approximately 255 hours at an average beam energy of 19 kW were devoted to the experiment. A flight path of 10.34 meters and a burst width of 10 ns were used.

The Harwell group[7] have pointed out recently an error in the spin assignment of the 153 eV resonance of Tm-169. This formed the motivation for the present experiment, and also emphasizes the importance of accurately assigning the resonance spins in this experiment. For this the low energy γ-rays in resonances have proved exceedingly valuable.

Although the low energy γ-rays following neutron capture do not contain significant primary radiation strength, they are important in providing indirect verification of the initial (resonance) state spin. This fact has been well established by many experiments, too numerous to quote here. The experimental fact was first noted by Draper and co-workers,[2] and explained by Huizenga and Vandenbosch[8] in terms of a simple cascade model involving sequences of dipole transitions. Subsequently this

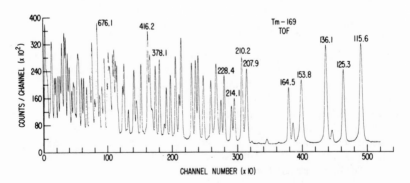

fig. 4.   Tm-169 Time-of-Flight Spectrum

model was refined by Poenitz[9] to include quadrupole transitions and energy dependent factors in the transition probabilities. Although a quantitative prediction of the population of a final state of given spin and parity by cascades from a resonance of given spin is not in general possible because of a lack of knowledge of such factors as multiplicities and quadrupole admixtures, it is an experimental fact that the low energy radiation exhibits intensity patterns precisely correlated with resonance spin. This fact provides an important tool for resonance spectroscopy. The low energy γ-radiation is relatively strong and free from Porter-Thomas fluctuation, since an averaging over many intermediate states has occurred.

Thulium can be taken to illustrate some of the problems of the method. A number of low lying states in Tm-170 have been identified by Sheline et al.[10] We have examined the intensity ratios of γ rays depopulating states of spins 0⁻, 1⁻, 2⁻ and 3⁻ as follows: 149(0⁻), 166(2⁻), 204(2⁻), 181(2⁻), 204(2⁻), 220(2⁻), 237(1⁻), 243(3⁻), and 311(3⁻). The simple argument leads us to examine ratios of γ-rays depopulating these states for 3⁻/1⁻ or 2⁻/0⁻ pairs, since for each pair the higher spin component should be sensitive to the resonance spin, while the lower spin component should be insensitive. The ratios 3⁻/1⁻ and 2⁻/0⁻ do show dependence on resonance spin; the ratios for thulium show a relative fluctuation of 50% from a spin 1 to a spin 0 resonance. A relative change of 1.5 to 1.0 is shown on fig. 5   for the average of 4 γ-ray pairs, 2 (3⁻/1⁻) and

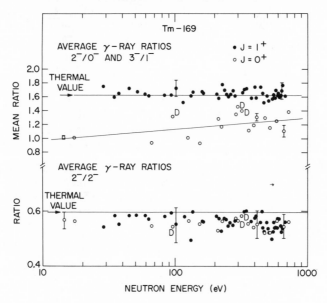

fig. 5.   Low Energy γ-Ray Ratios for Tm-169

2 $(2^-/0^-)$. For comparison the average of 2 $(2^-/2^-)$ ratios shows
no dependence on spin, as one expects. The ratio in Tm-169
display an interesting characteristic: the ratios characteristic
of spin 0 resonances rise with neutron energy. One might expect
that partly this is due to poorer resolution at higher energies;
the spin 1 resonance ratios, however, show no downward trend. A
more plausible explanation arises from the high thermal cross
section for Tm-169, 127 barns. The thermal capture cross section
is dominated by J =1 capture, hence thermal neutrons would give
rise to the γ-ray ratio characteristic of J=1 resonances. There
is no doubt that in a time-of-flight experiment using a pulsed
source that there are a certain fraction of thermalized neutrons
captured by the sample, and these thermal neutrons are favored
by a relatively high cross section relative to the resonance
neutrons. The Tm-169 time-of-flight spectrum of fig. 4 shows
rather clearly an increasing background with decreasing time-of-
flight. The behavior of the low-energy γ-ray ratios indicates
strongly that this increasing background is due to moderated or
thermalized neutrons.

On the basis of these ratios the following spin assignments
for thulium resonances is made   (Resonances used in determination
of correlation coefficients are underlined): 3.9(1), 14.5(0),
17.5(0), 28.8(1), 34.9(1), 37.6(1), 45.0(1), 50.8(1), 59.3(1),
63.2(1), 66.2(0), 83.6(1), 94.2(1), 95.7 (probably 0, incompletely
resolved from 94.2), 101.9(1), 115.6(1), 125.3(0), 132.3(1),
136.1(1), 153.8(0), 160.7(1), 164.5(1), 207.9(1), 210.2 (a doublet
1,0 near this energy), 214.1(1), 224.4(0), 228.4(1), 238.8(1),
243.8(1), 251.6(1), 260.6(1), 274.2(1), 283.8(1), 289 (1,0 doublet),
297.2(1), 319.9 (1,0 doublet), 352.1(1), 333.6(1), 346.8(1),
358.2(0), 378.1(1), 391(0), 400.5(1), 409.2(1), 416.2 (1,0
doublet), 441.4(1), 455.4(1), 459.9(1), 469(1), 472.8(0),
493.3(1), 512.9(1), 520.2(0), 542.8(1), 550.4(1), 557.4(1),
565.9(1), 573.9(1), 586.7(0), 592.4(1), 599.8(1), 607.8(1),
626.3(1), 631.8(1), 643(1), 659.5(0), 676.1(1), and 716(1). Note
that the important 153 eV resonance has now been assigned J = 0.

The ORELA experiments also show a positive correlation, as previously reported by Lone and Beer for resonances below 115 eV.
However, as the sample size increases to 17 resonances up to
300 eV, the average correlation coefficient for 15 final states
decreases to a point where it is only marginally significant
statistically. The effect is illustrated in the following average
coefficients.

$$\langle\rho\rangle_{\tilde{J}=1} \text{ (8 resonances, 15 γ rays)} = +0.16 \pm 0.10$$
$$\text{to 115 eV}$$

$$\langle\rho\rangle_{\tilde{J}=1} \text{ (17 resonances, 15 γ rays)} = +0.089 \pm 0.07$$
$$\text{to 283 eV}$$

It is also possible to investigate the J=0 resonances separately

$$<\rho_{\bar{J}=0} \text{ (5 resonances, 6 } \gamma \text{ rays)} = +0.15 \pm 0.22.$$
to 153 eV

Table I shows the coefficients for each γ-ray for the full set of 17 J=1 resonances as well as the correlation between reduced neutron width and the radiation width averaged over all 15 final states. It must be borne in mind, however, that the lowest energy resonance still has the highest reduced width of all resonances sampled; we have enlarged our sample by adding resonances of lesser widths. Hence the possibility exists that the correlation coefficient has been degraded by the addition of weaker resonances.

γ-Ray (eV)	ρ	γ-Ray (eV)	ρ
6594	-0.062	5945	+0.19
6556	+0.35	5911	+0.32
6445	-0.13	5900	+0.04
6389	-0.14	5809	-0.11
6375	-0.15	5737	+0.49
6356	-0.11	5730	+0.09
6003	+0.25	5684	-0.11
		5518	+0.41

$$\rho(\Gamma_n^o, \overline{\Gamma}_{\gamma i}) = +0.17 \ (P \cong 20\%)$$

TABLE I. Tm Correlation Coefficients

Although the magnitudes of correlations between neutron and radiative widths are small for heavy nuclei, this is not surprising in view of the small neutron widths and small level spacing characteristic of these nuclei. A much more favorable situation is to be found near the closed neutron shell at N=50. In this region the low-lying shell model states are S or D states, with large spectroscopic factors, and $\ell=1$ capture leads to strong transitions populating these single particle states.

The fact that, for these nuclei at low incident neutron energy, one finds oneself in the 3p giant resonance, suggests that the neutron resonant states in this region will produce radiative transitions describable in terms of a simple model in which the radiation arises from p → s or p → d single particle transitions.

Consider the case of Mo-92 and Mo-98 target nuclei; at Mo-92 the N=50 shell is closed and at Mo-98 the d 5/2 shell is filled. These target nuclei show considerable structure in the p-wave strength function. Rohr, Weigman and Winter[11,12] point out

that for Mo-98 a strong fluctuation in strength function exists:
up to 1 keV a value of $S_1 = 34 \pm 7 \times 10^{-4}$ is obtained, as compared to a value of $S_1 = 6.8 \pm 0.5$ from average cross section
measurement in the keV region. Figure 6 shows a time-of-flight
spectrum recorded in a 33 cc (GeLi) detector at ORELA for a
sample of 150 gms of Mo-98 oxide. The concentration of level
strength from 400-1000 eV is quite marked.

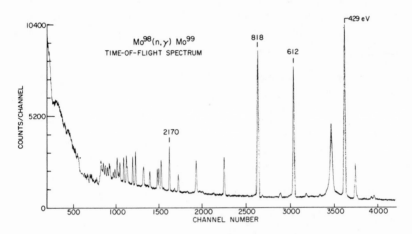

fig. 6. Mo-98 Time-of-Flight

The above experimental facts suggest the following questions:
Do the transitions from these anomalously strong $\ell = 1$ resonances
to the low-lying levels of largely single particle character obey
the statistical decay assumption   that the transition intensity
is independent of the nuclear parameters of the final state?   If
not, are they describable in terms of a simple model?

The first question can be answered simply by inspection.
Figure 7 shows the spectra from four very large resonances of
Mo-98; 429, 467, 612, and 818, as obtained at the BNL fast chopper.[13]
The 467 eV resonance is s-wave; the balance are p-wave resonances,
and the appearance of the spectrum is obviously different. The
strikingly similar ratios between lines 1, 3, and 5 for the 429
and 612 resonances and the dominant ground state transition in
each p-wave resonance makes it implausible to maintain that a
purely statistical decay is taking place. These observations
had already been made by the Geel group of Rohr, Weigman and
Winter[11] more than a year ago. To proceed further with a more quan-

titative analysis it was necessary to establish the spins of the
capturing states and to improve the quality of the resonance
spectra. Work carried out at the BNL fast chopper used angular
correlation techniques to establish that the 818 eV resonance
has spin 3/2 while the 612 and 429 have spin 1/2.   This explains
the qualitative differences between the spectra of the figure.
We can then proceed to evaluate these spectra in terms of the
simplest possible nuclear model: the valence neutron transition
model,[14] which ascribes the radiation to a single nucleon jump
outside a closed core:

$$\Gamma_{\lambda\mu} = \frac{16\pi/k^3}{9} \; \theta_\lambda^2 \theta_\mu^2 \left| \bar{e} \int_o^\infty dr \; u_\lambda r u_\mu \right|^2 \frac{\left| \langle j'I \, J_\lambda || Y^{(1)} || j''I \, J_\mu \rangle \right|^2}{2J+1}$$

This model is surprisingly well obeyed by the group of large p-
wave levels between 400 and 1000 eV. What is even more striking is the agreement between calculated and observed partial
widths. Figure 8   shows   a detailed calculation using
this model in the case of Mo-98. An effective charge
of eZ/A has been assumed for the neutron, and the radial integrals

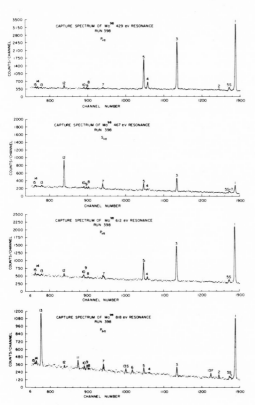

fig. 7.   Mo-98 Resonance
          Spectra

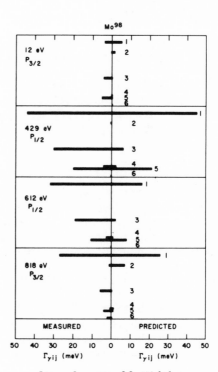

fig. 8.   Mo-98 Widths

have been taken from Lynn's calculation,[14] which employs a Woods-
Saxon potential. The agreement between observed and calculated
radiative widths is good for final states with large spectroscopic factors. For these cases the single particle effects
dominate the more complex reaction mechanisms.

An extension of the BNL chopper experiment was recently
undertaken in collaboration with the ORNL ORELA group
and appears as contribution 4.3 to this Conference by Cole, et al.

A detailed summary of the experimental results are presented
in Table II. Several spin and parity assignments are proposed in
this table, based on the transitions to known 5/2+ states in Mo-99.
The new experiment shows clearly that the observed spectral
regularities are associated with the anomalous strength function
peak below 1 keV. The detailed agreement with the model is
confirmed in this region, but at higher energies, where the
neutron reduced widths are smaller, the model fails to predict
the spectra accurately.

The startling success of the valence neutron model in predicting widths in molybdenum isotopes has stimulated interest in
examining strong transitions in neighboring nuclides, especially
near the 3p giant resonance. It is of considerable interest to
explore also the region of the 3s giant resonance for non-statistical effects. Some of these tests have been summarized by Bhat
et al. in contribution 4.1 to this conference. The case of
Sn-116, where additional information has recently been accumulated, serves to illustrate the successes and failures of this
model in predicting γ-ray spectra, and the types of measurements useful in this sort of search. In recent weeks measurements on the spins of the resonance in Sn-116 have been completed. Resonances at 147.9 and 632 eV have been examined at the
BNL fast chopper facility with detectors set at 135° and 90° to
the incident beam. From the theory of angular correlations, we
know that

$$I(90^{\circ})/I(135^{\circ}) = 10/7 \text{ for } J = 3/2 \rightarrow J = 1/2$$

$$I(90^{\circ})/I(135^{\circ}) = 8/11 \text{ for } J = 3/2 \rightarrow J = 3/2.$$

As shown in Fig. 9 both resonances display a non-isotropic
angular distribution, and are therefore p 3/2 resonances. From
the valence neutron model, evaluated for this case we know that

$$(p\ 3/2 \rightarrow s\ 1/2)/(p\ 3/2 \rightarrow d\ 3/2) \text{ is } \approx 15.$$

As is obvious from Fig. 9, the 147.9 eV resonance is in good
agreement with the model, the p 3/2 $\rightarrow$ d 3/2 transition being very
small, and the predicted width for the transition to the s 1/2
state is in good agreement with theory.

E_n (eV)	ℓ_j	I_γ (a)						$\gamma^2_{\lambda n}$ (eV) (b)
		1/2+ℓ=0 5926.9	5/2+ℓ=2 5829.4	3/2+ℓ=2 5575.9	1/2+ℓ=0 5400.9	3/2+ℓ=2 5379.1	5/2+ℓ=2 5310.9	
12	p 3/2	24.9± 1.5	3.4± 1.2	12.4± 1.5		31.1± 1.8	22.5± 5.2	122.6
401	p 3/2	23.8± 5.2	66.1± 9.3	23.5± 5.4		6.5± 4.6	0 ± 1.5	35.0
429	p 1/2	210.0± 5.0	0 ± 1.8	144.0± 3.0	0 ± 4.4	91.8± 2.5		1326.1
467	s 1/2	14.6± 2.8	0 ± 1.6	43.5± 2.6	20.3± 2.0			
612	p 1/2	204.0± 5.0	2.0± 2.2	119.0± 4.0	8.5± 2.3	57.1± 3.0	0 ± 2.0	755.3
818	p 3/2	125.0± 4.0	0 ± 2.0	15.2± 2.2	0 ± 2.2	5.9± 2.3	6.7± 2.2	601.9
1122	p 3/2	28.2± 5.4	0 ± 6.1	26.2± 6.1		4.8± 5.4	41.8± 6.8	52.0
1528	s 1/2	20.1± 6.5	0 ± 5.6	47.6±13.3	33.4± 7.7	9.0± 6.4	0 ± 6.4	
1920	p 3/2	4.0± 8.7	0 ±11.0	15.1± 9.5	30.1±11.9	0 ±12.0	49.9±12.7	24.4
2170	p 1/2	51.9± 7.3	4.4± 5.5	0 ± 5.5	18.6± 6.2	38.0± 7.3	8.4± 5.8	[248]
2460		34.4± 9.5	0 ± 8.1	57.9±11.0	6.6± 9.5	11.7±10.3	11.7±10.3	
2546	s 1/2	13.6±11.0						
2623	p(3/2)	50.6±12.5	17.1± 9.3	6.2±10.9		71.7±14.8		[58]
2950	p(1/2)	143.0±21.0		71.2±16.8	46.6±18.1			[47]
3260	p(1/2)	97.7±15.8		0 ±12.6				
3290	s(1/2)	48.9±22.4						
3792	p(1/2)	178.0±12.6	0 ± 7.8			16.5± 8.3	12.1± 8.3	[385]
4013	p 3/2	21.6± 9.9	113.7±14.0	29.7± 9.3			35.6± 9.9	
4471	s 1/2							
4571	p(1/2)	149.0±33.0	82.3±12.9	3.3± 9.9	108.2±15.2		7.3± 9.9	309.3
4842	p 3/2	69.4±13.6		17.2±10.0				269.4
Valence Model								
	p 1/2	1.0	0.0	0.14	0.05	0.47	0.0	
	p 3/2	1.0	0.27	0.02	0.06	0.05	0.02	

(a) = photons/1000 captures

(b) = gΓn from ref.

[ ] = assuming Γγ = 150 meV

TABLE II. Mo-98 γ-Ray Intensities (Normalized to 429 and 612 eV resonances, in BNL work)

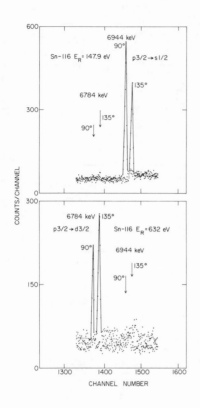

fig. 9. Capture spectra from a Sn-116 target at 148 and 632 eV, 90° and 135° positions.

On the other hand, the p 3/2 → d 3/2 .transition is intense in the 632 eV p 3/2 resonance, while the transition to the s 1/2 state is weak. The reduced width for these resonances is about the same, and we expected the model to work equally well for both cases. It does not. Furthermore a strong M-1 transition is observed from the 111.2 eV s-wave resonance of Sn-116 to the d 3/2 state. This is an example of a $\Delta\ell = 2$ transition which is forbidden by the simple model.

In the region of the 3s giant resonance, the RPI group[15] has recently reported a strong correlation between neutron and total radiation widths for targets of Cr, V, and Ni, a value of $\rho = 0.8$ for 12 resonances in even-even nuclei being obtained. The total radiation widths calculated from the valence neutron model are not large enough to account for such large correlations; the calculated width is only about 1/3 of the observed value. It is of interest to compare the model against the photoneutron measurements of Jackson and Strait.[16] In the latter experiment what is measured is the quantity $\Gamma_{\gamma 0}\Gamma_n/\Gamma$, for most resonances this is nearly equal to $\Gamma_\gamma 0$, the ground state radiative width. In each case examined; Cr-52 Ni-60 and Fe-56, the valency model is accurate for the first several resonances, but at higher neutron energies, there is complete disagreement. The success of the model, in each instance, may be associated with a relatively narrow range of excitation energy in the compound nucleus, which suggests the influence of some form of intermediate structure. The Mo-98 results are clearly in this class, the others may be as well. The suggestion of a narrow intermediate structure in the neutron strength function which is correlated with average radiative transition strength was also advanced by Coceva et al.[17] for the case of In-115; however, here no correlations of individual radiative widths with either final state or initial state (resonance) reduced widths have been noted.

We may summarize the present experimental situation by emphasizing the existence of strong departures from the statistical theory near the 3p giant resonance, with somewhat weaker evidence for width correlations in the 3s and 4s giant resonances. The agreement of experiment with the predictions of the valence neutron model is striking for many nuclei in the 3p maximum; in other nuclei the model fails. There is some evidence that the width correlations are associated with fluctuations in the neutron strength function. A much clearer picture of where and how width correlations occur will await more detailed experiments with larger sample sizes. These experiments are now technically possible with the new powerful neutron sources at our disposal, and we may hope that their execution is now close at hand.

REFERENCES:

1. H. H. Landon and E. R. Rae, Phys. Rev. 107, 1333 (1957)

2. J. E. Draper, C. Fenstermacher, and H. L. Schultz, Phys. Rev. 111, 906 (1958); C. A. Fenstermacher, J. E. Draper, and C. K. Bockelman, Nuc. Phys. 10, 386 (1959)

3. O. A. Wasson, R. E. Chrien, G. G. Slaughter, and J. A. Harvey, Phys. Rev. (in press)

4. M. Beer, M. A. Lone, R. E. Chrien, O. A. Wasson, M. R. Bhat and H. R. Muether, Phys. Rev. Letters 20, 340 (1968); M. A. Lone, R. E. Chrien, O. A. Wasson, M. R. Bhat, and H. R. Muether, Phys. Rev. 174, 1512 (1968)

5. S. F. Mughabghab, R. E. Chrien, and O. A. Wasson, Phys. Rev. Letters 25, 1670 (1970)

6. S. F. Mughabghab, O. A. Wasson, G. W. Cole, R. E. Chrien and M. R. Bhat, Bull. Am. Phys. Soc. II 16, 496 (1971); also see contribution 4.5, to this Conference.

7. B. W. Thomas, contribution 4.10 to this conference.

8. J. R. Huizenga and R. Vandenbosch, Phys. Rev. $\underline{120}$, 1305 (1960)

9. W. P. Poenitz, Zeit. für Physik $\underline{197}$, 262 (1966)

10. R. K. Sheline, C. E. Watson, B. P. Maier, U. Gruber, R. H.
    Koch, O. B.Sultz, H. T. Motz, E. T. Jurney, G. L. Struble,
    T. V. Egidy, T. H. Elze, and E. Bieber, Phys. Rev. $\underline{143}$, 857
    (1966)

11. G. Rohr, H. Weigmann, and J. Winter, Nuc. Phys. $\underline{A150}$, 97 (1970)

12. H. Weigmann, G. Rohr, and J. Winter, "Neutron Capture Measure-
    ments and Resonance Parameters of Mo-Isotopes", submitted to
    Third Conference on Neutron Cross Sections and Technology,
    Knoxville, 1971.

13. S. F. Mughabghab, R. E. Chrien, O. A. Wasson, G. W. Cole, and
    M. R. Bhat, Phys. Rev. Letters $\underline{26}$, 1118 (1971)

14. J. E. Lynn, "The Theory of Neutron Resonance Reactions,"
    Clarendon Press, Oxford, England, 1968, p. 333

15. R. C. Block, R. G. Stieglitz, and R. W. Hockenbury, Proc. of
    Third Conference on Neutron Cross Sections and Technology,
    Knoxville, 1971.

16. H. E. Jackson and E. N. Strait, "E-1 and M-1 Radiative
    Strength in ^{53}Cr, ^{57}Fe, and ^{61}Ni from Threshold Photoneutron
    Cross Sections,' to be published.

17. C. Coceva, F. Corvi, P. Giacobbe, and M. Stefanon, Phys. Rev.
    Letters $\underline{25}$, 1047 (1970).

DISCUSSION

FALLIEROS   (Brown)
  Can you tell me again what you meant by the predicted values for the transitions in Mo^{92}?

CHRIEN
  Yes. One calculates, assuming that the Mo^{92} core is inert in ascribing the total radiation strength of the single particle jump of a neutron from a $p_{3/2}$ initial orbit to a final state which is $d_{3/2}$ or $s_{1/2}$; using the single particle radial wave functions as calculated for a Woods-Saxon potential and listed in Lynn's book, we can calculate the partial radiative widths. These agree fairly well with the measured values--simple, single particle transition strengths. Of course, what goes in there is also the product of the initial and final state reduced widths.

NAMENSON   (U.S. Naval Research Lab.)
  I would just like to comment that in an experiment which we did at the Naval Research Laboratory on slow neutron capture by Re^{187}, which was similar to that by Coceva, we also found evidence for a very narrow intermediate structure in the radiative widths of the γ decay.

SHELDON   (Lowell)
  I have two questions which don't have a direct bearing on the material you presented, but perhaps you can give me some guidance. They both concern the radiative capture mechanism. The first question is: the energy dependence of electric dipole radiation. It's been suggested that this transition probability should go as E^5 rather than E^3. I wonder if you have any comment on that.
  The second question concerns radiative captures in e.g., (p,γ) reaction on nuclides around A $\sim$88 or heavier, in which subsidiary resonances have been found about 3-5 MeV below the giant E1 resonance. The model proposed to account for these in only a semi-quantitative manner has been in terms of a spin-flip resonance. This is not altogether convincing. I wonder if you have any comment on that point.

CHRIEN
  As far as your second question, I think I will defer comment on that to Tony Lane who will probably cover that topic of these pygmy resonances in his talk. As far as your first question is concerned, Lowell Bollinger has found that in most instances, the E^5 dependence is closer to the truth than an E^3 dependence for the electric dipole transition. This is ascribed to the tail of the giant resonance which is supposed to be built on each excited state of the residual nucleus. We find, however, cases where the E^5 dependence does not seem to be accurate. We find some cases where E^3 is obeyed, for example. I'd say that the preponderance of the evidence favors E^5 at this point.

INVESTIGATION OF THE CAPTURE GAMMA-RAY SPECTRA FOR INDIVIDUAL

RESONANCES IN THE ^{169}Tm(n,γ) REACTION *

B. W. THOMAS

U.K.A.E.A., A.E.R.E., HARWELL, DIDCOT

BERKS., UNITED KINGDOM

INTRODUCTION

Recent studies of the high energy gamma ray spectra resulting from resonance neutron capture by ^{169}Tm have indicated the existence of non-statistical effects in the form of neutron width-radiation width correlations [1,2]. These experiments have investigated resonances up to 160 eV using both Ge(Li) and NaI detectors. In the light of current developments in this field, the results would be important if it were not for the limited range of experimental data and the possibility of false correlations. These could be the result of poor time of flight resolution [1] or incorrect spin assignments [1,2]. The latter is particularly important in the present case, since the resonance spins for s-wave capture are 0 or 1, with the result that neutron widths and partial radiation widths of spin 0 resonances are substantially enhanced. The most reliable resonance spin measurements to date are those of the Brookhaven group [3]. These assignments have been made by comparing the ratios of certain low energy capture gamma rays below 200 keV. The present experiment was carried out to check and extend the spin assignments and obtain a set of partial radiation widths up to 160 eV with improved time of flight resolution.

EXPERIMENTAL

The present data were obtained using a 4 cm^3 Ge(Li) detector at the 10 metre flight path of the Harwell 45 MeV Linac, with the Neutron Booster Target. The thulium sample used was 4 grams of metal in the form of two 25 millimetre diameter discs. Experimental data were collected by a two parameter system linked

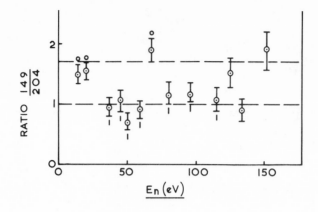

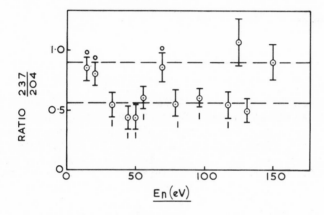

Figure 1   Ratios of low energy gamma rays (allocated J values from
           ref. 1)

to a PDP-4 computer which accepted both pulse height (gamma ray
energy) and time of flight (neutron energy) information for each
event. Gamma ray spectra for individual resonances have been
obtained in two runs, one for the low energy part below 1 MeV, the
other for high energy transitions above 1.5 MeV.

    The ratios of some low energy transitions below 250 keV were
used to assign the spins of resonances, and the results are in
complete agreement with those obtained from a similar experiment
[3]. Some ratios for gamma rays at 149 keV, 204 keV and 237 keV
are illustrated in figure 1. Additional spin assignments were
made for resonances at 125 eV (J=0), 136 eV (J=1) and 153 eV (J=0)
which differ from those obtained by cross section measurements
(BNL 325).

    The high energy spectra for 15 resonances (10 J=1,5 J=0) were

TABLE 1. Gamma-ray Intensities in ^{169}Tm(n,γ)^{170}Tm reaction (photons per 10³ captures)

E_n (eV)	3.9	14.4	17.5	34.8	38	44.8	50.7	59.2	65.8	83.4	94	115	125	136	153
	1⁺	0⁺	0⁺	1⁺	1⁺	1⁺	1⁺	1⁺	0⁺	1⁺	1⁺	1⁺	0⁺	1⁺	0⁺
$E_γ$ (keV)															
6594	1.45	16.40	0.71	0.00	1.72	1.30	0.89	1.20	2.11	-0.21	0.47	0.00	-0.14	3.00	1.65
6556	6.96	-0.14	0.06	0.03	0.56	0.95	-0.05	0.05	0.04	3.20	1.39	6.53	-0.14	0.86	0.27
6442	2.50	0.03	0.02	0.42	5.12	0.00	1.25	0.42	-0.10	1.15	1.42	2.42	0.27	0.71	0.82
6389	2.46	-0.19	0.00	2.34	1.54	2.24	0.89	1.59	0.20	1.10	0.00	1.40	0.39	-0.16	0.38
6375	0.97	-0.02	0.14	1.09	3.62	0.32	0.61	0.30	0.09	0.45	1.02	0.09	0.50	0.58	-0.11
6356	3.12	0.19	0.72	0.72	1.96	4.20	1.42	1.20	3.00	2.89	0.23	0.31	4.38	1.56	2.43
6003	1.63	10.05	0.30	-0.04	1.48	1.54	0.64	-0.03	-0.07	0.14	2.30	1.00	0.61	-0.13	3.36
5994	0.03	0.06	-0.02	0.46	-0.13	-0.17	0.42	0.08	1.42	-0.20	0.47	0.45	0.10	-0.13	-0.17
5945	5.45	0.32	0.71	0.50	3.03	0.17	1.30	1.15	0.50	1.75	1.08	3.41	3.40	0.91	4.75
5911	2.58	0.12	0.09	0.07	0.13	0.12	-0.16	0.26	0.10	0.07	0.83	0.31	0.29	0.76	-0.09
5900	0.23	0.35	0.19	2.92	1.70	0.40	0.16	0.31	0.33	0.74	2.64	0.12	-0.20	1.65	0.00
5858	1.28	0.18	0.08	2.78	2.33	4.32	0.00	0.19	0.30	0.27	2.14	0.72	0.46	-0.12	0.73
5810	-0.09	0.55	2.94	0.11	0.23	0.19	0.43	0.61	10.55	2.07	1.60	-0.05	1.78	0.24	8.25
5771	0.03	0.46	-0.01	0.25	1.08	0.11	0.11	0.18	0.03	1.08	0.41	-0.21	0.00	0.05	-0.46
5736 5730	4.10	-0.17	3.70	0.00	3.00	1.10	2.82	0.42	0.19	0.44	1.42	3.07	2.90	0.17	4.15
5684	0.04	-0.11	0.06	1.49	0.34	2.40	3.58	1.11	0.15	-0.12	1.26	0.55	0.10	3.25	0.19
5518	1.40	0.08	0.46	0.95	0.32	0.50	0.89	0.34	-0.11	0.34	1.23	0.27	0.33	0.77	0.34

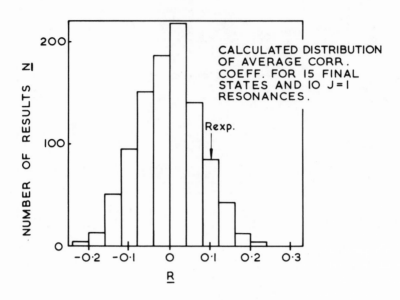

Figure 2   Distribution of average correlation coefficient from a
          Monte Carlo calculation

analysed for individual gamma ray intensities above 5 MeV.
Resonances were normalised to the summed spectrum above 1.5 MeV,
and absolute intensities obtained from a comparison of the 3.9 eV
resonance spectrum with published data [1].   The data, Table 1,
were in qualitative agreement with the spin assignments to the
extent that transitions from J=0 resonances to four known final
states of spin 0 or 2 were consistently weak.

Distribution of Gamma-ray Intensities

     The intensity distribution of transitions to 14 final states
from 10 J=1 resonances was investigated using a maximum likelihood
technique.   Gamma rays at 5994, 5771, 5736 and 5730 keV were
omitted from the analysis since they were either poorly resolved
or consistently weak (possible M1 transitions).   The stronger
transitions were assumed to be of E1 character and subsequently
corrected for an energy dependence on $E\gamma^3$.   The intensities were
found to follow a chi-squared distribution with 1.20±0.30 degrees
of freedom, which is consistent with the predicted Porter-Thomas
form.

Correlation of Gamma-ray Intensities with Reduced Neutron Widths

The gamma ray intensities of J=1 resonances were examined for a
possible correlation with reduced neutron widths using a relationship
of the form:-

$$
Corr(\Gamma_{n\lambda}^{0}, I_{\gamma\lambda f}) = \frac{\sum_{\lambda}(\Gamma_{n\lambda}^{0} - \overline{\Gamma_{n}^{0}})(I_{\gamma\lambda f} - \overline{I_{\gamma f}})}{\left[\sum_{\lambda}(\Gamma_{n\lambda}^{0} - \overline{\Gamma_{n}^{0}})^{2} \sum_{\lambda}(I_{\gamma\lambda f} - \overline{I_{\gamma f}})^{2}\right]}
$$

where $\overline{\Gamma_{n}^{0}}$ and $\overline{I_{\gamma f}}$ are averaged over resonances. The
transitions included in the analysis were identical to those of
previous authors [1] (i.e. transitions to final states observed
in d,p data). The average correlation coefficient for 15 final
states is 0.1, which contradicts previous evidence for a
correlation. This result is compared with a Monte Carlo
calculation, figure 2, which indicates that the probability of
obtaining a value greater than this is at least 14°/o.

* The paper was read by B. Patrick of A.E.R.E., Harwell.

REFERENCES
[1] M. Beer et al., Phys. Rev. Letters 20, 340 (1968).
[2] L. M. Bollinger, "Nuclear Structure, Dubna Symposium 1968"
     (IAEA Vienna, 1968) p. 317.
[3] M. R. Bhat et al., Brookhaven National Laboratory Report
     BNL 14729 (1970).

DISCUSSION

CHRIEN
    I just want to reemphasize the utility of using low-energy
gamma rays for spin assignments. I forgot to mention in my talk
that we have assigned approximately 70 resonance spins in thulium,
and I might say it's done with ease. It's very easy to do it.
It used to be, when I started out in the neutron business, that
one couldn't measure the spin and it was easy to get a neutron
width. Now, I think it's easier to measure the spin in many cases
than it is to obtain a neutron width. I feel we know the spins
in thulium better than we know some of the neutron widths, and
the lack of knowledge of neutron widths is a severe handicap in
our analysis. I think that maybe some of the others who have
worked in this--maybe Claudio Coceva could back me up. It's a
fine way to find a doublet, because when you find a point that's
between your two lines which represent low energy γ ray rates
you can be pretty sure that there's some trouble in resolving the
resonances.

BEER (M.A.G.I.)
    I would like to point out that the original BNL paper did not
include the 153 eV resonance. The large correlation coefficient
between partial radiation widths and reduced neutron widths was
primarily due to the 3 eV resonance; but in the more recent work,
which includes more resonances, the correlation coefficient goes
down. However, these resonances that have been included have low
reduced neutron   widths. What one would really need, then, to
resolve the question would be resonances with large reduced neutron
widths. Whether that can ever be done is a question.

LONE (Chalk River)
    One comment on the spin assignments on the basis of the low-
energy γ-rays. The method seems to work for rare earth nuclei,
but in Hg^{198} this method is not reliable. Two resonances in
Hg^{198} (23 eV and 90 eV) have 1/2+ spin and parity, but the population
of the first, second, and third excited states in Hg^{199} is quite
different for these resonances. This work has been reported at
the Houston 1970 APS meeting.

DECAY PROPERTIES OF NUCLEAR LEVELS POPULATED BY THE (γ,γ') REACTION

R. Moreh and A. Wolf

Nuclear Research Centre, Negev, Beer Sheva, Israel

The existence of large positive correlations between partial radiation widths from (n,γ) resonance reactions and the (d,p) spectroscopic factors are known to occur in several nuclides [1,2]. These positive correlations are explained by assuming that the resonance level contains two particle, one hole, doorway state components. Moreover, the partial radiation widths of levels formed by resonance neutrons in (n,γ) reactions were found to obey a Porter-Thomas distribution.

In the present work, some evidence is presented for the existence of similar phenomena in resonance levels populated by the (γ,γ') reaction. The technique of using the (γ,γ') reaction for studying the energy levels of nuclei is well known [3]. Experimentally the incident γ beam was obtained by thermal neutron capture on some element placed near a reactor core. Details of the experimental technique for studying the decay properties of nuclear levels by the present method may be found elsewhere [3]. The idea in this work [4] is that a resonance scattering of γ rays is obtained only when a random overlap exists between at least one of the incident γ lines and one level in the nuclide to be studied. This places a severe limitation on the choice of any particular target; nevertheless, it was possible to study several nuclides using this technique. Because of the similarity between the (γ,γ') reaction and the (n,γ) reaction using resonance neutrons, it was only natural to try and find whether the (γ,γ') reaction exhibits similar phenomena to that of (n,γ). In fact, the total radiative widths of bound resonance levels populated by the (γ,γ') technique were found to be of about the same magnitude [3] as those obtained by (n,γ). In addition the anomalous photon intensities [5] in the mass region A=180 to 208 were also found to occur in the (γ,γ') reaction [6] on Tl203 and Tl205. We hereby deal with other phenomena observed using the (γ,γ') reaction.

1. CORRELATIONS BETWEEN (d,p) AND (γ,γ') REACTION WIDTHS

There are several experimental difficulties in making a comparison between (d,p) and (γ,γ') reaction widths. First, the number of nuclides for which such a comparison may be applied is very small, the reactions (γ,γ') and (d,p) involve two different targets A and A-1 both of which are not always stable. The (n,γ)-(d,p) comparison however, is always possible because the same target is used in both experiments. Second, the scattered signal in a (γ,γ') process should be strong enough to render the study of the scattered photon intensities practically feasible. In spite of the above difficulties, four cases were found (Table 1) where such a comparison was possible.

Table 1

Final Nucleus	Resonance energy (MeV)	No. of final states	$\zeta\left[I_{\gamma f}, I_{dpf}\right]$	Percentile of $\zeta = 0$
Ni62	7.646	16	0.49	97
Cd112	7.632	12	0.54	96
Sn120	7.696	10	0.59	96
Sm150	8.998	–	strong	–

The experimental γ intensities for Ni62 and Cd112 were obtained in the present work by using iron capture γ rays [3] while the Sn120 and Sm150 γ intensities were obtained using nickel capture γ rays [7]. The (γ,γ') reduced widths $I_{\gamma f} = \Gamma_{\gamma f}/E^3$ from Ni62, Cd112, Sn120 and Sm150 were compared to the corresponding proton strengths $I_{dpf} = (2J_f+1)S_{\ell j}$ in the (d,p) reaction on Ni61, Cd111, Sn119 and Sm149. This is shown in Fig.1 and the results of the correlation coefficients and other data are shown in Table 1; it is very significant that a positive ζ was obtained although the probability of it being consistent with zero is not very small. It should be noted that the ground states of the four nuclides are 0+; the spins and parities of the resonance levels were determined by angular distribution and polarization measurements of the elastically scattered radiation [3] and were found to be 1-. Only E1 transitions to known 0+ and 2+ low-lying states were included in the calculation of ζ . In Ni62, the γ transitions were compared with $\ell =1$ protons [8] while in Cd112 and Sn120, $\ell =0$ and $\ell =2$ protons [9,10] were considered. In Sm150, it was not possible to obtain a quantitative measure of ζ because the spectroscopic factors and the spins were not reported for all final levels in the Sm149(d,p) reaction [11]. However, a striking correspondence between γ and proton intensities in Sm150 was observed; the collective group of levels below 1.6 MeV were weakly populated by both reactions, while the higher group of levels possessing single-particle properties were strongly excited by both reactions.

A simple minded picture suggests itself for explaining this correlation. The resonance level in Ni62 may be viewed as a particle-hole

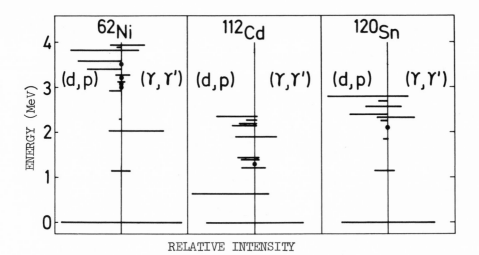

Fig. 1. A comparison between (γ,γ') and (d,p) strengths for Ni62, Cd112
        Sn120. The dots along the vertical lines indicate levels of the
        same spin and parity for which no transitions were observed.

El excitation of neutrons such that one of the paired $2p_{3/2}^2$ neutrons in
the ground state of Ni62 is excited to one of the orbitals $2d_{5/2}$, $2d_{3/2}$
or $3S_{1/2}$. The new system, which effectively consists of a Ni61 core +
$2d_{5/2}$ or $2d_{3/2}$ or $3S_{1/2}$ neutron will selectively prefer single-particle
neutron transitions of the form $2d_{5/2} \rightarrow 2p_{3/2}$, $2d_{3/2} \rightarrow 2p_{3/2}$,
$2d_{3/2} \rightarrow 2p_{1/2}$, $3S_{1/2} \rightarrow 2p_{1/2}$, $3S_{1/2} \rightarrow 2p_{3/2}$. The corresponding energies
are between 4 to 10 MeV and are comparable to the experimental transition
energies in the Ni62(γ,γ') spectrum. The intensity of such transitions
is dependent on the spectroscopic factor of the final states and hence
should be correlated to the proton intensities in the Ni61(d,p) reaction.
In Cd112 and Sn120, a similar situation holds; the last two neutrons are
paired to zero and assumed to be in a $S_{1/2}$ orbital. The last unpaired
neutron in Cd111 and Sn119 is also viewed to be in a $S_{1/2}$ orbit.
Therefore, the photoexcitation process is of the form $3S_{1/2} \rightarrow 3p_{3/2}$ or
$3S_{1/2} \rightarrow 3p_{1/2}$ while the de-excitation is of the form $3p_{3/2} \rightarrow 3S_{1/2}$,
$3p_{3/2} \rightarrow 2d_{3/2}$, $3p_{3/2} \rightarrow 2d_{5/2}$, $3p_{1/2} \rightarrow 2d_{3/2}$, $3p_{1/2} \rightarrow 3S_{1/2}$. Here again
the transition energies are of the order of 6 MeV. It should be noted
that more complicated configurations may contribute to the above
process. In Ni62 for example, excitations of the form $1f_{5/2} \rightarrow 2d_{3/2}$, or
$1f_{5/2} \rightarrow 2d_{5/2}$ are possible if the ground states of Ni62 and Ni61 are
assumed to contain the configurations $2p_{3/2}^2(2)1f_{5/2}^2(2),J=0$ and
$2p_{3/2}^2(2)1f_{5/2},J=3/2$ respectively.

2.   DISTRIBUTION OF PARTIAL RADIATION WIDTHS FROM BOUND LEVELS

The statistical distribution of partial radiation widths of unbound

Table 2

Final nucleus	Resonance Energy (MeV)	Number of final states	γ
Ni62	7.646	16	$0.4 \begin{smallmatrix} +0.6 \\ -0.2 \end{smallmatrix}$
Zn66	7.368	10	$1.1 \begin{smallmatrix} +0.9 \\ -0.3 \end{smallmatrix}$
As75	7.646	15	$1.0 \begin{smallmatrix} +0.5 \\ -0.5 \end{smallmatrix}$
Cd112	7.632	18	$1.0 \begin{smallmatrix} +0.5 \\ -0.4 \end{smallmatrix}$

levels populated by the (n,γ) reaction using resonance neutrons were extensively studied and found, for many nuclides, to obey a Porter-Thomas distribution [1,12]. In the present work, the statistical distribution of partial radiation widths from bound levels obtained from (γ,γ') reactions was studied and was found to be also consistent with a Porter-Thomas distribution. The isotopes studied are listed in Table 2. The experimental γ intensities for As75 were obtained using iron capture γ rays [3,13] in the same manner as Ni62 and Cd112; the Zn66 γ intensities were obtained using lead capture γ rays [14].

An attempt to determine the distribution of partial radiation widths for bound levels was done previously using a different approach. In this earlier work [7], the ratios of partial radiation widths, in 14 different nuclides, in the range $62 \leq A \leq 142$, were considered and the calculation was carried out under the rather extreme assumption that the decay properties of the resonance levels in all these nuclides are similar; the results were consistent with a Porter-Thomas distribution.

The present problem differs from other works in that the transitions considered originate from a single bound resonance to several final states. In (n,γ) work however, one generally considers transitions from several unbound resonance states to various final states. By using arguments similar to those applied in the (n,γ) work and assuming that the low-lying levels are uncorrelated, the distribution of El or Ml partial radiative widths from bound levels at high excitation is also expected to follow a Porter-Thomas distribution. However, these arguments should be applied with care because the excitation energies of the bound resonance levels, studied in a (γ,γ') reaction, are generally lower than those studied in the (n,γ) work. In fact, at much lower excitation energies, strong nuclear structure effects are expected to prevail and hence deviations from the Porter-Thomas distribution are expected to show up. In the calculation, only transitions to low-lying levels of known parity and spin and hence of the same multipolarity were included. The ground states of the nuclei Ni62, Zn66 and Cd112 are 0+ and the resonance levels in Ni62 and Cd112 are known to be 1-; only El transitions to known 0+ and 2+ low-lying levels in these two

nuclei were incorporated in the analysis. The resonance level in Zn66
is 1 (the parity of this level is unknown) and dipole transitions of the
same character to known 0+ and 2+ low-lying levels in Zn66 were considered.
The ground state of As75 is 3/2-, the resonance level is 1/2 (unknown
parity) and only dipole transitions to known 1/2- or 3/2- low-lying
levels were considered for the present analysis.

The statistical distribution of partial radiation widths in these
nuclei was investigated using a maximum likelihood method [12], that
gives the number of degree of freedom ν in a χ^2 distribution. This
method was described fully by Price et al. [15]; the same computer
program was employed in the present calculation. A Monte Carlo procedure
was applied to generate partial widths from a χ^2 distribution with
various values of the parameter ν , allowing it to assume non-integral
values; normally distributed experimental errors were added to these
widths. The results are listed in Table 2 and show that for the four
nuclides considered the ν values are consistent with a Porter-Thomas
distribution. It is interesting to note that two of the nuclides studied
here namely, Ni62 and Cd112 exhibit both phenomena i.e., a Porter-Thomas
distribution of partial radiation widths and a positive correlation
between (γ,γ') and (d,p) reaction widths. A similar behaviour was also
found for levels populated by resonance neutrons in the (n,γ) reaction[1].

REFERENCES

1. R.E. Chrien, K. Rimawi and J.B. Garg, Phys. Rev. C3 (1971) 2054.
2. S.F. Mughabghab, et al., Phys. Rev. Lett. 25 (1970) 1670.
3. R. Moreh, A. Wolf and S. Shlomo, Phys. Rev. C2 (1970) 1144.
4. B. Arad et al., Phys. Rev. 133 (1964) B684.
5. B.B. Kinsey and G.A. Bartholomew, Phys. Rev. 93 (1954) 1260;
   G.A. Bartholomew et al., Phys. Lett. 24B (1967) 47.
6. R. Moreh and A. Wolf, Neutron Capture γ-Ray Spectroscopy, Vienna,
   1969, p. 483.
7. Y. Schlesinger et al., Phys. Rev. C2 (1970) 2001.
8. R.H. Fulmer and A.L. McCarthy, Phys. Rev. 131 (1963) 2133;
   U. Fanger et al., Nucl. Phys. A146 (1970) 549.
9. P.D. Barnes, J.R.Comfort and C.K. Bockelman, Phys. Rev. 155 (1967)
   1319; J.A. Macdonald and H.D. Sharma, Nuc. Phys. A156 (1970) 321.
10. E.J. Schneid, A. Prakash and B.L. Cohen, Phys. Rev. 156 (1967) 1316;
    L.R. Noris and C.F. Moore, Phys. Rev. 136 (1964) B40.
11. R.A. Kenefick and R.K. Sheline, Phys. Rev. 133 (1964) B25.
12. L.M. Bollinger et al., Phys. Rev. 132 (1963) 1640.
13. R. Moreh and O.Shahal, Phys. Rev. 188 (1969) 1765.
14. N. Shikazono and Y. Kawaraski, Nucl. Phys. A118 (1968) 114.
15. D.L. Price et al., Nucl. Phys. A121 (1968) 630.

DISCUSSION

BOLLINGER (A.N.L.)

I want to start with a comment and end with a question. The nature of your experiment is such that you tend to select initial states that have unusually large ground state radiation widths. This being the case, it would seem to me that in order to make an unbiased analysis of the question of whether or not you have correlations, you should throw out the ground state. That being the case, I wonder if you were to do that whether your correlation coefficients that you obtain would not be substantially lower and therefore the evidence for correlation be less interesting.

MOREH

First, the (γ, γ') reaction is a biased process and our claim is that the reaction widths obtained from this biased process are correlated to the (d,p) reaction widths in the four cases studied. Second, if the ground state radiation width is omitted, the factor ρ is certainly reduced; I can't remember the exact numerical values of ρ. But now, it might be argued that one shouldn't omit the ground state transition only and may omit other transitions as well. When this was done, the value of ρ was found to increase again.

BOLLINGER

Well, that being the case, it seems to me you introduce a bias in the analysis which is not acceptable, because, had the ground state transition been weaker, you might well not have seen the resonance at all.

MOREH

One example is the Sm^{150} where the ground state transition is relatively weak (around 10%). I must emphasize that we are not claiming that such a (γ, γ')-(d,p) correlation should be positive in each case. It might well turn out that in many cases, these reaction widths are not correlated.

DAVID (Soreq)

I just want to comment on Dr. Bollinger's comment. In actual fact, this preferential choice of the γ_0 channel to excite these resonance states gives an average ground state transition something like three times the corresponding width to low-lying 2+ levels. It's a factor of three. This increased width is due to the pre-ferential excitation of these resonance levels.

THE 5.5 MeV ANOMALOUS GAMMA RADIATION IN ^{205}Tl(n,γ)^{206}Tl

E. D. Earle, M. A. Lone, G. A. Bartholomew
Atomic Energy of Canada Ltd., Chalk River

B. J. Allen
Australian Atomic Energy Commission and
Oak Ridge National Laboratory

G. G. Slaughter and J. A. Harvey
Oak Ridge National Laboratory

Introduction

An enhancement at 5.5 MeV in the γ-ray spectra following thermal and fast neutron capture in the mass region 180<A<208 represents a significant departure from the statistical model and has been interpreted in terms of doorway states[1,2]. This anomaly approaches a maximum at Tl and we have studied its magnitude and variation with neutron energy in the ^{205}Tl(n,γ)^{206}Tl reaction. The low lying level structure of ^{206}Tl is well known (fig. 1). These states have been observed in the (d,α), (t,α) and (d,p) reactions[3-5]. The level scheme calculated by Kuo and Herling[6] is generally in good agreement with experiment and consequently we have a knowledge of the various proton-hole, neutron-hole amplitudes of many of these states. Such data are invaluable for analysis of the capture γ-ray spectra.

The data were obtained in various experiments at the Fast Chopper Facility at the NRU reactor at CRNL and at the Oak Ridge Electron Linear Accelerator. Three measurements have been made at the latter facility. 1)High resolution (6 keV at 7.6 MeV) capture γ-ray spectra were obtained for the strong resonances up to 30 keV with a 36 cc Ge(Li) detector at a flight path of 10m. 2)Capture cross section measurements were made with Total Energy Detectors (T.E.D.)[7] at 40 meters with ∿0.2% resolution. This measurement yielded resonance energies up to 220 keV and the neutron and radiative widths of the stronger resonances. The ratio of capture yield above and below

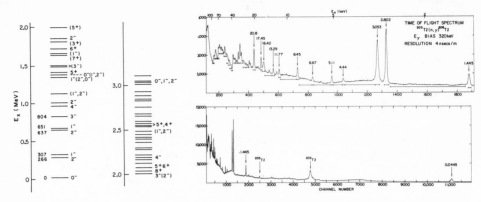

Fig. 1 Level scheme of   Fig. 2 Time of flight spectrum at the
^{206}Tl   10m station

4 MeV γ-ray energy which was found to be an indicator of the presence of the 5.5 MeV anomaly was also measured. 3)A transmission
measurement was made at a flight path of 80 meters. This measurement permitted definite s-wave assignments to be made to those resonances exhibiting a strong interference with the potential scattering.

Resonance Capture Gamma Rays

Capture γ-ray spectra were obtained by setting windows on resonances in the time of flight spectrum (fig. 2). The top curve
shows an expanded plot of the high energy region. The γ-ray spectra with the best statistics were those for the 45 eV, 1.44, 2.80
and 3.05 keV resonances. From transmission measurements Thomas et
al[8] concluded that the 45 eV resonance is probably p-wave; the
present transmission data show that the 1.44 keV resonance is also
probably p-wave. The two resonances at 2.80 and 3.05 keV are definitely s-wave as they both exhibit strong interference shapes in
the transmission measurement. These conclusions are reinforced by
the observation of transitions to 3⁻ final states from the 45 eV
and 1.44 keV resonances. Transitions to 3⁻ states from s-wave resonances would be M2(E3) while from the p-wave resonances they would
be M1(E2).

We next consider a characteristic difference in the γ-ray spectra (fig. 3) of these s- and p-wave resonances. The second escape
peaks are identified by arrows. Others are first escape or full
energy peaks. Primary transitions are recognized from the change
in their energy with the incident neutron energy or as transitions
to known levels (fig. 1), and are shown with the arrow under their
respective peaks. Peaks identified with arrows above the curve may
be due to primary or secondary transitions. From the known final
state spins we conclude that all the transitions above 4.5 MeV observed from the s-wave resonances are E1, whereas most of the transitions above 4.5 MeV from the p-wave resonances are M1(E2). The
strong peaks below about 1.5 MeV are due to secondary transitions.

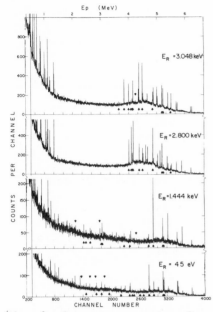

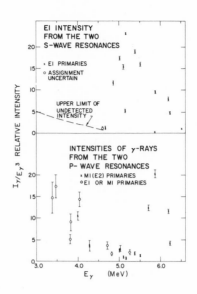

Fig. 3 Gamma ray spectra from four ^{205}Tl resonances. Arrows represent second escape peaks.

Fig. 4 Gamma ray reduced intensities from the data in fig. 3.

The reduced intensities of the γ-rays from these resonances, normalized to the total capture in each resonance, are plotted in fig. 4. The upper limit of undetected intensity is taken to be equal to twice the standard deviation of the data points at the position of the γ-ray peak. The 5.5 MeV anomaly is clearly seen in the s-wave resonances where strong transitions are observed only between 5 and 6 MeV. While the p-wave resonances have several intense high energy transitions there are no strong transitions in the 5 to 5.5 MeV region and many transitions are observed below 5 MeV. Thus the anomalous radiation results from E1 transitions. In the case of the 3 keV doublet, only one E1 transition is unobserved above 4.5 MeV, (i.e. at 5.853 MeV). This unobserved primary would be to an excited proton-hole state[6] at 0.652 MeV. Below 4.5 MeV no E1 transitions are observed although there are final states available with the appropriate spin and parity (fig. 1). The calculation of Kuo and Herling[6] predict eight states with spin 0^-, 1^- or 2^- between 3 and 4.5 MeV excitation energy. No transitions to these states have been observed from the 3 keV doublet. We estimate, by a Monte Carlo calculation, a 0.03% chance of this being due to Porter-Thomas fluctuations.

Resonances at higher neutron energies also exhibit the 5.5 MeV anomaly (fig. 5). Some of these resonances are known to be s-wave. P-wave resonances have not been definitely identified above E_n = 5 keV. The 5.853 MeV transition to the excited proton hole state is strong in many of these spectra including those from known s-wave resonances.

Capture Yield Ratios

By setting a bias at 4 MeV, capture γ-ray yields above and below this energy can be obtained from high resolution measurements
with the total energy detectors at 40 meters. The observed ratio (R)
for the p-wave resonance at 1.44 keV was 0.87 ± 0.07; for the s-wave
resonances at 2.80 and 3.05 keV, the ratios are 1.31 ± 0.04 and 1.32
± 0.04 (fig. 6). From this result we conclude that R is a sensitive
indicator of the presence of the anomaly in a resonance γ-ray spectrum. The T.E.D. measurement can then chart the presence and variation with neutron energy of the anomaly for many more resonances.
The errors in R increase with neutron energy because of poor statistics, and possibly for this reason a number of intermediate ratios
are observed. Nevertheless all the definite s-wave resonances as
assigned from the transmission measurement show large values for the
ratio up to a neutron energy of 220 keV. Thus the 5.5 MeV anomaly
definitely persists in strong s-wave resonances over this energy
range. High ratios are also observed for seemingly narrow and symmetric resonances and for a number of resonances not observed in
transmission. These may be p-wave or narrow s-wave resonances. At
present, in the energy range 0.045 to 220 keV, we can identify 19 s-
wave resonances as having the 5.5 MeV anomaly and 3 p-wave resonances
as not having it.

Resonance Parameters

Ninety resonances were observed in the capture cross section
measurement with the T.E.D. in the range 1.2 to 220 keV. The resonance parameters obtained from the analysis of these data are shown
in fig. 7. Most resonances with Γ_n greater than the resolution func-

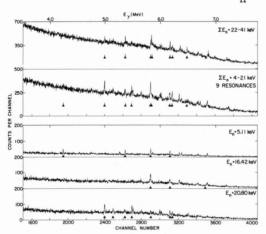

Fig. 5 Gamma ray spectra from
high energy resonances.

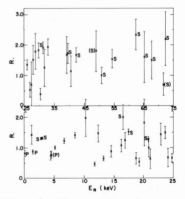

Fig. 6 The ratio of capture
yield above and below 4 MeV
gamma ray energy as measured
with the total energy detectors for resonances up to
75 keV.

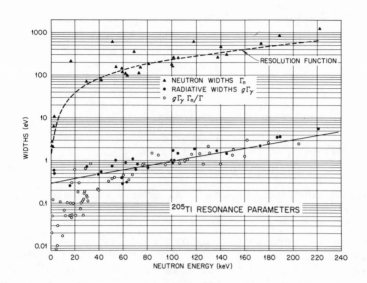

Fig. 7 Resonance parameters derived from the T.E.D. measurements.

tion are identified as s-wave from the transmission measurement, and it seems reasonable to assume that many of the others are also s-wave.

Sufficient data are available to test for correlations between Γ_n^o, Γ_γ and R. However, the uncertainty in the resonance spin, J, makes interpretation difficult. Instead of Γ_γ we use $g\Gamma_\gamma$ where g = 1/4 and 3/4 for s-wave resonances with J = 0 and 1 respectively. No correlation is found between R and $g\Gamma_\gamma$ or between R and Γ_n^o for 19 s-wave resonances but a correlation of 0.64 is found between Γ_n^o and $g\Gamma_\gamma$ for these resonances. However, this sample is heavily weighted with resonances with large neutron widths. By a Monte Carlo calculation we find that there is a 1% chance that this value is consistent with Porter-Thomas fluctuations.

Another observation is the energy dependence of $g\Gamma_\gamma$ in the neutron energy range 20 keV to 220 keV. Empirically this energy dependence can be expressed as

$$g\Gamma_\gamma (ev) = \exp(-0.7 + 0.01\ E(keV))$$

with a 50% error on the numerical coefficients. Such a strong energy dependence is difficult to explain on the basis of present statistical theory.

Reaction Mechanism

An analysis of resonance data has indicated the existence of a 5.5 MeV anomaly in s-wave resonances, and an energy dependence in $g\Gamma_\gamma$ for s- and possibly p-wave resonances to E_n = 220 keV. For an understanding of the reaction mechanism we now consider the information available on the configurations of the low lying states of ^{206}Tl.

^{205}Tl ground state w.r.t. ^{208}Pb

$$^{205}\left|\psi_{gs}\right) = \alpha_1\left|s_{\frac{1}{2}}^{-1}; \ p_{\frac{1}{2}}^{-2}\right) + \alpha_2\left|s_{\frac{1}{2}}^{-1}; \ p_{\frac{1}{2}}^{-1} f_{5/2}^{-1}\right) + \alpha_3\left|s_{\frac{1}{2}}^{-1}; p_{\frac{1}{2}}^{-1} p_{3/2}^{-1}\right)$$

$$+ \ \beta_1\left|d_{3/2}^{-1}; \ p_{\frac{1}{2}}^{-2}\right) + \beta_2 \ \text{-----}$$

$^{2.80}_{3.05}$ resonances are $^{205}\left|\psi_{gs}\right)$ + s neutron

$$\Big\downarrow \text{E 1}$$

$$^{206}\left|\psi\right) = \alpha_1'\left|s_{\frac{1}{2}}^{-1}; p_{\frac{1}{2}}^{-1}\right) + \alpha_2'\left|s_{\frac{1}{2}}^{-1}; f_{5/2}^{-1}\right) + \alpha_3'\left|s_{\frac{1}{2}}^{-1}; p_{3/2}^{-1}\right)$$

$$+ \ \beta_1'\left|d_{3/2}^{-1}; p_{\frac{1}{2}}^{-1}\right) + \beta_2' \ \text{-----}$$

Fig. 8   Dominant configurations in ^{205}Tl ground state and ^{206}Tl low lying states.

These states can be represented in terms of 1 proton-hole and 1 neutron-hole with respect to ^{208}Pb (fig. 8). Recent experimental and
theoretical work[3-6] has led to estimates of the α' and β' components. Similarly the low lying states of the ^{205}Tl target can be represented in terms of 1 proton hole and 2 neutron holes. The α terms
contain a $s_{1/2}$ proton-hole, the β terms a $d_{3/2}$ proton-hole. For the
ground state the α_1 component is expected[9] to be about 0.9.

The simplest neutron interaction is that of valency capture[10].
In this model the $s_{1/2}$ neutron jumps directly into a $p_{1/2}$ hole without disturbing the other nucleons. As the ^{205}Tl ground state is
mainly α_1, then only low lying states in ^{206}Tl with large α_1'
should be populated strongly and these would be 0^- and 1^-. However
we find transitions to 2^- states and to 1^- states with small α_1'
components, and hence we conclude that the valence capture model is
inadequate.

The next step is to assume the excitation of neutron doorway
states formed when the incoming neutron interacts with the target,
creating a hole in the $3p_{3/2}$ or $2f_{5/2}$ shell and filling a $p_{1/2}$ hole,
effectively increasing the α_2 and α_3 components. Then E1 transitions to any final state with a large α' component can occur. This
picture is consistent with the spectra from the 3 keV doublet where
the correlation between $\Sigma \left|\alpha'\right|^2$ and the partial γ-ray width is 0.3
and where no transition is observed to the state with a large β'
(fig. 4).

The correlation coefficient between reduced intensities to 14
final states from the 2.80 and 3.04 keV resonances is 0.15. This
small correlation indicates that more than one doorway state is
responsible for the 5.5 MeV anomaly and that the components of the
participating doorway states are different in these two resonances.
The γ-ray spectra from other s-wave resonances which contain the
5.5 MeV anomaly show a strong transition to the excited proton-hole

state at 0.652 (fig. 5) which has a large β'. This shows that, in addition to the neutron hole doorway states already discussed, proton-hole doorway states also contribute to the 5.5 MeV anomaly.

Conclusions

Evidence for an enhancement of $\sim$5.5 MeV gamma rays is found for strong s-wave resonances over an energy region of 2 to 220 keV. These results suggest the presence of a neutron capture mechanism not in accord with the statistical theory. As a result of a detailed knowledge of the configuration of low lying levels in ^{206}Tl, it appears that the effect cannot be explained on the basis of direct or valence capture but is consistent with a capture mechanism involving doorway states built on the ground and excited states of ^{205}Tl. The energy dependence of the radiation widths is not understood but may also be a consequence of this capture mechanism. Our measurements do not permit definite conclusions concerning the p-wave resonances.

We wish to thank Drs. R.L. Macklin and F.C. Khanna for many helpful discussions.

References

1) G.A. Bartholomew, Proc. Int. Conf. Neutron Capture γ-ray Spectroscopy, Studsvik 1969, I.A.E.A. (1969) 553 and references therein.

2) A. M. Lane, Annals of Physics 63 (1971) 171.

3) M.B. Lewis and W.W. Daehnick, Phys. Rev. c 1 (1970) 1577.

4) P.D. Barnes, E.R. Flynn, G.J. Igo and D.D. Armstrong, Phys. Rev. C 1 (1970) 228.

5) J.R. Erskine, Phys. Rev. 138 (1965) B851.

6) T.T.S. Kuo and G.H. Herling, NRL Memorandum Report 2258 (1971).

7) R.L. Macklin and B.J. Allen, Nucl. Instr. and Meth. 91 (1971) 565.

8) G.E. Thomas, L.M. Bollinger and R.E. Coté, private communication.

9) F.C. Khanna, private communication.

10) J.E. Lynn, Theory of Neutron Resonance Reactions (Clarendon Press, Oxford 1968) p. 326.

DISCUSSION

NEWSON   (Duke)

You said that you thought that there might be more than one doorway state there. Now, the analysis of the doorway state in the neighborhood of Pb^{208} is given by Divadeenam in a contributed paper to this conference. He believes that this is a single particle neutron coupled to a vibrating Pb^{208} core, or a core similar to that. He thinks that the total strength of the two-particle one-hole states is less than the three-particle two-hole states. If that analysis is correct, you've got lots of holes and particles to deal with.

CORRELATIONS BETWEEN PARTIAL WIDTHS OF DIFFERENT CHANNELS

Anthony M. Lane

Atomic Energy Research Establishment

Harwell, U.K.

Correlations of partial widths constitute an intriguing part of the more general subject of non-statistical effects in the isolated resonance regions of nuclear reactions. This subject is one of the growth points of present-day nuclear physics. It may well be the counterpart for the 1970's of analogue states for the 1960's. There are many challenges to the experimenter and theorist, for instance: Are the astonishingly high ($\geq$ 90%) correlations observed between (d,p) and thermal (n,γ) reactions for masses 40-60 characteristic of direct or compound nucleus mechanisms for the (n,γ) process, or of both? Is "anomalous capture" evidence for a pygmy resonance in the E1 absorption pattern of low-lying nuclear states? If so, what is the physical nature of this resonance? Despite much progress, and an impressive (and bewildering) accumulation of data, such basic questions as these do not yet have definite answers.

First the theoretical background will be sketched, then a review will be given of the data, and finally the current state of theoretical understanding will be discussed.

1. CORRELATIONS IN THE CONTEXT OF OTHER NON-STATISTICAL EFFECTS

The "statistical model" to nuclear reactions comes from a specialisation of the "compound nucleus model" of Bohr (1936). This model regarded a reaction as proceeding in two distinct stages through the formation and decay of intermediate "compound" states of all particles. The existence of such states at low bombarding energies in many reactions was confirmed by observation of sharp resonances, some of them being very striking such as the Gd(n,γ)

271

reaction where the peak-to-valley ratio at a resonance is 10^4 or more. In a region where compound states λ are separated in energy, i.e. their widths Γ_λ are less than their spacing, Kapur and Peierls (1938) showed that the scattering matrix has the form of a sum of resonance amplitudes:

$$S_{cc'} = A_{cc'} \sum_\lambda \frac{\gamma_{\lambda c}\gamma_{\lambda c'}}{E_\lambda - E - \frac{i}{2}\Gamma_\lambda}$$

c and c' label the ingoing and outgoing channels, and E_λ are the resonance energies (eigenvalues). $\gamma_{\lambda c}$ is the reduced width amplitude of resonance λ for channel c and can be expressed as an overlap $\langle \lambda | c \rangle$ of λ on a "channel state" $|c\rangle$ containing the product of wave-functions of the channel fragments and that of a suitable relative motion. $A_{cc'}$ takes account of uninteresting factors such as penetrabilities $P_c P_{c'}$.

This formula is the quantitative embodiment of the Bohr model. The specialisation of it for the statistical model, as developed by Bethe, Weisskopf and Ewing, and Hauser and Feshbach (1938-1952), is obtained by assuming that (1) $\gamma_{\lambda c}\gamma_{\lambda c'}$ have random signs, so that $\sum_{\lambda \text{ in } \Delta E} \gamma_{\lambda c}\gamma_{\lambda c'} \approx 0$ for levels in any given energy range ΔE containing many levels. (2) that $\frac{1}{\Delta E} \sum_{\lambda \text{ in } \Delta E} \gamma_{\lambda c}^2$, the "strength function", is independent of energy (and of c to a large extent). These extra assumptions mean that, for evaluation of $S_{cc'}$ or S_{cc} in a given energy range, all terms λ with E_λ outside the range can be dropped; in other words, there is negligible background from distant levels.

Before 1950, this extreme view seemed very reasonable. It was already known from the (n,γ) reactions that the values of the overlaps $\langle \lambda | c \rangle^2$ for neutron channels in certain rare earth nuclei were as small as 10^{-7} indicating that the neutron entrance channel state was only a fraction 10^{-7} of typical compound states, i.e. these states are exceedingly complicated, containing 10^7 or more components. Correspondingly a channel state $|c\rangle$ is "dissolved" or shared out amongst a vast number of compound states. At first sight, this makes assumptions (1) and (2) appear very reasonable perhaps inevitable. In fact, however, while the statistical model constitutes an acceptable expression of the high complexity of compound states, it is not implied by this complexity. A large part of the development of nuclear reaction theory since 1950 has grown from this realization, which has been forced on us by the experimental denial of all the major implications of the statistical model. These are (i) absorption cross-sections and transmission factors independent of energy (apart from penetration effects). This follows directly from (2). (ii) negligible "direct" cross-sections, where these are identified with any distant level background in $S_{cc'}$.   This follows directly from (1).   (iii) negligible correlations[+] between partial width amplitudes of different channels. This also follows from (1).

(i) was denied by the Barschall giant resonances in total neutron cross-sections in 1952 while (ii) was denied by the discovery of strong "direct" parts in many cross-sections since 1952: (p,p'), (d,p), (p,γ), etc.

In fact, (ii) and (iii) are closely related.   Essentially, the direct part of $S_{cc'}(E)$ is $\int dE\, s_{cc'}(E')/(E'-E^+)$ where $E^+=E+i\varepsilon$ and $s_{cc'}(E') \equiv \dfrac{1}{\Delta E} (\displaystyle\sum_{\lambda \text{ in } \Delta E \text{ at } E'} \gamma_{\lambda c}\gamma_{\lambda c'})$.   The numerator of the correlation is essentially the Im. part of the same integral, or $\pi\, s_{cc'}(E)$.   Such integrals are commonly met in physics and are usually such that the real and imaginary parts are comparable. Certainly the existence of direct cross-sections implies that correlations must occur, at least at some energies.   Thus the problem of understanding correlations for two channels is essentially identical to that of understanding the direct cross-section between them.

Now let us enquire whether assumptions (1) and (2) are distinct or not.   Although they are mathematically distinct, physically they are often close together, so that the violation of one is expected to lead to the violation of the other.   For example, the Barschall results are taken as evidence that the mean free path against collision of an incident nucleon inside a target nucleus is quite large.   This means that the incident nucleon may not only leave the nucleus without a collision but may also leave after a single collision with appreciable probability.   The first process corresponds to elastic scattering with single particle resonance effects implying violation of (2);   the second process corresponds to direct inelastic scattering implying violation of (1).

Now let us elaborate this picture a little with ideas going back to Bloch, Brown-de Dominicis, and Feshbach, Kerman and Lemmer. Suppose that the target states can be classified in a hierarchy 0p-0h, 1p-1h, 2p-2h etc. which is such that the Nth rank needs N interactions with the incident particle for its excitation. Consider inelastic scattering by a nucleon to a state ϕ_n of the Nth rank from the ground (0p-0h) state.   If the bombarding energy is near an incident single particle state u_ℓ the initial channel and

---

[+]The usual definition of the correlation ρ of two sets of quantities a_i and b_i is $\rho \equiv \sum_i (a_i-a)(b_i-b)/(\sum_i(a_i-a)^2 \sum_i(b_i-b)^2)^{1/2}$ where a,b are the mean values.   Since the mean value over λ of $\gamma_{\lambda c}$ is presumably zero, the partial width amplitude correlations is $\rho_{cc'} \equiv (\sum_\lambda \gamma_{\lambda c}\gamma_{\lambda c'})/((\sum_\lambda \gamma_{\lambda c}^2)(\sum_\lambda \gamma_{\lambda c'}^2))^{1/2}$.

near u_ℓ, in the final channel then $|c\rangle$ is the product $|\phi_0 u_\ell\rangle$ where ϕ_0 is the target ground state, and:

$$S_{cc'} = \Sigma \frac{\gamma_{\lambda c}\gamma_{\lambda c'}}{E_\lambda - E - \frac{i}{2}\Gamma_\lambda} \sim \langle\phi_0 u_\ell | \frac{1}{H-E} | \phi_n u_{\ell'}\rangle$$

writing $\frac{1}{H-E} = \frac{1}{H_0-E} + \frac{1}{H_0-E} V \frac{1}{H_0-E} \cdots$ where H_0 is the Hamiltonian without the target-nucleon coupling V, then the leading term in $S_{cc'}$ is that of Nth order in V. Now let us set up a set of states k with the properties that (1) they include the initial and final states $\phi_0 u_\ell$, $\phi_n u_{\ell'}$, (2) they include all other states up to and including the Nth rank (3) they diagonalize H. Noting that $S_{cc'} \sim \Sigma_{kk'} \langle\phi_0 u_\ell|k\rangle\langle k|\frac{1}{H-E}|k'\rangle\langle k|\phi_n u_{\ell'}\rangle$ for any set k that includes the channel states c and c' we find $S_{cc'} = \Sigma_k \frac{\langle\phi_0 u_\ell|k\rangle\langle k|\phi_n u_{\ell'}\rangle}{E_k - E - \frac{i}{2}\Gamma_k^\uparrow} +$

$\Sigma_{kk'} \frac{\langle\phi_0 u_\ell|k\rangle}{E_k - E - \frac{i}{2}\Gamma_k^\uparrow} \langle k|V\frac{1}{H-E}V|k'\rangle \frac{\langle k'|\phi_n u_{\ell'}\rangle}{E_{k'} - E - \frac{i}{2}\Gamma_{k'}^\uparrow}.$ The last

term gives the effect of levels of higher rank, including ultimately the fine structure λ. If we now average over an interval ε larger than the spacings of the levels of rank (N+1), using

$$\langle S_{cc'}\rangle_\varepsilon = S_{cc'}(E+i\varepsilon)$$

and if the interaction of levels be via the higher rank states is not too large, then the main effect of the second term in $S_{cc'}$ is to add a mixing width $\Gamma_k^\downarrow$ in the first term changing $\Gamma_k^\uparrow$ to $\Gamma_k = \Gamma_k^\uparrow + \Gamma_k^\downarrow$. If the levels k are not strongly overlapping then the last term is relatively small, and, for a given energy region, $S_{cc'}$ is dominated by a few levels k in the first term. In the case of really isolated levels, $\Gamma < D$, then one level dominates and, in the vicinity of a level k: $\Sigma_{\lambda \text{ in } \varepsilon} \gamma_{\lambda c}\gamma_{\lambda c'} \sim \frac{1}{\pi}\text{Im } S_{cc'} \sim$

$$\langle c|k\rangle\langle\phi|c'\rangle\left(\frac{\frac{1}{2\pi}\Gamma_k}{(E_k-E)^2 + \frac{1}{4}\Gamma_k^2}\right) \quad \text{(for c=c' and c}\neq\text{c')}.$$ These equations

reflect the feature that the partial widths of states λ in this vicinity arise through the "doorway" state k which is common to c and c':

$$\gamma_{\lambda c} = \langle\lambda|k\rangle\langle k|c\rangle \left(\Sigma_{\lambda \text{ in } \varepsilon} \langle\lambda|k\rangle^2\right) = \frac{\frac{1}{2\pi}\Gamma_k}{(E_k-E)^2 + \frac{1}{4}\Gamma_k^2}.$$

Obviously $\rho_{cc'} = 1$ in these circumstances.

We can use this picture to make assertions about correlations for a general pair of channels c,c'. Correlations and direct reactions will not necessarily be large merely because channels c,c' separately show non-statistical effects (i.e. violations of (1)).

Such effects show only that the separate channels have doorway
states (whose presence causes the observed cross-section
modulations), but this is insufficient in general for violation
of (2).   For this, we need the further property that there exist
doorway states which are common to the two channels, and that these
states be not strongly overlapping.   Such "common doorways" may
occur at any ranks in the hierarchies for c and c'.   If the rank
is low, then the correlations will extend over a wide energy range;
if high the range will be small.   (Of course, for a given
isolated common doorway state, any "inner doorways" (i.e. of higher
ranks) inside this state are automatically common doorways too).

     From this discussion, it follows that the problem of
understanding observed violations of (1), direct reactions or
correlations, is essentially that of identifying common doorway
states.   It may be objected that standard theories of direct
reactions appear not to involve such concepts.   Actually, they are
there implicitly.   The DWBA description of inelastic scattering to
a 1p-1h state ϕ_1 of the target is equivalent to the doorway picture
in which there are two common doorways, viz. those obtained by ad-
mixing $\phi_o u_\ell$ to $\phi_1 u_{\ell'}$ in perturbation theory.   The coupled channels
version is equivalent to two doorways of rank 2p-1h obtained by
diagonalising H between $\phi_o u_\ell$ and $\phi, u_{\ell'}$.   In a more complete
description, which may sometimes be necessary to take proper account
of collective effects, H would be diagonalised in all total states
containing target 1p-1h states besides these two states;   this
would result in more doorways of the same rank (2p-1h).   In any
event, justification for the DWBA or more refined approaches is
that the mixing of the original two doorways (between themselves
and to the others) via more complex states (i.e. of higher ranks
3p-2h etc) is small, i.e. they are not broadened to the point where
they overlap strongly.[+]

The Analogue State: A Classic Example
Of a Common Isolated Doorway

     It is amusing to note that, even if Barschall resonances, direct
reactions, and correlations did not exist and that the statistical
model were valid in general, there might still be exceptions to
it in the form of analogue states.   As is well-known, an analogue
state, because it corresponds to a low-lying state, is much
simpler than other states ($T_<$-states) near it.   In particular, its
natural decay widths for open allowed channels (p,p',γ,etc) will
be much greater than those for the $T_<$ states, roughly by the ratio
of level spacings for the two systems (sometimes reduced by (2T+1).
It follows that for any such channel the analogue will have width

---

[+]In practice this may be violated, at least for low partial waves,
because the spreading of the 2p-1h doorways exceeds their spacing.
The success of DWBA can then only be explained in terms of the
dominance of higher partial waves, and those waves not having single
particle states near the given energy.

significantly greater than the total of states with which it
strongly mixes provided that the analogue do not overlap.   Thus
the isolated analogue is a perfect example of a common doorway
not just for two but in general for several channels.   From the
work at Duke University, there is evidence that the analogue is
a (common) doorway for certain p,p' and γ channels, but no
convincing evidence that it is a significant doorway for isospin-
forbidden channels n and α.

The widths for p,p' and 5 γ channels are measured near the
analogue at $E_p \sim 2.00$ MeV in the $(p,p),(p,p')$ and (p,γ) reactions
on ^{54}Cr.   8 fine structure levels between 1.985 and 2.012 are
involved.   One finds for the p,p',γ_0 channels that the correlations
$\rho(\Gamma_{\lambda c}, \Gamma_{\lambda c'})$ are:

$$(c,c') = (p,p'), \quad \rho = 0.83$$

$$(c,c') = (p,\gamma_0), \quad \rho = 0.81$$

$$(c,c') = (p',\gamma_0), \quad \rho = 0.58$$

If the background levels had no intrinsic strength at all, all of
these figures would be 1.00.   (Furthermore, it would also follow
that the correlation for fixed λ,λ' over all channels c should also
be 1.00).   The large correlations show that the analogue accounts
for the majority of the widths of levels in its vicinity for the
p,p' and γ_0 channels.   This is really a consistency check since
this fact that the analogue is the dominant doorway for the
channels is already known from the strong peak in their strength
functions.

There is no real correlation between p,p',γ_0 channels and
the $\gamma_1 \ldots \gamma_4$ channels.   Apparently the analogue is a less important
contributor to the latter widths;   this is supported by the fact
that the γ_0 widths are 4–5 times larger than the others.

There are certain hints that there may be further effects in
the results.   For instance, the $\gamma_1 \ldots \gamma_4$ channels are quite
strongly correlated amongst themselves $(<\rho(\gamma_i,\gamma_f)>_{if} = 0.56)$,
essentially because one of the 8 levels dominates for all four
widths.   In fact, by regarding this level as the main doorway for
lines $\gamma_1 \ldots \gamma_4$ with the analogue as main doorway for p,p',γ, all
the main observed features of channel–channel correlations follow.
The level–level correlations have also been evaluated over
channels $\gamma_0 \gamma_1 \ldots \gamma_4$.   The mostly large positive values arise from
the fact that, for four levels, the γ_0 widths dominate strongly
(and partly dominates for a fifth level).   These trends also
follow from the two–doorway picture by assuming that the analogue
is a stronger doorway for γ_0 than the other is for $\gamma_1 \ldots \gamma_4$.   It
would be very nice to have data on f.s. levels further from the
analogue, so that one can directly determine the background widths
and correlations.

2. DATA ON NON STATISTICAL EFFECTS IN (n,γ) AND RELATED REACTIONS

There is a vast accumulation of data under this heading.[+]
However the great majority of it involves E1 radiation and 3 kinds
of effect:

(i) so called "anomalous capture", this is a bump in the high
energy part of the photon spectrum in some (n,γ) reactions.   This
bump is a violation of assumption 1, which implies smooth,
Maxwellian-type spectra.   It is seen both at thermal and resonance
energies.

(ii) correlations of intensities to various final states between
(n,γ) and (d,p) reactions on the same target.   Most reported cases
involve thermal (n,γ) data, but some resonance (n,γ) studies also
show correlations.

(iii) partial width correlations at (n,γ) resonances between neutron
and gamma widths.

Another feature of the majority of data is that it is
confined to certain incident orbital angular momenta in certain mass
regions.   The rule is that most non-statistical effects are found
when the initial and final states in the (n,γ) process occur in the
vicinity of single particle states.   (In practice, only initial s
and p states are observable), viz: A=35-65(3s → 2p), A=90-112
(3p → 3s,2d) and A=136-207 (4s → 3p).   The last region is actually
split into three parts by the deformation from A=150-185.

type of effect	s.p. trans.	3s → 2p	3p → 3s,2d	4s → 3p
(i)	thermal	25[32-65]]*	impossible	3[[138-142]],8[[185-206]]
(i)	res.	6[43-63]	3[92-98]	4[[185-206]]
(ii)	thermal	11[32-60]]**	impossible	3[138-142]]
(ii)	res.	1[55]	4[92-111]	****
(iii)	res.	6[50-60]	3[92-98]	3[163-173]]***[206]

negative

[+]References would be impossible to list.   The writer compiled a
list of data available on request.   Most recent data comes from BNL
ANL and AECL.   The older data on anomalous capture and the new data
on (γ,γ') from the Israel AEC complete the main sources.

The Table shows the number of cases of positive non-statistical effects and the mass-regions in which they occur. A double bracket denotes that one or more nuclei beyond this point have been studied and no effects found. Occasionally, inside the quoted ranges, cases of zero effect are reported: ^{*57}Fe(n,γ); $^{**}56$Fe(n,γ); $^{***}163$Dy$(2^+$ resonances$)$, ^{165}Er, ^{171}Yb, ^{169}Tm(0^+); ^{173}Yb(3^+);**** no effects in 6 nuclei near A $\sim$ 165 and 200. Here is some supplementary information on the three kinds of effect:

(i) <u>Anomalous Capture.</u> In the mass region 40-65, the high-energy part of the spectrum (say in the top 2 MeV or roughly $E_\gamma > 6$ MeV) is at least 50% and often more than 75% of the total capture. Beyond A=73 it is $\leqslant$ 15%. In the region 185-205, the high-energy part (say $E_\gamma > 4.5$ MeV) is of order 30% compared to $\sim$ 3% in the rare earths! There is some evidence that the main part of the anomaly is primary E1 lines; further that the bump is not an effect of final level density but due to a variation with final state energy of the E1 transition probabilities. If one makes the "Brink hypothesis" and assumes that the E1 absorption cross-section of various low-lying states is similar, then the anomalous bump implies the existence of a corresponding peak (or "pygmy resonance") in the absorption cross-sections at about the same photon energy as the bump. A detailed analysis of the ^{197}Au(n,γ) and $(d,p\gamma)$ data has demonstrated this in detail. Also in ^{197}Au, the pygmy resonance has been directly observed as a peak at 6 MeV in the (γ,γ_0) elastic scattering excitation curve. (In ^{203,205}Tl, the bump appears in the (γ,γ') spectrum when E_γ is fixed near $E_\gamma \sim$ 7 MeV).

Attempts to probe further the nature of the anomalous capture for the nucleus ^{197}Au has lead to confusion rather than illumination. $(d,p\gamma)$ studies and the (n,γ) studies up to $E_n = 4$ MeV have shown that the anomaly occurs for excitations other than the threshold region, and for incident partial waves other than $\ell=0$. However, the ^{197}Au$(n,n'\gamma)$ reaction at $E_n \sim$ 7 MeV which excites the threshold region in ^{197}Au shows no anomaly. (It is true that anomalous capture is mostly observed in odd-N compound systems, while ^{197}Au is even-N. However the observation of the anomaly in ^{202}Hg, ^{196}Pt seems to rule out an odd-even effect). Furthermore γ spectra from μ capture and $(p,p'\gamma)$ on ^{197}Au show no anomaly of the (n,γ) type. To confuse things further, these spectra show a new bump at 4.1 MeV. The ultimate in confusion has been set up in attempts to follow the anomaly towards ^{206}Pb, ^{208}Pb, where, from (γ,γ_0) curves, no bump occurs.

(ii) <u>(d,p)-(n,γ) Correlations.</u> These are often extremely strong and there is no question of the correlation being a random effect. The best cases are about ten even-even targets in the mass-range 40-60, where the correlations involve up to 15 states and are $\geqslant$ 0.8 and often $\geqslant$ 0.9. (Also two odd nuclei show a correlation $\geqslant$ 0.5, which is equally large when one allows for the two channel

spins).   The main hazard in evaluation (d,p)-(n,γ) correlations is
the possible existence of a spurious effect due to gross overall
energy dependences in the two sets of intensities.   Direct
inspection of the two sets of intensities enables one to see if a
large correlation genuinely arises from individual correlated pairs,
or from overall trends. (The large correlations in ^{165}Ho and ^{166}Er
can be dismissed on this basis).

Certain mass-regions and incident partial waves show zero
correlation, e.g. s-wave neutrons for $A \sim 100$ or 190.   Indirect,
but strong evidence for no correlations in some nuclei comes from
the success of the Argonne average-resonance capture-spectrum
analysis, which implies that transition probabilities are
independent of final state, except possibly for gross trends (e.g.
s-wave E1 transitions in the regions 105-120, 156-195).

(iii)   <u>Resonance Correlations</u>. These are not as convincing
statistically as the (d,p)-(n,γ) correlations.   The original case,
^{169}Tm(n,γ), has been changed considerably by inclusion of more
resonances, and re-assignment of resonance spins.   Present data
shows no correlation of statistical significance for five 0^+ or
eleven 1^+ states.   However, the largest Γ_n and Γ_γ for the 1^+
states occur in the same resonance;   further, both values are much
larger than the others.   This fact cannot be ignored.   One can
get a large correlation from this occurrence by changing the
definition (e.g. to one with square instead of linear weighting).
Other cases (^{163}Dy, ^{173}Yb) have since been reported, but there is
no real guarantee that these will not suffer the same fate as ^{169}Tm,
although they are more convincing on the face of it.   (17 resonances
of spin 3^+ in ^{167}Dy(n,γ) show $\rho=0.35$ for 5 highest energy E1 lines,
$\rho=0.22$ for 22 E1 lines;   8 resonances of spin 2^+ show no effect).
A convincing case is ^{206}Pb(n,γ), but, being near at a closed shell
and having a resonance-spacing of the order of 10^5 eV instead of
10 eV, it is rather special;   also the neutron width involved is a
significant fraction of the single particle value ($\sim 5\%$).

Several instances of correlations of Γ_n and <u>total</u> radiation
widths are reported for the mass-range 50-60.   Since the radiation
width is dominated by a few lines at the high-energy (anomalous
bump) end of the spectrum, the correlation effectively arises from
these few lines. (An important implication of such a correlation is
that the resonance tails in thermal capture are more likely to be
dominated by those occasional distant resonances with large Γ_n than
one would otherwise expect).   In this region, ^{55}Mn is reported to
show both effects (ii) and (iii), but this has recently been over-
shadowed by more impressive results on p-wave resonance capture on
^{92}Mo and ^{98}Mo.

To round off this general data survey, the following further
item should be mentioned.

(iv) (d,p)-(γ,γ') Correlations. Very recently (γ,γ') studies on
62Ni, leading to 0^+, 2^+ states via a single 1^- state at about 7 MeV
excitation have found a correlation with (d,p) intensities on ^{61}Ni;
similarly in ^{112}Cd, ^{120}Sn with $\ell=0,2$(d,p) intensities, and in ^{150}Sm
with $\ell=1,3$(d,p) intensities. Since the initial and final states
usually involve two channel spins, presumably uncorrelated, the
maximum possible correlation is not 1, but something like $1/\sqrt{2}$
as in (d,p)-(n,γ) correlations on odd targets.

Finally, here is a list of outstanding questions, posed by the
existing data and which have been at best incompletely answered.

1. In targets with $40 < A < 60$, where large thermal(d,p)-(n,γ)
correlations occur (e.g. ^{42}Ca), do the correlations occur at (n,γ)
resonances? There are 10 cases in even nuclei with $\rho > 0.8$ at
thermal energy. The capture cross-sections varies from 0.4b in ^{54}Cr
to 8.3b in ^{43}Ca, and this large variation makes it likely that
resonances are involved in the (n,γ) process. Note that the (γ,γ')
results mentioned above could be useful here.

2. Why does the (d,p)-(n,γ) correlation disappear abruptly at
A=61, and the anomalous capture weaken sharply at A=65, then fall
off with some fluctuations to disappear at about A=72? If the
effects are associated with the $2p_{3/2}$ and $2p_{1/2}$ final states, one
might expect little sudden change in view of the quite strong
mixing of these orbits with the $f_{5/2}$ one.

3. Is the anomalous bump in the range A=185-207 due to primary
radiation? (probably yes). Does it show an odd-even effect?
(probably not). Is it centred on those levels accounting for the
$3p_{1/2}$ and/or $3p_{3/2}$ states as revealed by $\ell=1$ (d,p) intensities?
(probably about 1 MeV higher). Is the anomalous part of Γ_γ
correlated with Γ_n (as seems to be so for $50 < A < 60$)? Why does
it not show up in (n,n'γ), (p,p'γ)and in capture spectra? Why
is there no bump in (γ,γ$_0$) excitation curves on ^{206}Pb, ^{208}Pb?
What is the 4.2 MeV bump seen in (p,p'γ) and μ capture spectra?

3. THEORY OF PARTIAL WIDTH CORRELATIONS
AND RELATED EFFECTS IN (n,γ) REACTIONS

In Chapter 1, we saw how correlations arise from the
existence of common doorways. For neutrons, the nature of the
initial (or entrance channel) doorway is clear, viz. the single
particle state $\phi_0 u_\ell$. Inner doorways correspond to increasing the
number of 1p-1h excitations of ϕ_0. For E1 photon absorption on a
given low-lying final state, even the nature of the initial doorway
for the threshold region (say E_γ=4-7 MeV) is not clear. The
identification of this doorway is an obvious pre-requisite to that
of finding a common doorway with the neutron channel. The first
candidate to consider is the giant resonance, centred 7 or more MeV

above threshold.   Dissipative effects spread this state into the
threshold region, and the resulting E1 strength contributes to the
observed strength.   For low-lying states where no non-statistical
effects are observed, this source may well be the dominant one
(e.g. low-lying states of both parities in A=70-90;   states of
negative parity in A=90-120, etc.).   However, for other states,
the observed non-statistical effects strongly suggests that another,
more local, source dominates.

If the data consisted only of a resonance correlation with
a single final state, then the common doorway could be a
complicated "inner" doorway appropriate to the final state, whose
effects occur only over a narrow energy range (say $\sim$ 10 keV).   The
fact that several correlations occur, of both types (ii), (iii),
suggests that the photon doorway is a fairly single structure
affecting a considerable energy range (say $\sim$ 1 MeV).   This general
conclusion is supported by the doorway inferred from the non-
statistical effects in the s-wave (n,γ) resonances for $40 < A < 60$ and
in the p-wave resonances for $92 < A < 98$.   In both regions, there is
evidence that all three classes (i),(ii),(iii) of non-statistical
effects occur.   For ^{92}Mo and ^{98}Mo, both $(d,p)-(n,\gamma)$ and resonance
correlations are observed.   Resonance correlations imply that a
large part of widths $\Gamma_{\lambda f}$ has the product form $\Gamma_{\lambda n} a_f$, while the
$(d,p)-(n,\gamma)$ correlations imply that a large part^n ^{f} has the form
$S_f b_\lambda$ where S_f is the spectroscopic factor.   When both occur we
infer that $\Gamma_{\lambda f}$ contains a term $\Gamma_{\lambda n} S_f$, i.e. $<\lambda|\phi_o u_\ell>^2 <\phi_o u_{\ell\prime}|f>^2$
where $\ell, \ell\prime$ are the ℓ- values of the neutron in the initial and final
states.   This is just the "valence model", which says that
radiation widths are controlled by single particle transitions of
the added neutron in the presence of the ground state of the target:
$\Gamma_{\lambda f} \sim <\lambda|\phi_o u_\ell>^2 <u_{\ell\prime}|d|u_\ell>^2 <\phi_o u_{\ell\prime}|f>^2$.   This diagnosis is
immediately supported by three further facts: it is consistent with
the anomalous bump which occurs near the single-particle energy
$E_\gamma \sim (\epsilon_\ell - \epsilon_{\ell\prime})$;   the observed absolute values of $\Gamma_{\lambda f}$ are consistent
with calculations using this formula, both for $A \sim$ 95 and $A \sim$ 55
(S.F. Mughabghab, private communication; A.M. Lane, Ann. Phys. 63,
171 (1971), Table 1);   for $A \sim$ 95, states f of $s_{1/2}$ and $d_{5/2}$ type are
observed, and their relative widths are consistent with the
calculated ratio of the single particle $3p \rightarrow 3s$, $3p \rightarrow 2d$ transitions.
This discussion implies that (for the above mass ranges) the state
$\phi_o u_\ell$ serves as a doorway state not only for the neutron channel,
but also for photon channels to the various final states in the
region of states $\phi_o u_{\ell\prime}$.

This result is an extraordinary one.   It means that the doorway for photon channels can be much simpler thay one has any right
to hope, and this situation immediately raises several challenges to
the theorist:

(a)   For a typical ground state we know that the giant E1 resonance at $E_\gamma \sim$ 16 MeV takes away the vast majority ($\sim$ 90%) of E1 strength from the uncoupled single particle transitions in the region 6–11 MeV.   How is it possible for some individual transitions to retain their strength undisturbed by this massive redistribution? What property separates them from the others?   Does the effect occur only for the valence particles in non-closed shell systems. Is it an effect for odd nuclei only?

(b)   Even if a few single particle transitions retain their strength and carry all significant E1 strength to the threshold region, why should they do so only in the presence of the core ground state? Corresponding to final states of type $\phi_0 u_\ell$, with strong E1 widths to higher excited levels λ that share state $\phi_0 u_\ell$, there will be final states of type $\phi_n u_\ell$, with strong E1 widths to levels λ sharing $\phi_n u_{\ell'}$, where ϕ_n is any lowlying excited state.   According to standard beliefs of optical model theory, states like $\phi_0 u_\ell$ and $\phi_n u_\ell$ are shared out into states λ over a considerable energy spread (say $\geqslant$ 2 MeV), implying that E1 radiation from states λ should be affected by several core-excited states $\phi_n u_\ell$ as well as $\phi_0 u_\ell$.   In fact it would be incomprehensible if it always happened that those single particle transitions that retain their strength did so in the presence of the core ground state.   It is thus encouraging that there exists some definite evidence that excited core states can also contribute in certain situations.   Unlike correlation effects (ii) and (iii), effect (i), the anomalous bump, does not specifically refer to the core ground state but presumably reflects the occurrence of all transitions like $\phi_n u_\ell \rightarrow \phi_n u_{\ell'}$, irrespective of core state ϕ_n.   Thus we expect the observation of (i) to be more widespread than (ii) or (iii), especially in those mass regions when $\phi_0 u_\ell$ is below threshold by a significant amount.   This happens in the range 185–205, where $\phi_0 u_{4s}$ is bound by 1–2 MeV;   here effects (ii) are zero, while effects (i) are at their strongest. (A smaller effect occurs for 60–65).   Thus the problem of explaining the relative unimportance of excited core transitions $\phi_n u_\ell \rightarrow \phi_n u_{\ell'}$, applies only when $\phi_0 u_\ell$ is near or rather above threshold, where at least $\phi_0 u_\ell \rightarrow \phi_0 u_{\ell'}$ is expected to be the most important single transition.

Detailed Calculations and Theoretical Speculations Relevant to the Valence Model

Calculational Evidence on the Dominance of Certain Single Particle transitions $\ell \rightarrow \ell'$ and for $\phi_0 u_\ell$ being a Common Doorway.

This comes from diagonalisation of 1p-1h states of spin 1- in the closed shell systems ^{60}Ni and ^{208}Pb, using zero-range forces (J.M. Soper, to be published).   In both nuclei, it is found that a significant amount of E1 strength remains in the low energy region after diagonalisation, and that this is strongly correlated with neutron components p^{-1}s.   In ^{60}Ni, the E1 strengths consists of 4%

of total $B^{\uparrow}(E1)$ at 6.5–8 MeV and 9% at 9–10 MeV, and this agrees roughly with observed values of radiation widths in this mass region.   In ^{208}Pb, about 4% is concentrated at 5.5–6.5 MeV;   this is about the amount indicated by anomalous capture in A< 208. Unfortunately the ^{208}Pb result is of interest only as a model since ^{208}Pb itself is observed to have < 1% in this region.   (Finite range forces give ≤ 1%, agreeing with this, e.g. S.M. Perez, Phys. Lett. 33B, 317 (1970)).

Calculational Evidence on the Existence of a Shell–Opening (or an Odd–Even) Effect.   Very lately, the first diagonalisation for the huge basis (up to 300 states) for an odd nucleus has been reported, viz. for the $2p_{1/2}$ proton nucleus ^{89}Y (J.D. Vergados and T.T.S. Kuo, Nucl. Phys. A168, 225 (1971).   The 1p–1h energies for the $p_{1/2} \to 3s_{1/2}$, $2d_{3/2}$ transitions are 7.2, 8.7 MeV.   The total $B^{\uparrow}(E1)$ after diagonalisation in 6–10 MeV is about the same in ^{89}Y as in the closed shells ^{88}Sr, ^{90}Zr, (Phys. Lett. 35B, 93 (1971)), so we conclude that there is no dramatic effect.   There is a hint of a mild effect in the fact that the $\frac{1}{2}- \to \frac{3}{2}+$ strength is mostly at 6–7.4 MeV, while the $\frac{1}{2}- \to \frac{1}{2}+$ and closed shell strengths are mostly at 7.4–10 MeV.

Theoretical Speculations on the Lack of Excited Core Contributions to Radiation Widths (especially for s–wave resonances in A=40–60; p–wave resonances in A=90–112).   This is perhaps the greatest and most persistemt mystery in the area of non-statistical effects. Let us focus on the p–wave (n,γ) reactions in the Mo isotopes. Spin $\frac{3}{2}^-$ occurs for channel states $\phi_0 u(p_{3/2}), \phi_n(0)u(p_{3/2}), \phi_n(2) u(p_{3/2})$, $\phi_n(2) u(p_{1/2}), \phi_n u(f_{7/2})$, etc. where $\phi_n(J)$ are excited target states of spin J.   The lowest excited states of ^{98}Mo are: $0^+, 0.73$; $2^+, 0.79, 1.43, 1.76$; $4^+, 1.5$ MeV.   One estimates about 6 states involving u(p) to occur within 2 MeV above threshold where $\phi_0 u(p_{3/2})$ is centred.   The state $\phi_n u(f_{7/2})$ is probably close to or just below threshold.   If one is guided by optical model beliefs on the spreading of states (typically over regions 2W ~ 3 MeV near threshold), then resonances λ should contain significant components of all these states.   Note that the final states absolutely must contain excited target state components because there are 2 or 3 states of each spin, $\frac{1}{2}+$, $\frac{3}{2}+$, $5/2+$.

From the ^{208}Pb calculation above, we may reasonably hypothesize that there is a distinction between f → d, and p → s,d transitions in the sense that the former lose their E1 strength to the giant resonance, while the latter retain theirs.   This enables us to ignore the presence of components $\phi_n u(f)$ in the resonances λ, but leaves the crucial question: how can we justify the neglect of components $\phi_n u(p)$.   Recently some light has appeared on this question from an unlikely source, viz. threshold anomalies in (d,p) reactions (A.M. Lane, Phys. Lett. 33B, 274 (1970)).   It was shown

that there is a strong compression effect operating on the spreading
of the 3p state in the threshold region.   Instead of the normally
expected spreading width of $2W \sim 3$ MeV, the 3p state has a width of
0.7 MeV.   The physical reason is very simple: A 3p state in a
Saxon-Woods well at threshold has 89% of its normalisation integral
beyond the inner turning point.   Thus mixing <u>matrix elements</u> of
$\phi_o u(3p)$ to other states are reduced by a factor $\sqrt{1-0.89} = \frac{1}{3}$ from
the normal value. (This would mean that the spreading width is
reduced by a factor 1/9, except that the act of spreading is to
push part of the state away from threshold where the effect is less.
The net effect on spreading is a factor $\sim \frac{1}{4}$).   Now let us consider
the case of mixing of $\phi_o u(3p)$ to a state which also contains u(3p)
like $\phi_n u(3p)$.   The mixing matrix element between the two is now
reduced <u>by two factors,</u> one for each state.   Thus, if each were
near threshold, the matrix element would be reduced by up to a
factor 1/9, <u>so that the mixing between the two states is much less
than normally expected.</u>   This fact is vital for understanding the
remarkable absence of excited core state effects in (n,γ) reactions.

     This simple picture may be suitably refined.   Evidently, as
presented, it fails for positive energies since the normalisation
integral is then 100% from the outside region.   Further, one may
object that, even for a negative energy, one should explicitly
include the nearby p-wave continuum in the calculation, and this may
have a drastic effect, and possibly restore the normal situation.
For coping with these problems, the R-matrix theory seems ideal.
First, by its very nature, it copes with the second issue automatically.   On the first issue it shows that the normalisation
effect persists at positive energies;   apart from a small range
above zero energy, the recipe is that the normalisation of the 3p
state should be taken to the outer turning point.   Consider the
mixing of $\phi_o u(3p) = \Phi_o$ with Φ_q (q=1...N) then the Hamiltonian
matrix for the (N+1) state problem is as usual except that $(E_o - E)$ is
replaced by $E_o + \Delta_o - E$, where Δ_o is the shift factor $\Delta_o = \gamma_o^2 (S(E_o) - S(E))$.
Making the linear approximation, $(E_o + \Delta_o - E) \approx (E_o - E)$
$(1 + \gamma^2 (\frac{dS}{dE})_{E_o}) = $ (say) $(E_o - E) N_o^2$.   The matrix is thus effectively
the usual one, but with all matrix elements M_{on} reduced to $M_{on} N_o^{-1}$.
N_o^2, which is < 1, is roughly the fraction of normalisation inside
the inner turning point relative to that over the whole region inside the outer turning point for the state u(3p).   This also
applies to any element in the basis Φ_q involving a 3p state.   Notice
that, near threshold, where the normalisation changes rapidly, the
linear approximation may be poor unless $(E - E_o)$ is small;   for the
state $\phi_n u(3p)$ the diagonal element is then $(E_n - E) + \gamma_o^2 (S(\varepsilon_{3p}) - S(E - e_n))$
where $E_n = e_n + \varepsilon_{3p}$, the sum of the core and 3p state energies (the
latter being defined as the energy for the natural boundary
condition: $(u'/u) = S(\varepsilon_{3p})$).

     Let us now apply the same ideas to s-wave resonances in the

region A=40–60.   Again there are many states involving excited cores that would normally be expected to be involved in the resonances; for spin $\frac{1}{2}$+, these are of types $\phi_o(0)u(3s), \phi_n(1)u(3s), \phi_n(2)u(d_{5/2})$. Assuming that d → p transitions lose their strength to the giant resonance, the problem is to justify the neglect of states of type $\phi_n u(3s)$.   The above p-wave theory may now be applied to this situation.   There is one essential difference, viz. that $S(E-e_o)$ for positive $(E-e_o)$ and s-waves is zero.   This means that for levels above threshold (resonances) no reduction in matrix element arises for the state $\phi_o u(3s)$.   However the matrix elements of states $\phi_o u(3s)$ are reduced, because the channels energies $\varepsilon = (E-e_n)$ for excited target states ϕ_n are negative and $S(\varepsilon) \neq 0$ for negative ε. Thus we again find a reduction in effective mixing matrix elements between $\phi_o u_\ell$ and $\phi_n u_\ell$; unlike the p-wave case, it arises from the normalisation of $\phi_n u_\ell$ only, rather than from both states.   Nevertheless it is an important effect, which operates over a larger energy-range than for the p-wave case.   If K is the channel wave-number of a negative energy channel then $S = -K$ $a\gamma_o^2$, where a is radius, and $N_o^2 = 1+\theta^2/Ka$, where $\theta^2 = \gamma_o^2 ma^2/\hbar^2$.   Setting $\theta^2 = 3.5$, a=5f, for channel energy –0.25, –1, –10 MeV $N_o^2 = 7.7, 4.3, 2.1.$

Finally, note the following implications:

(a)   the reduction effect on effective matrix elements is independent of the energy of the single particle state relative to threshold.

(b)   when the state $\phi_o u(3s)$ is centred on threshold (near A=55) it is the most important component in levels λ;   the reduction of the coupling to states $\phi_n u(3s)$ makes its dominance even _more_ marked. When it is below threshold, as at A ⩾ 60, a state $\phi_n \overline{u(3s)}$ is the largest contribution to λ;   the reduction of coupling to $\phi_o u(3s)$ now makes the latter even _less_ involved in λ.   _This reversal may help to explain the sudden disappearance of correlations at A=60._   When $\phi_o u(3s)$ is above threshold (at A ~ 45), it is the dominant contributor to levels λ amongst $\phi_n u(3s)$ since it is the nearest. The reduction in matrix element increases this dominance, agreeing with the observed fact that correlations are very strong at A ~ 45.

(c)   The new (γ,γ') data refers to states λ below threshold.   For these states, the neutron channel energy is negative, so the matrix elements between $\phi_o u(3s)$ and $\phi_n u(3s)$ are now reduced by a double factor (as in the previous p-wave case).   This helps to explain the strong $(d,p)-(\gamma,\gamma')$ correlations in ^{58}Ni for a state λ bound by 3 MeV [contribution 4.12].

DISCUSSION

NEWSTEAD
    What would be the effect of a closed shell on the spreading width?

LANE
    Could you be more explicit?

NEWSTEAD
    Yes. Newson tells me he finds that as one approaches a closed shell, the spreading width increases. I wondered if this fits in with what you were saying.

LANE
    This is the spreading widths of one of these simple states, when the ground state, ϕ_0, is a closed shell?

NEWSTEAD
    Yes. I think his case is actually more complicated.

NEWSON
    The particular case is this doorway state which we found in Pb^{208} plus a neutron. We believe that this is a neutron in a single particle state around a vibrating, excited core. We have done this measurement for Bi^{209}, Pb^{208}, Pb^{207}, Pb^{206}. We have a little smell of it in thalium. The upshot is that the spreading width of the doorway state for Pb^{208} is zero. The whole strength appears as one resonance. It's about 50 kilovolts for Pb^{207}, and going either way it jumps to 200 kilovolts. It seems to be going up very fast, and we speculate then that when you get a few more nucleons away from double magic, the spreading width becomes comparable to the imaginary term in the optical model, and you just don't see any of the effects of this particular kind of intermediate structure. I don't want to give my whole talk of tomorrow; but then taking this as a starting point, I looked as near as possible, to other double-magic combinations, most of them lighter, and there seems to be pretty fair evidence in all of them that the same thing is happening.

LANE
    I take it, in the case of Pb^{208} the small value is not due to the trivial reason that there are just no other levels around.

NEWSON
    There's only one level in a range of a couple of million volts.

LANE
    So it couldn't spread.

NEWSON
        There's no place to spread to.

LANE
        There seems to be a general sentiment that an increase away
from closed shells could be expected.

EARLE   (Chalk River)
        Your statement that we report evidence for correlations in
$T\ell^{206}$ was not quite what I meant to convey. Our data showed no
correlations between two of the pairs of variables studied and for
the third pair the correlation is uncertain because the data that
we have at present is somewhat selective.

LANE
        I had in mind something else you said that the γ-ray widths
for the anomaly as a whole might be correlated with neutron widths.
Or did I misunderstand that?

EARLE
        That's the third pair of variables that we studied, and we
feel, at the present time, because of the selective nature of the
data we have, that correlation might not be significantly different
from zero.

LANE
        Right. But the nominal value you quoted was 0.64, but that
may disappear in time, following some reasonable order. I hope it
disappears in this case. It's the first time I've been in this
position.

FALLIEROS
        I'm not sure I followed you in the case of Pb^{208}. You mentioned
that Soper does get some strength around 6 MeV, and that Kuo and
Brown do not. I'm not sure I understand what zero means in this
case. I've heard Dr. Chrien say before what he needed for his
single-particle transitions from the p-wave strength function. The
radiative strength was a very small fraction of the Weisskopf
value. Are we dealing with these small numbers, or what exactly
do we mean?

LANE
        Yes. To give you numbers, the single particle s to p transition
is about 1% of the sum rule in this nucleus. Soper finds this
energy to be about 6 MeV, and it is correlated with the $p \rightarrow s$
transition. Its strength is about 4% of the sum rule, backing up
the statement that I made that the single-particle transition actually
seems to pick up some strength. Brown-Kuo and all the other
authors who used finite ranges get well less than 1/4%. Experimentally,
the value is of the order of 1/4% or less.

MAHAUX (Liege)

Do I understand correctly that you propose that the threshold
effect is contained in the shift functions and in that case it's
surprising, because I thought that the shift functions were smooth
functions of energy through threshold, except for s-wave neutrons.
Is that correct?

LANE

Yes, that is true. The pictures I drew were for an s-wave
case. Although the drawing was rough, some of the curves were
discontinuous. They reflected this discontinuity for s-waves.
For other cases, the functions are continuous, but the effect is
there and very strong. For the p-wave case, this loss of normalization effect is, as I said, 90% at zero energy, and several MeV from
threshold it chops to something like 30%, so it's quite a strong
localization around threshold.

CHRIEN

I wonder if the threshold effect you mention is not responsible
for the lack of agreement of the predictions of the valence model
and the results of Jackson and Strait from photoneutron measurements.
In each one of three isotopes (Cr-52, Ni-60, Fe-56) the valence
neutron model predicts accurately the ground state radiative width
for the lowest energy resonance in each case. As one proceeds to
higher energies, the agreement is very poor.

LANE

Yes. This threshold effect I've been talking about is one that
would apply over an interval of hundreds of kilovolts. It offers
no explanation at all for anything as localized as a few kilovolts.

CHRIEN

I believe this is 100 kilovolts or so in the case of these
Ni isotopes. Jackson can correct me if I'm wrong on that.

NEWSON

I'd like to ask a stupid question. What do you mean when you
use the word threshold? That's not the neutron escape energy or
the inelastic scattering threshold, is it?

LANE

The neutron escape threshold.

NEWSON

So neutron escape energy is your threshold.

LANE

Yes; simply that.

NEWSON

     I see. So you say you expect funny things when the neutron
is coming out with low energy.

LANE

     Yes, for resonances near threshold, which means the resonances
we look at in practice. There are special things in the composition
of the resonance states and the actual nature of the resonance
states; the admixtures they have of the various channel states are
singular. They're very special because of the threshold effect.

BERMAN

     We've made a number of measurements on the total integrated
strength from the giant dipole resonance in medium and heavy nuclei.
Typically, you get 1.1 or 1.05 times the dipole sum rule in the
integrated crossection up to 30 MeV, beginning with the neutron
threshold. In the mass 90 region we've measured 5 isotopes, and
here typically we get about 0.9 of the dipole sum rule, about
15-20% below what is typical for heavier nuclei. Did I understand
you correctly when you said you expected 20% of the E1 strength
to lie between 5 and 10 MeV, which would be below the (γ, n)
threshold in Y^{89} and Zr^{90}?

LANE

     Yes. First, a matter of definition. I was speaking of the
nonenergy weighted strengths; you're probably talking of the
energy weighted one. No? All right. I am quoting the calculation
of Kuo and Vergados. It is a fact that their calculation gives
about 20% of the total strength between 5 and 10 MeV, which is of
course in violent contrast with the results by the same authors for
Pb^{208}, where the fraction is less than 1/4%, in that kind of energy
range. How that is explained, I don't know.

In Search of Non-statistical Effects in Neutron Capture*

Said F. Mughabghab

Brookhaven National Laboratory, Upton, N. Y. 11973

The reaction mechanism in radiative slow neutron capture has been dominated by the statistical compound nucleus model in which the mode of decay of the compound state is independent of its mode of formation. Because of the complexity of the initial state, the partial radiative amplitudes $\Gamma_{\gamma ij}^{1/2}$, where i and j designate the initial and final states respectively, are normally distributed with zero mean. In most cases, such a notion is born out by experiments carried out on heavy weight nuclei[1]. However, in recent years extensive theoretical and experimental investigations have been devoted for a search of simple reaction mechanisms. In order of increasing complexity, these are (1) The direct potential or hard sphere capture in the off resonance region in which the incident S-wave neutron is scattered by the boundary of the nuclear surface into a low-lying P state, and in the process electric dipole radiation is emitted. (2) The channel capture of Lane and Lynn[2] or valence neutron model of Lynn[3] in which the S or P wave neutron is scattered via the resonant state into a low-lying orbit, radiating γ rays. (3) The semi-direct process of Brown[4] or doorway states formations of Estrada and Feshbach[5], whereby an incoming neutron scatters in the target nucleus creating a 2P - 1H state. Subsequently a particle and hole combine to yield enhanced γ radiation. Lane[6] has shown that $\Gamma_{\gamma ij}$'s are correlated with the reduced neutron widths of the initial state. In this category, the giant dipole resonance, which is considered as a coherent superposition of 1P - 1H state, is included.

One can then write for the transition amplitude

$$\Gamma_{\gamma ij}^{1/2} = C_1 \; \Gamma_{\gamma ij}^{1/2} \Big|_{SP} + C_2 \; \Gamma_{\gamma ij}^{1/2} \Big|_{dS} + C_3 \; \Gamma_{\gamma ij}^{1/2} \Big|_{cn} \tag{1}$$

*Research supported by U. S. Atomic Energy Commission.

The contribution due to valence neutron transitions for electric dipole radiation[3] is given by

$$\Gamma_{\gamma ij}(E1) = \frac{16\pi\,k^3}{9}\; \Theta_i^2\; \Theta_j^2\; (\bar{e})^2\, r^2\, S \qquad\qquad (2)$$
SP

In this model it is assumed that interaction between the valence neutron and the core is negligible, i.e. the single particle strength is not coupled to the giant dipole resonance.

The doorway state contribution is

$$\Gamma_{\gamma ij}^{1/2} \;\alpha\; E_\gamma^{3/2}\, M_{dj}\, \Gamma_{ni}^{o\,1/2} \qquad\qquad (3)$$
ds

where M_{dj} is the matrix element connecting the doorway state, d, with the final state, j. In general, the last term in Eq. 1 which describes compound nucleus process dominates, so that it obscures any effects due to simple reaction mechanisms. The partial radiative width due to valence neutrons (Eq. 2) provides us with a guide as to what nuclei may be investigated for possible non-statistical effects. In addition, data of S and P wave neutron strength functions can serve to be very helpful in pointing out that regions of non-statistical effects for S-wave neutron capture are around A = 50 and 160, and P-wave neutron capture around A = 100 and 220.

At first let us consider non-statistical effects generated by the occurrence of doorway states. The conditions for the observation of doorway states are such that (1) the single particle states are located just near the neutron separation energy, (2) nuclei with large reduced neutron widths are to be considered, and (3) the matrix element connecting the doorway state with the final state is large. These conditions have been satisfied in target nuclei Dy^{163} and Yb^{173}. Experimental evidence[7,8] for correlations between partial radiative widths and reduced neutron widths have been reported for these isotopes.

In what follows I would like to discuss at length the influence of single particle effects in the neutron radiative process. Let us explore the existing available data to find out the success and the failure of the valence neutron model.

A. Thermal Neutron Capture

The thermal absorption cross section of Ni^{58} is dominated by a bound level at En = -28.5±5.0 kev and its Γ_n^o is well determined by Bilpuch et al[9] and Garg et al[10]. It is clear then that the thermal γ ray intensities will be dominated by the tail of the bound level, and as a result the partial radiative widths and intensities can be calculated on the basis of Eq. 2. The calculated partial radiative widths are listed in the second column of Table I. Also included in Table I are the γ ray energies and measured thermal γ ray intensities. If one considers a $\Gamma_\gamma = 0.55$ eV (compound nucleus formation) as obtained from the study of the systematics[12] and as calculated by the procedures outlined in ref. 13, one derives a total radiative width of 10.4 eV for the bound level. With the aid of this value, the partial radiative widths are converted into intensities. With the exception of 7696 keV γ ray, the agreement between experimental and calculated intensities is remarkable. An independent support for the validity of Γ_γ comes from a computation of the thermal absorption cross section. One obtains a value which is in close agreement with measurements.

Similar procedure as that outlined for Ni^{58} is applied for the thermal γ ray intensities[14] of Cr^{50}. The results are summarized in Table II.

	Table I				Table II		
E_γ(keV)	$\Gamma_{\gamma ij}$(eV) Cal.	I Cal.	I Thermal[11]		E_γ(keV)	I Cal.	I Thermal[14]
8998.9	5.57	534	490		8516	496	310
8533.4	2.45	235	210		8488	217	250
8121.0	0.48	46	48		7362	143	130
7696.0	1.37	131	17		6368	54	50
					6133	90	80

B. S-Wave Neutron Capture

For unbound S-wave neutron resonances with large reduced widths, it would be expected that doorway state contributions (Eq.3) would compete strongly with single particle effects. Let us investigate neutron capture in mass region of the 3S giant resonance. Particularly neutron radiative capture in the 7.8 keV resonance of Fe^{54} is considered. Partial radiative widths are calculated for final low lying states with significant single particle strength and the results are listed in Table III. A total radiative width of 1.13 eV is deduced for this resonance. Since about 40% of the thermal absorption cross section contribution is due to the 7.8 keV resonance, it would be expected that the thermal γ ray intensities[15] resemble those of the 7.8 keV resonance. As shown in the last two columns of Table III, there is reasonable agreement between the two sets of values. However at higher excitation energies, the model does not reproduce the experimental results.

Table III

Γ_γ(keV)	$\Gamma_{\gamma ij}$(mV)	I_γ(7.85)	I_γ(thermal)[15]
9927	571	506	650
8855	197	174	120
6820	47	42	21

C.   P-Wave Neutron Capture

The valency neutron model remarkedly accounted[16] for the observed γ ray intensities in P-wave neutron capture in $Mo^{92,98}$. Such a success motivated us to search for other cases in the 3P giant resonance and particularly in the Zr isotopes. Previous γ ray spectra measurements carried out by Rimawi[17] at the BNL fast chopper facility showed that the spectrum of the Zr^{96} 302 eV resonance is dominated by two strong transitions as shown in Fig. 1.

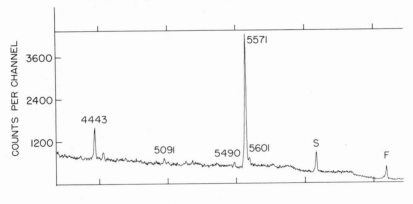

Fig. 1

This suggested that this resonance is P-wave in nature. Such a conclusion is supported by the small value of the thermal absorption cross section which can be understood only in terms of a P-wave interpretation of this resonance. The radiative partial widths are computed[12] and agreement with experimental values is found to be remarkable if one accepts a $P_{1/2}$ assignment for this resonance. This excellent agreement is attributed to the large reduced neutron widths of the initial and final states, in which case single particle effects dominate over other complex processes. A total radiative width of 555 mV is derived for this resonance which is at variance with an experimental[18] value of 220±50 mV. Note that the experimental value of Γ_γ favors a spin assignment of $P_{3/2}$. It would be of great interest to measure the spin and total radiative width of this resonance in order to clarify the situation.

Table IV

Isotope	E_n (keV)	$\Gamma_\gamma(P_{1/2})$ (mV)	$\Gamma_\gamma(P_{3/2})$ (mV)	Γ_γ(measured)[19] (mV)	Preferred Spin
Zr-90	7.249	454	441	400 ± 80	$P_{1/2}, P_{3/2}$
"	8.823	344	332	500 ± 80	$P_{1/2}, P_{3/2}$
Zr-92	4.160	493	390	500 ± 50	$P_{1/2}$
"	6.910	2330	1555	600 ± 80	$S_{1/2}$
Zr-94	5.800	783	544	650 ± 50	$P_{1/2}, P_{3/2}$
Zr-96	0.302	555	283	220 ± 50[18]	$P_{1/2}$

In addition, one arresting feature of the reported[19] radiative widths of $Zr^{92,94}$ resonances is the large spread of values. If one proceeds to calculate Γ_γ of resonances having large radiative widths on the assumption that these are $P_{1/2}$ or $P_{2/3}$ resonances, one obtains the results listed in Table IV. With the exception of the resonance at 6.91 keV, there is no evidence[19] to indicate that these resonances are formed by S-wave neutron interactions. Then the fact that the 6.91 keV resonance has a radiative width of 600±80 mV leads to the conclusions that M1 transitions in this resonance are enhanced. A similar situation exists in the 3.17 keV resonance of Mo^{92}.

D. M1 Radiative Strength in S-Wave Neutron Capture

One of the least understood problems in neutron capture investigations is the nature of the enhanced M1 radiative strength. This is in part due to the scarcity of M1 data. Some of the questions that can be raised are the following: Do valency neutrons play a significant role in M1 transitions? Are the $\Gamma_{\gamma ij}$'s correlated with the reduced neutron widths of the initial and/or final states?

One suggestion relating to the question of enhanced M1 strength is that it is due to collective cooperative phenomenon such as pigmy giant resonance located at energy of 7-8 MeV. In an effort to shed some light on this interesting problem, M1 transitions due to S-wave neutron capture in the even-even Mo isotopes have been studied[20]. Fig. 2 displays the γ ray spectrum due to neutron capture in the 3.17 keV resonance ($S_{1/2}$) of Mo^{92}. As shown, the spectrum is dominated by one γ ray feeding the first excited state ($S_{1/2}$) of Mo^{93}. This transition is of the type $S_{1/2} \rightarrow S_{1/2}$ and conforms to the ℓ selection rule of M1 transitions in the valence neutron model. Its partial radiative width is 68±8 mV. In addition, the partial radiative width due to neutron capture in the 467 eV resonance ($S_{1/2}$) and feeding the ground state ($S_{1/2}$) of Mo^{99} is

4±1 mV. Quantitative calculations of M1 partial radiative widths
in terms of the valency neutron model fails to reproduce these experimental values.

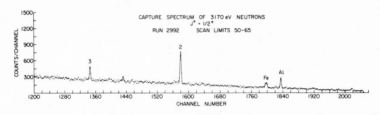

Fig. 2

   One mechanism in which enhanced M1 strength could be achieved
is through the excitation of 2P-1H state. The following model is
proposed in mass region A = 100. The $S_{1/2}$ valence particle excites
a $g_{9/2}$ particle into a $g_{7/2}$ orbit, thus creating a 2P-1H configuration as shown in Fig. 3. The $g_{9/2}$ particle interacts with the
$g_{7/2}$ hole through the magnetic dipole operator, emitting M1 radiation, while the incoming $S_{1/2}$ neutron goes into an $S_{1/2}$ orbit.
It is of interest to point out that the energy difference between
the $g_{9/2}$ and $g_{7/2}$ single particle states is 7.2 MEV. This is about
the same energy as that observed for Mo^{92}. This model would require that M1 partial radiative widths be correlated with reduced
neutron widths of particularly the initial state. This is remarkably supported by the M1 data of Mo^{92} and Mo^{98}. In addition this
model accounts for the large radiative width of the 6.9 keV resonance in Zr^{92}, predicting a value of 470 mV. It is stressed that
much further experimental work in the study of high energy M1 radiation remains to be done in order to verify these ideas.

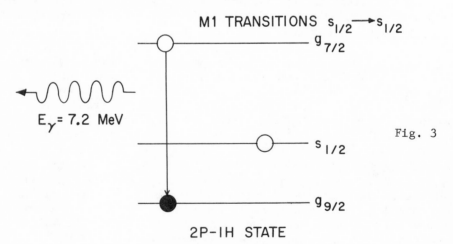

Fig. 3

In conclusion it has been shown that the valence neutron model accounted remarkably well for a large body of data. However, the doorway state mechanism is invoked in order to explain enhanced M1 radiation. It is hoped these findings would stimulate further theoretical and experimental investigations in neutron capture.

References

(1) L. M. Bollinger in Experimental Neutron Resonance Spectroscopy Ed. J. A. Harvey Academic Press p. 235, 1970 and references therein

(2) A. M. Lane and J. E. Lynn, Nucl. Phys. 17, 586 (1960).

(3) J. E. Lynn, The Theory of Neutron Resonance Reactions (Clarendon Press, Oxford, England, 1968) p. 333.

(4) G. E. Brown, Nucl. Phys. 57, 339 (1964)

(5) L. Estrada and H. Feshbach, Ann. Phys. 23, 123, 1963.

(6) A. M. Lane, Phys. Letters 31B, 344 (1970).

(7) S. F. Mughabghab, R. E. Chrien, and O. A. Wasson, Phys. Rev. Lett. 25, 1670, 1970.

(8) S. F. Mughabghab, O. A. Wasson, G. W. Cole, R. E. Chrien, and M. R. Bhat, Bull. Am. Phys. Soc. 16, 496 1971. See these Proceedings also.

(9) E. G. Bilpuch, K. K. Seth, C. D. Bowman, R. H. Tabony, R. C. Smith, and H. W. Newson, Ann. Phys. 14, 387 (1961).

(10) J. B. Garg, J. Rainwater, and W. W. Havens, Jr., Phys. Rev. C, 3, 2447, 1971.

(11) L. N. Bystrove, Z. A. Rudak, E. T. Firsov as reported in Nucl. Data, ed. K. Way Section A, Vol. 3, No. 4-6 (1967).

(12) S. F. Mughabghab, Third Neutron Cross Sections and Technology Conference (March 1971) Knoxville, Tennessee, Paper III 5.

(13) S. F. Mughabghab and R. E. Chrien, Phys. Rev. C, 1, 1850 (1970).

(14) G. A. Bartholomew, E. D. Earle, M. R. Gunye, Can. J. Phys. 44, 2111 (1966).

(15) S. E. Arnell et al and Fissov et al as reported in Nuclear Data Set A volume 3, No. 4-6 (1967).

(16) S. F. Mughabghab, R. E. Chrien, O. A. Wasson, G. W. Cole, and M. R. Bhat, Phys. Rev. Lett. 26, 1118 (1971).

(17) K. Rimawi Thesis, N. Y. State University at Albany, 1970 (unpublished).

(18) J. Morgenstern. CEA-R-3609. Previous measurements showed that Γ_γ = 370±90. Comp. Rend. 254, 4009 (1962).

(19) Z. M. Bartholome, R. W. Hockenbury, W. R. Moyer, J. R. Tatorczuk, and R. C. Block, Nucl. Sci. and Eng. 37, 137 (1969) and private communication.

(20) S. F. Mughabghab, R. E. Chrien, O. A. Wasson, M. R. Bhat, and G. W. Cole. These Proceedings.

DISCUSSION

PATRICK

I should like to point out that we have studied the inverse reaction, that is (γ,n), and looking at the neutron coming out in the molbdenum mass region. In quite a number of elements, we have found a large resonance in the neutron crossection at about 7.8 MeV gamma ray energy. This, we believe, is an M1 giant resonance. This occurs in Sn, which we have published, and also in Mo, Zr, and there's one or two others that we have examined.

MUGHABGHAB

I'm sorry. I wasn't aware of that work.

HANSEN   (C.E.R.N.)

It is interesting to note that a $g_{9/2}^{-1}$ $g_{7/2}$ pgymy resonance can also be observed in the high-energy beta decay in the mass 100 region. The data came from the study of delayed-photon emission from Xe^{115} for which the beta strength function shows a resonance at about 5.5 MeV with about 1/20 of the total strength available.

CHANNEL CORRELATION EFFECTS FOR FRAGMENTED ANALOG STATES

G. E. Mitchell
North Carolina State University and Triangle Universities
Nuclear Laboratory*
E. G. Bilpuch, J. D. Moses, W. C. Peters and N. H. Prochnow
Duke University and TUNL*

In this paper we present results from several different experiments illustrating the variety of channel correlation phenomena which occur in proton, neutron, and gamma decay of fragmented analog states. Fragmented analog states provide a set of fine structure resonances whose character is at least partially understood, and for which correlations are expected between partial widths of various decay channels.[1] In simplest terms, correlations are expected between those channels which are enhanced by the analog state.

All experiments discussed below were performed in the TUNL 3 MV Van de Graaff laboratory. Overall energy resolution of 300–400 eV was obtained using thin solid targets (1–2 $\mu g/cm^2$) of enriched Ti, Cr, and Ni isotopes. Protons were detected with surface barrier detectors, neutrons with BF_3 counters, and gamma rays with a 3 x 3" NaI(Tl) crystal and an 80 cc Ge(Li) detector.

Inelastic proton decay from an analog state is often enhanced (when the parent state contains the suitable configuration--excited target $\otimes$ neutron). In this event the analog is a doorway for both the elastic and inelastic channels, and the partial widths in these channels should be correlated. Fig. 1 shows elastic and inelastic excitation functions[2] for ^{48}Ti, and a multi-level, multi-channel fit to these data. The cluster of levels near 2.95 MeV is the analog of the 18th excited state of ^{49}Ti. Fig. 2 shows the reduced widths near this analog for the elastic and both inelastic channels (channel spins 3/2 and 5/2). There is a high correlation between the p_0 and p_1 widths, although much of the correlation is due to one very strong resonance. As expected, the widths are not correlated away from the analog.

299

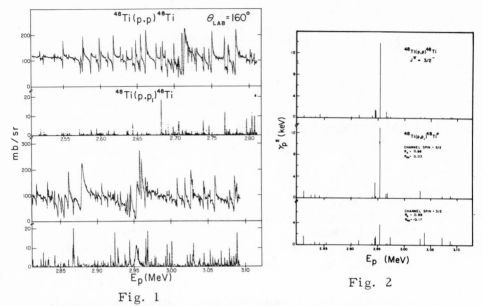

Fig. 1

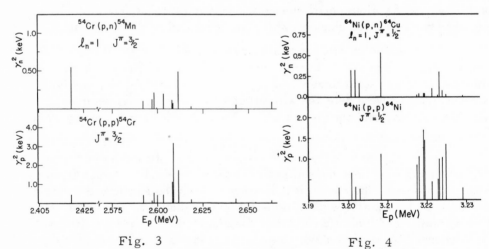

Fig. 1 Elastic and inelastic data for ^{48}Ti. The solid line is a fit
to the data.

Fig. 2 Elastic and inelastic reduced widths for the analog of the
18th excited state of ^{49}Ti.

Fig. 3   Fig. 4

Fig. 3 Proton and neutron reduced widths for the analog of the
3rd excited state of ^{55}Cr.

Fig. 4 Proton and neutron reduced widths for the analog of the
1st excited state of ^{65}Ni.

According to accepted theories of analog states the neutron widths are not directly enhanced by the analog state. (Some enhancement may occur through mixing with the anti-analog, but this effect is expected to be small.[1]) There is little direct experimental evidence on this question.

In the present experiments, proton and neutron partial widths of analog state fine structure resonances were measured.[3] Measurements of resonance widths away from the analog state proved impossible, since the elastic resonances were observable only when enhanced by the analog state. Lacking information on the background, we test for (strong) enhancement by determining the linear correlation coefficient between neutron and proton partial widths. Two analog states were studied: the analogs of the third excited state of ^{55}Cr and the first excited state of ^{65}Ni. Figs. 3 and 4 show the measured neutron and proton reduced widths in the vicinity of these analog states. The linear correlation coefficients are $r \simeq 0.1$ for the ^{54}Cr analog and $r \simeq -0.1$ for the ^{64}Ni analog. Since these values are consistent with statistical behavior, we conclude that there is no evidence for enhancement of the neutron widths of these analogs.

Fig. 5 shows the elastic and capture excitation functions including the fine structure of the analog of the ground state of ^{55}Cr. The capture data shown in Fig. 5 are obtained from NaI(Tl) spectra by summing over the energy regions indicated. The solid line through the elastic data is a multi-level fit.[4] The widths obtained from this analysis are shown in Fig. 6. This fine structure distribution is highly unusual, having most of the strength at the extremes of the distribution. Since the $3/2^-$ resonances not strongly enhanced by the analog state are too weak to be resolved in the elastic scattering, there is essentially no information on the background widths in this energy region.

In order to obtain detailed information on the decay of these resonances, spectra were measured at the resonance energies with a Ge(Li) detector.[5] Strong transitions are observed to the ground state of ^{55}Mn (this is the strongest transition), and to excited states at 1.524, 2.251, 2.564, and 3.037 MeV. Except for the ground state ($J\pi = 5/2^-$), these levels are all populated via $\ell = 1$ transfer in the ^{54}Cr (^{3}He,d) reaction, and presumably indicate the (fragmented) anti-analog state. The (^{3}He,d) strengths are shown in Fig. 7, along with the relative strengths (divided by E_γ^3) for gamma transitions to these levels.

Fig. 8 shows the decay of the eight fine structure resonances to the five final states most strongly populated in ^{55}Mn. For the resonances numbered 2, 6, 7, and 8, the ground state transition dominates. For the other resonances, no clear decay pattern is evident. Fig. 9 shows the relative strengths for elastic, inelastic, and five capture channels for all eight reso-

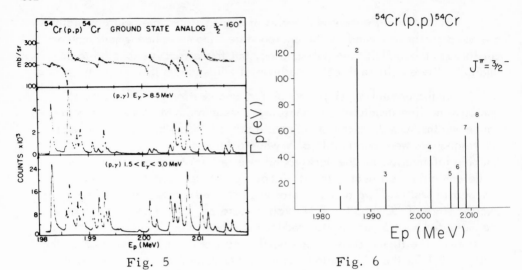

Fig. 5   Fig. 6

Fig. 5   Elastic and capture data for the analog of the ground state
        of ^{55}Cr. The solid line for the elastic data represents a
        fit, while the line through the gamma ray data is merely to
        guide the eye.

Fig. 6   Elastic widths for the 8 fragments of the analog.   The
        resonances are numbered for later reference.

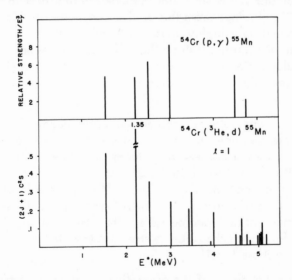

Fig. 7

Fig. 7   (^{3}He,d) strengths to excited states of ^{55}Mn and relative
         strengths of gamma ray transitions to those states (summed
         over the 8 fragments of the analog).

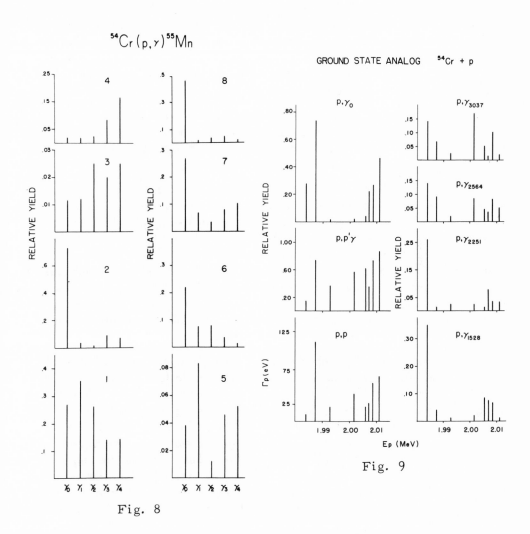

Fig. 8

Fig. 8 Decay patterns for the 8 fragments of the analog. The
subscripts 0, 1, 2, 3 and 4 refer to the ground state and
excited states at 1.528, 2.251, 2.564 and 3.037 MeV,
respectively. Note that the decay of resonances 2, 6, 7
and 8 is dominated by the ground state transition.

Fig. 9 Relative strengths for elastic, inelastic and gamma ray
channels. Note the similarity of the patterns for p, p_1
and γ_0 channels.

nances. The elastic, inelastic, and γ_0 strengths are all correlated. This is qualitatively apparent in Fig. 9, and quantitatively the linear correlation coefficients are large and statistically significant. Further, it is interesting to note that the γ_{1528} and γ_{2251} widths have a similar pattern which is qualitatively different from the elastic pattern, and that the γ_{2564} and γ_{3037} widths also appear to be correlated, although not as strongly. The interpretation of these correlations is discussed below.

The correlation between the elastic proton and inelastic widths is another example of the enhancement of inelastic widths by the analog, as in the ^{48}Ti data discussed above. The correlations between γ_{1528} and γ_{2251} and between γ_{2564} and γ_{3037} are unexplained at present.

The strong gamma rays (other than γ_0) are presumably analog to anti-analog state transitions. M 1 analog to anti-analog transition strengths in the s-d shell are predicted to be on the order of Weisskopf units, and such transitions are observed.[6] In nuclei with an unfilled $f_{7/2}$ proton shell, however, these transitions are observed to be inhibited[7,8] by a factor of $\sim 10 - 100$. Summing our data over the eight resonances and four final states (using preliminary absolute efficiency measurements for the Ge(Li) detector), we find an analog to anti-analog strength of about a tenth of a Weisskopf unit, assuming that the transitions are pure M 1. The analog to anti-analog gamma widths are not correlated with the elastic widths.

The correlation between the elastic and γ_0 channels, and the strength of the γ_0 transition, can be explained tentatively as follows: The parent state (the ground state of ^{55}Cr) beta decays to the ground state of ^{55}Mn with a log ft value of 4.95. There should be a corresponding M 1 gamma decay from the analog state to the ground state of ^{55}Mn. Using the prescription given by Hanna,[9] the strength Λ(GT) of the ground state beta branch corresponds to a gamma width of about 0.3 - 0.4 eV. Summing over the eight fragments of the analog state, we obtain $\Gamma_\gamma \sim 1$ eV, assuming that the transition is pure M 1. Thus it appears that the analog of the ground state of ^{55}Cr is a doorway for the γ_0 channel as well as for the elastic and inelastic channels.

The authors would like to thank Dr. A. M. Lane for extremely valuable advice and comments, Prof. L. C. Biedenharn, Prof. D. Robson and Dr. R. Chrien for valuable discussions, and Dr. J. C. Browne and Dr. G. L. Morgan for help in performing some of these experiments.

* Work supported in part by the U. S. Atomic Energy Commission
1. A. M. Lane, Isospin in Nuclear Physics, ed. D. H. Wilkinson (North-Holland, Amsterdam, 1969) p. 509
2. N. H. Prochnow, H. W. Newson, E. G. Bilpuch and G. E. Mitchell (to be published)

3. J. D. Moses, J. C. Browne, H. W. Newson, E. G. Bilpuch and G. E. Mitchell, Nucl. Phys. A168, 406 (1971)
4. J. D. Moses, H. W. Newson, E. G. Bilpuch and G. E. Mitchell (to be published)
5. W. C. Peters, E. G. Bilpuch and G. E. Mitchell (to be published)
6. P. M. Endt, Nuclear Isospin, eds. Anderson, Bloom, Cerny and True (Academic Press, New York, 1969) p. 51
7. S. Maripu, Phys. Letters 31B, 181 (1970)
8. H. V. Klapdor, Phys. Letters 35B, 405 (1971)
9. S. Hanna, Isospin in Nuclear Physics, ed. D. H. Wilkinson (North-Holland, Amsterdam, 1969) p. 591

DISCUSSION

BERMAN   (L.R.L.)
     Another type of channel-resonance effect concerns the correlation
of $\Gamma_{\gamma 0 \lambda}$ and $\Gamma_{n\lambda}^0$ for s-wave ($J^\pi = 1/2^+$) states in Cr^{53} and Fe^{57}. We
have examined the threshold photo-neutron crossections for these
nuclei and compared them with the Duke neutron data. For eleven
$1/2^+$ states in Fe^{57}, the correlation coefficient was calculated to
be 0.14; for nine $1/2^+$ states in Cr^{53} it was 0.07. Therefore,
for both cases, the c.c.'s are not significantly different from
zero; indicating the channel-resonance terms are small compared with
the compound-nucleus terms.

JACKSON   (Argonne)
     Since the question of the correlation between the ground
state radiation width and the reduced neutron width seems important,
we have done essentially the same analysis, and I'll report the
results Friday morning. We have looked with threshold photoneutrons at the compound nuclei, Cr^{53}, Fe^{57}, and Ni^{61}. In each
case, we see correlation coefficients which are quite small and
we feel indicate no evidence for a significant correlation. In
fact, if you combined the three nuclei, as was done in an earlier
experiment due to Block and presented as evidence for a strong
correlation in the total radiation width, we find a correlation
coefficient which is less than 0.1 and indicates that there is no
very strong correlation.

CHRIEN
     I hate to keep repeating myself, but if you look at the
contribution of Bhat to this conference, in each of these isotopes
the corresponding neutron targets would be Cr^{52}, Ni^{60}, and Fe^{56}, and
one uses the valence neutron model to predict the ground state
radiation widths. For the first resonance in each isotopes, the
agreement is remarkably good. As one goes up, the agreement
disappears. This is certainly an odd coincidence, if indeed it is.

JACKSON
     It was suggested that perhaps this threshold effect might
explain this effect, but I don't believe it will. For example,
if you look at the Ni data you find that within 100 kilovolts of
the state you feel shows a strong correlation there is another state
with a strong reduced width, and we see no strong corresponding
radiation width. I don't think that is a possible explanation.
Monahan and Elwyn proposed doorway states in Fe^{56}+n at 400 and
800 keV. We find no strong correlation between these structures
and our averaged threshold photo-neutron crossection.

SESSION V

AVERAGE RESONANCE PARAMETERS

Chairman   - W. W. Havens, Jr., Columbia University
Scientific Secretary - B. J. Allen, Oak Ridge National Laboratory

STRENGTH FUNCTIONS AND INTERMEDIATE STRUCTURE

Henry W. Newson

Triangle Universities Nuclear Laboratory and Duke University

Durham, North Carolina

About five years ago at the neutron conference in Antwerp,[1] I reviewed the status of strength functions. We were plagued by some instrumental difficulties which, however, have been cleared up mostly by the efforts of Divadeenam[2] and Bilpuch and this happy development was reported about a year later at the Washington neutron conference.[3] As a result of this improvement in reliability, we decided that a further effort on average cross sections would lead to enough new and accurate data to make a considerable improvement in the optical model interpretation of cross sections and hopefully to obtain a set of parameters which would reproduce the experimental facts accurately over most of the Periodic Table. Figure 1A shows the most optimistic interpretation of the labors involved in the theses of Pineo,[4] Divadeenam,[2] Tabony[5] and of earlier work.[5] In this slide we have not broken up the cross section curves into their s- and p-wave components but have simply plotted the cross sections averaged from 300 to 650 keV. One sees immediately that the experimental points bring out only the strong p-wave peak at A = 90 and the two strong s-wave peaks (not well resolved) in the rare earth region. The theoretical calculations by the models of Buck and Perey[6] and of Tamura[7] are also shown. One sees a fairly good fit to experiment in the upper curves but, looking at the data at the lower one we see immediately that there are distortion parameters needed in Tamura's program and these parameters are not very well known either experimentally or theoretically.[8] The upper limit of the parameter lies in the rotational region of the rare earths and one sees that the solid line is in reasonably good agreement with the experimental points while the dashed line is somewhat low, but the two curves seem to bracket the data. On the other hand, as might be expected, the lower limit of

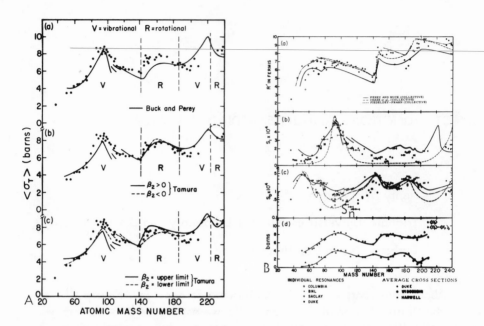

Fig. 1A. Calculated cross sections averaged from 300 to 650 keV and compared to measurements (mostly Duke and Wisconsin). The results are nearly
the same for averages from 0 to 650 keV. Fig. 1B. Strength functions compiled in 1970. Values depending on a few resonances have been omitted.
Essentially the same figure was discussed in detail at Antwerp.[1]

the distortion parameter gives the better fit in the vibrational region around
A = 90. While this is reasonably good the catch is that, in addition to all
the other more or less free parameters of the optical model, this treatment
now includes some distortion parameters which, while measurable in principle, are not in fact very well known experimentally; nor do the techniques now available appear adequate for a better determination in order
to remove the ambiguity in Fig. 1A. Figure 1B shows the more conventional
interpretation of the data in which we have broken the average cross section up into s- and p-wave strength functions and s-wave phase shifts. The
bottom curve in Fig. 1B shows the total resonance contribution to the cross
section which sometimes drops to very small values, particularly in the
neighborhood of mass number 50. It is for this reason that we do not see a
very clearcut peak in Fig. 1A at the well known s-wave giant resonance
around A = 50. These data in Fig. 1B include not only our own average
cross section data but also individual resonance data for which a reasonably large number of resonances have been analyzed. Again the fit is fairly
good near the strength function peaks, but we still see the well known tendency for a low s-wave strength function[9] particularly in the neighbor-

hood of tin but actually spreading nearly all the way between the p-wave giant resonance at A = 90 and the s-wave giant resonance near A = 140.

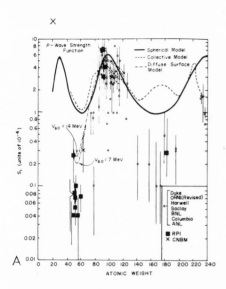

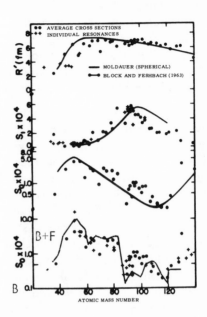

Fig. 2A. A plot of p-wave strength functions including recent anomalously high (X) and low (RPI) measured values.[10,11] This erratic behavior we attribute to the effects of doorway states. Fig. 2B. Shows much the same data as in Fig. 1B but emphasizes theoretical attempts to account for anomalously low s-wave strength functions.

In Fig. 2A we show other discrepancies reported by the hosts of this conference. In this case, the p-wave strength function in the neighborhood of the peak of the s-wave at A = 50 drops practically to zero, according to recent work at RPI,[10] and seems to be well below reasonably predicted (Fig. 1B) optical model minima. Another discrepancy is shown in this slide where the Albany[11] and Columbia workers find a very strong p-wave strength function between A = 19 and 27 Both of these recent results may be of instrumental origin, but my feeling is that many discrepancies are probably due to a fundamental shortcoming of the optical model as suggested some time ago by Block and Feshbach.[9]

Figure 2B shows much the same strength function data as shown in Fig. 1 with various attempts to account for the low s-wave strength function between mass numbers 90 and 140. The curve of Block and Feshbach[9] based on the assumption of 2p-1h doorway states is certainly an improvement over the calculations that are shown in Fig. 1B, but this better agreement is purchased at the cost of the introduction of even more free parameters into the optical model which is now approaching the complexity of a slowly converging infinite series which needs a new term for each new experiment. Indeed Moldauer[12] has shown, also by suitable adjustment of the parameters of the optical model, that s-wave strength functions in this region can be accounted for without recourse to the 2p-1h states. Thus the explanation of these low s-wave strength functions is still in question. It will be the object of the remainder of this talk to attempt to demonstrate clearcut cases of doorway states.

About the time of the Antwerp conference we announced a recently discovered doorway state,[13] observable as a single resonance in the neutron cross section of Pb^{208} which appeared to break up into a number of levels when Pb^{206} was bombarded. This argument was considerably strengthened by the fact that an accurate sum rule was observed between the width of the single (parent) resonance of Pb^{208} and the dozen daughter resonances in Pb^{206}. For a long time we feared that this might be the only demonstrable example of doorway states which affect neutron cross sections. (The isobaric analogue states which were discussed by Ed Bilpuch[14] yesterday can be considered to be doorway states which affect only the proton cross sections, but he also reports a few which do not appear to have an isobaric parent.) In order to find other cases, we attempted to increase the range of our measurements so as to resolve resonances in Pb^{206} + n at higher energies where it was known that there were s-wave resonances of Pb^{208}. While we succeeded[15] in getting rather good ($\Delta E \gtrsim 2$ keV) resolution in the region between 1.5 and 2 MeV, it was not sufficient to identify and measure the width of s-wave resonances in Pb^{206} at this high energy. However, we were able to make a considerable improvement in our curve fitting technique as illustrated in Fig. 3A. This is an example of shape analysis of the total cross section of natural strontium[15] at the highest resolution which we have been able to achieve with Li(p,n) neutrons. The relatively sparse points which appear in most places are the experimental total cross section measurements and the closer points, which appear as solid lines in the flatter regions, are the calculated cross sections. This figure is actually a photograph of an oscilloscope trace from which the calculated and measured cross sections could be compared as the analysis proceeded. This is not our best example since, in that case, the experimental and calculated points would be indistinguishable on the scale of this figure. The next step in fitting the largest peak on the curve is to type in a higher assumed E_0 for that resonance and consequently move the calculated peak horizontally to get better agreement with experiment. This operation could be performed without any other readout than visual observation of the oscilloscope. Thus, as the experimenter fits his data, he can vary parameters and check the agreement by observing the oscilloscope but without any readout until he considers that the fit is optimum. This improvement in technique is simply a

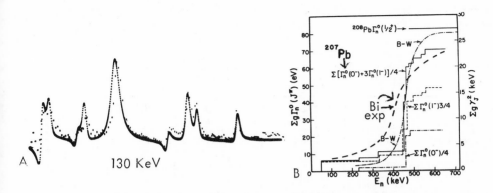

Fig. 3A. Photograph of oscilloscope trace during fitting process of natural strontium cross section curve. The closer points are calculated from an R-matrix code and the widely spaced ones are experimental. The total energy range is about 200 keV. Fig. 3B. Interpretation of the observed s-wave resonances of Pb^{207} and Bi^{209} (dashed) showing the apparent effect of a doorway state previously found in Pb^{206} and Pb^{208} + n. The lower two histograms show the separated effects of the $1^- = I^\pi + 1/2$ and $0^- = I^\pi - 1/2$ where $I^\pi = 1/2^-$ for Pb^{207}, and the upper histogram gives the weighted sum which is compared to Pb^{208}.

matter of convenience, but it made the process of curve fitting practical in some cases which had previously been given up as hopeless. This method was worked out by Fred Seibel [16] as part of his thesis and later refined by Pineo.

Figure 3B shows the results of analysis of the cross sections of the odd nuclei nearest to doubly magic Pb^{208}. The upper horizontal line shows the reduced width of the single strong s-wave resonance in Pb^{208} at about 500 keV. The dot-dash $\int$-shaped curve is the integrated Breit-Wigner formula and shows what one would see if the spreading width were about equal to the escape width and if the latter were about the same as the neutron width of the single resonance in Pb^{208}. Note the total width ($\Gamma_d \equiv \Gamma^\uparrow + \Gamma^\downarrow \sim 100$ keV) is about the distance between 1/7 and 6/7 of the saturation value. The solid histogram [17] gives the weighted sum of the two s-wave channels obtained from Pb^{207} + n. One sees that the histogram rises steeply between 400 and 500 keV and that there is a relatively long flat tail which shows that there is an appreciable spreading width, $\Gamma^\downarrow$ about the same size as the escape width, $\Gamma^\uparrow$. The sketched-in part of the figure shows the experimental results [17] of the analysis of Bi^{209} + n. The integral-shape of this (dashed) experimental curve is clearcut in the lower energy part and there is still curvature present in the upper portion which shows a definite tendency to flatten out near the horizontal line corresponding to the $1/2^+$ resonance in Pb^{208}. In this case, the escape width $\Gamma^\uparrow$ seems to be in reasonable agreement with that found with Pb^{208}, but the spreading or damping width, $\Gamma^\downarrow$, is much greater than that of Pb^{207}, as may be seen in the table.

Table I. Estimated Parameters of Doorway States (keV)

Target	Tl^{205}	Pb^{206}	Pb^{207}	Pb^{208}	Bi^{209}
E_d	---	~475	~450	500	~350
$\Gamma_d = \Gamma^\uparrow + \Gamma^\downarrow$	($\gtrsim$ 400)	~250	~100	58	~300
$\Gamma^\uparrow = \Sigma\Gamma_n$	(~ 50)	58	~ 50	58	~ 50
$\Gamma^\downarrow$	($\gtrsim$ 350)	~190	~ 50	0	~250

In addition to the parameter from Fig. 3B, Table I includes those from our earlier interpretations of the Pb^{206} cross section.[13] This doorway state has been discussed theoretically by Divadeenam and Beres[18]. We also have data on natural thallium which looks as if we could find a doorway state like that of Bi if we had a sufficiently thick sample of separated Tl^{205}.

In Fig. 3B we have not only shown evidence that the doorway state affects the neutron scattering of odd as well as the even-even isotopes of lead, but (particularly in the case of bismuth) our interpretation now depends on cross sections which include a much larger number of resonances than the dozen which were found[13] in Pb^{206} + n. This makes it very unlikely that the ∫-shaped pattern of the latter is a statistical accident. Later we will show that in molybdenum, apparent intermediate structure in even-even targets disappears in an odd target where resonances are more numerous.

In spite of the crudity of the estimates of spreading width, $\Gamma^\downarrow$, in Table I, we see a very definite trend: As we go away from doubly magic Pb^{208}, the spreading widths increase rapidly. If this trend continued, $\Gamma^\downarrow$ would soon approach several MeV (the imaginary part of the optical potential) and hide the intermediate structure completely. Thus it would appear that doorway states, narrow enough to be easily recognizable, are most likely to be detected in other doubly magic targets, and will be found less often as holes or particles are added to the closed shells. If the spreading widths, $\Gamma^\downarrow$ become much larger than those in Table I, one would observe a nearly linear stairstep plot and the strength function (which, of course, is proportional to the slope) would be a constant which one could then compare with optical model predictions such as those in Fig. 1B. Doorway states are indicated when the strength function is not constant over the energy range of the measurements. Since any relatively flat portion of the stairstep curve must be between narrow doorway states, one is more likely to observe a flat curve at low energies followed by a sharp increase at

higher energies. However, it will happen occasionally that there will be
a doorway state near $E_n = 0$ in which case the low energy strength function will be very high at low energies and decrease as the neutron energy
increases. It is also possible (particularly when the spreading width is
large) to find a stairstep curve which is nearly linear throughout the range
of the measurements. We illustrate examples of all three cases in Fig. 4A,

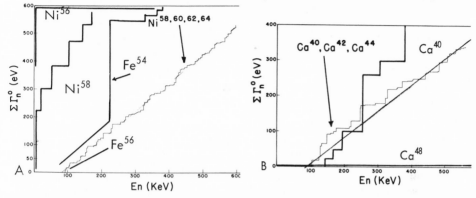

Fig. 4A. Apparent doorway state effects near $_{28}Ni^{56}_{28}$ (calculated 2p-1h histogram[19]). The experimentally measured resonances of Ni^{58}, Fe^{54}, and Fe^{56}
are compared to a composite curve for all even Ni isotopes which ap-
proximates the average (optical model) strength function in this mass region. The strong, low energy resonance in Ni^{58} is actually at negative
neutron energy.[20]   Fig. 4B. A similar comparison of doubly magic Ca^{40}
and Ca^{48}. Predicted 2P1h state of Ca^{40} are near 250 and 600 keV

which also includes (upper left hand corner of the Figure) a prediction by
Divadeenam[19] of a two-particle one-hole (2p-1h) doorway state near zero
neutron energy for a Ni^{56} target. Ni^{56} might be called "triply" magic
since $N = Z = 28$ and when $N - Z = 0$ we have the effects of a closed
shell since every nucleon is strongly attracted by the three others which
share the same orbit. Unfortunately, Ni^{56} is a short-lived radioactive nucleus which is not available in sufficient quantity to measure a neutron total cross section. However, we can compare the calculation with the
nearest neighbors for which measurements are available. The results of the
Fe^{56} measurements[20] are shown in the lower part of the curve where they
are compared with the total reduced widths of the four even-even Ni
isotopes. This composite curve was drawn some time ago in order to obtain an average strength function for the Ni isotope to plot in Fig. 1B. The
average number of neutrons for the composite curve is 33, far enough from
the magic number to make observation of any doorway state effects unlikely.
Fe^{56} is about equally far away since it has two neutrons more and two protons less than the magic numbers. Thus the spreading width for Fe^{56} should
be similar to that of the composite curve and indeed we see no signs of in-

termediate structure in either one. However, Fe^{54} has only two protons less than double magic and Ni^{58} has two neutrons more. These two stair-step curves[20] do indeed show a steep rise of roughly the strength predicted by the calculations. The fact that the steeply rising regions of these two nearly doubly magic nuclei are displaced from each other is to be expected since the different nuclear radii will cause a shift of the single particle states by about the observed amount.

The closer we approach a double magic nucleus the more likely we are to see at least a portion of the ʃ-shaped pattern, but going farther away, as in the case of Fe^{56}, the intermediate structure is often hidden by wide spreading widths. There is actually evidence[21] for a doorway state in Fe^{56} at higher energies than those emphasized in the Figure.

Fig. 4B is a similar slide comparing a composite stairstep curve com-bining the reduced widths of three calcium isotopes[20] to doubly magic[17] Ca^{48} and triply magic[20] Ca^{40} (N = Z = 20). We see again that the stair-step curve for Ca^{40} rises much more rapidly than the composite curve after a somewhat wider flat region at lower energies. The steep part of the rise falls between two predicted doorway states indicated by arrows in the fi-gure.[19] (In this case Divadeenam did not estimate the widths.) While the apparent effects of the doorway states are less marked here than near Ni^{56}, it is again clear that whatever effect there is in Ca^{40} tends to smooth out when we go to targets farther from the magic numbers. For a target of Ca^{44} (farthest calcium isotope from double magic) we do   see a linearly in-creasing stairstep plot from 0 to 600 keV. The case of Ca^{48} is very striking; we have been able to find no s-wave resonances up to 1.4 MeV. Divadee-nam and Beres[18] predict a 2p-1h s-wave doorway state much higher than the range of the experimental measurements. Ca^{48} is probably an extreme case of doorway states in magic nuclei with narrow spreading widths and spaced widely enough that the strength function (even when averaged over 600 keV) is anomalously low like those of natural tin (Z = 50) isotopes pro-posed by Block and Feshbach.[9] See Fig. 1B. One should also look for doorway states in light targets where N = Z inasmuch as this combination should behave like a closed shell in the heavier nuclei. Looking back over the neutron resonance data, one does indeed find definite signs of this ef-fect in fluorine, potassium, chlorine, phosphorous, magnesium and alumi-num. The best example is natural silicon for which N = Z = 14 correspond-ing to the closure of the $d_{5/2}$ shells. We were privileged to analyze the data taken by the Bureau of Standards group[22] using their electron linac as a pulsed neutron source. This data started at about 500 keV and it was an-alyzed up to the inelastic scattering threshold. A preliminary interpretation is shown in Fig. 5. One sees the p-wave strength concentrated between 500 and 1000 keV. The possible presence of d-waves at the higher neutron energies throws some uncertainty into the analysis but does not spoil the

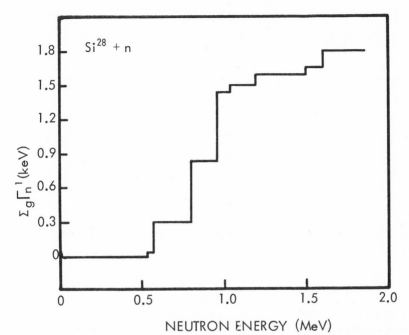

Fig. 5. Interpretation of the p-wave resonances in natural silicon (92% Si[28]) based on National Bureau of Standards Measurements between 0.5 and 2.0 MeV. Earlier measurements([20]) showed that the p-wave strength below 0.5 MeV is small.

steep rise followed by a flattening in the p-wave strength.

Finally, we were anxious to find evidence for a doorway state in another heavy nucleus besides those near Pb[208]. The best candidates were Ba[138] and Sr[88]. Both have a magic number of neutrons but the latter has the strong closed subshell at 38 protons, and a level density suitable for analysis of individual resonances. The barium resonances[15] unfortunately turned out to be too closely spaced to be readily analyzable, but we are able to fit the cross section curve of strontium[23] which is shown in Fig. 6. It is interpreted in Fig. 7 where there is a very decided change in slope, indicating a large increase in p-wave strength function, at about 300 keV where there are two extremely strong resonances which almost double the strength of the stairstep curve. There are also indications of the flattening to be expected at the top of the ʃ-shaped curve but the analysis in the highest energy region is only preliminary. This figure also compares the experimental data to the calculations of Divadeenam and Beres, which are based on both two-particle one-hole[18] doorway states and also on vibrating core + single particle[18] neutron states. It is perhaps most interesting that both models predict that a considerable fraction of the single particle strength of the giant optical model resonance is concentrated in the region

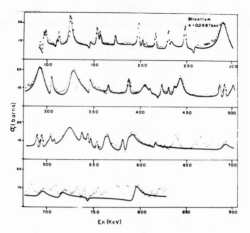

Fig. 6. The total cross section of natural strontium (83% Sr^{88}) fitted with an R-matrix code[23] (Preliminary).

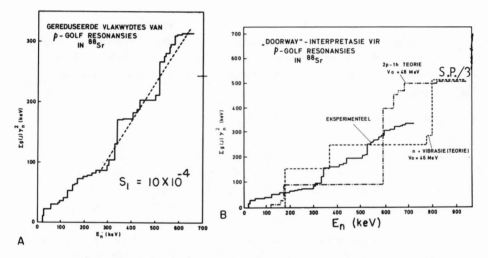

Fig. 7A. The weighted sum (Preliminary) of the reduced p-wave widths in Sr^{88}. The strength function in usual neutron units of 10^{-4} (eV) is three times the value in the figure, i.e., the slope of the dashed line is about 10×10^{-4}. Fig. 7B. The same data as 7A compared to the predictions of both 2p-1h and n + vibrating core doorway states. The flat regions at the upper right correspond to about 1/3 of the single particle strength (Wigner Limit).

of the measurements. Thus both the $\int$-shape and the amplitude of the experimental curve indicate doorway states in this region which correspond roughly with the calculations.[18]

Zr^{90} might have been an even better candidate for this experiment but is known that it's level density is so high that the resonances would have been even more difficult to analyze than those in Ba^{138}. What one therefore needs is a method of locating doorway states in averaged cross sections, a procedure which is probably necessary to identify the doorway states in tin which were predicted by Block and Feshbach. Figure 8 indicates the difficulties in attempting to identify doorway states in averaged cross section curves. Figure 8A shows low resolution cross sections of several molybdenum isotopes.[2] There are many fluctuations which can cer-

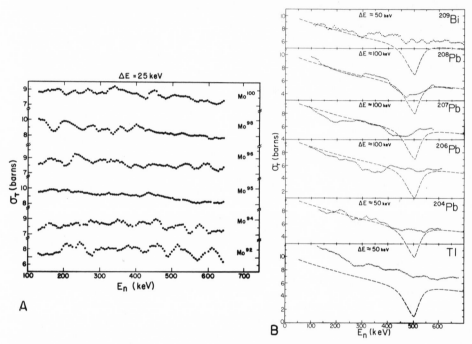

A

B

Fig. 8A. Total cross sections of six molybdenum isotopes measured under nearly identical conditions and then averaged over the same energy interval (ΔE) to insure identical effective resolution. Fig. 8B. Total cross sections of Pb^{208}, Pb^{207}, Pb^{206}, and natural bismuth and thallium. They were averaged to simulate partial resolution of the resonances. In each case the average experimental curves are compared with what would have been observed with perfect resolution on Pb^{208} if no p- or d-wave resonances were present.

tainly be called intermediate structure, and one would be tempted to attribute them to doorway states similar to those found in lead. However, in the Mo^{95} curve (which includes about ten times as many resonances as any of the others) the fluctuations have practically disappeared, and they must be attributed to lack of a significant number of resonances in the averaging interval. It will be remembered that the odd targets near Pb^{208} (particularly Bi) behaved differently. In Fig. 8B one sees the effect of averaging cross sections in the neighborhood of Pb^{208} where the doorway state structure is definite, but in these averaged curves one sees fluctuations little more impressive than those of the even–even Mo isotopes. It is evident that one cannot rely on simple inspection of partially resolved cross section curves to identify intermediate structure. However, since we had data and consistent calculations for Sr^{88} it seemed most interesting to attempt to understand an odd nucleus, Y^{89},[24] in that region just as we did near Pb^{208}. While the measured cross section itself had no obvious structure in the neighborhood of the predicted doorway state, we found that by taking the measured value of the s–wave potential scattering at low energy (where there are no resonances) and then extrapolating it to higher energies by the relation $\sigma_p^o = 4\pi\lambda^2\sin^2 kR'$, a σ_p^o curve was predicted which followed the experimental curve for Y^{89} rather closely up to several hundred keV. When this extrapolated potential scattering curve was subtracted from the measured[24] cross section we found indeed that a peak appeared at about 500 keV (Fig. 9) near a strong predicted[18] $3/2^-$ 2p–1h state. The resonances of Y^{89} at this energy would not be resolvable by any technique now

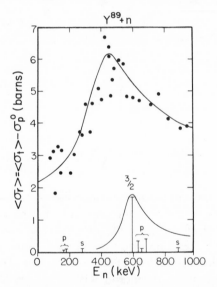

Fig. 9. The s–wave potential scattering, σ_p^o is extrapolated from low energies and subtracted from an unresolved total cross section[24] curve of Y^{89}. The difference is compared schematically to the contribution to the cross section to be expected from the strongest predicted[18] 2p–1h state if the spreading width $\Gamma^\downarrow= 0$. Weaker predicted 2p–1h states are indicated as vertical lines with flags. Poor agreement in absolute cross section can be ascribed to error in the extrapolation of σ_p^o.

known or even contemplated, but it appears that a predicted doorway state for Sr^{88} is affecting the slightly heavier Y^{89}. Analogous treatment of differential cross section curves may make it possible to identify doorway states in other unresolved resonance spectra.

In conclusion, it seems quite likely that doorway states and other intermediate structure play some part in almost any neutron cross section, but spreading widths (which so far can not be predicted with any certainty) are usually so large that the total widths of the doorway states overlap. Consequently, except very near doubly magic nuclei, it is usually impossible to separate these effects from those described by the optical model.

References

1.   H. W. Newson, Nuclear Structure Study with Neutrons (Antwerp, 1966).

2.   M. Divadeenam, Ph.D. Thesis (1967), Duke University (unpublished). M. Divadeenam and H. W. Newson, Bull. Am. Phys. Soc. 12, 106 (1966).

3.   H. W. Newson, Neutron Cross Sections and Technology Conference (Washington, 1968).

4.   W. F. E. Pineo, Ph.D. Thesis (1970), Duke University (unpublished). W. F. E. Pineo, M. Divadeenam, E. G. Bilpuch, and H. W. Newson, Bull. Am. Phys. Soc. 15, 568 (1970).

5.   K. K. Seth, R. H. Tabony, E. G. Bilpuch, and H. W. Newson, Physics Letters 13, 70 (1964).
        K. K. Seth, R. H. Tabony, E. G. Bilpuch, and H. W. Newson, Comptes Rendus du Congres International de Physique Nucleaire, 1964 (Paris).
        R. H. Tabony, Ph. D. Thesis (1966), Duke University (unpublished).

6.   B. Buck and F. Perey, Phys. Rev. Letters 8, 446 (1962). The total cross sections versus A curve was generated by using the strength functions and R' given in this reference.

7.   T. Tamura, ORNL Report No. 4152 (Coupled Channel Computer Program JUPITOR I). Revs. of Mod. Phys. 37, 679 (1965). Computor Code JUPITOR I was used at Oak Ridge by M. Divadeenam and at Duke by W. F. E. Pineo.

8.   P. H. Stelson and L. Grodzins, Nuclear Data 1, 427 (1965).

9*.   B. Block and H. Feshbach, Ann. of Phys. 23, 49 (1963).

10.   R. C. Block, private communications.

11.   J. B. Garg, U. N. Singh, J. Rainwater, W. W. Havens, Jr. and S. Wynchank, these proceedings.
        U. N. Singh, J. B. Garg, J. Rainwater, S. Wynchank, M. Slagowitz, and H. Liou, these proceedings.

12.   P. Moldauer, Nucl. Phys. 47, 65 (1963).

13.   J. A. Farrell, G. C. Kyker, Jr., E. G. Bilpuch, and H. W. Newson, Phys. Letters 17, 286 (1965).

14.   E. G. Bilpuch, these proceedings (Section 2).

15.   J. G. Malan, W. F. E. Pineo, M. Divadeenam, E. G. Bilpuch, and H. W. Newson, unpublished.

16.   F. Seibel, Ph.D. Thesis (1968), Duke University (unpublished).

17.   F. Seibel, E. G. Bilpuch, and H. W. Newson, Ann. of Phys. (in press).

18.   M. Divadeenam and W. P. Beres, these proceedings (Section 8).
        W. P. Beres and M. Divadeenam, Phys. Rev. Letters 25, 596 (1970).
        M. Divadeenam, W. P. Beres, and H. W. Newson, Ann. of Phys. (in press).
        M. Divadeenam, E. G. Bilpuch, and H. W. Newson, Bull. Am. Phys. Soc. 16, 495 (1971).

19.   M. Divadeenam, unpublished.

20.   C. D. Bowman, E. G. Bilpuch, and H. W. Newson, Ann. of Phys. 17, 319 (1962).
        J. A. Farrell, E. G. Bilpuch, and H. W. Newson, Ann. of Phys. 37, 367 (1966).
        E. G. Bilpuch, K. K. Seth, C. D. Bowman, R. H. Tabony, R. C. Smith, and H. W. Newson, Ann. of Phys. 14, 387 (1961).

21.   A. J. Elwyn, and J. Monahan, Nucl. Phys. A123, 33 (1969).

22.   R. B. Shwartz, R. A. Schrack, and H. T. Heaton, Bull. Am. Phys. Soc. 16, 495 (1971), and private communication.

23.   W. F. E. Pineo, E. G. Bilpuch, H. W. Newson and J. G. Malan, Bull. Am. Phys. Soc. 16, 495 (1971).

24.   Wisconsin measurements and BNL 325.

*The calculations in Fig. 2B attributed to $(B + F)^9$ are actually due to Feshbach, Kerman and Lemmer.

DISCUSSION

BERMAN
      In the absence of Denys Wilkinson, I think we should compliment
Henry on his elegant set of slides. Secondly, I have a request
to hear once again, an explanation why the energy of the doorway
state should be nailed to the neutron-separation energy rather
than to the ground-state energy of the compound nucleus.

NEWSON
      To coin a phrase, I'm glad you asked that question, because
the next slide will provide the answer. (Refer to the slide page
325.) Our best case of what we claim to be a doorway is seen
in Pb^{207} and Pb^{209} compound nuclei. A long time ago K.Way pointed
out that if you took these two level schemes and fitted them together
at the neutron escape energy, the single particle levels in Pb^{209}
seem to line up fairly well with the sign of single particle levels
in Pb^{207}. In particular, the $1/2^+$ level which is what we want to
line up properly looks pretty good here. There is newer and
better data that seems to be consistent with this. If you add a
neutron to Pb^{208} it's obvious you can get single particle strengths
with the entering neutron being a valence neutron. Pb^{208} cannot
have a single hole state. The closest you can come to that in
Pb^{208} is a 1p-1h state. Whereas the ground states in Pb^{207} are
1h states. There are a whole set of levels here in Pb^{207} that have
no connection with any possible state in Pb^{208}--they have entirely
different spins and parities, and there must be some displacement
of this kind in order to line up the single particle levels. You
can get a single particle level here by adding a neutron to Pb^{206}
and the lowest single particle level would be a $9/2^+$ here, just
as the ground state of Pb^{209} is $9/2^+$. Attempting to translate this
to low energies and to compare the levels by their spins and
parities and so forth will get you nowhere at all. The ground
state is $9/2^+$ for Pb^{209} and $1/2^-$ for Pb^{207} and so forth. You
must take account of the fact that when you say I've got a bunch
of levels which I think are the daughter products of the doorway
state you are comparing levels of the same spin and parity. So
to make that comparison, you ought to be able to compare levels of
the same spin and parity near the ground state. I'm glad you asked
that question. I hope I satisfied you.

FULTZ (Livermore)
      Henry, I looked at your Si^{28} results there, and it looked
like you had kind of a step in your strength function. Could you
tell me roughly what excitation energy that would correspond to
if one were to come in with, say, a gamma ray and make the Si^{29}
nucleus by the inverse reaction.

NEWSON
      I really don't know what the excitation energies are. The

neutron escape energy, of course, is usually around 8 MeV and I
haven't looked at it any closer than that.

FULTZ
     Anyway, what I wanted to say is, have you done Mg^{24}?

NEWSON
     Yes, but not very well.

FULTZ
     What did you find there?

NEWSON
     Well, we found a p-wave level at 80 kilovolts, and another
one at a few hundred and that's about all we've done. We find this
to agree with some of the effects, that is almost no s-wave
strength in Mg^{24}, and the strength is distributed erratically.
One thing I would like to get some of my friends, who have much
better methods of measuring than we do (at least that's what they
tell me), to get them to do some of these things like natural
magnesium which is mostly Mg^{24} and work these things out up to a
million volts or higher, as we did for Si.

FULTZ
     We did the inverse reaction Mg^{26} (γ,n) Mg^{25} and we looked
at the ground state transitions and made the same plots you did,
and this is because we were able to analyze all the resonances
right through the giant resonance. We also found a step in our
strength function at about 16 MeV and we interpreted this as the
onset of the giant resonance. (Phys. Rev 4C. 149, 1971) It
would be interesting to see whether the inverse reactions produce a
break in the strength functions at the same energies as is observed
in the (γ,n) crossection for nuclei which exhibit this effect.

NEWSON
     It would have been better if you'd done Mg^{25}.

FULTZ
     We did Mg^{25}, but we were unable to resolve the resonances,
because they were just so tightly packed; and unfortunately we
were unable to analyze it.

NEWSON
     When you get down to where the neutron and proton numbers
are the same, you have something like a double magic effect again
There's a certain amount of gap because of the symmetry effect.
All through that region what data there is, and it's neutron
(and it's very scratchy), looks as if this effect would be coming
in--probably much stronger. Certainly in Si^{28} it's much stronger

than it is in S^{32}. I wouldn't be a bit surprised to see a similar
pattern to Si in all the magnesium isotopes. We do know there's
a big p-wave resonance at about three or four hundred kilovolts
in all the magnesiums. There was not time to review all the data,
particularly when it got scratchy-I decided not to bother with it.

JACKSON

We are actually studying the $Si^{29}(\gamma,n)$ crossection in
threshold for the neutron measurements, and in fact we see a strong
concentration of radiative strength that correlates very well with
the staircase function that you presented in your slide. It looks
like there is evidence for a strong correlation between the two
experiments. This is $Si^{29}(\gamma,n)$. In answer to Stan Fultz's
question, in (γ,n) reaction, the corresponding state would appear
at around an excitation of 10 MeV.

NEWSON

I see. I did want to make one more remark that the Livermore
people, Berman et al, did do some (γ,n) work on Pb^{207} that
seemed to be consistent with our work with Pb^{206} plus a neutron.

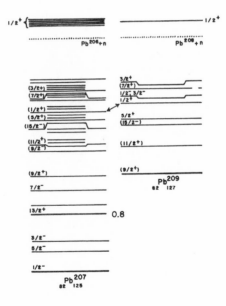

AVERAGE TOTAL NEUTRON CROSS SECTIONS

OF HEAVY ELEMENTS AT 2.7 keV *

W. Dilg and H. K. Vonach

Technische Universität München

München, German Federal Republic

Average total cross sections of 31 heavy elements
(A>100) at 2.7 keV mean neutron energy have been measured
at the Munich Research Reactor FRM. The data were used to
study the s-wave strength function in the mass range of
the 4s giant resonance.

1. METHOD

Transmission measurements were carried out with Na
resonance neutrons, selected from the pile epithermal
spectrum by means of a resonance scattering method {1}.
The principle is illustrated by Fig.1: The detector only
counts neutrons, which have twice been scattered from thin
layers of NaF. The first layer ("source") is placed in the
center of the FRM tangential beam tube, the second one is
placed close to the detector in the well collimated beam
arising from the "source". In both steps neutrons with
energies near the 2.85 keV Na resonance are strongly preferentially scattered. The effective spectrum thus obtained
is expected to be centered near (100-200 eV below) the resonance energy and to have a width about 500 eV. The integral contribution to the counting rate due to off-resonance
neutrons is about 15%. It is further reduced by a factor
∿3, as additionally a (thick) Na-filter difference technique is applied. The quoted estimates on the beam quality
are results of different transmission measurements on Na,
B and Cl, in order to check the spectrum used.

2. MEASUREMENTS

By this energy selection method a high count-rate and correspondingly high statistical accuracy was achieved (sample-out counts $\sim$ 800/sec). Thus very thin samples of the heavy elements could be used (n $\approx$ 0.001 - 0.004 atoms/barn), for which finite-thickness corrections are small. The average cross sections are then rather immediately connected with the observed transmissions $<T> \approx 1 - n<\sigma>$.

The experimental problem, of accurately determining transmissions in the range 95 - 99%, was solved by the following procedure: Sample-in and -out measurements were alternated in short periods (2 - 5 min) during the whole data collection time (4 - 20 hours per sample). This eliminates small variations in the beam intensity and detecting equipment. The average transmissions were thus reproducible within counting statistics ($\Delta<T>/<T> \approx 0.05 - 0.2\%$).

3. RESULTS

Small corrections for resonance self-protection effects were applied, starting from the thin-sample-expansion

$$n \cdot <\sigma> \;=\; \ln \left(\frac{1}{<T>} \cdot (1 + \frac{n^2}{2} \cdot (<\sigma_c^2> - <\sigma_c>^2) + ..) \right) \qquad . \quad (1)$$

Herein σ_c = compound nucleus part of the cross section. With the usual assumptions estimates of the quadratic term

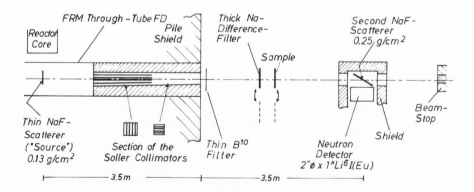

Fig.1: Schematic representation of the experimental arrangement. The Soller collimation system eliminates radiation, which directly arises from the core and the walls of the tube.

are easily obtained for the elements of interest, if even-
even and odd mass isotopes are distinctly treated and the
effects of both Doppler-broadening and strong overlap (in
some of the odd mass isotopes) are considered. The corrections exceeded in no case a few per cent.

Table 1 lists the measured cross sections, the "effective" number of resonances (equ. 4, 7) in the energy interval of the measurement, the values of the potential scattering radius R' and the p-wave strength function S_1 used in
the evaluation of the s-wave strength function, and the
s-wave strength functions calculated from the measured
cross sections. The calculation was based on equ.2, assuming only contributions of potential scattering and s- and
p-wave resonances to the total cross section

$$<\sigma> = 4\pi R'^2 + 2\pi^2 \lambda^2 \cdot (E/1eV)^{1/2} \cdot (S_o + 3k^2R^2 \cdot S_1) \quad . \quad (2)$$

As far as possible experimental values R' were used to subtract the potential scattering. The p-wave contribution,
calculated from the listed estimates of S_1, is very small
except in a few cases (Rh, In, Sn), where it amounts up to
20% of the s-wave resonance contribution.

The errors quoted for S_o take into account uncertainties due to (1) the measured cross sections, (2) the corrections (R', S_1), (3) the limited number of levels in the
energy range $\Delta E \simeq 500$ eV covered in our measurement. The
resulting errors are mostly dominated by this latter uncertainty, which is for a single isotope

$$\Delta S_o/S_o = (2.5/N)^{1/2} \quad , \quad (3)$$

with $$N = \Delta E/\overline{D} \quad . \quad (4)$$

For elements with several isotopes the strength function
calculated from the measured cross section is an average
over the values for the corresponding isotopes

$$S_o = \sum_i c_i \cdot S_{o i} \quad (5)$$

(c_i=isotopic abundance of isotope i). Each $S_{o i}$ has an uncertainty due to the finite number of resonances averaged
as given by equ.3. Combining these errors and assuming all
$S_{o i}$ to be approximately equal we obtain for this case

$$\Delta S_o/S_o = (2.5/N_{eff})^{1/2} \quad , \quad (6)$$

with $$1/N_{eff} = \sum_i c_i^2 \overline{D}_i / \Delta E \quad . \quad (7)$$

Thus the presence of even-even isotopes with wide spacing considerably decreases the "effective" number of levels. The worst cases, i.e. elements with N, $N_{eff} < 10$ (Te, La, Ce, Nd, Hg), are listed separately in the table and were omitted in the evaluation of S_0.

4. DISCUSSION

4.1 Cross Sections: To our knowledge there are only a few published cross sections in the low keV region to be directly compared with ours. Data from BNL 325 (1958) and ref. {2} are in general considerably lower than ours, probably due to the use of thick samples. Agreement is found with sporadic more recent measurements (Ag: {3}, Re: {4}).

Fig.2 shows the mass dependence of our total cross section results and also the cross sections predicted by the Duke s- and p-wave fit parameters {5,6} determined from total cross section measurements at higher energies (3-650 keV). With three exceptions (Ce, La, Nd), caused by the extremely small effective number of resonances, our data show the expected smooth gross structure in the dependence on the mass-number. Within errors they are in agreement with the extrapolated Duke data.

4.2 Strength Functions: In Fig.3 our values of the s-wave strength function are compared with previous results. Most of our data are only moderately accurate due to the relatively small energy interval used. Yet our results clearly show and thus confirm the previously observed features of the s-wave strength function: (1) the minimum at A=120, (2) the splitting of the 4s giant resonance, (3) a remarkable difference against optical model calculations {7} in the mass range near A=165. This was interpreted {8} as the appearance of an additional weak maximum caused by spin-orbit interaction {9} .

As shown in Fig.3, our results generally agree within errors with previous data both from resolved resonance and average cross section work. Only for some elements with small level spacing (Re, Eu, Ta) somewhat higher results have been obtained with the latter mehtod. This is seen in the comparison of data from refs. {4} ($Re^{185\ 187}$), {11} ($Eu^{151\ 153}$), {10} (Ta), with those of refs. {4} (Re_{nat}), {12} (Eu_{nat}), {6} and ours.

For the rest of the elements our cross section results at a few keV are in agreement with the known averaged s-wave resonance parameters, which mostly had been deduced from measurements at lower energies.

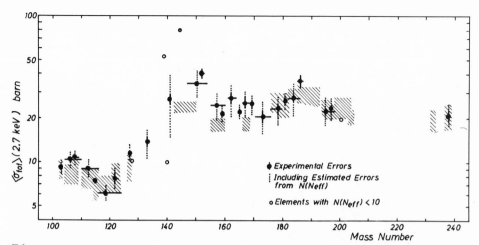

Fig.2: Measured cross sections versus mass number.
Shaded: Cross sections at 2.7 keV, calculated from the
Duke s- and p-wave parameters, ref.{5,6}.

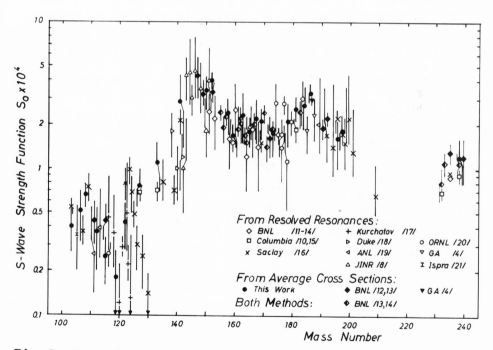

Fig.3: Experimental s-wave strength function for A > 100.

TABLE 1: SUMMARY OF RESULTS

Element		$\langle\sigma_{tot}\rangle$ at 2.7 keV (barn)	N (N_{eff})	data used in the evaluation: R' from ref.	$10^4 \cdot S_1$	$10^4 \cdot S_0$
Rh		9.2±0.4	25	{16}	4 ±2	0.40±0.14
Pd		10.1±0.5	(25)	{6}	3 ±1.5	0.52±0.20
Ag		10.7±0.4	(80)	{3}	2.5±1.5	0.66±0.13
Cd		9.0±0.4	(15)	{5}		0.37±0.17
In		7.4±0.3	80	{5}	2.5±1.5	0.25±0.09
Sn		6.1±0.3	(13)	{5}	2 ±1.5	0.18±0.09
Sb	(a)	7.7±0.8	(50)	{5}	2 ±1.5	0.43±0.15
I	(b)	11.4±0.7	40	{5}	1.5±1.5	0.76±0.22
Cs	(c)	13.7±1.2	25	6±.5fm (d)		1.1 ±0.4
Pr	(a)	27 ±3	10	{16}		2.9 ±1.5
Sm		34.6±1.4	(35)	7.5±.5fm (d)		3.4 ±0.9
Eu	(a)	40.4±3	(800)	7.5±.5fm (d)		4.1 ±0.5
Gd		24.6±1	(35)	{14}		2.3 ±0.7
Tb	(b)	21.2±1.5	130	7.5±.5fm (d)		1.7 ±0.3
Dy	(a)	27.8±2.5	(25)	{14}		2.3 ±0.8
Ho		22.1±1	80	7.5±.5fm (d)		1.8 ±0.4
Er	(a)	25.8±2.6	(35)	{14}		2.2 ±0.7
Tm	(a)	24.7±2	70	{16}		2.1 ±0.5
Yb		20.5±1.2	(20)	{14}		1.8 ±0.6
Hf	(c)	23.3±2.5	(30)	{6}		2.1 ±0.7
Ta	(a)	26.6±2	110	{6}		2.6 ±0.5
W		27.4±1.7	(20)	{6}		2.7 ±0.9
Re	(b)	36 ±3.5	(220)	{5}		3.3 ±0.5
Pt		22.6±1.2	(18)	{16}		1.6 ±0.6
Au		23.8±1	30	{16}	1.5±1.5	1.8 ±0.5
U	(a)	21.0±2	30	{6}	2.4±0.3	1.2 ±0.4

Elements, for which N(N_{eff}) < 10:

Te	(a)	10.2±1	(∿4)
La	(a)	53 ±3	∿7
Ce		10 ±1	(<2)
Nd	(a)	80 ±5	(∿5)
Hg	(a)	20 ±1.4	(<2)

Remarks:

(a) measured oxide, using σ(O)=3.8b. (b) measured powder.
(c) measured CsCl and HfB_2, using σ(Cl)=1.6b and σ(B)=6.5b, as obtained in transmission on CCl_4 and B_4C.
(d) estimated from the trend of experimental R' values.

 * The paper was given by H. Vonach.

5. REFERENCES

{1} The experimental set up, previously used to produce
    130 eV Co resonance neutrons by the same method, has
    been described in W.Dilg, H.Vonach, Z.f.Naturfor-
    schung 26a (1971)442.
{2} D.Gayther, K.Nicholson, Proc.Phys.Soc.70A(1957)51.
{3} R.Chrien, Phys.Rev.141(1966)1129.
{4} S.Friesenhahn et al., J.Nucl.Energy 22(1968)191.
{5} R.Tabony, K.Seth, Ann.Phys.46(1968)401.
{6} K. Seth et al., Phys.Lett.13(1964)70.
{7} B.Buck, F.Perey, Phys.Rev.Lett.8(1962)444.
    A.Jain, Nucl.Phys.50(1964)157.
{8} L.Pikelner, Dubna Symposium(1968)349.
    E.Karzhavina et al., Sov.J.Nucl.Phys.7(1968)161.
{9} Y.Elagin et al., Sov.Phys.JETP 14(1962)682.
{10} J.Desjardins et al., Phys.Rev.120(1960)2214.
{11} D.Goldberg et al., BNL 325, 2.edition, supplement
    No.2(1966).
{12} D.Hughes, Physica 22(1956)994.
    More recently, M.Harlow et al., Washington Conf.(1968)
    p.837, reported an average total cross section $\approx 80b$
    for Eu at 1 keV, which yields $10^4 S_0 \approx 5$.
{13} D.Hughes et al., Phys.Rev.Lett.1(1958)461.
{14} S.Mughabghab, R.Chrien, Phys.Rev.C1(1970)1850.
    R.Chrien et al., Phys.Lett.24B(1967)573.
{15} J.Garg et al., Phys.Rev.137(1965)547.
    J. Garg et al., Phys.Rev.134B(1964)985.
{16} R.Ribon et al., Nucl.Phys.A143(1970)130.
    R.Alves et al., Nucl. Phys.A134(1969)118.
    S.de Barros et al., Nucl.Phys.A131(1969)305.
    H.Tellier et al., Comptes Rendues 269B(1969)266.
    R.Alves et al., Nucl.Phys.A131(1969)450.
    J.Morgenstern et al., Nucl.Phys.A123(1969)561.
    J.Morgenstern et al., Washington Conf.(1968)867.
    A.Michaudon et al., Nucl.Phys.69(1965)545.
{17} Y.Adamchuk et al., Sov.J.Nucl.Phys.3(1966)589.
    G.Muradyan et al., Sov.J.Nucl.Phys.8(1969)495.
{18} E.Bilpuch et al., Ann.Phys.14(1961)387.
    H.Newson et al., Ann.Phys.8(1959)211.
{19} H.Jackson, L.Bollinger, Phys.Rev.124(1961)1142.
{20} J.Harvey et al., ORNL-4513(1970)62;ORNL-3778(1965)38.
{21} C.Coceva et al., Antwerpen Conf.(1965)contribution 72.

DISCUSSION

MICHAUDON   (Saclay)
    You obtain the values of the strength functions for s-waves
from only one measurement point and it seems you have to subtract
the contribution from the p-wave strength function and the potential
scattering. Could you give a figure for the amount of this
correction, say, in the worse case around A=100 where the s-wave
strength function is small?

VONACH
    The maximum correction for p-wave strength was about 20%.
The potential scattering subtractions were, of course, greater
20-30%, but as these potential scattering radii are known rather
accurately, they don't produce much error. The errors of the R_p
are included in the error I showed in the slides.

RAINWATER   (Columbia)
    You said that the samples are thin, but can you make it
quantitative--what were the n values for the samples?

VONACH
    We used samples with transmissions between 95 and 99% effectively,
and corresponding to 0.001 to 0.004 atoms per barn.

RESONANCE PARAMETERS AND THE OPTICAL MODEL[*]

P. A. Moldauer

Argonne National Laboratory

Argonne, Illinois 60439, U.S.A.

INTRODUCTION: THE OPTICAL MODEL

We begin with a schematic review of the place of the optical model[1] in reaction theory. Like all quantum mechanical computations the calculation of nuclear cross sections involves the three distinct steps shown in Fig. 1. First one must determine the Hamiltonian of the system, in this case the nuclear forces. Second, one must solve the Schrödinger equation for the wave function. In this instance only the asymptotic part of the wave function is needed. It is specified by the elements of the S-matrix. Finally, one uses the wave function to calculate the expectation values of the required observables. In this case the observables are cross sections (in units of $\pi \lambdabar^2$) which are bilinear functions of the S-matrix elements (or a linear function in the case of the neutron total cross section).

The same three steps also apply to model calculations. The optical model Hamiltonian is defined by a complex potential $V_c + iW_c$ in each channel c specified by a set of asymptotic quantum numbers (reaction fragments, orbital and spin angular momenta). Step 2 yields a complex optical model phase shift with negative imaginary part which in Step 3 defines an optical model total, shape elastic and absorption cross section (or transmission coefficient). The connection between theory and the model can be made at any one of the three stages of calculation. Here we define the optical model phase shift as specifying an appropriate energy average $\bar{S}_{cc}$ of the corresponding diagonal S-matrix element.

There is also a generalized coupled channels optical model, which in Step 2 yields the complete average S-matrix. But there

335

will be no time to discuss this here. It will be assumed throughout that the average S-matrix is diagonal.

It is important to keep track of the number of independent
parameters involved. In that regard the optical model is somewhat
strange. It starts out with a complex potential that is ordinarily
specified by six independent parameters for each channel (though
commonly required smoothness properties as functions of energy,
spin, and fragment mass, effectively reduce this number). These
six parameters specify two independent components of the complex
phase shift or S, which in turn yields only one directly observable
cross section, the average total cross section. The shape elastic
and absorption cross sections are directly observable only in the
limit of very many open channels.

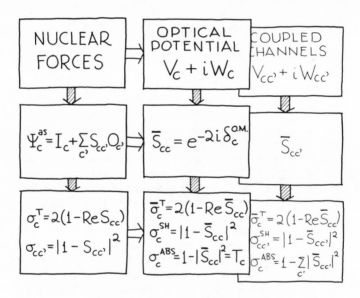

Fig. 1 The optical model.

How can the two optical model phase shift parameters be connected with two directly observable quantities? That requires additional information or assumptions regarding the nature of resonances
or fluctuations. With such additional resonance information, the
optical model can be made to yield information about average resonance parameters, average reaction cross section and even fluctuation statistics.

We shall examine the connection of the optical model with three
distinct resonance formalisms: The compound nucleus model, R-matrix
theory, and the theory of S-matrix resonance poles. Each of the
three resonance formalisms will be seen to generate its own characteristic set of resonance parameters. Each of these is related in

its own way to the optical model parameters and to observable cross section features.

OPTICAL MODEL AND THE COMPOUND NUCLEUS

The earliest and simplest resonance theory is Bohr's compound nucleus model,[2] which is specified by a set of unstable (compound) states each having an energy E_μ and a partial with $\Gamma_{\mu c}$ for decay into channel c. The sum of the partial widths is the total level width Γ_μ and the resulting S-matrix elements have the energy dependence shown in Fig. 2. From this energy dependence one calculates the familiar resonance cross section consisting of a series of Breit-Wigner resonances.

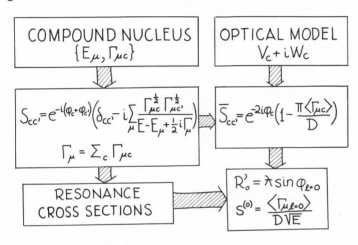

Fig. 2 Compound nucleus and optical model.

The average spacing D of compound levels and the distributions of spacings and of widths and their correlations are discussed in the statistical theories of complex spectra[3,4] under the assumption that the compound states are relatively stable, that their lifetimes are long enough to achieve statistical equilibrium in the nuclear phase space. The results of these theories are discussed in other sessions in this conference.

The connection with the optical model is made by calculating the energy averages of the diagonal S-matrix elements. As shown in Fig. 2, the two compound nucleus parameters which the optical model specifies are the background scattering phase shift ϕ_c and the average partial width to average level spacing ratio $\langle \Gamma_{\mu c} \rangle /D$. In the case of low energy s-wave neutrons these two parameters are customarily given as an equivalent scattering radius R' and a strength function $s^{(o)}$. Both of these parameters are directly measurable and provide sensitive tests for optical potential parameters, even where the absolute precision of the strength function data is not high.[5] For example, the depth of the strength function minimum determines the radial width of the surface imaginary potential. The mass number of the minimum determines the radial position of W. The strength function maximum and R' determine mainly the magnitude and radial size of V.

Strength functions have also been defined at higher energies and for higher partial waves. But these definitions are not quite unique. A more direct parameterization is obtained by use of the channel transmission coefficient (or absorption cross section) T_c, defined on the right side of Fig. 3. The transmission coefficient is less than or equal to unity, as it must be, for all values of $\Gamma_{\mu c}/D$ according to the quadratic relation given in Fig. 3. This formula reduces to the familiar $T_c = 2\pi\langle\Gamma_{\mu c}\rangle/D$ for $T_c \ll 1$.

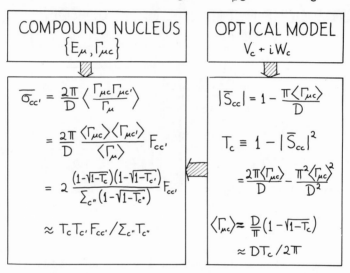

Fig. 3 Compound nucleus and optical model, continued.

In addition to the cross section resonance structure, the compound nucleus model permits one to calculate the average reaction cross section. The first formula was derived by Bethe (Fig. 3 left).[6] It was rewritten in terms of the average partial widths and a width fluctuation correction factor $F_{cc'}$ by Lane and Lynn.[7] This, in turn can be expressed in terms of the optical model transmission coefficients. In the limit where all $T_c \ll 1$ one gets the familiar Hauser-Feshbach formula[8] with width fluctuation correction.

As was pointed out already by Bethe all these derivations for $\overline{\sigma}_{cc'}$ are valid only when all $T_c \ll 1$. In fact, the whole compound nucleus picture breaks down for $2\pi\Gamma_\mu > D$. The mean period of motion or return time of a compound state is h/D. (Therefore a long return

time implies small spacings, not small widths as is often incor-
rectly stated). The mean lifetime of a compound state is $\hbar/\Gamma_\mu$.
(A small width is due to a surface barrier preventing re-emission
after one period). Therefore $2\pi\Gamma_\mu << D$ implies that the compound
state lives through many periods of its motion, that statistical
equilibrium is established in the compound system. On the other
hand $2\pi\Gamma_\mu > D$ means that the compound state decays before it has completed even one period of its characteristic motion. The state has
no time to be fully formed and no equilibrium is established. The
simple compound nucleus model then breaks down, and so does the
applicability of the statistical theories of compound resonance
parameters.

OPTICAL MODEL AND R-MATRIX THEORY

To get around this breakdown of the compound nucleus model,
Wigner invented a theoretical trick.[9] He put the compound system
into an impenetrable theoretical container which shuts off all decay. The resulting spectrum of stable R-matrix states is characterized by level energies E_μ and wave function coefficients $\gamma_{\mu c}$,
as shown in Fig. 4. The channel radii and boundary conditions that
specify the theoretical container determine channel parameters χ_c
and L_c^o. The more or less free choice of channel radii and boundary
conditions defines an infinite set of possible R-matrices for each
Hamiltonian.

By carrying out the theoretical process of removing the container and permitting the system to decay one obtains the S-matrix

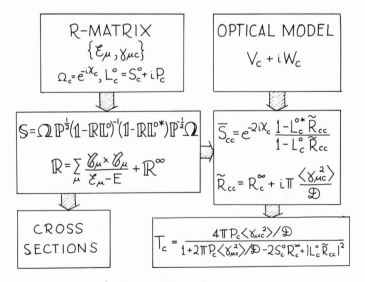

Fig. 4 R-matrix and optical model.

elements in terms of the R-matrix and container parameters as indicated in matrix notation in Fig. 4. From the S-matrix elements one again calculates cross sections and all their properties in the usual manner.

To relate this to the optical model, one must again average the diagonal S-matrix elements over energy. The result of this is indicated on the right side of Fig. 4. For each R-matrix container specification, the optical model specifies the two R-matrix parameters $\langle \gamma_{\mu c}{}^2 \rangle / D$ and R^∞. D is the mean spacing of the levels E_μ. The term R^∞ is added to the R-matrix when the sum is extended over only a finite portion of the infinite R-matrix level spectrum. It accounts for the contribution of the omitted distant levels. The averages $\langle \gamma_{\mu c}{}^2 \rangle$ and D are meaningful only as averages over limited energy ranges and therefore R^∞ is always required when considering such averages. The full expression for the transmission coefficient in terms of R-matrix parameters is also shown in Fig. 4. Again, as required the range of T_c is between 0 and 1. The numerator of this expression is the familiar result for $T_c \ll 1$. A graphical description of the relationship between S_{cc} and R-matrix parameters is shown in Fig. 5 for the special boundary condition case $S_c^0 = 0$. The second half of the circle is symmetric to the first half with R^∞ negative. By changing boundary conditions and channel radii, one can change the values of $\langle \gamma_{\mu c}{}^2 \rangle$ and R^∞ which correspond to a given optical model phase shift.

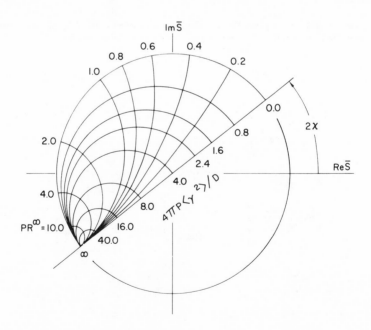

Fig. 5   Relation between R-matrix and optical model parameters.

Since the R-matrix states are completely stable and in equilibrium, one may expect that the statistical theories of complex spectra apply to them. One should therefore be able to use statistical
theories to calculate the mean level spacing D of R-matrix states
and the distributions and correlations of level spacings and wave
function coefficients $\gamma_{\mu c}$. Together, these statistical theories and
the optical model determination of R^{∞} and $\langle \gamma_{\mu c}^2 \rangle / D$ provide the basis
for a complete statistical description of cross sections, also in
situations where compound nucleus concepts are invalid. The procedure for this is outlined in Fig. 6. The Optical model determines
the average S-matrix which gives total cross sections (and in the
coupled channel case direct reaction cross sections) directly. From
the real and imaginary parts of the average S-matrix one calculates
the two R-matrix parameters R^{∞} (or $R^{\infty}_{cc'}$, in the coupled channel case)
and $\langle \gamma_{\mu c}^2 \rangle / D$. Together with level density and statistical theory
this specifies the R-matrix statistically, which in turn yields the
statistical S-matrix from which one can calculate statistical prop-

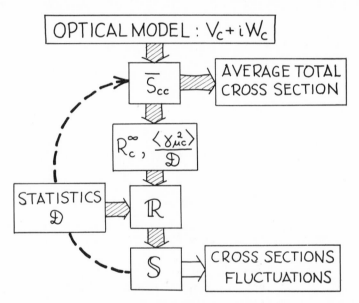

Fig. 6 Procedure for statistical cross section calculations.

perties of cross sections such as their averages, fluctuation properties, etc. By averaging the statistical S-matrix to obtain the
optical model S one checks the whole procedure.

This method can be used to determine optical model parameters
from cross section fluctuations. Fig. 7 shows high resolution measurements of neutron scattering by Titanium (top). For comparison
two statistical cross sections are shown (bottom) which were calculated by the method of Fig. 6 and averaged over 2 keV to match the
2 keV experimental resolution. The two statistical curves were

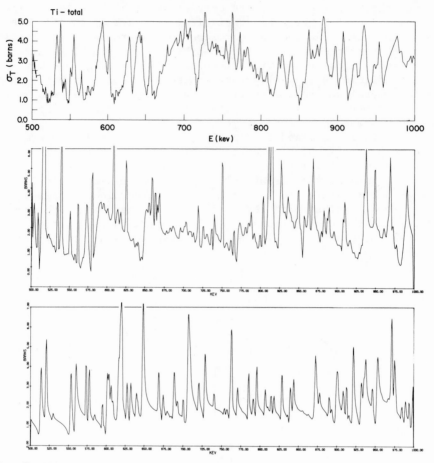

Fig. 7   Experimental and statistically calculated total neutron cross
         sections for Titanium from 500 to 1000 keV.

obtained from different optical models which yield about the same
average cross sections but, as can be seen, very different kinds of
resonance fluctuations. One of the calculated curves, looks qualitatively very much like the experiment.[10]

OPTICAL MODEL AND S-MATRIX POLES

The artifical boundary condition dependent R-matrix parameters
E_γ and $\gamma_{\mu c}$ have no simple direct connection with cross section features except in the isolated resonance limit. To obtain theoretical
parameters that are more directly related to resonances and fluctuations, it is useful to write the S-matrix in the form of the pole
expansion shown in Fig. 8.[11] In the domain of validity of the compound nucleus picture the pole expansion of Fig. 8 becomes the com-

pound nucleus S-matrix of Fig. 2. But in more general situations the numbers of pole parameters (E_μ, Γ_μ, complex $g_{\mu c}$) is twice the number of independent parameters that specify the unitary symmetric S-matrix. It is therefore essential that this pole expansion be obtained from either an R-matrix or a related K-matrix which contains only the requisite number of dynamically independent parameters. The connection between R or K-matrix parameters and S-matrix pole parameters is in general very complex, even as regards their averages.

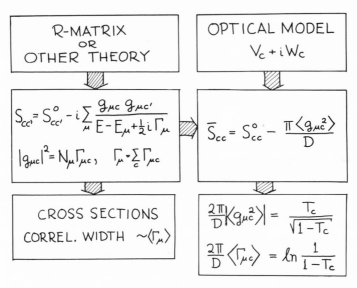

Fig. 8   S-matrix poles and optical model.

Cross section features are related to pole parameters. For example the well-known cross section correlation width[12] Γ_{corr} is related to the Γ_μ and the relative magnitudes of reaction cross sections to various channels are related to the $g_{\mu c}$.

Again we calculate the average diagonal S-matrix elements to obtain the connection with the optical model (Fig. 8) with a result that is reminiscent of the compound nucleus formula. As in the compound nucleus model, the average partial width to spacing ratio is uniquely related to the optical model transmission coefficient.[13] That relation which is also given in Fig. 8 reduces to the formula of Fig. 3 in the compound nucleus limit. The same is true of the formula for G_c, the average of $g_{\mu c}^2$.[14] These two relations are plotted in Fig. 9 (top curves), and compared with the compound nucleus formula of Fig. 3 (middle curves) as well as with the linear extrapolation $T_c = 2\pi \langle \Gamma_c \rangle /D$ of the isolated resonance limit (bottom curves).

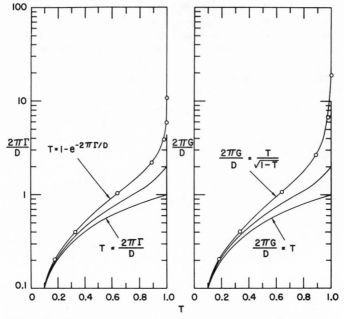

Fig. 9 Relations of average pole parameters to optical
model transmission coefficients.

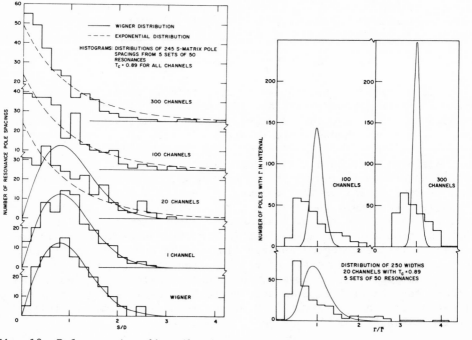

Fig. 10 Pole spacing distribution.   Fig. 11 Pole width distribution.

When $\Gamma > D$ the pole parameters represent at best the properties
of a highly unstable non-equilibrium system, their statistical properties would have to be discussed by means of non-equilibrium statistical mechanics. In order to be able to utilize the standard
equilibrium statistical results we have to derive the statistical
properties of the poles from the statistics of the equilibrium R-
matrix states. The resulting S-matrix pole parameter statistics are
known only qualitatively from numerical studies.[14]

With increasing energy, the average pole density D must be expected to lag progressively further behind the R-matrix level density $\mathcal{D}$ obtained from statistical theory. This is a consequence of
the cumulative effects of the level shift to which each open channel contributes.

The Wigner repulsion of pole spacings gets washed out as Γ/D
increases. This is apparent from the histograms in Fig. 10 which
show the progressive change of statistically calculated pole spacings
from a Wigner distribution to the uncorrelated exponential distribution as the number of competing strongly absorbed channels is increased from 1 to 300.

In the same calculations the distribution of total widths,
shown by histograms in Fig. 11, were found to be much broader than
the naively expected χ^2-distribution indicated by the continuous
curves. It is interesting to note that these width distributions
have marked peaks at their lower limits, which generally agree with
the total width that would be obtained from $T_c = 2\pi \langle \Gamma_\mu \rangle_c /D$. This
appears to be reason why both observed and numerically generated

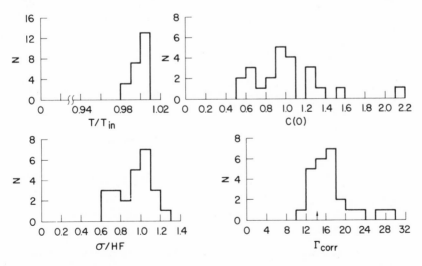

Fig. 12   Distributions of average cross sections and fluctuation
          parameters.

correlation widths are generally smaller than the theoretical average widths of Fig. 8.

Finally, one can calculate average reaction cross sections from
the S-matrix in Fig. 8. The formula depends on the details of the
pole parameter statistics as well as on the optical model phase
shifts.[16] The question of how well the final result approaches the
Hauser-Feshbach formula has not been solved generally. But numerical studies indicate that in the limit of many open channels average
reaction cross sections do follow the Hauser-Feshbach prediction.
On the other hand, the numerically generated average reaction cross
sections exhibit a very large dispersion of values. This is illustrated in Fig. 12 which concerns a calculation of the kind outlined
in Fig. 6 with 25 competing channels most of them having T_c = .99
and 240 levels. The calculated transmission coefficients $\hat{T}$ are all
very close to the input transmission coefficients T_{in}. But the calculated average cross sections are widely distributed about the
Hauser-Feshbach values. Similarly the cross section r.m.s. fluctuations (or autocorrelation coefficients) C(0) have a wide distribution
about the expected value of unity. The measured correlation widths
cluster in a wide range of values about $\Sigma T_c D/2\pi$ which is indicated
by an arrow. Clearly for large Γ/D average reaction cross section
and fluctuation measurements cannot be expected to yield very reliable optical model or other dynamical information.

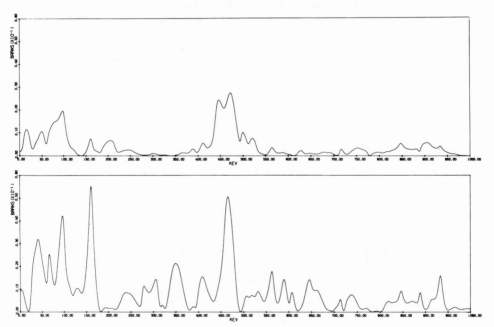

Fig. 13 Intermediate structure in statistically generated cross
         sections.

The large dispersion of widths, and particularly the long tail of the width distribution implies the occasional occurence of large cross section bumps which may be interpreted as intermediate structure.[17] The fact that such intermediate structure may be consistent with the statistical picture does, of course, not preclude a study of its more specific dynamic origins. An example of apparent entrance channels intermediate structure in two statistically generated reaction cross sections is shown in Fig. 13. Both curves come from the above mentioned 25 channel calculation.

CONCLUSIONS

The optical model, together with an appropriate resonance formalism, provides important information on average resonance parameters, average reaction cross sections, and cross section fluctuations. The compound nucleus model which gives the simple resonance picture breaks down when resonances are no longer isolated. Then one has to rely on either R-matrix theory whose parameters are governed directly by the statistical laws of complex spectra but are only very indirectly related to observables; or one may use S-matrix pole parameters which are more directly related to observations, but are subject to complicated and inadequately known correlations and statistical distributions. Certain averages of both types of resonance parameters are directly and fairly simply related to the optical model phase shift.

REFERENCES

*   Work performed under the auspices of the U. S. Atomic Energy Commission.
1.   H. Feshbach, C. E. Porter and V. F. Weisskopf, Phys. Rev. 96, 448 (1954).
2.   N. Bohr, Nature 137, 344 (1936).
3.   T. Ericson, Advances in Physics 9, 425 (1960).
4.   C. E. Porter, Statistical Theories of Spectra: Fluctuations, Academic Press (New York, 1965).
5.   P. A. Moldauer, Nucl. Phys. 47, 65 (1963).
6.   H. A. Bethe, Rev. Mod. Phys. 9, 69 (1937).
7.   A. M. Lane and J. E. Lynn, Proc. Phys. Soc. (London) A70, 557 (1957).
8.   W. Hauser and H. Feshbach, Phys. Rev. 87, 366 (1952).
9.   E. P. Wigner and L. Eisenbud, Phys. Rev. 72, 29 (1947).
10. E. Barnard, J.A.M. de Villiers, D. Reitmann, P. A. Moldauer, A. B. Smith and J. Whalen, to be published.
11. J. Humblet and L. Rosenfeld, Nucl. Phys. 26, 529 (1961).
12. T. Ericson, Ann. Phys. (New York) 23, 390 (1963).
13. P. A. Moldauer, Phys. Rev. 157, 907 (1966); 177, 1841 (1969).
14. P. A. Moldauer, Phys. Rev. Lett. 19, 1047 (1967).

15. P. A. Moldauer, Phys. Rev. <u>171</u>, 1164 (1968).
16. P. A. Moldauer, Phys. Rev. <u>135</u>, B642 (1964).
17. H. Feshbach, A. K. Kerman and R. H. Lemmer, Ann. Phys. (New York) <u>41</u>, 230 (1967).

DISCUSSION

SINGH   (SUNY/Albany)

The neutron strength function for orbital angular momentum is defined in terms of the optical model transmission coefficients, and the penetration factors. In the denominator of the expression on page 90 in the paper by Elwyn et al.-Phys. Rev. 133 B80, 1964, there is a $\sqrt{E}$ term. I am just curious to find out at what neutron energy the strength functions calculations should be made.

MOLDAUER

For s-wave strength functions, the answer is simple. At energies until you start deviating appreciably from the $\sqrt{E}$ dependence you keep averaging. This deviation will happen in different places in the optical model. In some places it doesn't happen for a long time, in others it happens within a few tenths of kilovolts. The reason that I prefer the transmission coefficient to the strength function is because, for p-waves and higher waves, the strength function is not uniquely defined. You must also specify at least the radius in addition to the strength function. That's not too serious a matter, but I think it is just as easy to talk about transmission   coefficient as a parameter. Then you do not have to worry about the energy region in which the parameter is appropriate. It's appropriate to any region in which you can get a large enough average. I'm sorry. I didn't really answer your question.

SINGH

Suppose we have an experimental p-wave strength function, derived on the basis of the averaging of resonances from say, up to 200 keV. In the optical model calculation, if we take the neutron energy at 50 keV or 100 or 150 keV, we get different strength functions for elements such as K or Cℓ.   In the paper by Perey and Buck in 1962, they made some optical model calculations, for even-even nuclei at an energy of 40 keV, which is below the 2^+ excitation energy for all the nuclei considered.

MOLDAUER

I really have no other comment. Maybe we can discuss specific cases, but I don't really think there is one overall rule that is followed.

PEREZ

I have some comments regarding the distribution of the poles of the S-matrix. Dr. de Saussure and myself have studied this problem for the few channel case. We found that for the two level case and high degree of level interference, the Wigner repulsion disappears and we get the theoretical distribution obtained by Hwang. However for the same degree of interference and a population of 100 levels, the Wigner level spacing distribution is preserved to a high degree of accuracy. We do not have the same results for a large number of channels as you do. Do you have any comments on this?

MOLDAUER

I think I'm not quite sure how to interpret the distribution of two level matrices. I think I would make the same comment that I made Monday. There are interesting things one can get from such calculations but I'm not too sure that the distribution law is one of them.

All of the averages discussed in my paper are local averages. This means that the averaging interval (or resolution functions) covers only a small part of all resonance terms. Experimental averages, such as those generated by instrumental resolution functions, are local averages, and so are averages considered in statistical theory, optical model theory, etc. In contrast to this, the sum rules mentioned by Feshbach in his paper are derived from global averages that involve all the resonance terms that there are. When transmission coefficients approach unity, the two kinds of average give very different results.

COMPOUND NUCLEUS EFFECTS IN ^{24}Mg+α ELASTIC SCATTERING AT 20-23 MeV

A.G. Drentje and J.D.A. Roeders

Nuclear Physics Accelerator Institute,

University of Groningen, The Netherlands

Abstract. Excitation functions and angular distributions for ^{24}Mg+α elastic scattering show at extreme backward angles important contributions of compound nucleus reactions. Several reaction mechanisms are discussed to explain the angular distributions.

The present experiments have been started to study the mechanism of elastic α scattering from ^{24}Mg and of the inelastic scattering processes leading to the low lying excited states in ^{24}Mg, in particular to the 3$^+$ state. We measured the energy dependence of the cross sections at a number of scattering angles (94°, 150°, 160° and 164°; energy resolution 30 keV) with energy steps of 100 keV between 19 and 27 MeV (lab. system). In addition we measured complete angular distributions at 13 selected bombarding energies around 20, 21 and 23 MeV with an energy resolution of 100 keV.

The elastic excitation functions are shown in fig. 1. Clearly visible is the irregularly varying fine structure due to compound nucleus effects. Also broad enhanced structures in an excitation function can be discerned, for instance around 20.3 MeV (94°) and 21.2 MeV (160° and 164°), and also at higher energies. From the elastic and inelastic excitation functions the coherence width Γ has been obtained by three different methods of analysis: normal autocorrelation functions, trend-reduced autocorrelation functions, and the peak-counting method [1,2]. The value of Γ, found by averaging the values from the three methods, increases from 81 to 116 keV in the bombarding energy range 17.7-27 MeV.

The arrows in fig. 1 indicate the selected bombarding energies at which angular distributions have been measured [3]. Fig. 2 shows three elastic distribution functions, obtained at 21.00, 21.37 and

21.55 MeV. These three examples feature the characteristics of all
elastic angular distributions:

a) one observes $d\sigma(\theta)/d\sigma_R(\theta) < 1$ at most angles except in the extreme
   backward region where this ratio can be greater than 1, sometimes
   reaching values of 10 or more, for instance at 21.37 MeV;

b) the position of the maxima and minima at forward angles ($\theta < 80^\circ$)
   is nearly independent of the energy, but the position of the
   extrema in the backward part ($\theta > 120^\circ$) changes with energy;

c) at some energies the modulations in the angular distributions
   between 90° and 160° are very deep; this effect is very pronounced at 21.37 MeV, and the maxima are rather strong.

These observations indicate that direct reaction processes are
accompanied by important compound nucleus contributions. In order
to wash out statistical fluctuations in the cross sections average
angular distributions were determined over energy intervals
$\Delta E \simeq 800$ keV, which amounts to about 8 times the coherence width Γ.
At the nominal energies of 20, 21 and 23 MeV the available five,
four and four distributions, respectively, have been averaged.
It appears that these distributions have their backward maxima at
about the same positions; the backward oscillations have a smaller
period than the forward oscillations.

We have tried to fit the averaged angular distributions with
optical model calculations. A four parameter potential with equal
real and imaginary geometry was used. The analysis was started with
five sets of parameters found by McFadden and Satchler [4]) to give
the best fits for $\theta \leq 120^\circ$ at 24.7 MeV. It turns out that at 23 MeV
reasonable fits can be obtained to the forward angle cross sections
as well as to the cross sections in the entire range of scattering
angles. However, at 21 and 20 MeV good fits could be obtained for
the forward angular region only, but their quality deteriorates
with decreasing energy. The best fit at 21 MeV is shown in figure 3;
the parameter values differ a little from those in ref. 4. The
dashed curve in figure 3 gives the calculated angular distribution
extrapolated from the fit at forward scattering angles; it may be
remarked that the pattern near 120° is out of phase with the observed
angular distribution.

Energy averaged compound nucleus cross sections may be calculated with the Hauser-Feshbach formula. In the case of elastic
scattering of spinless particles this formula reads:

$$< \frac{d\sigma}{d\Omega}(\theta) > = \frac{1}{4k^2} \sum_J A^J(\theta) \frac{(T^J)^2}{\sum_{c''} T^J(c'')} W \tag{1}$$

where $A^J(\theta)$ is a geometrical factor, W the Moldauer correction
factor accounting for entrance and exit channel interference effects,

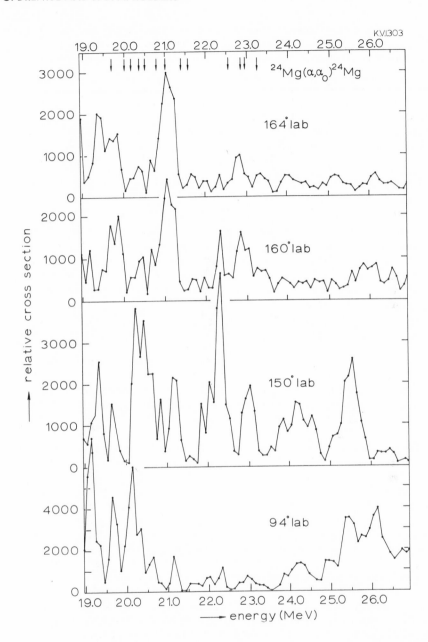

Fig. 1. Excitation functions for ^{24}Mg+α elastic scattering measured with 30 keV energy resolution and 100 keV step size. Angular distributions have been measured at energies indicated by the arrows.

the T^J's transmission coefficients; the summation in the denominator is over all possible exit channels. The transmission coefficients were obtained from the OM fit to the forward angle cross sections and are shown in fig. 4. It can be seen that important compound nucleus formation occurs for J values up till J = 10.

For the present experiments one deals with a highly excited compound nucleus $^{28}Si^*$ (excitation energy $\mathcal{E}_x$ about 28 MeV). In this situation the mean width Γ_J of the participating levels is larger than their mean spacing D_J [1]). For calculating purposes the following expression, derived by Eberhard et al. [5]), is very useful

$$\frac{1}{2\pi} \sum_{c''} T^J (c'') = \frac{\Gamma_J}{D_J} \simeq \frac{\Gamma_o}{D_o} (2J+1) \exp\{-J(J+1)/C\} \tag{2}$$

where Γ_o/D_o is the width to spacing ratio of spin zero levels in the compound nucleus; C is an adjustable parameter, which is about equal to $2\sigma^2$, σ^2 being the spin distribution parameter of levels in the residual nucleus. The formula holds if the CN decay is dominated by the emission of only one kind of particle. In our case this must be the alpha particle, since

a) in the summation over J in (1) only a narrow range of J-values is important, namely $7 \le J \le 11$,

b) within that J-range the compound nucleus decays mainly via α-particle emission, because protons and neutrons cannot carry away such high angular momenta. This is a consequence of the negative Q-values for the reactions (α,n) and (α,p), -7.2 MeV and -1.6 MeV respectively.

Calculations were performed with D_o = 2.3 keV and C = 21.2. Furthermore we take $\Gamma_o = \Gamma$, since Γ_J changes very little with J in the range of important J values; Γ = 98 keV at 21 MeV. With these values of the parameters the ratios Γ_J/D_J following from Eberhards formula (cf. fig. 4) agree rather well with the Γ_J/D_J values calculated by Roeders [1]) by means of the so-called backshifted Fermi gas model [6]). The same values of Γ_o/D_o and $C \simeq 2\sigma^2$ correctly predict the inelastic scattering cross sections to the 3^+ state (5.22 MeV) and 4^+ state (6.00 MeV) of ^{24}Mg [7]), and the observed energy dependence of the coherence width.

We now apply relation (2) and calculate the compound nucleus angular distribution, assuming no interference between entrance and exit channels because of the many competing channels and overlapping levels (hence W = 1). The result is shown in figure 3. It is striking that a number of pronounced maxima and minima appear which, just as in the OM calculations, have about the correct position at large backward angles.

It is concluded that at energies of 20-23 MeV compound elastic scattering accounts for 50% or more of the extreme backward scatte-

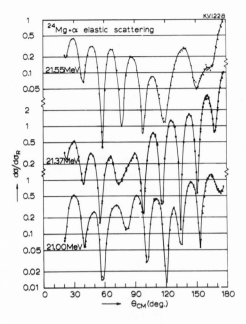

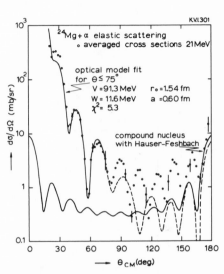

Fig. 2. Angular distributions for ^{24}Mg+α scattering measured with 100 keV energy resolution. The curves are guides to the eye.

Fig. 3. Energy averaged experimental cross sections at nominal energy of 21 MeV. An OM fit for $\theta \leq 75^\circ$ and its extrapolation is shown. The Hauser–Feshbach curve has been calculated assuming that the compound nucleus ^{28}Si* decays via α-particle emission only.

ring; the same conclusion is also reached in the analysis of the cross section fluctuations. Deviations from the Hauser-Feshbach predictions at backward angles between 90° and 150° are observed at all energies; they are strongest at 21 MeV. These deviations might have been smaller when more angular distributions were available in determining the average distribution over the 800 keV interval. Because of the small number of angular distributions used, the experimental distribution at 21.37 MeV, showing pronounced oscillations with high maxima at backward angles, strongly affects the average distribution at 21 MeV. The strong oscillation of the 21.37 MeV distribution may be explained by an enhanced density of compound nucleus levels with one particular spin value, namely J = 10. On the one hand this follows from the location of the backward minima in the 21.37 MeV angular distribution; they coincide within 1 degree with zero's of $P_{10}(\cos \theta)$. On the other hand, the exponential form in (2) and the values of the transmission coeffi-

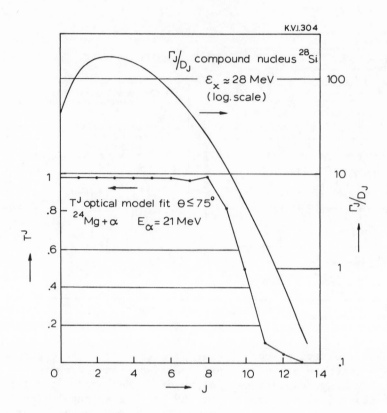

Fig. 4. Optical model transmission coefficients from
the fit shown in fig. 3, and the $(\Gamma/D)_J$ ratio calculated with Eberhards expression (2).

cients in fig. 4 show that levels with J = 10 contribute very
strongly to the compound nucleus cross sections.
The remaining deviations from the HF calculations cannot be ascribed
to direct reactions. If direct reaction processes were dominating
also at backward angles, one should expect a regular repetition of
bumps in the excitation functions (fig. 1), which is not observed.
Our conclusion may be that the angular behaviour between 90° and
150° indicates the presence of reaction types intermediate between
pure direct reactions (dominating at forward angles) and compound
nucleus processes (dominating at extreme backward angles).

We wish to thank Professor Dr. H. Brinkman, and Drs. A. van
der Woude and L.W. Put for their contributions to this paper and
the experiments; we are indebted to the directors of the Free University Cyclotron and the Philips Cyclotron for using these
machines.

This work is part of the research programme of the University of Groningen and the "Stichting voor Fundamenteel Onderzoek der Materie" (F.O.M.), the latter being financially supported by the "Nederlandse Organisatie voor Zuiver-Wetenschappelijk Onderzoek" (Z.W.O.).

References:

1. J.D.A. Roeders, Thesis, Groningen, 1971.
2. J.D.A. Roeders, L.W. Put, A.G. Drentje and A. van der Woude, Nuovo Cim. Lett. 2 (1969) 209.
3. A.G. Drentje, Thesis, Groningen (1971) (in press).
4. L. McFadden and G.R. Satchler, Nucl. Phys. 84 (1966) 177.
5. K.A. Eberhard, P. von Brentano, M. Böhning and R.O. Stephen, Nucl. Phys. A 125 (1969) 673.
6. H.K. Vonach, A.A. Katsanos and J.R. Huizenga, Nucl. Phys. A 122 (1968) 465.
7. A.G. Drentje and J.D.A. Roeders, Phys. Lett. 32B (1970) 356.

DISCUSSION

RICHTER   (Heidelberg and Bochum)
Have you looked into the possibility that the optical model
itself gives you a backward peaking? There are mechanisms which
could give rise to backward peaking other than the compound nucleus.
Could you comment on that?

DRENTJE
You are right. That's what I didn't discuss. In the optical
model, one could build in an L absorption term. Such an L-dependent
absorption gives rise to backward enhancement. However, in the
excitation functions, such an optical model backward enhancement
should predict a more or less regular appearance of smooth varying
maxima and minima. That's in contradiction with what we observe.
From that, we conclude that the observed backward effects are not
direct effects.

PEREZ
I am no expert in this at all, but there is a paper by McVoy
(Phys. Rev., March, 1970) in which he explains some of this backward
oscillations by introducing a set of Regge poles. Do you have any
comments on this? I've looked on that, too. These Regge poles are
present, in the case of very weak absorption, for instance, in the
case of heavy ion scattering, and I've tried to find them also in
this case. I can find such a Regge pole when I take the imaginary
potential zero or nearly zero. However, with the given value of the
imaginary potential, they are not present.

STRUCTURE IN THE (n, n') CROSS SECTIONS

OF Se ISOTOPES

E.S. Konobeyevsky, R.M. Musayelyan,

V.I. Popov and I.V. Surkova

Institute for Nuclear Research, Moscow, USSR

Inelastic scattering of low energy neutrons can give a valuable information on statistical properties of resonance parameters. In a review given at the Conference on Nuclear Structure Study with Neutrons held in Antwerp in 1965 Ferguson [1] has discussed a few examples demonstrating the importance of width fluctuation corrections in theoretical description of average cross sections. It has been shown in a number of publications that these corrections calculated under an assumption of uncorrelated widths improved significantly an agreement of experimental results with theoretical cross sections expressed in terms of optical model transmission coefficients. In some cases, however, experimental cross sections have appeared too high compared to predictions based on the assumption of uncorrelated widths. An essential discrepancy of this kind has been observed by the authors [2,4] and by Lister and Smith [3] for the first 2^+ levels of even-even isotopes of Ge. It was suggested in refs. 2-4 that large experimental cross sections observed in that case could be accounted for by channel coupling effects.

The purpose of the experiments discussed in this paper was to obtain further information on inelastic neutron scattering in the mass region $A \approx 70 - 80$. Excitation fuctions of the (n,n') reaction leading to the first 2+ levels of ^{76}Se, ^{78}Se, ^{80}Se and ^{82}Se were measured in the energy region from

a threshold to ≈ 1.5 MeV. Gamma-rays following
neutron scattering were detected in a ring geometry
using a Ge(Li) detector with a volume of 74 cm³. A
scattering sample contained about 300 g of natural
Se. The energy resolution in the experiment was about
10 keV.

Experimental cross sections versus mean center-of-
mass energy are shown in fig. 1. Errors of measured
gamma-ray yields leading to uncertainties in energy
dependence of the cross sections are order of 5 per
cent or less. Absolute cross section values have
errors not higher than 20 per cent. One must take into
account that experimental points have not been
corrected for contribution of transitions from second
2^+ levels. In the case of [76,78]Se, which have second 2^+
levels lying at 1217 and 1310 keV respectively, the
correction could be of the order of 20 per cent at
the highest energies.

The main feature of the data shown in the figure
is that the cross sections obtained for all four
nuclei are on the average considerably higher than
theoretical values based on usual statistical
assumptions and the optical model. This is illustrated
by the curve (1) which was obtained using the Moldauer
formula [5] for a fluctuation cross section. The
parameters $\langle \theta_c \rangle$ entering into the formula are related to transmission coefficients by an expression:

$$\langle \theta_c \rangle = (2/Q_c) \left[1 - (1 - Q_c\, T_c)^{1/2} \right]$$

In calculations the transmission coefficients of a
Saxon-Woods potential [6] were used, the parameter
Q_c being taken equal to 1. Width fluctuation corrections were calculated using an assumption of uncorrelated widths with Porter-Thomas distribution.
Evidently, Hauser-Feshbach cross sections with the
same transmission coefficients will be closer to
experimental points but still an appreciable difference will remain particularly in the case of [76]Se and
[78]Se.

A qualitative explanation of these results that
can be given as well as in the case of Ge is that we
encounter here with strong effects of channel coupling.
The physical nature of this coupling can be similar
to that leading to an increase of s-wave neutron
strength functions in the mass region A ≈ 60 - 80 in

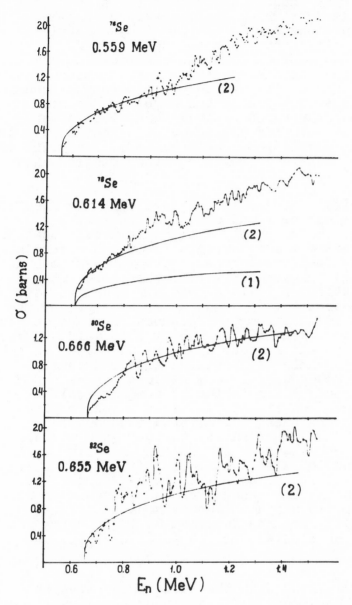

Fig. 1 Comparison of experimental and calculated cross sections.

comparison with theoretical values for a spherical
potential. As well known this behaviour of s-wave
strength functions can be explained by a particle-
vibration coupling of a neutron-core system.
Apparently, the effect of a particle-core coupling
on average inelastic scattering cross sections can
be not only due to its influence on strength function
values but also because it can give rise to strong
correlations of resonance parameters. The latter
must be taken into account in calculations of a width
correlation correction. Besides, in an analysis of
experimental data a correlation cross section must be
added.

In principle, a relative role of width fluctuation effects can be found out from an analysis
considering simulteneously data on inelastic scattering cross sections and strength functions determined,
for example, from total cross section measurements.
In the energy region close to a threshold, when
an approximation of well-separated resonances can be
applied, theoretical fluctuation cross sections are
simply related to strength functions of entrance and
exit reaction channels. For energy intervals not very
large one can make further approximation used in total
cross section analysis, assuming that a strength
function is proportional to a reduced strength function and a penetration factor of a square-well
potential. One can expect that in the mass region
considered, when energy of emitted neutrons is low
enough, the main contribution to inelastic scattering
will be given by s-wave neutrons in the exit channel.
This is supported with a behaviour of experimental
cross sections at energies $\leqslant$ 100 keV. A fluctuation
cross section at these energies can be considered
depending on two strength function parameters, a
d-wave reduced strength function S_2 in an entrance
channel and an s-wave strength function S_0 in an
exit one.

Calculations with the factorised strength
functions showed that a fit to experimental data near
thresholds under an assumption of uncorrelated widths
could be obtained only with S_2 and S_0 values significantly larger than expected for these nuclei. The
curves (2) in the figure correspond to $S_2 = S_0 =$
4×10^{-4}, nuclear radius being taken equal to
$R = 1.45 \ A^{1/3}$ F. P-wave strength functions were

assumed rather arbitrarily to be $S_1 = 2 \times 10^{-4}$.
Contribution of partial cross sections with $l > 1$
in an exit channel can be neglected. Experimental s-wave strength functions of Se isotopes [7] and other
nuclei in this mass region vary usually from
1×10^{-4} to 2×10^{-4}. Though there are no definite
experimental data on p- and d-wave strength functions,
one can hardly expect higher values for them. We
can conclude therefore that coupling definitely takes
place for entrance and exit channels with $l = 2$ and
$l = 0$ respectively, leading to strong correlations
of corresponding partial neutron widths. As to p-wave
channels, no definite conclusion of that kind can be
made at the moment. It should be mentioned that
strictly speaking an approximation of isolated
resonances can not be applied at energies of a few
hundred keV above thresholds. Therefore, the curves
(2) don't give of course a realistic description of
cross sections in large energy intervals, and have been
drawn mainly to demonstrate some difference in the
experimental data for different isotopes.

Quantitative description of neutron inelastic
scattering with inclusion of channel coupling could
be made, for example, in terms of a model with a
particle-vibration interaction. Then the discussed
here channel-channel correlations can be interpreted
as an effect of intermediate states of a particle-
phonon nature having large contributions of the
components with a core in ground and one-phonon
states and a neutron correspondingly in 2 d- and
3 s-states.

In this connection a question can be asked
to what extent the observed cross section fluctuations
can be considered as an intermediate structure
related to intermediate states. A mean fluctuation
width determined directly from the experimental data
was found to be close to experimental energy resolution. In the attempt to reveal a structure with
larger widths a statistical analysis was carried out
using an approach similar to that applied by Carlson
and Barschall [8] in their analysis of total cross
sections. Mean square fluctuations of experimental
cross sections averaged over energy intervals varying
from ≈ 10 keV (the experimental resolution) to
≈ 100 keV were compared with statistical theory
predictions in the limit of well-separated resonances.
An approximate theoretical expression for a mean

square fluctuation was obtained using several simply-
fying assumptions. It was supposed that a width
fluctuation effect could be approximately taken into
account by putting a width fluctuation factor in the
Moldauer cross section equal to 1. This leads to the
Hauser-Feshbach formula with transmission coefficients
proportional to strength functions. Contributions
from partial waves with $l > 1$ in an exit channel were
neglected as well as dependence of strength functions
on total angular momentum. Assuming further that
spin dependence of level densities is given by a
factor $2 J + 1$ one comes to an expression for a mean
square fluctuation containing a level density
parameter and two parameters determined by distributions of level widths and spacings. It was found that
mean square fluctuations of the experimental cross
sections vary with averaging intervals according to
a statistical model expectation for intervals
100 keV. Assuming certain distribution laws for
widths and spacings it is possible now to evaluate a
level density parameter from the fit to the relative
mean square fluctuations determined from the experiment. Some ambiguity of this estimation, in addition
to approximations mentioned earlier, arises from the
uncertainty of strength function values. Taking the
Porter-Thomas and Wigner distributions for widths
and spacings and optical model strength functions,
following average spacings for $J = 1/2$ were obtained:

^{77}Se - 0.29 keV, ^{79}Se - 0.46 keV, ^{81}Se - 0.5 keV,

^{83}Se - 3.2 keV.   These values can be compared with
experimental s-resonance spacings obtained in ref. 7
from neutron resonance studies:

^{77}Se - 0.7 keV, ^{79}Se - 1 keV, ^{81}Se - 1.2 keV.

Taking into account the difference in excitation
energies of the compound nuclei, level spacing
values estimated here are too low compared to
spacings measured directly, the difference being given
by a factor 2. With all the approximations made in
the analysis this can not be considered as a
serious disagreement.

We come to the conclusion that the cross
section fluctuations with a width of the order of
100 keV or less observed in this work appear to be
of a statistical nature and give no evidence of an
intermediate structure of such a width.

A possible interpretation of the data presented here in terms of a model with intermediate states seems to require the states with a comparatively large spreading width of at least a few hundred keV.

REFERENCES

1. A.T.G. Ferguson, Proc. Intern. Conf. on Nuclear Structure Study with Neutrons, Antwerp (North-Holland Publ. Co., Amsterdam, 1966) p. 63.
2. E.S. Konobeyevsky, R.M. Musayelyan, V.Y. Popov and I.V. Surkova, Izv. Ak. Nauk, ser. fiz. 33 (1969) 1753.
3. D. Lister and A.B. Smith, Phys. Rev. 183 (1969) 954.
4. E.S. Konobeyevsky, R.M. Musayelyan, V.I. Popov and T.V. Surkova, Yad. Fizica 14 (1971) 14.
5. P.A. Moldauer, Rev. Mod. Phys. 36 (1964) 1074.
6. G.I. Marchuk and V.E. Kolesov, Application of Numerical Methods for Neutron Cross Section Calculations Atomizdat, Moscow, 1970.
7. H. Maletzky, L.B. Pikelner, I.M. Salomatin and E.I. Sharapov, Preprint P3-3956, Dubna, 1968.
8. A.D. Carlson and H.H. Barschall, Phys. Rev. 158 (1967) 1142.

S- AND P-WAVE STRENGTH FUNCTIONS AND THE SPIN, ISOSPIN

AND SPIN-ORBIT DEPENDENCE OF THE OPTICAL POTENTIAL

C.M. Newstead, J. Delaroche and B. Cauvin

Département de Physique Nucléaire

Centre d'Etudes Nucléaires de Saclay, France

1. SPIN DEPENDENCE OF THE S-WAVE STRENGTH FUNCTION. It was suggested by Feshbach [1] that the optical model potential might contain a spin-spin term of the form I.σ (where I and σ are respectively the target and projectile spin operators) which would give rise to different scattering for the various possible channel spins. One manifestation of this potential would be a spin dependence of the s-wave neutron strength function S_o. The effect would be more prominent in high spin nuclei.

To search for spin dependence of S_o we have assigned the compound nucleus spins J = I $\pm$ 1/2 for some 200 s-wave resonances in ^{143,145}Nd, ^{147,149}Sm and ^{165}Ho using the Saclay 65 MeV Electron Linear Accelerator as a pulsed neutron source in conjunction with the gamma-ray multiplicity method [2]. The strength functions obtained on the basis of these spin assignments are given in Table 1. We also include in this table recent results for the rare earth region obtained by other laboratories with the methods indicated in the table.

In order to determine the magnitude of the strength of the spin-spin potential V_{ss} we have carried out a series of coupled channels optical model calculations using Tamura's code Jupitor I [3], with a real spin-spin potential of the form $V_{ss}f(r)$I.σ where $f(r)$ is taken as the Saxon-Woods form factor. The potential parameters employed in the calculations and the strength function predictions are summarized in Table 1. We find that the experimental results are well described by a spin-spin strength $|V_{ss}| < 0.5$ MeV.

Measurements carried out by Wagner et al [4] with 350 keV polarized neutrons incident on a polarized ^{165}Ho target indicate that the difference in the measured total cross-section with parallel and anti-parallel spin orientations ($\sigma_{\uparrow\uparrow} - \sigma_{\uparrow\downarrow}$, the so-called "spin-spin effect") is consistent with a value for V_{ss} between $-$ 130 and

+ 280 keV. Similar experiments were carried out on ^{165}Ho by Fisher et al [5] and Kobayashi et al [6] with conflicting results. Recently Healey et al [7] have reported a "spin-spin effect" in ^{59}Co which suggests a value of V_{ss} = – 0.4 MeV below 1 MeV neutron energy and V_{ss} = + 0.17 MeV from 1 MeV to 8 MeV. Clearly considerable work remains to be done in the determination of the sign and magnitude of V_{ss} as well as its possible energy dependence.

An interesting point that emerges from the present study is that for certain nuclei a repulsive spin-spin potential is required while for others an attractive term is needed. It is tempting to speculate on the origin of the phenomenological spin-spin potential and its sign. A detailed understanding of this as well as the radial dependence f(r) would require microscopic calculation. It would appear that the spin-spin interaction may be derived from the spin dependent part of the effective two body interaction assuming that only the extra nucleons outside the core contribute to the target spin. The form factor f(r) would thus be determined by the sum of the convolution of the wave functions of these extra nucleons and the radial dependence of the effective interaction. Clearly f(r) and V_{ss} would vary with the target. The sign of the interaction would depend on the particular mixture of spin dependent forces in the effective two body interaction.

2. *S-WAVE STRENGTH FUNCTIONS AND ISOSPIN DEPENDENCE OF OPTICAL POTENTIAL*. Recent measurements of the s-wave neutron strength function S_0 carried out at Saclay for the isotopes of tellurium [11] and at Oak Ridge for the isotopes of tin [12] have shown a remarkable systematic decrease of S_0 with mass number A for each isotopic chain. Similar but somewhat less well defined trends has been found by the Brookhaven group in the isotopes of Dy, Er and Yb [13] and by the Oak Ridge group in the isotopes of Hf [14].

This behaviour of S_0 with A is quite contrary to the predictions of the conventional optical model which indicates a gradual rise from the minima of the mass 100 region to the shoulder of the 4S size resonance in the one case and similarly an increase from the minima in the splitting of the 4S size resonance to the shoulder of the heavy mass component. The effect would appear to occur in the vicinity of minima but it may be that it is masked in the presence of maxima by the rapid variation of S_0 with A. To date the best known example is the behaviour of the tellurium isotopes where extensive measurements for some 400 resonances have enabled the determination of quite accurate values of S_0 [11].

We have investigated the possibility that this effect is correlated with the asymmetry parameter (N–Z)/A by carrying out a series of coupled channel optical model calculations with potentials containing complex asymmetry terms. The Lane potential $4V_1(T.t)/A$, where T and t are respectively the isospin of the target and projectile, is well established for the real potential strength [15]. Inclusion of such a term in the optical potential does little to improve the comparison between prediction and experiment for $S_0(A)$

even if twice the normally accepted strength $V_1 = 30$ MeV is employed.

It seems plausible to assume that the imaginary potential strength may also have an isovector component. The first suggestion of this came from Perey [16] who found that a diffuse absorptive potential with a strength $W_D = W_0 + W_1(N-Z)/A$, where $W_1 = 48$ MeV, gave a good description of 9-22 MeV proton elastic scattering angular distributions for a wide range of nuclei. Since then a number of investigations of proton scattering, particularly for chains of isotopes, have shown the need for such a term in the imaginary potential. The matter is by no means settled however since other studies find no evidence for W_1.

Satchler [17] pointed out that to be sure this term was really isovector in nature rather than a fortuitous occurence signifying only a correlation with neutron excess or even $A^{1/3}$, it was necessary to demonstrate that (as a consequence of T.t) the absorptive potential strength for neutrons decreased with mass number as $W_D = W_0 - W_1(N-Z)/A$. The relatively poor quality of neutron scattering data (compared to its proton counterpart) has prevented an unambiguous conclusion from being reached on this point.

The exciting thing about the present study is that it provides for the first time the required evidence for the behaviour of the imaginary neutron potential. As can be seen in Figure 1 the experimental results for the isotopes of tellurium and tin are well described by just such an absorptive potential strength which decreases with increasing mass number with $W_1 = 45$ MeV for Sn and $W_1 = 63$ MeV for Te. These strengths are of the same order of magnitude as the value found by Perey [16], but are of course employed in the expression with opposite sign.

In the rare earth region it is also possible to describe the experimental results for those isotopes which show a decrease with mass number. However it requires about 5 times stronger the above isospin potential strength to do this. As pointed out by Greenlees et al [18] it is the volume integral of the potential which is the important parameter not merely the strength. For a surface peaked absorptive potential the volume integral is proportional, to first order, to the product W_1 a, where a is the diffuseness. Since diffuseness is now thought to increase as one goes to heavier nuclei and may even have a dependence for a given isotopic chain of the form $a = a_0 + a_1(N-Z)/A$, it may quite possibly be that a proper account of the variation of diffuseness will decrease the value of W_1 required to describe the experimental results in the rare earth region.

3. *P-WAVE STRENGTH FUNCTIONS FROM AVERAGE TOTAL CROSS SECTION MEASUREMENTS*. The determination of the p-wave neutron strength function S_1 from analysis of the average total neutron cross section over a wide energy range has the advantages of automatically including a large number of resonances (which minimises the effect of Porter-Thomas fluctuations of neutron widths) and eliminates the necessity of identifying the orbital angular momentum ℓ of each

resonance which is an extremely tedious and time consuming process.

Essentially the method consists of separating the various partial wave contributions to the average total cross section by virtue of their different energy dependence. Each partial wave cross section depends on two parameters the distant level parameter R_ℓ^∞ and the strength function s_ℓ. In particular the average collision function over many levels $\bar{U}_\ell$ is related to the R-function $(R_\ell = R_\ell^\infty + i\pi s_\ell)$ by the expression,

$$\bar{U}_\ell = e^{-2i\phi_\ell}(1 - \hat{L}_\ell^* R_\ell)/(1 - \hat{L}_\ell R_\ell)$$

where ϕ_ℓ is the hard sphere phase shift and $\hat{L}_\ell = \Lambda_\ell + iP_\ell$ is the logarithmic derivative of the incoming wave function at the nuclear surface with Λ_ℓ the modified shift factor and P_ℓ the penetrability factor for the ℓth partial wave.

The partial wave cross sections are obtained from the well known expressions :

$$\sigma_{cn}(\ell) = (2\ell+1)\pi\lambda^2(1-|\bar{U}_\ell|^2) \; ; \; \sigma_{se}(\ell) = (2\ell+1)\pi\lambda^2|1 - \bar{U}_\ell|^2$$

While at low energy the compound nucleus cross section $\sigma_{cn}(\ell)$ depends only on s_ℓ and the shape elastic cross section $\sigma_{se}(\ell)$, only on R_ℓ^∞, at higher energy each of the partial cross sections depends on both R_ℓ^∞ and s_ℓ.

The procedure is to determine the value of the R_ℓ^∞ and s_ℓ by making a least squares fit of the theoretical total cross section $\sigma_t = \Sigma\{\sigma_{cn}(\ell) + \sigma_{se}(\ell)\}$ to the measured average total cross section. It is necessary to make the plausible assumptions that $s_3 = s_1$ and $R_2^\infty = R_3^\infty = 0$ since these quantities make little contribution to the cross section below 1 MeV and cannot be determined from the fit.

The neutron total cross sections of ^{165}Ho and ^{239}Pu were measured from 10 keV to 1 MeV neutron energy using the Saclay 65 MeV Electron Linear Accelerator as a pulsed neutron source in conjunction with a 200 meter flight path. The experimentally determined cross sections are shown in Figure 2 along with the least squares fits and resulting partial wave cross sections. The ^{239}Pu result is to be considered preliminary as measurements are continuing.

The s-wave strength functions obtained from the average analysis of the present study are in good agreement with the values $S_0 = 1.77 \pm 0.14 \times 10^{-4}$ for ^{165}Ho [11] and $S_0 = 1.25 \pm 0.10 \times 10^{-4}$ for ^{239}Pu [19] obtained from analysis of many individual resonances.

The value $S_1 = 1.63 \pm 0.25$ obtained for ^{165}Ho is of some interest as it confirms the rotational splitting of the 4P size resonance predicted by the collective model. Most other published results for S_1 in this mass region were obtained from average capture measurements which must be viewed with some scepticism because of the difficulties involved in the analysis. The ^{239}Pu result of $S_1 = 2.33 \pm 0.35$ would also seem to be in better agreement with the collective model and to confirm previous results of this magnitude for the neighbouring nuclei ^{235}U and ^{238}U.

We have carried out a series of coupled channels optical model calculations for S_0 and S_1. As can be seen in Tables 2 and 3 the predictions are in good agreement with the experimental results for both s- and p-wave strength functions. The potential employed for the ^{165}Ho calculation is the same as that given in Table 1 and includes a spin-spin term. It is interesting to note that the experimentally observed spin splitting of the s-wave strength function is described with the same optical potential.

The spin-orbit coupling strength V_{so} = 7.5 MeV is of the same order of magnitude as that used in shell model calculations. There was no need to employ a stronger value of V_{so} and indeed in the strength function minima the predictions of S_1 are insensitive to V_{so}. Calculations (with V_{so} = 7.5 MeV) were also carried out for other nuclei in this mass range. No evidence for a subsidiary bump in the minima of the 4S size resonance was found as might be expected due to spin-orbit coupling of the d-wave channel.

We wish to thank Dr. A. Michaudon for helpful advice and discussion.

REFERENCES

1. H. Feshbach, Ann. Rev. Nucl. Sci., 1958, 8.
2. C. Coceva, F. Corvi, P. Giacobbe and G. Carraro, Nucl. Phys., 1968, A117, 586.
3. T. Tamura, Rev. Mod. Phys., 1965, 37, 679.
4. R. Wagner, P. D. Miller, T. Tamura and H. Marshak, Phys. Rev., 1965, 133, B29.
5. T. R. Fisher et al., Phys. Rev., 1967, 157, 1149; Nucl. Phys., 1969, A130, 609.
6. S. Kabayashi et al., J. Phys. Soc Japan, 1967, 22, 368.
7. D. C. Healey et al., Phys. Rev. Letters, 1970, 25, 117.
8. M. Asghar, P. Ribon, E. Silver and J. Trochon, Nucl. Phys., 1970, A145, 549.
9. S. F. Mughabghab, R. E. Chrien and O. A. Wasson, Phys. Rev. Letters, 1970, 25, 1670.
10. O. A. Wasson, S. F. Mughabghab and R. E. Chrien, B.A.P.S., 1971, 16, 496.
11. H. Tellier and C. M. Newstead, Proc. Third Conf. Neutron Cross Sections, Knoxville 1971.
12. T. Fuketa, F. A. Khan and J. A. Harvey, B.A.P.S., 1963, 8, 71.
13. S. F. Mughabghab and R. E. Chrien, Phys. Rev., 1970, C1, 1850.
14. T. Fuketa and J. A. Harvey, Report ORNL-3778 (1964).
15. A. M. Lane, Phys. Rev. Letters, 1962, 8, 171.
16. F. G. Perey, Phys. Rev., 1963, 131, 745.
17. G. R. Satchler, Nucl. Phys., 1967, A91, 75.
18. G. W. Greenlees, G. J. Pyle and Y. C. Tang, Phys. Rev., 1968, 171, 1115.
19. H. Derrien - A. Michaudon, Paris Conf. 1966; Private Communication, 1971.

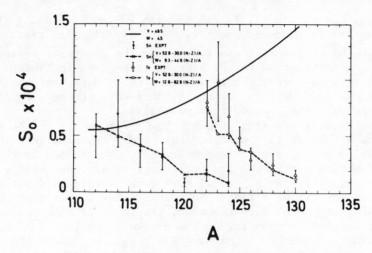

Fig. 1 - EXPERIMENT AND CALCULATION FOR S_o.

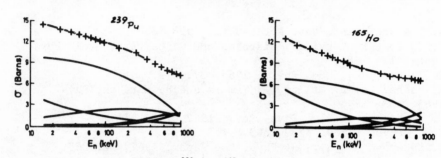

Fig. 2 - LEAST SQUARES FIT TO ^{239}Pu AND ^{165}Ho MEASURED $<\sigma_t>$ TO OBTAIN S_o AND S_1.

Table 1 - COMPARISON OF THEORETICAL AND EXPERIMENTAL VALUES FOR THE SPIN DEPENDENCE OF S_o

| TARGET (I^π) | OPTICAL MODEL CALCULATIONS | | | | | | EXPERIMENTAL RESULTS | | | |
	V_{ss}	V	W	β_2	$S_o(J^-)$	$S_o(J^+)$	$S_o(J^-)$	$S_o(J^+)$	J Assign. Method	Reference
^{143}Nd (7/2$^-$)	- 0.1	48.45	1.95	0.01	9.03	2.59	8.4 ± 4.3	4.1 ± 2.2	γ-RAY MULTIPLICITY	PRESENT STUDY
^{145}Nd (7/2$^-$)	0.1	47.95	1.37	0.01	0.55	2.26	0.64 ± 0.34	1.70 ± 0.75	γ-RAY MULTIPLICITY	PRESENT STUDY
^{147}Sm (7/2$^-$)	0.5	48.01	14.0	- 0.03	2.36	3.04	3.09 ± 1.90	5.04 ± 3.40	γ-RAY MULTIPLICITY	PRESENT STUDY
^{149}Sm (7/2$^-$)	0.3	47.67	8.0	- 0.03	3.77	5.48	3.28 ± 1.80	7.18 ± 2.40	γ-RAY MULTIPLICITY	PRESENT STUDY
^{155}Gd (3/2$^-$)	0.5	47.47	10.0	0.30	1.99	2.04	1.38 ± 0.73	1.72 ± 0.60	γ-RAY MULT. + SCATT.	ASGHAR et al[8]
^{157}Gd (3/2$^-$)	0.5	47.16	8.0	0.29	1.92	1.97	2.94 ± 1.56	2.26 ± 0.88	γ-RAY MULT. + SCATT.	ASGHAR et al[8]
^{163}Dy (5/2$^-$)	0.5	46.99	2.0	0.24	0.88	1.26	0.54 ± 0.26	1.24 ± 0.40	LOW ENERGY γ-RAYS	MUGHABGHAB et al[9]
^{165}Ho (7/2$^-$)	- 0.25	47.50	3.0	0.30	2.42	1.35	2.50 ± 0.50	1.45 ± 0.3	γ-RAY MULTIPLICITY	PRESENT STUDY
^{167}Er (7/2$^+$)	- 0.5	47.13	4.0	0.30	3.31	1.55	5.95 ± 3.15	1.76 ± 0.76	LOW ENERGY γ-RAYS	WASSON et al[10]
^{177}Hf (7/2$^-$)	0.5	47.11	4.0	0.25	1.38	3.75	1.34 ± 0.58	3.40 ± 1.20	γ-RAY MULTIPLICITY	COCEVA et al[2]

Geometry set for all calculations : r_o = 1.25 f, a = 0.65 f, b = 0.47 f. Potentials strengths in Mev. S_o in units of 10^{-4}.

Table 2 - OPTICAL MODEL CALCULATIONS OF STRENGTH FUNCTIONS

TARGET	V	W_D	V_{so}	B_2	a	b	r_o	$S_o \times 10^4$	$S_1 \times 10^4$
^{165}Ho	47.5	3.0	7.5	0.30	0.65	0.47	1.25	1.82	1.61
^{239}Pu	47.5	5.0	7.5	0.33	0.65	0.47	1.25	0.98	2.32

Table 3 - EXPT. VALUES

$S_o \times 10^4$	$S_1 \times 10^4$
1.66 ± 0.24	1.63 ± 0.25
1.15 ± 0.16	2.33 ± 0.35

DISCUSSION

BLOCK

I'd like to point out a measurement that was done at R.P.I. several years ago by Bartolome   et   al., on the tungsten isotopes near mass 180, and we too get a low p-wave strength function of about 15-20% of the black nucleus value, in agreement with the low value for erbium obtained at Columbia, but in disagreement with your high value for holmium. Another point I'd like to make: How do you take into account inelastic scattering in your fitting of, say, the plutonium average crossection? I notice you go up to 1 MeV, and I didn't see any notation for inelastic scattering.

NEWSTEAD

Yes. This is not important here, because we're considering the total crossection, and the inelastic component is included in the compound nucleus contribution. It doesn't change the value of the strength function.

MOLDAUER

I think I should include a comment about the strength function minimum known for so many years but studiously ignored. Suddenly, everybody is interested in it now. The point that I made a number of years ago is that when people first started with the surface absorption optical model, they said let's make a one fermi wide surface peak and this was thrown out of a hat, because one is a beautiful number, and nobody had really thought about it. Most data are not sensitive to this number. In fact, the strength function minimum is almost the only piece of data which is sensitive to this number. It is sensitive to this because the absorption takes place in the region of the wave function node. Therefore, how much absorption you get depends on how wide in radius the absorptive region is. By making it narrower and taller, you can keep the absorptive properties in other regions the same but decrease the absorption in the region of the strength function.

NEWSTEAD

Yes. I'm quite familiar with your phenomena of fringe absorption. Because of lack of time, I could not discuss it. That will certainly give you the low values you describe, but it won't give the dependence with neutron excess that I showed in the tellurium and tin isotopes. The full interpretation of the anomolous minima is still not clear. Whether it is partly due to fringe absorption is an open question. It is physically hard to understand, I think you must admit, why the absorption takes place so far outside the nuclear surface, in such a sharp band. I don't think that this is just the effect of the Pauli Principles.

MOLDAUER

I don't think it's fringe absorption. It's just the width of the absorptive peak.

NEWSTEAD
     Why is it so small?   What is the physical interpretation?

MOLDAUER
     You have to calculate the finite nucleus in order to do that.
There are some old calculations which I've seen which use local
density approximations and things like that which do suggest that
it's considerably narrower than the one fermi that is traditionally
used. That's a complicated question which I don't think one can
answer in a qualitative manner.

NEWSTEAD
     Yes. It would be interesting to extend your calculations into
the deformed region to see the effect of fringe absorption in the
minimum of the 4-s splitting, to see whether it would give the low
values some people are quoting for the s-wave strength function.

COCEVA
     Concerning the spin dependence of the s-wave strength function,
I saw that our data on Hf^{177} was quoted as supporting this dependence.
However, we have recently extended our measurements on Hf^{177} and
carried out new measurements on Hf^{179}. The new results show that
the s-wave functions for the two spin states are equal for both
isotopes, within our experimental errors.

NEWSTEAD
     So your new measurements show no spin dependence in Hf. That
is just another example of the phenomena described by Dr. Lane--
the hazards of this business of predicting things.

VONACH
     I have not quite understood your explanation on this changing
of sign of the spin-spin potential between neighboring elements.
Could you please comment?

NEWSTEAD
     I did not really give an explanation--rather a bit of hand-
waving. What I said was that in order to know the sign, the strength,
and indeed the shape of the spin-spin potential, one would have to
do a microscopic calculation. These quantities would depend very
much on the particular mixture of spin-dependent forces between the
projectile and the nucleons outside the core. The spin appears in
several parts of the phenomenological two-body potential.
It would depend on which term dominated as to what the sign would
be.

FALLIEROS

Your simple formula $W\sigma|\text{matrix}|^2\rho$ would seem to imply that if W decreases with N-Z/A for neutrons, the level density ρ should do the same. Could you comment on that?

NEWSTEAD

It would appear that the level density ρ should include a $t \cdot T$ term. The really interesting question, it seems to me, is the microscopic interpretation of the imaginary iso-vector potential.

SESSION VI

NUCLEAR LEVEL DENSITY

Chairman   - K. J. LeCouteur, Australian National
           University
Scientific Secretary - K. Ratcliff, State University of New York
           at Albany

A SURVEY OF NUCLEAR LEVEL DENSITY THEORIES

Claude BLOCH

CEN Saclay - BP n°2 , 91 - Gif-sur-Yvette, France

The theory of nuclear level density is about as old as nuclear physics itself. Actually, many features of the models used in present days are already found in the famous article published by H.A. Bethe in 1937[1]. To a considerable extent, the theory of nuclear level density rests upon models which describe nuclei as made of independent   objects, which may be nucleons or quasi-particles. In such a model, the evaluation of the level density is essentially a combinatorial problem.

Although any independent particle or quasi-particle model is a considerable simplification of the physical reality, it cannot be said, even to-day, that the full consequences of such a simple model have been completely explored. The reason is twofold. On one hand, the experimental data on level density is still rather poor. This is because nuclear levels may be individually detected only in small regions of energy (accessible, for instance by slow neutron capture). In other regions of energy the measurable cross-sections depend not only on the level density, but also on matrix elements or partial widths, which are also unknown. On the other hand, even a simple independent particle model of nuclear level density involves a large number of parameters, namely the positions of the individual single particle levels. This is why in most calculations, additional assumptions have been introduced in addition to the independent particle model. In the crudest form, this leads to the so-called continuous approximation, but several kinds of improvements have been proposed as we shall recall below.

A general feature of level density theories is their extreme sensitivity on the parameters entering the underlying model. This is due to the fact that the level density is, roughly speaking, an

379

exponential function of energy and of the parameters. It is therefore often possible to fit experimental data with very crude models,
by an appropriate choice of parameters which corrects the deficiencies of the model. In particular, the influence of the nucleon-
nucleon interactions seems, so far, hard to detect in the observed
level densities. Nevertheless, a number of authors have explored
the use of the sophisticated methods developed in statistical mechanics for the treatment of systems of interacting particles.

In all the treatments which I have mentioned so far, the single
particle energies are considered as a known ingredient of the theory.
This is justified by the fact that for spherical or slightly deformed nuclei, the actual resolution of the one-body Schrödinger equation is a simple matter. Recent experimental evidence on fission and
isomeric states has drawn a lot of attention on the highly deformed
nuclear states. The determination of single particle levels becomes
then a difficult problem justifying the construction of an appropriate theory which I shall mention in the end.

I – THE LEVEL DENSITY OF AN INDEPENDENT PARTICLE MODEL : THE BASIC COMBINATORIAL PROBLEM

It may be useful to recall here very briefly the main ideas in
the treatment of the basic combinatorial problem[2].

Let us assume for simplicity that we have only one kind of
particles, which may occupy a set of single particle levels of energy ε_i . Let $n_i = 0, 1$ be the occupation numbers of these single
particle levels. If we consider a system of A particles, the number of levels for the whole system having an energy equal to E is
the number of solutions of the two equations

$$A = \sum_i n_i \quad ,$$

$$(1)$$

$$E = \sum_i n_i \varepsilon_i \quad ,$$

where the n_i are supposed to take the values 0 and 1 in all
possible ways. In order to solve this combinatorial problem, one
introduces two selecting variables $\alpha \ \beta$, and the generating func-
tion

$$Z(\alpha \ \beta) = \left(1 + e^{\alpha - \beta \varepsilon_1}\right)\left(1 + e^{\alpha - \beta \varepsilon_2}\right) \cdots = \prod_i \left(1 + e^{\alpha - \beta \varepsilon_i}\right) \quad . \quad (2)$$

For actual calculations, it is more convenient to evaluate instead
of Z , the function

$$\Phi(\alpha\beta) \quad = \quad \log Z(\alpha\beta) \quad = \quad \sum_i \log \left(1 + e^{\alpha - \beta\varepsilon_i}\right) \quad . \qquad (3)$$

Next it is easy to convince oneself that the number of solutions
of the equations (1) is equal to the coefficient of $e^{-\alpha A + \beta E}$ in
the expansion of $Z(\alpha \beta)$. This coefficient, and therefore the level
density may be obtained in the form of the Fourier integral

$$\rho(A \ E) \ = \ \frac{1}{(2i\pi)^2} \int_{-i\infty}^{i\infty} d\alpha \ d\beta \ \ e^{-\alpha A + \beta E + \Phi(\alpha \ \beta)} \quad , \qquad (4)$$

where Z has been expressed in terms of Φ by means of (3).

The formal expression (4) is useful only if there is a way to
evaluate the integral. This may be done asymptotically for large A
and E by means of the <u>saddle-point method</u>. This last ingredient
of the theory rests upon the fact that the integrand in (4) has a
sharp maximum at the saddle-point, defined as the point where it
is stationary. This point is determined by the conditions

$$\frac{\partial\Phi}{\partial\alpha} = A \quad , \quad \frac{\partial\Phi}{\partial\beta} = -E \quad \text{(saddle point).} \qquad (5)$$

Finally, by expanding the exponent in (4) up to second order in α
and β , the integrals may be carried out elementarily and one
obtains the expression :

$$\rho(A \ E) \ = \ e^S \Big/ 2\pi \sqrt{D} \ , \qquad (6)$$

where :

$$S \ = \ (\Phi - \alpha A + \beta E) \quad , \quad D \ = \ \begin{vmatrix} \Phi''_{\alpha^2} & \Phi''_{\alpha\beta} \\ \Phi''_{\alpha\beta} & \Phi''_{\beta^2} \end{vmatrix} \quad , \qquad (7)$$

calculated at the saddle-point.

II - APPLICATIONS AND EXTENSIONS

The basic derivation recalled above has been the object of many discussions, applications and generalizations.

A. Evaluation of the Function $\Phi(\alpha\ \beta)$

Any evaluation of level density clearly starts from the function $\Phi(\alpha\ \beta)$ and an approximation must be introduced in order to be able to use the general expansion (3).

The simplest treatment consists in replacing the distribution of the single particle levels by a continuous distribution of density $\rho(\varepsilon)$. This defines the <u>continuous approximation</u> for $\Phi(\alpha\ \beta)$:

$$\Phi(\alpha\ \beta)\ \ =\ \ \int\ d\varepsilon\ \rho(\varepsilon)\ \text{Log}(1 + e^{\alpha - \beta\varepsilon})\ . \tag{8}$$

For all nuclear excitations occurring in practice, the nucleus remains a highly degenerate Fermi system. This corresponds to <u>large values</u> of α and β . The significant part of the integral is then concentrated in a small region surrounding the Fermi energy $\varepsilon_F \pm \alpha/\beta$. Denoting by ρ_F the single particle level density considered as a constant in that small region, one obtains the famous expression

$$\rho(A\ E)\ \ =\ \ e^{a\sqrt{U}} \bigg/ U\ \sqrt{48}\ ,\ a = \pi\ \sqrt{\frac{2\rho_F}{3}}\ \ , \tag{9}$$

where $U = E-E_o$ is the excitation energy measured from the ground state.

The continuous approximation (9) gives correctly the general behaviour of level densities, but it can certainly not be trusted for precise numerical calculations.

As an extreme opposite approach, it was proposed[3] to retain the exact single particle states in the expansion (3). When α and β are large, the number of significant terms around the Fermi energy is small and the actual calculation is reasonably simple, especially for light nuclei. This <u>totally discrete approximation</u> has the drawback that it does not provide a simple explicit formula for $\rho(A\ E)$, but it proved very accurate when tested against an exact enumeration of the levels of Ne^{20} . An interesting feature which appeared in these calculations is the <u>shell effect</u>. When a nucleus has an incomplete outer shell, many nuclear levels may be formed with no excitation energy (in the pure independent particle model where residual interactions are neglected) by recoupling the par-

ticles in the incomplete shell. This increases the level density
in a way which is felt quite high up in energy.

Between these two extreme approximations, several methods have
been proposed in order to retain some of the basic simplicity of
the continuous approximation, while taking at least approximately
into account the discrete nature of the single particle spectrum.
They are based on the Euler-Mac-Laurin summation formula for ap-
proximating a discrete sum involving an analytic function $f(n)$
by an integral

$$\sum_{i=0}^{N} f(i) = \int_{0}^{N} f(x)\ dx + \frac{1}{2}[f(N)+f(0)] + \frac{1}{12}[f'(N)-f'(0)] + \ldots (10)$$

This expansion may be used for the evaluation of the discrete sum
occurring in (3) provided the levels ε_i may be expressed in terms
of an analytic function φ, for instance as $\varepsilon_i = \varphi(i)$. The particularly simple case where φ is a constant, giving rise to equally spaced levels has been considered by Rosenzweig[4].

A generalization due to Kahn and Rozensweig[5] is obtained by
taking for the single particle levels a finite bunch of g levels
repeated periodically, with a spacing d between bunches

$$\varepsilon_{ik} = \frac{i}{d} + \eta(k) \quad (i = 0,1,2,\ldots ; \ k = 1,2,\ldots,g). \quad (11)$$

This level scheme may be interpreted as a succession of equally
spaced shells, each shell consisting of g levels.

Under these assumptions, and provided the correction for the
discrete summation in (3) is limited to the terms retained in (10),
the calculation of the level density for the whole system may be
carried out as simply as in the continuous approximation. It yields
an expression identical with (9) where $\rho_F = g/d$, and U is replaced by an effective excitation energy U^* given by

$$U^* = U + \frac{g}{12} - \frac{1}{2g}\left(n - \frac{g}{2}\right)^2 + \sum_{k=1}^{n} \eta(k) + \frac{1}{2}\sum_{k=1}^{g} \eta^2(k). \quad (12)$$

Here n is the number of nucleons in the last, partially filled
shell. Such a displacement of the ground state energy is a common
feature of many of the improvements of the simple continuous approximation (9).

These ideas are pushed further in several communications to
this Conference. Thus Hiroshi Baba[6] presents a treatment of a

similar model in which the shells are degenerate, corresponding to $\eta(k) = 0$. By applying the procedure of the totally discrete approximation he obtains an expression very similar to that of Kahn and Rosenzweig, except that it contains an extra-factor approximately equal to the binomial coefficient

$$\begin{pmatrix} g \\ n \end{pmatrix} . \tag{13}$$

This coefficient is precisely the number of ways of distributing n particles in g levels, which is the number of states of the system having a vanishing excitation energy.

Böhning[7] reports a similar calculation, in which he obtains, besides an energy shift of the type (12), a modification of the parameter a occurring in (9), which becomes now energy dependent. According to Vonach, Schantl and Böhning[8], the resulting expression may be fitted very well to experimental level density data for nuclei ranging from $A = 30$ to $A = 200$.

B. The Saddle-point Approximation

It has been shown, a long time ago[3], by comparing the totally discrete approximation to an exact counting of the levels in the case of Ne^{20}, that the saddle-point approximation is actually quite accurate, except at very low energy. Therefore, most of the attempts for improving the simple expression (9) rely on the saddle-point approximation. In communication 6.7 to this conference, however, Böhning[7] indicates the possibility of calculating the integral occurring in (4), for special forms of the function Φ, without making use of a saddle-point approximation. He obtains then, instead of the exponential function occurring in (9), a slightly more complicated elementary function. The resulting expression has no singularity at zero energy. According to Böhning, it is numerically quite accurate.

C. Distribution of Angular Momentum and Isospin

It is not difficult to introduce neutrons and protons, or which amounts to the same the isospin, and also to evaluate the density of levels with a given angular momentum.

Consider for instance the introduction of an isospin $t_{zi} = \pm 1/2$ for each single particle state. In complete analogy with the argument of section I, the number of levels of A nucleons, at the energy E and having the total isospin T_z is equal to the <u>number of solutions</u> of the system of three equations

$$A = \sum_i n_i \quad , \quad E = \sum_i n_i \varepsilon_i \quad , \quad T_z = \sum_i n_i t_{zi} \quad , \tag{14}$$

where the occupation numbers n_i take again the values O and 1
in all possible ways. The continuation of the argument is quite similar to the discussion given above, except that instead of two
selecting variables α, β one must now introduce three variables
α, β, γ. Similarly, the density of levels having a given projection
of the angular momentum along the z axis can be evaluated by introducing a fourth selecting variable.

To a very good approximation, in the case of angular momentum,
for instance, one obtains a Gaussian distribution for the projection M along the z axis

$$\rho(M) \sim e^{-M^2/2\sigma^2} , \tag{15}$$

where σ is the so-called <u>spin cut-off factor</u>. The density of levels
with given angular momentum J is easily deduced from the density
of levels with given projection of angular momentum along the z
axis, by writing

$$\rho_J = \rho(J+1) - \rho(J) \sim (2J+1) e^{-(J+1/2)^2/2\sigma^2} . \tag{16}$$

The value of the spin cut-off factor depends on which of the approximations discussed in section A is used for evaluating the sum
over single particle states occurring in the function Φ . The spin
cut-off factor is intimately related to the moment of inertia of the
nucleus. When the continuous approximation is used, one obtains for
the moment of inertia the rigid body value. It is well known that
this does not fit experimental data, and therefore, numerous improvements of the theory have been proposed, which cannot be discussed here for lack of space[3][9].

D. Introduction of some special types of two-body interactions

The expression (16) may be generalized in a way which amounts
to replacing the rotational group associated with angular momentum
by other groups. For instance, in the supermultiplet approximation,
each level is associated with a given representation of the four
dimensional unitary group, and may therefore be characterized by
three quantum numbers conventionally denoted by P , P' and P" .
In the Elliot model, each level is associated with a representation
of SU3 and characterized by two corresponding quantum numbers. It
is possible to obtain a formula very similar to (16) giving the distribution of levels with given quantum numbers characterizing the
representation of the invariance group underlying the model. The
resulting expression has the same form as (16), except that $(J+1/2)^2$
is replaced by the Casimir operator of the group, and 2J+1 by
the dimension of the representation of the group[3][9].

As an application of such distributions, one may evaluate
the effect on the nuclear level density of some special types of
two-body interactions having the property that the energy shift
which they produce is proportional to a Casimir operator. Starting
from the partial densities of levels with given quantum numbers,
for which the Casimir operator has the same value, one calculates
the total level density by shifting each of these partial densities
by the corresponding amount, and then summing all the resulting partial densities[3][9].

E. Density of states with given numbers of particles and holes

In view of the discussions of intermediate structures associated with states having given numbers of particles and holes, it
is interesting to derive expressions, for the partial level densities of such states. The general method is the same as that discussed in section I. One needs here three selection variables
associated with energy, and the numbers of particles and holes.
For sufficiently large numbers of particles and holes, it is possible to introduce a continuous approximation[10]. For small numbers of particles and holes, however, the saddle-point approximation
is not good. When these numbers are very small i.e. of the order of
a few units, it is possible to carry out explicitly the integrations
without using the saddle-point approximation or any other approximation. This yields very accurate expressions which are, however
somewhat complicated[10]. Various approximations applicable in
intermediate situations have been discussed by Böhning[11] and
Williams[12].

III - LEVEL DENSITY FOR SYSTEMS OF INTERACTING PARTICLES

The level density $\rho(A\,E)$ is directly related to the grand
partition function $Z(\alpha\,\beta)$ defined in thermodynamics by

$$Z(\alpha\,\beta) \;=\; \mathrm{Tr}\; e^{\alpha A - \beta H} \;=\; \int dA\;dE\;\rho(A\,E)\;e^{\alpha A - \beta E} \;, \qquad (17)$$

where α is the chemical potential and β the inverse temperature.
Actually the argument of section I may be regarded as the statistical
mechanics of a system of independent particles, the expression (4)
being the inverse relation of (17). A considerable amount of work
has been done in statistical mechanics for evaluating $\Phi(\alpha\beta)=\log Z(\alpha\beta)$
in the presence of interactions, for instance in the form of dia-
gram expansions. The relation (4) yields then corresponding evaluations of the level density, taking into account the interactions.

Such a program was tested a long time ago[13] in the simple case of interacting nucleons in a p-shell. Comparison of the statistical theory with exact levels gave very good results. A number of authors have since applied similar ideas to more realistic models[14]. As already mentioned in the introduction, the uncertainties both in the experimental results and in the parameters entering the theory have prevented any real application of these more sophisticated theories to the interpretation of actual data.

A special mention should be made of the pairing interaction. The corresponding system is equivalent to an ensemble of non-interacting quasi-particles, and the grand partition function may therefore be evaluated by the simple method described in section I. Along these lines, Moretto[15] discusses, in a contribution to this conference, the influence of angular momentum on the pairing correlations, and obtains the critical temperature for the superconductivity phase transition as a function of angular momentum.

Next, I would like to report on some recent work referring to a slightly different subject, which however illustrates very well the application of the methods of statistical mechanics to the theory of level density. It concerns random Hamiltonians.

This concept is introduced as a simple simulation of complicated residual interactions. Random Hamiltonians have been considered particularly in connection with the statistics of level spacings, but I shall mention here only the problem of level density. This problem arises, for instance, if one considers the effect of the residual interactions between nucleons within an incomplete shell. In the absence of interactions, all the levels obtained by coupling these nucleons in all possible ways are degenerate. The residual interactions produce a distribution of the levels around the unperturbed energy, the actual calculation of which implies the evaluation of many matrix elements and a (usually) large diagonalization problem. Although such calculations have often been performed for particular nuclei, it is interesting to simulate the residual interactions by random matrices in order to obtain general shapes of the level distributions.

In the early work on random matrices[16] all elements of the matrix were considered as independent random variables, with the same, usually Gaussian distribution. The resulting level density follows the famous semi-circle law first derived by Wigner. This model ignores, however, the very important fact that the residual interactions are essentially two-body interactions. Monte-Carlo calculations based on the diagonalization of many test matrices[17] obtained by considering random two-body interactions have shown that the resulting level distribution differs markedly from the semi-circle law, and is actually consistent with a Gaussian shape.

A. Gervois[18] has investigated this type of problem using purely analytical methods, based on the perturbation theory of statistical mechanics. Considering a shell of g degenerate states, containing n nucleons, she treats the interaction V between these nucleons as a perturbation. If all energies are measured from that of the shell, the unperturbed Hamiltonian H_o vanishes, and the expansion of $\Phi(\alpha \beta)$ reads

$$\Phi(\alpha \beta) = \Phi_o(\alpha \beta) + \sum_{p=1}^{\infty} \frac{(-\beta)^p}{p!} \langle V^p \rangle_C \quad , \tag{18}$$

where Φ_o is the expression of log Z in the absence of interaction given by (3) which reduces here to

$$\Phi_o = g \log (1+e^{\alpha}) \quad . \tag{19}$$

The brackets in the expansion denote the expectation value in the sense of statistical mechanics[19] and the symbol C indicates that connected diagrams only should be taken into account. For instance at the second order

$$\langle V^2 \rangle_C = \sum_{rsmn} f_r^+ f_s^+ f_m^- f_n^- \mid \langle rs|v|mn \rangle \mid^2 \quad , \tag{20}$$

where $\langle rs|v|mn \rangle$ denotes the matrix element of the two-body interaction between single particle states r, s and m, n whereas f^+ and f^- are the Fermi statistical weight factors for particle and hole lines, respectively, which (since $H_o = 0$) reduce here to

$$f_m^+ = 1 - f_m^- = 1 / (1+e^{\alpha}) \quad . \tag{21}$$

Statistical assumptions are now introduced on the matrix elements of the two-body interaction. The $\langle rs|v|mn \rangle$ are considered as independent random variables, having all the same probability distribution. The precise form of this distribution does not matter much, but it is convenient to assume symmetry, so that the expectation value of any odd power of a matrix element vanishes. Finally, one makes the limiting assumption of a very large shell, $g \to \infty$. As for the number of nucleons in the shell, two cases are of particular interest :

a) "Normal" density (n/g finite and different from 1). It is then easy to show that the dominant term, in the limit $g \to \infty$, is the quadratic term p = 2 of the expansion (18). This gives a quadratic dependence of $\Phi(\alpha,\beta)$ as a function of β . Substitution

of the corresponding expression into the integral (4) yields for
$\rho(A\ E)$ a <u>Gaussian distribution</u> in E . It is important to note
that this result is obtained under the assumption that

$$\left(\frac{n}{g}\right)^2 \left(1-\frac{n}{g}\right)^2 \sum_{rsmn} \overline{|\langle rs|v|mn\rangle|^2} \sim \sum_{rsmn} \overline{|\langle rs|v|mn\rangle|^2} \qquad (22)$$

is finite in the limit g $\rightarrow \infty$.

    The argument for the selection of dominant diagrams may be
presented as follows, taking for instance the assumption of a large
system implying conservation of momentum at each interaction. In
(22) we have then 3 independent summations, and the matrix element
of v must be of order $g^{-3/2}$ if this expression is finite. For
a connected diagram with p interactions, the number of independent
summations is p+1 and the total power of g in front of its contribution is therefore p+1 $-3p/2 = 1-p/2$. It is negative except
for p = 2 (the diagram p = 1 gives a vanishing contribution).
    b) <u>Extreme low density</u> (n or g-n = 2). When n is finite,
corresponding to a small number of particles (g-n finite corresponds to a small number of holes and the discussion is the same
provided the role of particles and holes is exchanged) the statistical factor f^- is a small quantity and the dominant diagrams
are those containing the smallest number of hole lines, namely the
<u>ladder diagrams</u> (Fig.1). The product of matrix elements of v for
such a diagram is of the form

$$\langle m_o n_o|v|m_1 n_1\rangle \langle m_1 n_1|v|m_2 n_2\rangle \ldots \langle m_k n_k|v|m_o n_o\rangle \qquad . \qquad (23)$$

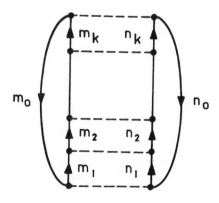

Figure 1

Its expectation value vanishes under the statistical assumptions formulated above unless equality relations between intermediate states are introduced in such a way that (23) becomes a product of even powers of individual matrix elements. For instance, a non-vanishing expectation value is obtained for

$$\langle 0|v|1\rangle \langle 1|v|2\rangle \langle 2|v|1\rangle \langle 1|v|3\rangle \langle 3|v|1\rangle \langle 1|v|0\rangle \quad , \qquad (24)$$

where, for simplicity, the intermediate states $m_i n_i$ have been indicated simply by i . The scheme of identical intermediate states is conveniently represented by a diagram such as Fig.2 (for the product 24).

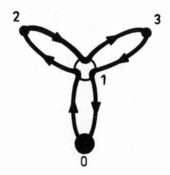

Figure 2

One must still determine the dominant family of these diagrams. It is easy to see that each identification of intermediate states, by reducing the number of summation variables introduces a small factor $1/g$. The dominant diagrams are therefore obtained by introducing the smallest possible number of these identifications. The corresponding diagrams are (rooted) trees, each branch of which is a double line. It may be shown that all of them give contributions of the same order, under the assumption that

$$\frac{1}{g^2} \sum_{r\,s\,m\,n} \overline{|\langle rs|v|mn\rangle|^2} \qquad , \qquad (25)$$

remains finite as $g \to \infty$. Note that this condition corresponds to (22) when n is a fixed number independent of g . The resummation of all the tree diagrams may be carried out and yields for $\rho(A\ E)$ the semi-circle law.

It may seem surprising that the low density limit requires the resummation of an infinite family of diagrams, whereas in the

case of the normal density, there is a single dominant diagram.
The reason is that in the low density limit, according to (25) as
compared with (22) one is led to take a larger interaction, the trace
of the square of which goes to infinity as g^2, instead of remaining finite
as in the normal density case. This is why diagrams with many interactions become important in the low density limit, whereas they
are negligible in the normal density case.

When n is larger than 2 but still much smaller than g ,
it may be shown that additional diagrams must be taken into account,
since the introduction in a particle line of an insertion, for
instance, of the type represented on Fig.3 does not change the
order of magnitude of the contribution. The level distribution has
then a shape intermediate between the semi-circle law and a Gaussian.

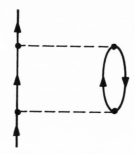

Figure 3

IV - THE DISTRIBUTION OF SINGLE PARTICLE LEVELS
IN A HIGHLY DEFORMED POTENTIAL WELL

The determination of the single particle energy levels ε_i is
quite difficult even by numerical methods when the potential well
has a large deformation of the type appearing in the study of fission or the nuclear isomeric states. As a first step it is interesting to consider the case of a deformed infinite square well. The
problem amounts then to finding the distribution of eigenvalues for
the wave equation

$$(\Delta + k^2) \, \psi(r) = 0 \quad , \tag{26}$$

where the wave number k is related to the single particle energy ε
by $\varepsilon = k^2$, assuming that $\hbar^2/2M = 1$, with one of the following
two boundary conditions

$$\psi(r) = 0 \ \ (A) \quad \text{or} \quad \partial\psi/\partial n = 0 \ \ (B) \tag{27}$$

on the nuclear surface S.

The determination of the asymptotic distribution for the solutions of the eigenvalue problem (26)(27) is a famous mathematical
problem which was solved, in first approximation, by H. Weyl in 1911.
This early result was then elaborated, and the following asymptotic
expansion in powers of $1/k$ was obtained

$$\rho(\varepsilon) \approx \frac{Vk}{4\pi^2} \begin{matrix} - \\ + \end{matrix} \frac{S}{16\pi} + \frac{1}{12\pi^2 k} \int_S d\sigma \; \frac{1}{2} \left(\frac{1}{R_1} + \frac{1}{R_2} \right) + \ldots , (28)$$

where V is the nuclear volume, S the area of the nuclear surface and R_1, R_2 its main curvature radii. In this expression,
the signs $-$ and $+$ refer respectively to the boundary conditions
(A) or (B).

The asymptotic expansion (28) gives a smooth expression for
the single particle level density, which does not include any shell
effect, or in other words, which applies only when the wavelength
is very small compared to nuclear dimensions. This condition is
clearly not satisfied at the useful single particle energies, which
are of the order of the Fermi energy. The asymptotic evaluation of
the density of eigenvalues for the wave equation with boundary conditions on a surface of arbitrary shape has recently been extended
in such a way that it becomes applicable to nuclei[20] . The main
ideas are as follows.

A. Definition of a Smoothed Density

Since the exact density is a sum of δ-functions, for which an
asymptotic evaluation, strictly speaking, has no meaning, it is
necessary to replace it by a smoothed density. By introducing a
smoothing function of the Lorentz type, we may define it as

$$\rho_\gamma(\varepsilon) = \frac{\gamma}{\pi} \int \frac{d\varepsilon'}{(\varepsilon-\varepsilon')^2 + \gamma^2} \; \rho(\varepsilon') , \qquad (29)$$

where $\rho(\varepsilon')$ is the exact density, and γ the smoothing width.

B. Introduction of a Green Function

Next we consider the Green function $G(r \, r', z)$ for the wave
equation

$$(\Delta + z) \, G(r \, r', z) = - \delta(r-r') , \qquad (30)$$

and one of the boundary conditions (27). It is easy to show that
the smoothed density (29) is simply expressed in terms of the Green

function by

$$\rho_\gamma(\varepsilon) \; = \; \frac{1}{\pi} \; \text{Im} \int_V d^3r \; G(r \, r \,, \varepsilon + i\gamma) \;. \qquad (31)$$

C. The Multiple Reflection Expansion

The next step is the evaluation of the Green function $G(r \, r', z)$. We start from the Green function of the wave equation (30) for the infinite space

$$G_0 \; = \; \frac{e^{ik|r-r'|}}{4\pi|r-r'|} \;, \qquad (k = k_r + ik_i = \sqrt{\varepsilon + i\gamma}). \qquad (32)$$

It may then be shown that the Green function G for the volume V is given by the expansion

$$G(r \, r') = G_0(r \, r') \mp 2 \int_S d\sigma_\alpha \; \frac{\partial G_0(r \, \alpha)}{\partial n_\alpha} \; G_0(\alpha \, r')$$

$$+ (\mp 2)^2 \int_S d\sigma_\alpha \; d\sigma_\beta \; \frac{\partial G_0(r \, \alpha)}{\partial n_\alpha} \; \frac{\partial G_0(\alpha \, \beta)}{\partial n_\beta} \; G_0(\beta \, r')$$

$$+ (\mp 2)^3 \int_S d\sigma_\alpha \; d\sigma_\beta \; d\sigma_\gamma \; \frac{\partial G_0(r \, \alpha)}{\partial n_\alpha} \; \frac{\partial G_0(\alpha \, \beta)}{\partial n_\beta} \; \frac{\partial G_0(\beta \, \gamma)}{\partial n_\gamma} \; G_0(\gamma \, r')$$

$$+ \, \ldots \qquad (33)$$

where the signs $-$ and $+$ correspond again to the boundary conditions (A) or (B). This expansion may be described as representing the propagation of a wave from an initial point r' to a final point r with any number of reflections on the boundary surface S.

Substitution of the multiple reflection expansion (33) into (31) yields a formal expansion for the smoothed level density. For discussing this expansion, we note that the wave number k occurring in the Green function G_0, according to (32) has a positive imaginary part k_i proportional to the smoothing width γ. As a function of $|r-r'|$, G_0 has a range of the order of $1/k_i$. If we denote by R a typical nuclear dimension, we may now distinguish two cases :

a) The smoothing width γ is so large that $k_i R \gg 1$. The range of the function G_0 is then very small compared with nuclear dimensions. It follows that the expansion (33) gives significant contributions only when the points r, α, β, γ, $\ldots$ are all near one another. By evaluating the resulting contribution from the

successive terms, one recovers precisely the expansion (28). This
corresponds to the situation where the smoothing width has been
chosen sufficiently large to wipe out all oscillations in the density of eigenvalues. One obtains then a completely smooth dependence
on the energy.

   b) <u>The smoothing width is small</u> : $k_i R \lesssim 1$. In this case, the
propagation of the wave described by G_o is no longer restricted
to small distances, and waves may travel back and forth across the
whole volume V . It is nevertheless possible to find an asymptotic
evaluation of the level density, under the condition that the energy is large.

   More precisely, if k_r is the real part of the wave number k,
we now assume that $k_r R \gg 1$. We notice then that the integrand
in any of the terms of the expansion (33) contains a rapidly varying
exponential of the form

$$e^{ik_r \ell} = e^{ik_r [|r-\alpha| + |\alpha-\beta| + \ldots + |\zeta-r|]} \quad , \quad (r' = r), \quad (34)$$

where ℓ is the <u>total length travelled by the wave</u>. The remainder
of the integrand is a slowly varying factor. In the limit of large
k_r , each term of the expansion (33) may be evaluated asymptotically by the <u>principle of stationary phase</u>, according to which important contributions come only from the neighbourhood of configurations
of the points r (inside V) , α, β, ... (on the surface S) such
that the <u>length ℓ is stationary</u>. Two such configurations are represented on Fig.4.

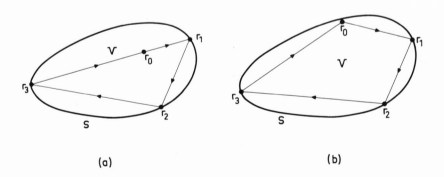

(a)   (b)

Figure 4

In the case a) , the point r is on the straight line between the
first and last reflection points, and there is a mirror reflection
at each point α, β, ... on S . It can be shown that these configurations give the largest contributions. The case b) corresponds to

a "boundary" contribution. The point r is near the boundary S
of the volume V and there is a mirror reflection at all points.
Detailed evaluation shows that the corresponding contribution is
smaller than in case a). In first approximation, these boundary
contributions may be omitted.

Let us denote by L the length of a stationary polygon of
the type described above. The magnitude of its contribution depends
on its degeneracy, defined as follows. For an isolated stationary
polygon, we say that the order of degeneracy is $q = 0$. In the
case of a surface with an axis of symmetry, any regular polygon
in a plane orthogonal to the axis and to the surface S is stationary. The order of degeneracy is then $q = 1$, since rotations
around the axis produce a one-parameter family of equivalent stationary polygons. In the case of a sphere, the stationary polygons
will be the regular polygons in the diametrical planes. The order
of degeneracy is then $q = 3$ (the orientation of the plane depends
on 2 parameters, and within a given diametrical plane the polygon
may still undergo a one-parameter rotation), except for the flat
polygons reducing to a repeated diameter, in which case the degeneracy is only $q = 2$, corresponding to the two parameters defining the orientation of the diameter.

It is easy to show by application of the principle of stationary phase, that the contribution of a stationary polygon of length
L having an order of degeneracy q is the real part of a quantity
proportional to

$$\left(\frac{1}{k}\right)^{1-q/2} e^{ikL} . \tag{35}$$

Thus we obtain oscillations in the level density. For an isolated
polygon, the amplitude of the oscillations is proportional to 1/k.
In the case of the sphere, for all polygons, except repeated dia-
meters, the amplitude is proportional to $\sqrt{k}$.

In order to verify the existence of these oscillations, we
have performed numerical experiments for the case of the sphere.
First of all, starting from the exact levels given for the sphere
by the zeroes of Bessel functions of half-integer order, we have
evaluated the Fourier transform of the level density expressed as
a function of k , in order to see if peaks associated with the
lengths of the stationary polygons really show up. The result is
shown on Fig.5, which gives the modulus of the Fourier transform
of the level density evaluated by taking all eigenvalues up to
$k = 37$. The arrows represent the lengths of the regular polygons
given for a sphere of unit radius by

$$L(p,t) = 2p \, \text{Sin} \, \frac{t\pi}{p} , \tag{36}$$

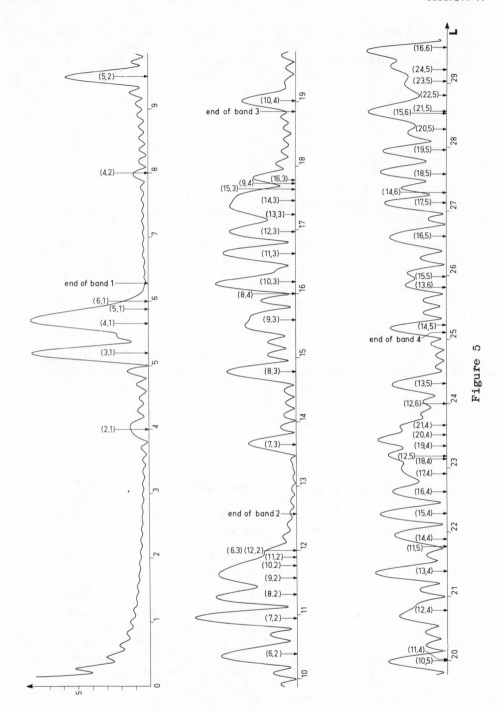

Figure 5

where p denotes the number of sides of the polygon and t the
number of turns it makes around the center. For instance, (4,2)
represents a diameter described four times, whereas (7,3) represents
a starred heptagon making three turns around the center. It is seen
that a large number of regular polygons appear very clearly.

The real purpose of the theory, of course, is to provide a
simple evaluation of the level density. It is necessary therefore
to take into account all contributions up to a given order. In
the selection of dominant contributions, the imaginary part k_i
of the wave number k plays a role by providing an exponentially
decreasing factor $e^{-k_i L}$, which favors the paths of smallest
lengths. It appears, however, that paths tend to form bands. For
instance, in the case of the sphere, the polygons for a given t ,
and a number of sides p going to infinity have a length which
converges toward $2\pi t$, the length of t circles. It may be shown,
however, that although the energy dependence of the contribution
(35) does not depend on the number of reflections p , the numerical coefficient in front of this contribution goes to zero as p
goes to infinity, in such a way that the total contribution of a
band converges slowly.

Fig.6 represents several evaluations of the level density for
a sphere. The thin curves are the smoothed densities evaluated
from the exact levels for different values of the smoothing. The
thick curves are obtained by taking the sum of the volume and surface terms in the expression (28), and a certain number of oscil-
lating terms of the type described here.

For a large value of the smoothing width ($k_i = 0.5$) the density oscillations are quite damped. They exhibit nevertheless very
clearly an interference pattern between two sinusoidal oscillations.
The corresponding thick curve has been calculated by taking into
account only the contributions of the regular triangle and square,
and it is seen that the interference between the corresponding two
contributions reproduce the density extremely well. The same holds
for a smaller smoothing width corresponding to $k_i = 0.2$. The
oscillations are then much larger, but the simple curve obtained
by taking into account only the triangle and the square still reproduces very well the exact curve except in the region where destructive interference takes place between the corresponding two
contributions. There, it is not surprising that contributions from
other polygons become important.

Finally, for a small smoothing width $k_i = 0.1$ the oscillations have a much more complicated shape, and it was necessary in
order to reproduce the exact curve to take into account 12 polygons :
the diameter repeated twice, the first six polygons with $t = 1$,
and the first five polygons with $t = 2$. The agreement is then
very good between the exact and the approximate curves. It is

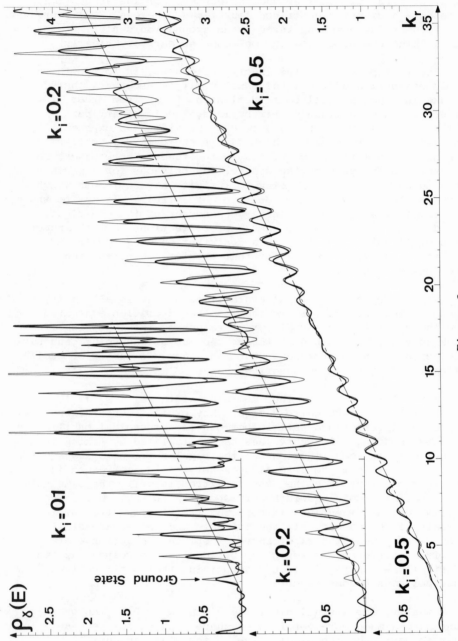

Figure 6

interesting to note that the shape of the first oscillations is no
longer sinusoidal, but ressembles very much a Lorentzian. Actually
the first oscillations correspond to individual **levels,** the ground
state and the first few excited states. Higher up, the density
oscillations correspond, on the contrary to a bunching of the levels. It is remarkable, however, that any asymptotic theory valid,
in principle, for large energies extends to such low energies that
the ground state may be located.

Of course, the sphere cannot be considered as a very good
example of the highly deformed surfaces for which the theory was
constructed. Actually, Mr. Bonche at Saclay has undertaken a
calculation of the fission potential barrier including shell effects
calculated according to this theory. This is underway and I cannot
yet present any result. On the other hand Mr. Bonche has also
studied the extension of the method to a slightly different problem, namely the evaluation of the nuclear density. One must therefore keep r constant in the expression (31) (instead of integrating

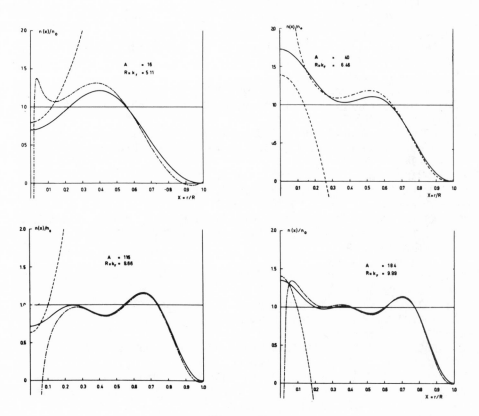

Figure 7

over V) and integrate over energy from 0 to the Fermi energy. The
resulting quantity is the sum of the squares of the single particle
wave functions for which the energy is smaller than E_F , taken at
the point r. Again, asymptotically one must consider the stationary
paths starting and ending at r (no stationarity required at r) and
involving any number of reflections on S. The preliminary results
presented on Fig. 7 have been obtained by taking simply the one-
reflection paths: going from r to S and back along a diameter. The
calculations carried out for several values of the number of particles A yield the dash and dot curves. The full curves give the
exact density. Near the center, the one-reflection stationary paths
tend to become degenerate and this gives rise to a divergence.

A special evaluation near the center, taking properly into
account this degeneracy yields the dotted curves.

REFERENCES

1) - H.A. BETHE; Rev. Mod. Phys. 9, 69 (1937)
2) - G.E. UHLENBECK and van LIER; Physica 4, 531 (1937)
3) - C. BLOCH; Phys. Rev. 93, 1094 (1954)
4) - N. ROSENZWEIG; Phys. Rev. 108, 817 (1957); Nuovo Cimento
      43, 227 (1966)
5) - P. KAHN and N. ROSENZWEIG; Phys. Rev. 187, 1193 (1969)
6) - HIROSHI BABA; Communication 6.2
7) - M. BÖHNING; Communication 6.7
8) - M. VONACH, R. SCHANTL and M. BÖHNING; Communication 6.6
9) - See, for instance, I. Kanestrøm; Nucl. Phys. 83, 380 (1966)
      A109, 625 (1968)
10)-C. BLOCH in Summer School of Theoretical Physics, Les Houches
      1968 ed. C. de Witt and V. Gillet (Gordon and Breach, New
      York 1969)
11)-M. BÖHNING; Nucl. Phys. A152, 529 (1970)
12)-F.C. WILLIAMS; Nucl. Phys. A166, 231 (1970)
13)-R. BALIAN; Nucl. Phys. 13, 594 (1959)
14)-V. CANUTO; Nuovo Cimento 28, 742 (1962) M. SANO AND S. YAMASAKI;
      Prog. Theor. Phys. 29, 397 (1963) D. ZIVANOVIC; Fizika, 2,
      177 (1970)
15)-L.G. MORETTO; Communication 6.1
16)-See, for instance, M.L. Mehta, Random Matrices, Acad. Press,
      New York (1967)
17)-J.B. FRENCH and S.S.M. WONG; Phys. Lett. 33B, 449 (1970)
      O. BOHIGAS and J. FLORES; Phys. Lett. 34B, 261 (1971)
18)-A. GERVOIS; Communication 6.14
19)-C. BLOCH and C. DE DOMINICIS; Nucl. Phys. 7, 459 (1958)
      C. BLOCH in Studies in Statistical Mechanics, ed. J. de
      Boer and G.E. Uhlenbeck, (North-Holland) (1965)
20)-R. BALIAN and C. BLOCH; Ann. of Physics (in the press) see also
      R. Balian and C. Bloch; Ann. of Physics, 60, 401 (1970); 64,
      271, (1971).

DISCUSSION

KAPOOR (BARC)

Strutinsky defines the uniform single particle distribution by averaging with the Gaussian distribution function the Hermite polynomial coefficients. I see that you are using a different approach. Do you have any comments why you should do everything this way and not the other way? The second thing I would like to know is if you now have any program to calculate the nuclear shell correction energy using this smooth single particle level distribution, not just for deformed nuclei but also for spherical ones?

BLOCH

Let me first answer the first question. I do not like the Gaussian as a smoothing function very much because one of the tricks for deriving the expression of the smoothed level density in terms of the Green function is to close a contour in the complex plane as you very well know. If you take a Gaussian, you can't do that, because it goes to infinity on the imaginary axis. The Gaussian is therefore not a nice function to use in this kind of derivation, but I do not think that it matters very much whether you use a Gaussian or some other smoothing function. You may be worried by the possibility that all these results might depend on a particular choice of the smoothing function. Actually, one would like the results to be independent of any choice. We have been able to show that all the derivations can be repeated if, instead of the Lorentz shape, one takes a smoothing function which is a sum of any number of poles, chosen in such a way that it is real and goes nicely to zero at infinity. The Lorentz shape is obtained when one takes only two complex conjugate poles. But one can take any number of poles, and derive very similar formulas. That gives us confidence that the results are quite independent of the shape. With a Gaussian function, I don't know how to do the calculation, but I wouldn't expect the result to be significantly different. As to the second question--I must say that we have not as yet made any applications beyond what I have just mentioned, but we plan to do such calculations.

MORETTO (Berkeley)

Let me come back to the question of Dr. Kapoor. When you try to smooth out the curve of a Gaussian function, you immediately realize that there are some problems associated with it, namely, that you tend to alter the lowest derivatives. Therefore, you are supposed to introduce a correcting function that will be such as to retain the low order derivatives the way they are supposed to be. I wondered if you ever considered that problem with your Lorentzian smoothing.

BLOCH

No.

SCHARFF-GOLDHABER (BNL)
   On your last slide was the distribution given as a function
of E?

BLOCH
   I'm very sorry. I should have said that. The nucleon
density is plotted as a function of the radial distance for various
values of A.

SCHARFF-GOLDHABER
   Is this density supposed to be independent of charge?

BLOCH
   This is a very simplified model in which one assumes that
all the nucleons are identical and that one can put four nucleons
on each single particle level. It does not take into account any
coulomb energy. This is just a first schematic model.

WIGNER (Princeton University)
   I'm afraid I have two questions. The first one is very
trivial. If you have an average density proportional to the
energy and smooth it out with a density that is asymptotically,
$1/E^2$, don't you get an infinity at zero energy?

BLOCH
   The density is not proportional to the energy but to the root
square of the energy. Then the integral is convergent.

WIGNER
   That means it's convergent, but I am afraid that the density
at zero will be heavily influenced by what comes from a distance.
You are right. In more than a three dimentional space, it will
not be convergent.

BLOCH
   Actually, we extended our theory to any number of dimentions
and this is why we introduced smoothing functions of the type I
mentioned previously. One can then always take as smoothing function
a   meromorphic function which has a tail going to zero as one
over any power of the energy that one likes.

WIGNER
   The question really is, does the distribution at zero not get
much influenced by what comes from a distance?

BLOCH
   No.

WIGNER
    The second question is, I am always wondering what it means to have the level density for a deformed nucleus, because in the excited state the nucleus will have different deformations. What does it mean?

BLOCH
    This is undoubtedly a very deep question. It is of course very surprising that one can apply the independent particle model for fission, for instance, and be able to describe the fission process by considering that one has a potential which is becoming more and more deformed and ultimately ends up in two separate wells and still have independent particle motions throughout.

WIGNER
    That doesn't explain what the meaning of the calculation is.

BLOCH
    No, but there is a lot of experimental data that supports it. Now you are absolutely right: it remains a tremendous problem for the theoreticians to understand why the residual interactions do not destroy completely this very simplified picture. This is an entirely unresolved problem, and I cannot say anything about it.

LE COUTEUR
    Perhaps I might be able to make a comment on the question that Dr. Wigner asked. We really are concerned with particles in a boundary which is moving a little when the nucleus is excited. If you think of a nucleus following slow neutron capture where the excitation is about equal to the neutron binding energy, the temperature of that nucleus is already higher than the characteristic energy of the first quadrupole vibrations in an even-even nucleus. There will be some excitation of these vibrations. One is left with the question whether the amplitude of vibration is sufficient to move this spectrum very much. The estimates one can make suggest that at this energy the effect is not yet very big, but as the temperature rises, of course, it becomes more important. There are some calculations more recently of Dr. Moretta which show this effect in more detail, and at excitations of say 40 MeV where the temperature would be about twice as high. Then, of course, this Planck's factor for the excitation of the oscillation is very much bigger, and Moretta's results do show quite large effects. I might mention that this is a problem of great antiquity. The first calculations on this effect known to me were done in 1937 by Bagger, published in the Ann. de Physique. As the nucleus is excited, you have waves on the surface, the nuclear matter reaches out a little further and depresses the coulomb barrier. The effects he calculated at that time were rather small because at that time one had no reliable estimates of the frequency of these excitations, and, of course, no measurements, either.

DISTRIBUTION THEORY OF NUCLEAR LEVEL DENSITIES AND RELATED

QUANTITIES*

J. B. French and F. S. Chang

Department of Physics and Astronomy

University of Rochester, Rochester, New York 14627

1. INTRODUCTION

The concepts and methods applied in the usual theories of
nuclear level densities are so different from those used in studying nuclear properties in the ground-state domain that contacts between the two activities have been disappointingly few. In the
ground-state region the detailed nature of the effective Hamiltonian,
and the detailed results which flow from it, are paramount, and
the spectroscopist working there has not been much interested in
average properties such as level densities. In the neutron-thresh-
old region, and at higher excitation energies, averages (and fluctuations about them) have been of major interest. Since it is not
feasible to do conventional detailed spectroscopy in the extremely
large model spaces encountered at high excitation, the result has
been that the spectroscopist has mainly continued to work at low
energy, while the person whose interests lie higher has been com-
pelled to use quite different methods. These methods in particular
have required him to ignore the residual interaction effects (except insofar as they are taken account of by choice of single-particle energies and the use of BCS quasiparticles as the basic
fermions).

There have been two major consequences of this. On the one
hand no justification for ignoring the residual interactions has
really been given, nor have the parameters of the conventional
theory been evaluated in terms of more fundamental quantities; thus
the theory has been left without an adequate foundation. On the
other hand the great mass of level density information available

*Supported in part by the U.S.Atomic Energy Commission.

has not been used to study or to test the effective interaction.
There is more to the latter than appears on the surface. In recent
years the emphasis in low-energy spectroscopy has shifted away from
the use of interactions chosen to fit spectroscopic data, and towards using more fundamental interactions derived from free nucleon
scattering; in deriving effective interactions which operate properly
in small low-lying model spaces one must know or assume much about
the higher states. Moreover some of these states are "renormalized
away", a process whereby the states themselves are discarded but
their effects on low-lying states taken account of by modifying the
interaction; this process must be undone when the spectroscopist
wants to deal with the high-lying density. It is clear then that
the connections between the two domains, arising from the effective
interaction, are both interesting and important.

There are moreover many cases where one needs information about
states and interactions at higher energy of a less detailed nature
than would be provided by standard spectroscopy (if that could be
done) but more detailed or more accurate than given by the usual
combinatorial methods. We think here of such things as state occupancies, and the decomposition of the interaction into various parts
(hole-particle, pairing, multipole, etc.) which are encountered for
example in the study of compound-nucleus formation.

We can expect to make progress with these questions only if
there exists some principle which simplifies the at-first-sight
hopelessly complex problem of many interacting particles in a large
spectroscopic space. The central limit theorem supplies such a
principle. It has of course been used since the beginning in de-
composing the level density according to angular momentum; but we
shall need more general versions and quite different ways of applying the theorem. Making use of these we shall describe a procedure
for confronting the general problems described and constructing a
theory which takes account of orbital structures and residual interactions; it yields the level densities as a sum of partial densities
in which various symmetries (angular momentum, isospin, configurations, space-symmetry, etc) are specified. It can of course also
be applied to the case of non-interacting particles. In fact the
conventional theories for this latter case are essentially combinatorial in nature, this arising from the fact that single-fermion
energies are additive; one can extend the notion of combinatorial
analysis so that it encompasses the interacting-particle case as
well.

During the past few years the resulting so called "spectral
distribution" methods (1-3) have been used for a number of purposes
at low excitation energies. But although the distributions which
enter are themselves partial level densities, applications to level
densities at higher excitations have only recently begun. Besides
the fundamental effective-interaction problem mentioned above

(which is really ingrained in the physics) various technical problems of accuracy arise when one uses the very large spaces needed at high excitation and these are just now being dealt with. This report is then a preliminary one designed more to make clear the general procedures for calculating level densities and related quantities than to give final results for them.

2.   DISTRIBUTIONS AND THE CENTRAL LIMIT THEOREM

We shall deal always with a finite piece of the Fermi spectrum, i.e., with a finite set of single-particle states. If we start with N_1 of these then, for m_1 particles ($0 \leq m_1 \leq N_1$) we have $\binom{N_1}{m_1}$ states whose energies may span a large domain, (about 100 MeV for 12 particles in the (ds) shell, for which $N_1 = 24$). If N_1 and the effective H are appropriate, the lowest m_1-particle energies will be good approximations to the eigen-energies of the m_1-particle nucleus; but as we go up in excitation energy (say above 8 MeV in the (ds)^{12} case) the chosen N_1 becomes inadequate, the energies which we calculate become expectation values instead of eigenvalues, and the corresponding state density becomes a partial density. If now we extend the Fermi spectrum considered, say to $N_1 + N_2$ states, we may similarly partition the total number of particles, to m_1+m_2; these partitions then define a "configuration"; and similarly to $\Sigma N_i=N$ with $\Sigma m_i=m$. In this way we are of course representing the state density which rapidly increases with energy, as a superposition of partial (configuration) densities each of which spans a finite domain. If we had agreed in advance on the maximum excitation energy of interest we could have chosen N appropriately and thereby dispensed with the partitioning, and sometimes this is good to do. But the partial densities are themselves of real interest; they enable us moreover to derive the average Fermi occupancies; and when N is large we shall, by partitioning, be able to increase the accuracy with which the distributions can be calculated. It should be clear that we shall also need further decompositions of the level density (for example according to angular momentum) which do not correspond simply to a partitioning of the Fermi spectrum.

We assume that the complete "universe" for our problem is defined by N and that we have an effective Hamiltonian H which operates in the m-particle spaces defined thereby. Let Ψ_i, E_i be the Hamiltonian eigenfunctions and eigenvectors. Suppose that our interest is with a partial density $\rho_\alpha(m)$ defined by the m-particle states $\phi_\alpha(m)$ where α runs over the orthonormal set $\underset{\sim}{\alpha}$. The states ϕ_α need not be Hamiltonian eigenstates, and, more generally, the space spanned by $\underset{\sim}{\alpha}$ need not be a Hamiltonian eigenspace. We consider the intensities with which each ϕ_α occurs in each eigenstate Ψ_i and in this way are led to a partial density $\rho_\alpha(E)$; explicitly

$$\phi_\alpha = \sum_i B_{\alpha i} \Psi_i \qquad\qquad \rho_{\underset{\sim}{\alpha}}(E_i) = \sum_\alpha |B_{\alpha i}|^2$$

The distribution ρ_{α} is of course discrete but we shall often consider it as continuous; its p'th moment is seen to be

$$M_p(\alpha) = d^{-1}(\alpha)\int \rho_{\alpha}(E)\ E^p dE = d^{-1}(\alpha) \sum_{\alpha \epsilon \alpha} <\phi_{\alpha}^+ H^p \phi_{\alpha}> \equiv <H^p>^{\alpha}$$

where $d(\alpha)$ is the dimensionality of the set α, and we have written
both continuous and discrete forms. $<H^p>^{\alpha}$ is of course the expectation value of H^p averaged over α.

Instead of the <u>simple moments</u> we may consider the <u>central
moments</u>, μ_p, which are the moments for $H-\mathcal{C}(\alpha)$ where $\mathcal{C}(\alpha) = M_1(\alpha)$
is the centroid; the <u>cumulants</u> K_p (whose structure we shall see
shortly) are certain polynomials in the central moments up to order
p; like the μ_p these are homogeneous in H, and "p-linear" in the
sense that multiplying H by a factor λ multiplies K_p by $(\lambda)^p$. K_2,
which equals the second central moment, is the variance, $\sigma^2 = M_2 - (M_1)^2$,
or square of the <u>width</u> σ; the Gaussian distribution has as a defining property that <u>all</u> higher cumulants vanish. We introduce finally the scale-independent quantities K_p/σ^p the values of which fix
the <u>shape</u> of the distribution. K_3/σ^{3} and K_4/σ^4 are commonly called
the "skewness" and the "excess".

The most important and most striking feature of the distribu-
tions $\rho_{\alpha}(E)$ is that, for a wide range of α subsets, they are closely
Gaussian in form as long as the number of "active" fermions is
sufficiently large and the interactions are of low enough particle
rank (a k-body interaction has particle-rank k). The indications
indeed are that this is true if α corresponds to an irreducible
representation of U(N) or one of its subgroups (summed over equivalent representations if there are such); thus for example we would
expect Gaussian forms for the distribution of all m-particle states
(the group is U(N)), of all states of a given configuration (a
"direct-sum" subgroup), of all states with given T or spin-isospin
SU(4) symmetry (direct-product subgroups), or fixed angular momentum J, or fixed (J,T), etc.

We are not able here to give any general discussion of this
(and indeed the whole thing is by no means completely understood).
But the role of the central-limit theorem in yielding normality
will be clear when we consider the distribution over all states for
m non-interacting particles. If, moreover, m/N is small enough we
may ignore the Pauli principle "blocking effects" (which correct
for the fact that two particles may not occupy the same single-particle state) and then have that the m-particle density is given by
a convolution

$$\rho_m(E) = \int \rho_{m'}(E')\ \rho_{m-m'}(E-E')dE'$$

Things simplify if we introduce the Fourier transform $g(t) = \int \rho(E)e^{itE}dE$
and the further Maclaurin expansion, $\ln g(t) = \sum K_p(it)^p/p!$ Since
the transforms multiply under a convolution it is immediate that
the cumulants add. Thus we have easily that $K_p(m) = m\ K_p(1)$,

which is simply a generalization to higher order of the well-known
result that the variances σ^2, which are <u>quadratic</u> quantities in H,
add <u>linearly</u> when we combine independent systems. But the shape
parameters are K_p/σ^p which then, for p>2, decrease as particles are
added, being proportional to $m^{1-p/2}$; this in turn implies a trend
towards normality. The argument given here is of course that used
in deriving the elementary central limit theorem which asserts that
under repeated convolutions a function tends towards Gaussian. It
implies that information contained for example in the two-particle
matrix elements and spectrum is gradually lost as we add particles
to the system.

The same argument applied to the distribution for a fixed con-
figuration, defined by $\vec{m}\equiv(m_1,m_2,\ldots m_\ell)$ which represent the numbers
in the separate "orbits" $(N_1,N_2,\ldots N_\ell)$, also gives an additivity of
the cumulants, $K_p(\vec{m})=\Sigma\ m_i K_p(00..1_i..00)$ and a corresponding normality for large m. Similarly for example the additivity of M_J, the
z-component of angular momentum, implies that under appropriate
circumstances we shall have a Gaussian energy dependence for states
of fixed J, as well as the familiar (2J+1)×Gaussian dependence on
J for fixed E. The energy variation of the spin cut-off factor,
which we consider later, modifies the energy dependence as do also
departures from the {(2J+1)×Gaussian} form.

If we have two-body (or higher rank) interactions two compo-
nents $(\underset{\sim}{m}',\underset{\sim}{m}-\underset{\sim}{m}')$ of an m-particle system may not be regarded as independent; for the two-particle energies are not additive, combin-
ing two systems does not give a simple convolution, and the tendency towards normality as particle number increases is not all so
clear. It may however be shown, by making a decomposition of H
with respect to the U(N) group of transformations in the single-
particle space, that, in large-N spaces with large fixed particle
number, there is an <u>a priori</u> tendency for H to behave like a one-
body operator; specifically,ignoring the U(N)-scalar part of H,
one finds that the part of the interaction whose Young (column)
shape is [N-1,1], and whose form is (n-1)×(traceless one-body operator) where n is the number operator, grows more rapidly with particle number than the "irreducible two-body" H, and thus may be expected to dominate often for large particle number. But then the
system is similar to one for non-interacting particles and the
argument used above applies. There is actually a strong indication
that, much more generally than this, the approach to normality obtains for operators whose particle rank (or unitary rank) is much
smaller than the particle number, or equivalently when only a small
fraction of the particles interact simultaneously, so that the trans-
forming away of the interactions should not in fact be essential;
it is this feature, the source of which has been studied via the
equation ahead, which modifies the semicircular distributions found
in conventional random matrix ensembles in the direction of
Gaussian forms as we make particle-rank restrictions on the H's in

the ensemble(4).

The qualitative arguments above are incomplete and in any case
leave us unprepared for the remarkable accuracy with which distributions are often fitted by Gaussian forms. Moreover they do not
give us a satisfactory basis for quantitative calculations. Instead
we may, considering the distributions as defined by their moments,
explicitly evaluate the low-order cumulants taking account of Pauli
effects and residual interactions. If we consider only the centroid
and width we would assume Gaussian; any strong departure from normality would be revealed by the skewness and the excess. The technical problem encountered here is that of evaluating traces over
the subsets $\underset{\sim}{\alpha}$ of powers of H, for which conventional spectroscopy
is of course inadequate, the spaces often being extremely large.
General methods are however available when α defines an irreducible
representation of a U(N) subgroup, in which case also we expect an
asymptotically normal distribution so that only a few moments are
needed. In one set of cases the problem is solved by expressing
the density operators in terms of the Casimir invariants of the
group which defines $\underset{\sim}{\alpha}$; more complicated operators are encountered
in other cases, including that of fixed angular-momentum for which
we later indicate an approximate method. With the use of these
operators we can express the moments for m-particles as linear combinations of the moments for simpler systems. A typical result,
for a k-body operator which is unitarily irreducible and transforms
as [N-v,v] is that the even-order cumulants for the distribution of
all m-particle states, is given by

$$K_p(m) = \binom{N-m}{u+1} \binom{m}{u} (m-u) \binom{m-v}{k-v}^2 \sum_{t=v}^{u} \binom{N-t}{u+1}^{-1} \frac{(N-2t)(-1)^{t-u}}{(m-t)(N-m-t)} \binom{u}{t} K_p(t)$$

$$\longrightarrow \binom{m}{u}(m-u)\binom{m-v}{k-v}^2 \sum_{t=v}^{u} \frac{(-1)^{t-u}}{(m-t)} \binom{u}{t} K_p(t)$$

Here u=pv/2 and the second form gives the large-N limit, a comparison of the two forms giving then Pauli-principle blocking effects.
Involved here is a different kind of cumulant combination than that
encountered for non-interacting particles, but one which for higher-
order cumulants is by no means as easy to apply (since the "input"
cumulants themselves will then refer to relatively complicated systems). The unitary-group decomposition implied by the equation extends easily (3) to the fixed-configuration and configuration-isospin cases and yields a physically significant orthogonal separation
of the Hamiltonian.

The distributions constructed as above are of course partial
level densities. Their sum yields also distribution approximations
to the ground-state energies and the low-lying spectrum. If there
are no degeneracies the exact (cumulative) distribution function
has a staircase form taking on the values (k-1,k) at the k'th eigenvalue. A natural way then to recover these energies (Ratcliff (2))

is to take the energies at which the distribution function has values $(k-\frac{1}{2})$. If there are exact-symmetry degeneracies an obvious
modification is needed.

We comment finally on certain "stabilities" of the distributions. When m is large enough we have the standard convolution (or
multiplicative) stability, in that adding a particle does not change
the form of the distribution; more surprising is the _additive_ stability which arises from the relationship between representations
of a group and its subgroups. If a set of m-particle representa-
tions $\underset{\sim}{\alpha}_i$ of a group G combine to give a representation β for a
larger G' then, normality for each group tells us that the sum of
the approximate Gaussians for the α_i will produce a single Gaussian
(for β); an example of this, where the 45 Gaussians for the jj representations of $(ds)^{12}$ combine beautifully to a single one, is given
in reference (1).

3.   ILLUSTRATIONS AND APPLICATIONS

Agreeing that we are now able to calculate low-moment approximations to various partial densities let us proceed to a series
of applications. Some of these will be relatively simple, solvable
exactly by combinatorial methods for non-interacting particles or
by conventional spectroscopy when there are interactions; we consider these in order to get some idea of the accuracy of the methods
and of the problems to be encountered in more serious applications.
Besides the level densities we consider also such related quantities as orbit occupancies and the weights with which various parts
of the interaction contribute at various excitation energies.

3(a)   The Odd-Parity Even-Parity Density Ratio

Figure (1a), due to F. C. Williams (5), shows the 5-particle
5-hole density for 10 non-interacting particles distributed over
the single-particle spectrum described in the caption; the total
number of 5p-5h states is 63,504. The agreement with the Gaussian
distributions is excellent as it is also for the other distributions
with a fixed number of hole-particle pairs.

If the single-particle spectrum contains states of both parities then we expect that the combined level density (both parities
together) should have a Gaussian form. However, the fixed-parity
densities, though each is representable as a sum of Gaussians,
should not themselves be in general Gaussian. This is because
fixed-parity configurations, though corresponding to representations
of a U(N) subgroup, are not "complete" with respect to U(N). It is
clear that this should be true for non-interacting particles, but
it is also to be expected for interacting ones (the argument here

being that the formal role of parity conservation is simply to put
to zero some of the two-particle matrix elements, which does not
however affect the two-body nature of the interaction).

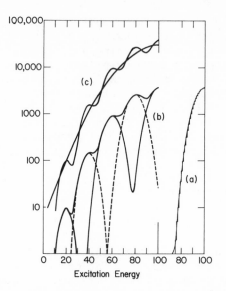

Fig. 1. The state density for 10 non-interacting particles distributed over two opposite-parity sets of 10 states, each with unit
spacing, separated by a 10-unit gap. The spectrum spans 0-200 and
is symmetrical about the mid-point. (a) compares the 5-particle 5-
hole exact and distribution densities; (b) gives the exact positive-
parity (dashed line) and negative-parity densities and their sum;
(c), which is displaced upwards by one cycle, gives the total density as calculated by the Gaussian distribution over all states,
and as a sum of fixed-configuration (hole-particle) Gaussians.

Figure (1b) shows the exact separate and combined distributions
for the same single-particle spectrum as in Fig. (1a) but where the
particle and hole states have opposite parity. Figure (1c) shows
the total density as calculated via the hole-particle distributions;
it agrees extremely well with the exact density of (1b). Shown
also is the two-parameter normal distribution which also fits well
though it does not of course reproduce the ripples which show up in
the finer distribution, and which are prominent because of the large
single-particle gap at the Fermi surface.

Figure 2 shows an example for 12 particles in the space generated by $(ds+f_{7/2})$, interacting by a Brown-Kuo interaction (6).
All configurations are considered, the number of states being 5.6
$\times 10^9$. The (d,s) single-particle energies are close to 0^{17} but two
values are given for $(\varepsilon_{7/2}-\varepsilon_{5/2})$. The larger, which derives from
averaging the interaction over 0^{16} closed shells, gives too large

a value (20 MeV) for the excitation energy of the lowest negative-parity state; because of the large gap it yields a large-amplitude oscillation in $\rho^{(+)}/\rho^{(-)}$. The smaller $\varepsilon_{7/2}$ value, which is chosen to give a proper negative-parity lowest state, produces a much smaller oscillation. It appears that, at energies where the "other-parity" states have well set in, the common equal-density assumption is liable to be good.

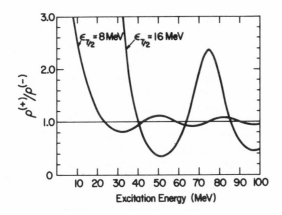

Fig. 2: The ratio of the even-parity to odd-parity state densities for $(ds+f_{7/2})^{12}$ with two values of the $f_{7/2}$ single-particle energy.

3(b)   The Formal Basis for Non-Interacting-Particle Densities

Suppose that the single-particle space consists of a set of spherical orbits with given single-particle energies. In the absence of residual interaction a spherical configuration consists of a set of degenerate states, which both move and spread under the influence of an interaction. The movement (of the centroids) can be taken account of by modifying the single-particle energies (1-3)

$$\varepsilon(\alpha) \Rightarrow \varepsilon(\alpha) + \frac{1}{2}(m_\alpha - 1)W_\alpha + \frac{1}{2}\sum_{\beta \neq \alpha} m_\beta W_{\alpha\beta}$$

where α, β are the orbits, W_α the average interaction between two particles in the same orbit, and $W_{\alpha\beta}$ similarly between two orbits; we may assume in any given case that these energies have already been taken account of in the non-interacting-particle model so that the significant question is about the effects of the spreading and the admixing of configurations in single eigenstates. A reasonable criterion for the adequacy of the simple model is that the widths

should not be large compared with the configuration spacings. In
(ds)^{12} we find however that, when reasonable interactions are introduced, the average spacing between the 45 jj configuration is
about 1 MeV while the corresponding configuration half-widths (σ)
are about 10 MeV; and similarly in other cases. Spherical orbits
then, at least in light and intermediate nuclei where calculations
have been made, do not give an adequate basis for a non-interacting
particle model. It is a difficult open question however whether in
these or other cases there might exist an adequate non-spherical
basis.

3(c) A Si^{28} Calculation with Realistic Interactions

Zuker (7) has constructed and diagonalized the 839-dimensional
matrix for (ds)^{12} with J=T=0, using the interaction of Kahana,Lee
and Scott (8). From the eigenvalues the cumulants have been evaluated as far as the fourth (it is probable that the accuracy is
sufficient to make this profitable), and the corresponding distributions constructed. The parameters and a comparison between the exact spectrum and those derived via the distributions with two, three
and four moments is given in Table 1. From the small values of the
skewness and excess we expect that the distribution must be very
close to Gaussian (appreciable departures only show up when one or
both of these parameters is larger in magnitude than ~.5). Table
1 shows how true this is, and shows moreover how much better we do
with third- and fourth-moment corrections. Much of the deviation
in fact comes from a few states at the extremes of the spectrum,as
we see from the reduction in the RMS error when the five lowest and
five highest states are omitted.

Table 1

Given are the R.M.S. differences between Zuker's exact J=T=0 (ds)^{12}
spectrum and the distribution spectra defined by two, three, and
four moments. In computing the "modified" error the five highest
and five lowest states are omitted. The dimensionality is 839, the
spectrum span is 88 MeV, and the ground-state energy is -104.9. A
negative ground-state error implies that the distribution ground
state is lower than the exact one. The centroid energy, width,
skewness and excess, are respectively -64.64 MeV,13.03 MeV,0.25,
and 0.20. The single-particle energies are those of 0^{17} with ε(5/2)
= 0.

No. of Moments	R.M.S.Error	Modified R.M.S.Error	Ground State Error
2	.81 MeV	.66 MeV	-2.0 MeV
3	.36	.15	+4.8
4	.24	.11	+1.5

Changing from two to three or four moments corresponds to making a "secular" change in the average density and the fact that this gives an improvement implies that the errors which show up in the two or three moment distributions are not in the nature of fluctuations (or "noise"). The pattern of the deviations makes it clear that this is correct; in the normal spectrum, levels 1-153 show a negative deviation, 154-724 a positive one, and 725-839 again negative with altogether only a half-dozen exceptions. Even with four moments we have a simple sign pattern so that we have not, even in this case, reached the irreducible errors which would be describable as "fluctuations".

The ground-state error which results from the Projected-Hartree-Fock calculation is about 7 MeV, so that with a few parameters we do much better than that. There being available for analysis no other matrix calculations of this size we are unable to say for certain whether the error found here is typical. We can however see one of the difficult problems facing us in our attempt to construct an absolute theory of level density, namely the location of a reference energy from which to measure the excitations. Experimentally of course one uses the ground state, and similarly in the theory for non-interacting particles (where one does of course introduce some ad-hoc corrections); but while from the standpoint of the spectroscopist an error of only a few MeV in the Si^{28} ground state is excellent, such an uncertainty in the reference energy produces a considerable error in the level densities. We can expect moreover that, when we increase by many orders of magnitude the vector-space dimensionality in order to deal with the densities encountered at high excitation the problem will be a great deal worse; the possible deterioration in very large spaces is in fact the second major technical problem, the experience being that a Gaussian approximation is quite untrustworthy more than 4σ away from the centroid. To deal with these problems we may, besides using finer distributions, also use an excited reference energy, reading off from the experimental data the difference between that energy and the actual ground state. We may determine an appropriate reference point by matching either the total number of states below that point or the density of states in its neighborhood.

3(d)   Level Densities for Fixed Isospin

The Zuker calculation described above is for fixed angular momentum and isospin. When we deal with heavier nuclei (with a neutron excess) it is almost essential to consider the densities for fixed isospin; on the one hand, only then can we find a reference point from which to measure our energies, while, on the other hand, the isospin decomposition considerably reduces the size of the vector spaces encountered and thus increases the accuracy. In fact fixed-isospin distributions are of value in testing the accuracy of the methods as a function of the dimensionality of the space, the relevant feature being the large variation of dimension with isospin; for example, in three-orbit (fp)-shell calculation we find, on using a Brown–Kuo type of interaction, good agreement with experiment for the isobaric mass difference, even though the results being compared come from spaces varying in size from 80 to 35,000.

Fortunately the technical problems of evaluating fixed-isospin centroids and widths are easily solved either in an isospin formalism (in which the direct-product nature of the isospin group makes things easy) or else in a p-n formalism (3) in which we deal with T_z and recover T by an elementary process of subtraction. In fact most of the applications so far made have involved specifying both configurations and isospin, the computing in all such cases being quite elementary. A typical example of the isospin decomposition, for the $(ds)^{12}$ partial density, is given in reference (3).

3(e)   Orbit Occupancies

The single-particle occupancies in a many-particle nuclear state are the simplest parameters which enter into the description of the multi-particle density. The ground-state occupancies are often directly measureable (in a spherical orbit representation) via stripping and pickup reaction sum rules, but at higher excitations they are of equal interest; they naturally enter for example in studies of compound nucleus formation. A plot of the occupancies versus the single-particle energies yields the profile of the Fermi surface. It is not surprising also that the occupancies are important parameters in the combinatorial theory of level densities.

To calculate the Fermi profile for a given set of single-particle states, in the presence of interactions or otherwise, is quite straightforward. For the occupancy of state i we isolate it from its companions in the single-particle space by the partition $N \to [N-1_i, 1_i]$ which, for m particles, defines two configurations $[\underset{\sim}{m}, 0_i]$ and $[\underset{\sim}{m}-1_i, 1_i]$;   then we have   as a function of the multi-particle energy E that

$$\langle n_i \rangle_E = \rho_{[\underset{\sim}{m}-1_i, 1_i]} (E) / \rho_{\underset{\sim}{m}} (E)$$

where $\rho_{\underset{\sim}{m}}(E)$ is the total density, the sum of the two configuration densities (for increased accuracy we may of course choose a further partitioning of the (N-1) states). Plotting $\langle n_i \rangle_E$ for fixed E against the single particle energy ε_i gives the Fermi profile. For non-interacting particles with an arbitrary single-particle spectrum $\langle n_i \rangle_E$ has of course a simple analytic form in terms of single-particle sums.

For a test of the procedure we show first, in Table 2, the

Table 2

Occupancies, at selected excitation energies, for 8 non-interacting particles in 16 single-particle states with unit spacing. The excitation energy span is 0-64. By a hole-particle symmetry both the exact and distribution occupancies satisfy $\langle n_i \rangle + \langle n_{(16-i)} \rangle = 1$ for fixed E, and $\langle n_i \rangle \rightarrow 1 - \langle n_i \rangle$ when $E \rightarrow 64 - E$. For each energy the first line gives the exact occupancies and the second the four-moment distribution occupancies. At the midpoint energy, E=32, both the exact and distribution occupancies are precisely $\frac{1}{2}$ for each state.

State #	1	2	3	4	5	6	7	8
$\langle n_i \rangle$:E=5	1.00	1.00	1.00	0.86	0.86	0.71	0.71	0.57
	1.00	0.98	0.94	0.88	0.81	0.73	0.64	0.55
$\langle n_i \rangle$:E=15	0.87	0.83	0.78	0.72	0.68	0.62	0.58	0.53
	0.87	0.82	0.78	0.73	0.68	0.63	0.58	0.53
$\langle n_i \rangle$:E=25	0.67	0.63	0.61	0.59	0.56	0.55	0.53	0.51
	0.66	0.64	0.61	0.59	0.57	0.55	0.53	0.51

exact and 4-moment distribution occupancies at three energies for 8 particles in 16 evenly spaced states. The total number of states is 12,870. A demanding test on the non-interacting-particle occupancies is that they should reproduce the multiparticle energy, via $E = \Sigma \langle n_i \rangle_E \cdot \varepsilon_i$. The two-moment distribution comes within 2 units of satisfying this when $10 < E < 54$ (the spectrum span is 0-64) while the much superior four-moment distribution is within 1 unit for $3 < E < 61$, and within 0.1 unit for $14 < E < 50$.

We would expect single-state occupancies to fluctuate from state to state when we have residual interactions, and then we might average over a few states to produce results comparable with those given by smooth distributions. On the other hand an averaging process is already built in, via the orbit degeneracy, when we analyze in terms of spherical orbits, and further averaging is unnecessary. Table 3 shows the excellent agreement (3) with the 3-orbit exact shell-model calculations, largely due to Wong (9), for the ground-state occupancies of some Ni and Cu isotopes. The interaction is Brown-Kuo and the distributions have fixed configuration and isospin.

In the (ds) shell Castel et al (10) have recently calculated Hartree-Fock proton and neutron occupancies and found them to compare well with experiment. For the same interaction the distribution occupancies also compare well, especially when the particle numbers are fairly large. In Mg^{26} for example the H.F. proton occupancies $d_{5/2}, s_{1/2}, d_{3/2}$ respectively are given as 2.9, 0.9, 0.2 and the same numbers are found for distributions; the H.F. neutron occupancies are 3.7, 1.3, 1.0 compared with 4.3, 1.3, 0.4.

Table 3

Fractional occupancy (%) of the single-particle orbits for ground
states of (f,p) nuclei, as predicted by the distribution method and
detailed spectroscopic calculations.

Nucleus	Distribution			Detailed Spectroscopy		
	$f_{5/2}$	$p_{3/2}$	$p_{1/2}$	$f_{5/2}$	$p_{3/2}$	$p_{1/2}$
^{60}Ni	2.0	40	13	5.5	38	8.3
^{61}Cu	3.5	48	18	6.0	47	13
^{61}Ni	6.2	44	18	10	48	17
^{62}Cu	4.8	55	26	5.2	55	24
^{62}Ni	8.3	49	27	9.3	46	30
^{63}Cu	8.9	58	32	8.7	59	32
^{64}Ni	17	50	48	22	48	40
^{65}Cu	19	61	45	20	60	43

The calculations which give the distribution occupancies produce them at all energies. For example we find for the states of
^{63}Cu, described as $(fp)^{23}$, that the proton and neutron occupancies
vary slowly and almost linearly with excitation energy. For $f_{7/2}$(n)
the occupancy falls regularly from 91% at the ground state to 85%
at 25 MeV; for the other neutron orbits the corresponding percen-
tages are $f_{5/2}$(30,46), $p_{3/2}$(88,80), $p_{1/2}$(67,67). For the protons
we find $f_{7/2}$(70,63), $f_{5/2}$(9,20), $p_{3/2}$(58,52) and $p_{1/2}$(28,36). Since
at higher energies the distribution occupancies are insensitive to
the ground-state energy we believe them to be accurately given even
in very complicated cases. There are many interesting features
about occupancies which we shall discuss at a later time.

3(f) More General Fixed-Energy Averages and Fixed-J Densities

The procedure of using the distributions to give the variation
of the occupancies with excitation energy can be used more gener-
ally with other operators. Provided that the spaces $\underset{\sim}{\alpha}$ are eigen-
spaces for O (so that in an eigenstate average we encounter no
matrix elements connecting different $\underset{\sim}{\alpha}$) we may write as a reasonable
form for the energy variation of an average quantity that

$$<0>_E = \sum_{\underset{\sim}{\alpha}} <0>^{\underset{\sim}{\alpha}} \cdot \rho_{\underset{\sim}{\alpha}}(E) / \sum_{\underset{\sim}{\alpha}} \rho_{\underset{\sim}{\alpha}}(E)$$

This has been used (3) to yield the variation with excitation energy
of the norms of the various parts into which the Hamiltonian naturally decomposes when we examine how it operates in a multiconfigura-
tion space (pairing, multipole, effective single-particle energy,
and so forth); it appears that this decomposition in particular
will be very useful in studying various kinds of higher energy
phenomena.

We consider this method here in order to yield the J-decomposition of the level densities, using for this purpose the J_z operators whose moments are calculable very simply, and then using the standard subtraction method to isolate the J values. Observe that the effects of the residual interactions are taken account of via the densities. A more accurate method for evaluating fixed-J densities would be by calculating the fixed-J Hamiltonian moments; but, because J does not enter via a direct-product group, it turns out that this is technically very much harder to do. Using the form above we are able to calculate very easily the fixed-energy moments for J_z; the odd-moments vanish, the second moment gives the spin-cut-off factor $\sigma_J^2(E)=<J_z^2>_E$, and the fourth cumulant, the excess, gives a correction to the standard Gaussian form; the latter turns out to be small whenever we have more than a few active particles (as it should by the general arguments of 2) and so the standard form is well justified. Since the spin-cut-off factor varies slowly with excitation energy, its values are not sensitive to residual errors in the ground-state energy and we believe that, in cases where the effective interaction, and hence the orbit occupancies, are well understood, they should be very accurate indeed.

Figure 3, which uses only the second J-moment shows the three orbit ($f_{5/2}$,$p_{3/2}$,$p_{1/2}$) level density (3) as calculated for Cu^{63} using a Brown-Kuo interaction, and compared with exact shell-model calculations of Wong (9). We stress that the distribution calculation is absolute in that it is calculated from a specification of the orbits and Hamiltonian matrix elements, the ground state being fixed by the same calculation; there are no free parameters of any kind. The agreement is excellent and could even be improved by taking account of the J-excess which, being negative, would slightly depress the peaks. We conclude that in smallish spaces such as we encounter here (the largest fixed-J matrix is about 300-dimensional) we have a simple and accurate theory of the level density.

3(g)   The Absolute Level Density in ^{63}Cu

Figure (3) gives a partial density for ^{63}Cu. It is clear that, if we wish to calculate the total density up to, say, 20 MeV, we must extend to the complete (fp) shell and allow excitations of a few nucleons from or into the adjacent shells. We report on the as yet incomplete results.

We consider the T=5/2 density only, this being the major component. The higher-T densities, which come in at 9 MeV, could be treated the same way or via the isobaric-spin cut-off factor. We express the results in terms of the level densities, the number of levels for fixed (m,T,T_z); fixed-J densities follow of course from ρ and $\sigma(J)$.

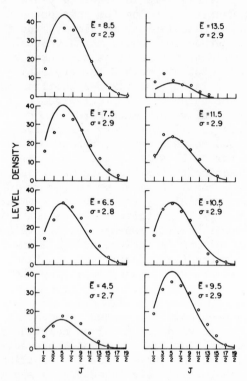

Fig. 3. The absolute 3-orbit T = 5/2 partial level density in ^{63}Cu (continuous curves) as derived from the configuration-isospin distribution with the calculated angular-momentum cut-off factor σ, compared with the exact shell-model results of Wong. Except at the lowest and highest energies, where 3-MeV bands are used, the exact counting is done in 1-MeV intervals.

We take the single-particle energies from A=41, the only real problem here being with $g_{9/2}$ which we take as 5.6 MeV above $f_{7/2}$, which may well be low. We use the 15-orbit B-K (6) interaction (all orbits up through (gds)) and, in the three- and four-orbit calculations, appropriate renormalizations of them. There are difficulties with the use of these interactions which we have not yet seriously considered. For example in the extension from three to four orbits (opening up $f_{7/2}$) we find a surprisingly large increase in the $f_{5/2}$ occupancy even at low excitation. Perhaps then the (four→three)-orbit renormalization is questionable. But conversely, since this occupancy change gives a large increase in the spin cut-off factor (from ~ 2.8 to ~ 4.5) we have an illustration of the argument made above that level density measurements (σ(J) in this case) can inform us about the interaction.

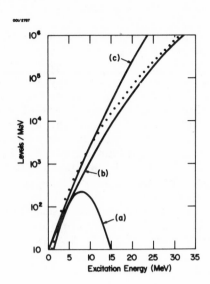

Fig. 4. The level density in ^{63}Cu described (a) as $(f_{5/2}p_{3/2}p_{1/2})^7$
(b) as $(fp)^{23}$, and (c) as $(fp)^{23}+(fp)^{22}g_{9/2}+(fp)^{21}(g_{9/2})^2$. In (c)
the ρ^-/ρ^+ ratio is 1.4, 1.5, 1.8 at 10, 15 20 MeV respectively.
For (a) $\sigma(J)\sim3.0$ at all energies; for (b) $\sigma(J)$ increases from 4.6
to 4.8 in the 5-20 MeV range; in (c) the values are 5.7 to 6.0.
Shown for comparison is $\exp(b\sqrt{E})$ with $b=2.9$ MeV$^{-\frac{1}{2}}$.

    Figure 4 shows the results as they presently stand. (a) con-
siders only the three orbits of Fig. 3; direct comparison with
shell-model is available, the ground-state error being less than
0.5 MeV. (b) extends to the complete shell and (c) admits one and
two particles in $g_{9/2}$, producing thereby states of both parities.
The dimensionalities $d(m,T,T_z)$ at the three stages are 1.0×10^4,
4.5×10^9 and approximately 10^{12}. In (b) the reference energy is
fixed by matching the observed number of negative-parity levels in
the 2.3-3.3 MeV band, and similarly for (c) with the total number of
levels (the ρ^+/ρ^- ratio comes out satisfactorily). The (a) curve
does not meet the others at the reference energy because three
orbits are inadequate near 3 MeV.

    The incomplete results underestimate ρ^+ by perhaps a factor
3-4 and give too large a value for $\sigma(J)$. A similar thing happens
with Ni^{63}. Including the other excitations will increase ρ^+ and,
since the orbits to be added are small, should decrease $\sigma(J)$; it
remains to be seen to what extent the deficiencies will be made up.

4. FINAL REMARKS

Our concern has been to demonstrate that spectroscopy at low excitation energies and level density and related quantities at higher can profitably be considered together. The density appears as a natural extension of the low-lying spectrum and the same concepts, orbits and effective interactions, enter into both. Occupancies, J and T cut-off factors, decompositions of H which are adapted to studies of compound-nucleus formation, and other such things can all be calculated. Extensions to deal with more complicated symmetries and with the distributions of various kinds of transition strengths, and more systematic applications, are also in order. And there are of course problems, for example formal ones about the source of normality and the extent to which it is achieved, technical ones about accuracy in large spaces, and more profound ones about the nature and role of the effective interaction. Some thought has been given to some of these things but a great deal remains to be done.

ACKNOWLEDGMENTS

We have benefited greatly from discussions about level densities with J. R. Huizenga and C. C. Lu. For supplying us with unpublished results and permitting us to quote them, we are indebted to F. C. Williams, S. S. M. Wong, and A. Zuker, and for help with the calculations to V. N. Potbhare. The first-named author also acknowledges the hospitality of Brookhaven National Laboratory where part of this report was written.

REFERENCES

1.   J. B. French, in "Nuclear Structure" (A. Hossain, et al, Eds.) North-Holland, Amsterdam (1967).
2.   J. B. French and K. F. Ratcliff, Phys. Rev. C3, 94 (1971); K. F. Ratcliff, Phys. Rev. C3, 117 (1971).
3.   F. S. Chang, J. B. French and T. H. Thio, Annals of Physics (in press).
4.   J. B. French and S. S. M. Wong, Phys. Letters 33B, 449 (1970); 35B, 5 (1971); O. Bohigas and J. Flores, Phys. Letters 34B, 261 (1971); 35B, 383 (1971).
5.   F. C. Williams, private communication.
6.   T. T. S. Kuo and G. E. Brown, Nucl. Phys. A114, 241 (1968), and private communication from T. T. S. Kuo.
7.   A. Zuker, private communication, and M. Soyeur and A. Zuker, to be published.
8.   S. Kahana, H. C. Lee and C. K. Scott, Phys. Rev. 185, 1378 (1969).
9.   S. S. M. Wong, Nucl. Phys. A159, 235 (1970).
10. B. Castel, et al, Nucl. Phys. A157, 137 (1970).

DISCUSSION

KAPOOR (BARC)
As you know, the numerical calculations of level density also
have been carried out to the calculation of nuclear or direct
counting entropy, starting from non-interacting single particle
level sequence. This has been done by various people. We have also
done it. Suppose now one takes into account the effective interaction, by means of some spreading function of the type you mentioned or by some other method which will remove the degeneracy.
Then suppose we are able to take this into account and generate a
new single particle level sequence which has the effective interaction built in. Then one does the same numerical calculations
which we have been carrying out. Do you think that the two results
will converge?

FRENCH
I think that's a great question, because it refers to
Section B, which I left out. What is the basis for the single
particle theory of level density? I don't, obviously, know the
answer to that, but I'll point out the following thing. When you
include the effective interactions, two things happen. First of
all, states move, and they spread. I think this is a quite
natural kind of notion--that if the movement is small and the spread
is very large, then you don't have a proper basis for single particle densities. On the other hand, if the spread is small compared
to the spacing, then you do have a proper basis no matter what the
movement is. If you take spherical orbits, it is absolutely clear
in light nuclei that the spreading dominates everything. In the
ds-shell, when one turns on the effective interaction, the JJ configurations in $(ds)^{12}$ may spread over ten configuration spacings.
In that case, with spherical orbits, I am absolutely convinced that
there's no formal basis for level densities in light nuclei.
We have not made calculations in heavy nuclei. That, however, does
not rule out the possibility that there may be a non-spherical orbit
basis in which things are a great deal better. That's a very
interesting question. That's not an easy question. If you think
you know what the basis is, if you have some notion of what the
basis looks like, then it's easy to make calculations. One is not
restricted here to spherical orbits. If you ask the formal question, what is the best possible basis, that's a hard question. For
spherical orbits and light nuclei, I think there is no basis at all.
That's my strong belief. For heavy nuclei, I am not sure--I doubt
it. For non-spherical orbits, perhaps. Does that answer your
question?

NEWSON(DUKE)
I'm a little bit surprised by your statement that you don't
think that you can account for single-particle effects in level

densities. If you go back a few years and look at the spacings you
get by looking at neutron resonances, you find a factor of 1,000
between the level density, say, of Ta and the level density of Bi.
This obviously is a shell effect. It gets confusing as you move
into light nuclei because of the difference between a shell and a
subshell. Certainly, in heavy nuclei there could be no question of
the qualitative effect of the individual particle model.

FRENCH

I'd like to point out that perhaps I meant a little bit more
by my remark. I'm not in a sense just talking about the level
density. I'm also talking about the nature of the levels. There
was a remark by Dr. Bloch, namely, that when you turn on the interactions, some states go up and some go down. On the average, it
doesn't make too much difference. I think that's absolutely right.
That's one part. When I ask is there a formal basis, I am really
asking about the nature of the states. That is, not only does the
combinatorics turn out right--the combinatorics are pretty stable
as regards to many changes--but I don't think that other things
which depend on the nature of the states we're counting will turn
out right. I made an absolute remark for light nuclei. That
doesn't mean it is absolutely true. We have not made calculations
for heavy nuclei, but I would doubt it. I myself don't see a conflict. I'm not only counting states. I'm also looking at the
nature of them. I believe a proper basis will not only tell you
how many, but will give you an idea of what kind.

NEWSON

We don't even know how many there are yet.

FRENCH

I well know that.

Experimental and Theoretical Nuclear Level Densities

John R. Huizenga

University of Rochester

Rochester, New York 14627[+]

I. INTRODUCTION

In this talk, I will deal mainly with the experimental aspects of the nuclear level density, emphasizing the various methods and techniques employed in the determination of nuclear level densities. The evaluation and analysis of experimental data have been performed traditionally on the basis of the Fermi gas independent-particle level density formula based on Bethe's theory[1,2,3,4]. Hence, in the first section of my talk I will discuss the data in terms of the conventional level density formulas,

$$\rho(U,I) = \frac{a^{1/2}}{24\sqrt{2}} \frac{\hbar^3}{\mathscr{I}^{3/2}}(2I+1)(U+t-\Delta)^{-2} \exp\{2[a(U-\Delta)]^{1/2} - [I(I+1)/2\sigma^2]\}$$

$$(1)$$

and

$$\rho(U,I) = \text{Const}(2I+1)(U+t-\Delta)^{-1/2}\exp\{[(U-\Delta)/T]-[I(I+1)/2\sigma^2]\}$$

$$(2)$$

For the Fermi gas formula given by Eq. 1, the parameters to be determined by experiment are the Fermi gas level density parameter $\underline{a}$, the energy shift $\underline{\Delta}$ and the moment of inertia $\mathscr{I}$. The thermodynamic temperature $\underline{t}$ is defined by an equation of state,

$$U-\Delta = at^2-t$$

$$(3)$$

[+]Work supported in part by the United States Atomic Energy Commission.

and the moment of inertia is related to the spin cutoff
factor σ by

$$\sigma^2 = \mathcal{J}t/\hbar^2 = 0.0137\ A^{5/3}\ (\text{MeV}^{-1})\ t\ (\text{MeV}) \qquad (4)$$

where a nuclear radius of $1.2\ A^{1/3}$F is assumed. In the
present treatment, the quantity Δ is an adjustable
parameter to account for pairing and shell corrections
and defines the energy of a fictive ground state with
respect to the actual ground state[5,6,7,8]). If low-
energy level density data is available, an extrapolation
of a linear plot of the square of the entropy S as a
function of excitation energy U gives a measure of Δ. The
constant temperature level density formula given by Eq.2
has been used extensively also to fit experimental data.

In the second part of my talk, I will discuss some
exemplary comparisons between experimental results and
theory. It is well known that the conventional level
density formulas are derived on the basis of simplifying
assumptions which are especially objectionable in certain
mass regions.

II. EXPERIMENTAL METHODS AND TECHNIQUES EMPLOYED TO
     DERIVE INFORMATION ON NUCLEAR LEVEL DENSITIES

1) Neutron Capture Resonances

A good deal of experimental information on nuclear
energy levels near the neutron binding energy has been
obtained by counting the observed resonances as a function of neutron energy. In this type of experiment, the
width Γ of each level must be less than the level spacing
D and the experimental resolution must be good enough
to resolve individual levels. Average level spacings
have been obtained from slow-neutron resonances for about
200 nuclei[9-11]). Hence, this type of measurement has
given us the most extensive information on level densities covering the entire range of A values across the
whole periodic table. In the case of s-wave neutron
capture on a zero spin target, the density of levels
of a selected angular momentum and parity (1/2+) is
obtained. The total level density ρ(U) is related to
the average spacing between 1/2+ levels $<d_{1/2+}>$ by,

$$\rho(U) = 2\sigma^2/<d_{1/2+}(U)> \qquad (5)$$

Although the technique of measuring neutron resonances
is an extremely important one in terms of level density
information, it is not without experimental difficulty.

Weak levels may be experimentally unobserved on occasion and some ambiguity in interpretation of levels may arise from the possibility of p-wave resonances. Although p-wave resonances are not expected to be observed for low energy neutrons in most mass regions, the possibility of observing such resonances exists for light nuclei and regions of A where a maximum exists in the p-wave strength function.

A number of authors[11-16] have analyzed neutron resonance data with level density formulas of the type given by Eq. 1. A recent compilation[16] of a values is shown in Fig. 1. The energy shift in this analysis has been assumed equal to the pairing energy values of Gilbert and Cameron[17] and a rigid-body moment of inertia has been used in the computation of the spin cutoff factors. In such measurements one is concerned whether levels with a single spin and parity value are truly representative of the total level density. Nuclear structure, for example, may cause oscillations of the positive and negative parity levels with energy. Calculations[18] of the ratio of positive to negative parity levels near the neutron binding energy for selected nuclei show that this ratio may vary considerably. The solid line in Fig. 1 represents a values given by a = A/8. One observes that a increases with A as expected theoretically. However, there are marked deviations from this smooth trend especially for A values near closed shells. For example, in the vicinity of the Z=82, N=126 shells, a values are more than a factor of two smaller than those of nearby nuclei. Such irregularities are associated with the low single-particle level densities near the Fermi energy for nuclei near closed shells. I shall return to this subject in the second part of my talk.

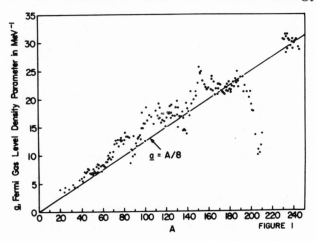

FIGURE I

The neutron resonance data give a measure of the level density at a single excitation energy. Hence, the level density parameters determined from the analyses of such experimental data

as described above are subject to error due to the
effects of nuclear structure which introduces an energy
shift in many of the nuclear ground states. This uncertainty in the effective excitation energy may introduce in some nuclei an error in the derived value of a
of up to 25%. It has been known for some time that the
conventional shifted Fermi-gas model was not adequate to
describe the experimental level density data between the
ground state and the neutron binding energy. In order
to improve agreement with experiment, a back-shifted
Fermi gas model has been proposed[5] in which the level
density parameter a and the energy shift Δ are treated
as energy-independent adjustable parameters. In this
model, the effective excitation energy is increased
since the fictive ground state of the Fermi gas lies
between the actual ground states of the even and odd-
odd nuclei. This model was initially used for light
nuclei[8] and an extension to heavier nuclei is given in
contributed paper 6.6.

2) Charged-Particle Capture Resonances

Information on level densities may be obtained from
charged-particle capture resonances in the same way
described previously by neutron capture resonances.
These measurements are restricted, however, to light
nuclei where the level spacing is large and the Coulomb
energy is small. From the density of resonances ρ_{res},
the total level density $\rho(U)$ is calculated from the
relation

$$\rho_{res} = \rho(U) \sum_{I,\pi} [(2I+1)/4\sigma^2] \exp[-I(I+1)/2\sigma^2] \qquad (6)$$

by summing over the spins and parities of the resonances.
Considerable error may arise due to the assumption made
about the ℓ waves contributing to the total number of
resonances.

As an example of level density information obtained
from this technique, we show in Fig. 2 the number of
levels in ^{28}Si in the vicinity of 12.5 MeV[19]. The
experimental data points result from a count of resonances for the ^{27}Al(p,γ), ^{24}Mg(α,γ) and ^{27}Al(p,α$_0$)
reactions[20].

Recently, proton elastic scattering experiments
performed with very good energy resolution have demonstrated the fine structure of analogue states in light
nuclei. This technique is described in a separate

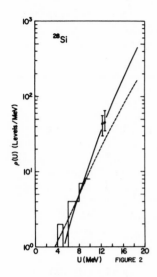

FIGURE 2

contribution to the conference. Since
the analogue state, $T_>$, mixes to some
degree with the $T_<$ levels of the
nucleus, the number of resonances one
observes is a function of the density
$T_<$ levels which have the same spin
and parity as the analogue state.
Information about levels in copper[21]
and cobalt[22] isotopes has been
published. Since the proton bombarding
energy is in the 2-3 MeV region,
there are very few p-wave resonances
observed except in the vicinity of
an analogue state. In the region of
an analogue state, the strength of
the $T_>$ state is shared with $T_<$ background states and the widths of the
$T_<$ states near the analogue are sufficiently increased such that a number
of them are observable. However, some caution must be
exercized in using the number of these p-wave resonances
for level density measurements since their density
(corresponding to 1/2- and 3/2- levels), even in the
vicinity of an analogue state, are too low relative to
the s-wave resonances (corresponding to 1/2+ levels).
If in some nuclei one could be assured of seeing all the
$\ell=0$ and 1 resonances, a comparison of the number of 1/2+
and 1/2- levels would give a measure of the ratio of
positive to negative parity levels which we mentioned in
section II-1.

3)   Counting of Resolved Levels in a Residual Nucleus
            Excited by Nuclear Reactions

      Nuclear reactions, such as the (n,n'), (p,p'),
(α,α') and (p,α) reactions, have been utilized to study
resolved levels in a large number of nuclei at relatively low excitation energies[23-33]. High resolution
magnetic spectrographs used in conjunction with Van de
Graaff accelerators are capable of giving overall resolutions of the order of 10 keV for energetic charged
particles. With such equipment, one is able to study
isolated levels up to excitation energies of approximately 5 MeV for nuclei around A=60. If this technique
is to be used for accurate level density measurements,
one must be assured that levels of all angular momenta
are excited and that the experimental energy resolution
is sufficient to see all levels as separate lines. Relative cross sections for populating levels of different I

values in compound nucleus reactions have been calculated
and show that only a very small percentage of levels are
expected to be missed if one chooses the reactions
properly. For example, the levels in ^{56}Fe have been
measured by both the (p,p') and (p,α) reactions. The
question of unresolved levels is a more difficult one.
In dealing with levels of mixed spin and parity, Wigner
proposed that these levels of different spin and/or
parity are not in any way correlated in position. The
resulting distribution of spacings from a sequence of
levels which is a superposition of sets of different spin
and/or parity has a shape intermediate between the
Wigner and the exponential distribution[34]. However,
for a spin dependent level density of both parities
given by eq. 1, the distribution of spacing approaches
the exponential distribution,

$$P(S/\bar{S}) = \exp(-S/\bar{S}) \tag{7}$$

even for rather small values of σ. An analysis of the
experimental spacing distribution for 622 levels with
average spacing $\bar{S} \geq 30$ keV in nuclei around A=50 is
consistent with an exponential spacing distribution[35].
Once one knows the spacing law for levels of all spins
and both parities and has an estimate of the experimental
energy resolution, it is possible to correct the level
density as a function of energy for levels missed due
to the finite energy resolution. At 5 MeV to 6 MeV, the
correction for a nucleus with A=50 may approach 50%.

4)   Measurement of Particle Spectra from Compound
                 Nucleus Reactions

The inclusion of angular momentum effects in compound nucleus reactions has been explicitly formulated
by several authors[36-40]. Such an angular and energy
dependent differential cross section for compound
nucleus reactions including angular momentum is given by,

$$\frac{d^2\sigma_{ab}(\varepsilon_b)}{d\varepsilon_b d\Omega_b} = \sum_{\substack{L=0 \\ \text{even}}}^{\infty} B_L(\varepsilon_b) \, P_L(\cos\theta) \tag{8}$$

The function $B_L(\varepsilon_b)$ is given by

$$B_L(\varepsilon_b) = \frac{(2I_a+1)^{-1}(2i_a+1)^{-1} \, k_a^{-2}}{4} \sum_{S_a,S_b,I_b,\ell_a,\ell_b,J}$$

(Continued from Page 430).

$$\frac{(-)^{S_a-S_b} T_a^{\ell_a}(\varepsilon_a) \, T_b^{\ell_b}(\varepsilon_b) Z(\ell_a J \ell_a J; S_a L) Z(\ell_b J \ell_b J; S_b L) \rho(U_b, I_b)}{G(J)} \quad (9)$$

and $G(J)$ is given

$$G(J) = \sum_{b'} \int_0^{U_b' max} dU_{b'} \sum_{\ell_{b'}=0}^{\infty} T_{b'}^{\ell_{b'}}(\varepsilon_{b'}) \sum_{S_{b'}=|J-\ell_{b'}|}^{J+\ell_{b'}} \sum_{I_{b'}=|S_{b'}-i_{b'}|}^{S_{b'}+i_{b'}} \rho_b(U_{b'}, I_{b'})$$

$$(10)$$

Integration of eq. 8 over all angles gives an energy dependent differential cross section which reduces to

$$\frac{d\sigma_{ab}(\varepsilon_b)}{d\varepsilon_b} = K \, \sigma_b(\varepsilon_b) \, \varepsilon_b \, \rho_b(U_b, I_b=0) \quad (11)$$

if one assumes a spin dependent level density with a $(2I+1)$ dependence. Use of eqs. (11) and (1) allows one to determine the Fermi gas level density parameter $\underline{a}$ from the slope of the straight line obtained from a plot of

$$\ln\{[d\sigma_{ab}(\varepsilon_b)/d\varepsilon_b](U_b+t_b-\Delta_b)^n/\varepsilon_b \, \sigma(\varepsilon_b)\} \text{ vs. } (U_b-\Delta_b)^{1/2}$$

$$(12)$$

where the parameter $\underline{n}=2$. It is well known that the implicit assumption of a $2I+1$ spin dependent level density needed to derive eq. 11 is unrealistic. Hence, the values of $\underline{a}$ determined from eq. 12 may be subject to considerable error.

A large number of reaction spectra have been analyzed with eqs. 11 and 12 with a variety[41-43] of values of $\underline{n}$ including 0, 5/4, 3/2 and 2. In order to in-

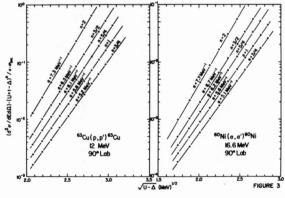

FIGURE 3

terpret the meaning of the level density parameter $\underline{a}$ obtained from eq. 12, we have analyzed[44] theoretical spectra generated with eq. 8 by the method of eq. 12. Some examples of the results are shown in Fig. 3. The slope technique gives a different value of $\underline{a}$ for each value of $\underline{n}$.

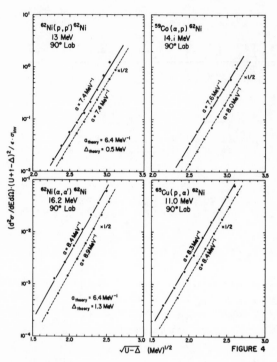

FIGURE 4

In order to obtain the correct value of $\underline{a}$, the values of $\underline{n}$ must be 7/4 and 1 for the $^{63}Cu(p,p')^{63}Cu$ and $^{60}Ni(\alpha,\alpha')^{60}Ni$ reactions, respectively. The value of $\underline{n}$ depends on angle also, especially for the (α,α') reaction.

The analyses of experimental and theoretical spectra by eq. 12 are shown in Fig. 4 for four different reactions[44] leading to the residual nucleus ^{62}Ni. The solid lines represent the experimental data. The spectra from the $^{62}Ni(p,p')^{62}Ni$ and $^{59}Co(\alpha,p)^{62}Ni$ reactions with n=2 give values of $\underline{a}$ which are smaller than the corresponding values from the $^{62}Ni(\alpha,\alpha')^{62}Ni$ and $^{63}Cu(p,\alpha)^{62}Ni$ reactions. The dashed lines are derived from theoretical spectra calculated with eq. 8 where $\underline{a}=6.4$ MeV^{-1}. It is interesting to note the analyses of the theoretical spectra with eq. 12 give slopes in excellent agreement with the experimental data. In all cases, the values of $\underline{a}$ derived from eq. 12 are too large (15% for the evaporated protons and 30% for the evaporated alpha particles). In order to derive the correct value of $\underline{a}$ one must use the proper value of $\underline{n}$ for each different reaction (see Fig. 3). Since most evaporation spectra have been analyzed with the approximate formula of eq. 12 and a variety of values for $\underline{n}$, one must expect the reported values of $\underline{a}$ to reflect these uncertainties.

Spectra from (n,n') reactions have been widely studied[45-48] and interpreted in terms of level density parameters. In some cases, the values of $\underline{a}$ derived from these experiments are a factor of two different from corresponding values of $\underline{a}$ derived from charged-particle spectra and neutron resonance data. The discrepancies appear to be related to both experimental problems and

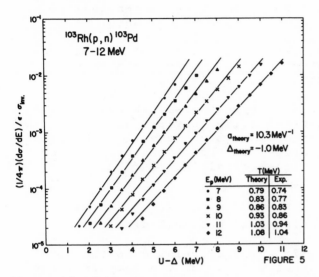

^{103}Rh(p,n) ^{103}Pd
7–12 MeV

$a_{theory} = 10.3\,\text{MeV}^{-1}$
$\Delta_{theory} = -1.0\,\text{MeV}$

E_p(MeV)	T(MeV) Theory	Exp.
• 7	0.79	0.74
■ 8	0.83	0.77
▲ 9	0.86	0.83
× 10	0.93	0.86
▼ 11	1.03	0.94
♦ 12	1.08	1.04

$(1/4\pi)(d\sigma/dE)/\epsilon \cdot \sigma_{inv.}$

U–Δ (MeV)

FIGURE 5

the method of data analysis[49]. Contribution 6.8 deals with (n,n') spectra. Analyses of theoretical neutron spectra from the ^{103}Rh(p,n)^{103}Pd reaction are shown in Fig. 5. The theoretical spectra are calculated with eq. 8 and analyzed with an equation similar to eq. 12 except that a constant temperature type level density is employed. The theoretical temperatures corresponding to different proton bombarding energies are compared to experimental data[50] analyzed by the identical equation. The theoretical and experimental temperatures increase in a very similar way as predicted by the Fermi gas model (the increase in the temperature with proton bombarding energy was cited previously as evidence[50] for the failure of the Fermi gas model).

5) Excitation Functions of Isolated Levels

As pointed out by Ericson[51], absolute cross sections for formation of isolated residual levels in compound nucleus reactions can be used to determine the values of the level density of the various residual nuclei formed in the reaction under consideration. As the cross section for formation of any particular level is governed by the competition of decay probability through this selected reaction channel to that for all other channels, the number of competing channels can be determined from the cross section of a single level. This number of effective competing reaction channels is directly related to the level densities of all the residual nuclei, the appropriate reaction Q values, and the transmission coefficients for the various emitted particles. Therefore, measurements of the absolute cross sections for isolated final levels as a function of bombarding energy give the energy dependence and absolute values of the level densities of the residual nuclei, whereas the usual statistical analysis of the spectra of

emitted particles gives only the energy dependence of
the level density and not the absolute values.

According to the statistical theory of nuclear reactions, the differential cross section for a reaction
leading to a final state with angular momentum I_B,
parity π_B and excitation energy U_B in the residual
nucleus B, can be expressed by eqs. 8-10 where the level
density in the numerator of eq. 9 is replaced by a single
level. The application of this technique to level density measurements is described in the literature[5]. Contribution 6.10 reports level densities of ^{54}Fe and ^{58}Fe
determined by this technique.

6) Cross Section Fluctuation Widths

The quantity $G(J)$ in eq. 9 is related to the total
width Γ_J and average spacing $D_{J,\pi}$ of the compound levels
at excitation energy U_C by[52]

$$\Gamma_J = (D_{J,\pi}/2\pi)G(J) \qquad (13)$$

With some simplifying assumptions it is possible to determine $\Gamma/D_{0,\pi}$ from thick target cross section measurements[53,54]. With a knowledge of Γ from fluctuation
measurements, information on the level density in the compound nucleus at a high excitation energy is obtained[5,6].
The level density of ^{60}Ni is shown in Fig. 6. In the
vicinity of nuclear shells, one would expect this technique to give larger values of <u>a</u> than other techniques

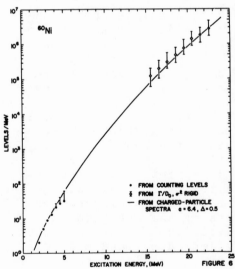

which measure <u>a</u> at low
excitation energy. A survey
of level density information
from fluctuation widths is
published[55]. Contribution
6.13 is an application of
the fluctuation technique.

7) Nuclear Reaction Times by the use of the Blocking Effect

Recently, measurements
have been reported[56-60] of
nuclear reaction times based
on the blocking effect of
charged particles in single
crystals. This technique
offers a unique method of

determining lifetimes of the order of 10^{-17} seconds.
Lifetime measurements[56,57] of ^{71}As and ^{73}As from proton
inelastic scattering on ^{70}Ge and ^{72}Ge have been interpreted[61] with statistical theory (see eq. 13). The decay
widths from the direct lifetime measurements are well
reproduced with reasonable level density parameters[61].

III. COMPARISON BETWEEN EXPERIMENTAL AND THEORETICAL RESULTS

    The level density formula used in the previous
section is based on the independent particle assumption
which allows one to write a nuclear partition function
in a simple form using single particle energies. Calculation of the conventional level density from the partition function involves the additional assumptions
a) that the single particle level scheme can be approxi-
mated by a continuous distribution, b) that the excited
nucleus can be treated as a degenerate Fermi system and,
c) that the Laplace transform can be inverted by the
saddle-point method. For actual nuclei assumption (a)
is certainly incorrect[62,63]. It has been shown that the
ratio of the level densities of two nuclei in which the
same degenerate shell model levels are filled to a different extent in the ground state may be large[64]. These
calculations have been extended to included periodic[65]
independent particle energy-level schemes and systems
with a perturbation[66] in the position of a single particle level. The Rosenzweig effect can be approximately
accounted for by a shift in the energy scale. The saddle-
point approximation introduces some error, especially
at low excitation energies.

    A number of research groups are performing theoret-
ical calculations of level densities based on various
approximations as exemplified by contributions 6.1, 6.2,
6.3, 6.7 and 6.11 to this conference. I wish now to
report briefly on state density calculations which we
have performed[67] for systems of non-interacting fermions
with single-particle energies given by Nilsson[68]. The
numerical computations were done on an IBM-360-65 computer with a recursion relation described previously[69].
The overall state density calculation consisted of three
sub-programs. First, a set of single-particle levels is
generated for a chosen nuclear deformation. Secondly,
the state density as a function of excitation energy is
computed numerically by use of recursion relations.
Thirdly, a smearing function is employed to smooth local
fluctuations in the state density.

Some results of the state density calculations
for a closed-shell nucleus ^{208}Pb are shown in Fig. 7.
The square of the entropy S (S is defined here as
ln[$\sqrt{48}$ U ω(U)]) is plotted as a function of the excitation
energy U. One sees a large deviation from the Fermi gas
prediction due to the shell structure. At excitation
energies larger than 35 MeV, an asymptotic relationship,
$S^2=4a(U-\Delta)$, is observed where the ground state shell
correction Δ is 9.6 MeV. With the same set of single
particle levels, the shell correction calculated by
Strutinsky's procedure is 10.6 MeV. The asymptotic
value of the level density parameter a = 23.3 MeV^{-1}, or
approximately A/9. The experimental level density information on ^{208}Pb is consistent with the results shown in
Fig. 7. If one fits the neutron resonance data with a
Fermi gas model[16] and a Δ = 1.2 MeV[17], one obtains
a = 10.3 MeV^{-1}. Treating the low-energy theoretical
state density in the same way gives a value of a in
remarkably close agreement (see insert of Fig. 7). The
theoretical state density of ^{208}Pb at low energies is well
fitted with a constant temperature model, a characteristic
of closed shell nuclei[47,71]. As can be seen from Fig.
7, not only is the level density parameter a dependent on
excitation energy but also the value of Δ, which means
that the reference surface itself depends on excitation
energy. In the high energy asymptotic region, the level
density parameter a has been determined from inelastic
scattering[72] to be somewhere between A/10 and A/7 (these
values are based on liquid drop masses with no shell and
pairing corrections). For intermediate excitation energies, it has been known for some time from Γ_f/Γ_n measurements, that a_f/a_n is approximately 1.3 for nuclei in the
vicinity of closed shells[73-75]. This result is consistent with calculations of the type shown in Fig. 7
for the two different deformations. As the excitation
energy increases, it is obvious that this ratio should
approach unity if the excitation energies are measured
from the liquid drop
reference surfaces. A
decrease in the a_f/a_n
ratio with excitation
energy is in better
accord also with the
higher energy experi-
mental data[76].

State density calcu-
lations for ^{242}Pu at
four deformations are

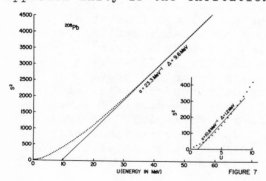
FIGURE 7

shown in Fig. 8.
Although the Δ
values differ
slightly in abso-
lute magnitude
from those obtained
by the Strutinsky
procedure[76],
their trend with
deformation is
the same. At a
deformation of
ε=0.21 (near the
deformation of
the ground state)
the asymptotic
level density
must be measured
from a reference
state 2 MeV above
the ground state
while for ε=0 the
reference state is
7 MeV below the
ground state. The
third deformation
(ε=0.41) is

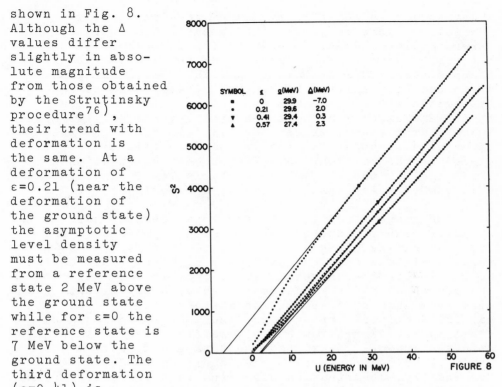

FIGURE 8

near the peak of the first barrier and the fourth deformation (ε=0.57) is in the second potential well.
Although levels through the N=12 oscillator shell were
included in the calculation, the decrease of the asymptotic value of $\underline{a}$ to 27.4 MeV^{-1} for ε=0.57 is evidence
for insufficient single particle levels in the calculation
for very large deformations. The results shown in Fig. 8
are important for an understanding of certain features
of fission. In the calculation of fission probability
by the transition state method, one requires a knowledge
of the minimum number of exit channels (for some specified energy) across the saddle point. The present calculations indicate that the minimum number of exit
channels will occur at the liquid drop saddle point already for medium excitation energies. This result agrees
also with fission fragment anisotropy data which for
medium excitation energies give the liquid drop saddle
point shapes[77-79] as postulated recently[80] (see also
contribution 6.11). The washing out of the shell effects
with energy has consequences wherever the statistical
properties of nuclei play a role.

Up to now we have made comparisons between experiment

and theory based on a Fermi gas of non-interacting
nucleons. Professor French has described at this conference a level density theory which takes account of
the nucleon-nucleon interactions. The dashed-line in Fig.
2 shows results[81] of this theory for ^{28}Si. The ground
state has been shifted by 3.5 MeV as suggested by recent
calculations[82]. Good agreement is obtained between
theory and experiment up to an excitation energy of about
10 MeV. The theory misses levels at 12.5 MeV but this
is expected insofar that only 1d-2s orbitals are included in the calculation. Very recent calculations
which include the 1f$_{7/2}$ levels give a level density
at 12.5 MeV which again agrees with experiment[83].

One property of the nuclear level density which is
of great general interest is its spin dependence. The
level densities of eqs. 1 and 2 contain this spin dependence in the spin cutoff factor σ. Early information
on the dispersion σ^2 was obtained from isomer ratio
measurements[84-89]. Information about σ may be obtained
also from angular distribution measurements[39,90] of
particles emitted into the continuum or to an isolated
final level. For a few nuclei, sufficient experimental
spin assignments[20] exists to estimate σ directly.
Other techniques have also been employed to determine σ.
For example, contribution 6.12 reports a value of σ from
a measurement of the relative probability of neutron
resonance capture in ^{179}Hf leading to spins 4 and 5.

A compilation of some values of σ is given in Fig.
9 (the excitation energy is approximately 8 MeV). The
solid line represents values calculated with the rigid-
body moment of inertia at an excitation energy of 8 MeV.
The triangles are from theoretical calculations[81,91]

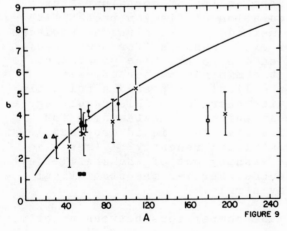

FIGURE 9

while all the other
points represent ex-
perimental values
(crosses, isomer,
ratios;[84-89] solid
circles and triangle,
angular distribution
measurements;[44,90]
open rectangle, neutron
resonance measure-
ments;[91] open circles,
analyses of levels of
known spins;[44] solid
squares, neutron cross
sections[92]). For
A$\leq$110, the σ values

are well represented by the eq. with a rigid-body moment
of inertia. The little information available for large
A indicates that σ is smaller than the value predicted
from a rigid-body moment of inertia. In making the above
comparison one should be aware that σ may fluctuate from
nucleus to nucleus due to the different available shell
model orbitals.

The author wishes to acknowledge the collaboration
of C. C. Lu, L. C. Vaz, F. C. Williams, Jr. and G. Yuen-
Chan in much of this work. Thanks are due also to J. B.
French and F. S. Chang for stimulating discussions. The
support of the Nuclear Structure Research Laboratory by
the National Science Foundation is gratefully acknowledged.

REFERENCES

1) H. A. Bethe, Phys. Rev. $\underline{50}$ 332 (1936); Rev. Mod. Phys.
    $\underline{9}$ 69 (1937)
2) C. Van Lier and G. E. Uhlenbeck,Physica $\underline{4}$ 531 (1937)
3) J. M. B. Lang and K. J. LeCouteur, Proc. Phys. Soc.
    (London) $\underline{A47}$ 585 (1954)
4) J. M. Blatt and V. F. Weisskopf, "Theoretical Nuclear
    Physics" (John Wiley and Sons, Inc., New York 1952)
    P.371
5) J. R. Huizenga, H. K. Vonach, A. A. Katsanos, A. J.
    Gorski, and C.J. Stephan, Phys. Rev. $\underline{182}$ 1149 (1969)
6) A. A. Katsanos, R. W. Shaw, R. Vandenbosch and D.
    Chamberlin, Phys. Rev. $\underline{C1}$ 594 (1970)
7) E. Gadioli and L. Zetta, Phys. Rev. $\underline{167}$ 1016 (1968)
8) H. Vonach and M. Hille, Nucl. Phys. $\underline{A127}$ 289 (1969)
9) Brookhaven National Laboratory Report BNL-325, II
    Edition Supplement No. 2 (1965-66)
10) J. E. Lynn, "The Theory of Neutron Resonance Reactions"
    Clarendon Press, Oxford (1968)
11) U. Facchini and E. Saetta-Menichella, Energia Nu-
    cleare $\underline{16}$ 54 (1968)
12) T. D. Newton, Canadian Journal of Physics $\underline{34}$ 804
    (1956); $\underline{35}$ 1400 (1957)
13) A. A. Ross, Phys. Rev. $\underline{180}$ 720 (1957)
14) E. Erba, U. Facchini, E. Saetta-Menichella, IL
    Nuovo Cimento $\underline{22}$ 1237 (1961)
15) D. W. Lang, Nucl. Phys. $\underline{26}$ 434 (1961); Nucl. Phys. $\underline{77}$
    545 (1966)
16) H. Baba, Nucl. Phys. $\underline{A159}$ 625 (1970)
17) A. Gilbert and A. G. W. Cameron, Can. Journal of
    Phys. $\underline{43}$ 1446 (1965)
18) M. Hillmann and J. R. Grover, Phys. Rev. $\underline{185}$ 1303 (1969)
19) H. K. Vonach, A. A. Katsanos and J. R. Huizenga,
    Nucl. Phys. $\underline{A122}$ 465 (1968)

20) For a compilation of such data, see for example, P. M. Endt and C. van der Leun, Nucl. Phys. A105 1 (1967)

21) J. C. Browne, H. W. Newson, E. G. Bilpuch and G. E. Mitchell, Nucl. Phys. A153 481 (1970)

22) D. P. Lindstrom, H. W. Newson, E. G. Bilpuch and G. E. Mitchell, Nucl. Phys. A168 37 (1971)

23) A. A. Katsanos, J. R. Huizenga and H. K. Vonach, Phys. Rev. 141 1053 (1966)

24) A. A. Katsanos and J. R. Huizenga, Phys. Rev. 159 931 (1967)

25) R. G. Allas, L. Meyer-Schützmeister and D. von Ehrenstein, Nucl. Phys. 61 289 (1965)

26) M. A. Grace and A. R. Poletti, Nucl. Phys. 78 273 (1966)

27) J. H. Bjerregaard, P. F. Dahl, O. Hansen and S. Sidenius, Nucl. Phys. 51 641 (1964)

28) G. Brown, A. MacGregor and R. Middleton, Nucl. Phys. 77 385 (1966) .

29) G. Brown, S. E. Warren and R. Middleton, Nucl. Phys. 77 365 (1966)

30) A. MacGregor and G. Brown,Nucl.Phys. 88 385 (1966)

31) M. Mazari, W. W. Buechner and A. Sperduto, Phys. Rev. 112 1692 (1958)

32) A. Aspinall, G. Brown and S. E. Warren, Nucl. Phys. 46 33 (1963)

33) A. Sperduto and W.W.Buechner,Phys.Rev. 134B 142(1964)

34) N.Rosenzweig and C.E.Porter, Phys.Rev. 120 1098 (1960)

35) J.R.Huizenga and A.A.Katsanos, Nucl.Phys. A98 615(1967)

36) A.C.Douglas and N. MacDonald,Nucl.Phys. 13 382 (1959)

37) L. Wolfenstein, Phys. Rev. 82 690 (1951)

38) W. Hauser and H. Feshbach, Phys.Rev. 87 366 (1952)

39) T. Ericson and V. Strutinsky, Nucl. Phys. 8 284 (1958); 9 689 (1959)

40) A.M. Lane and R.G.Thomas, Revs. of Mod.Phys. 30 257 (1958)

41) D. Bodansky, Annual Review of Nuclear Science 12 79 (1962)

42) E.Gadioli and L.Zetta, Nuovo Cimento 51A 1074 (1967)

43) A. V. Ignatyuk and V. S. Shorin, Soviet Journal of Nucl. Phys. 12 660 (1971)

44) C.C.Lu, L.C.Vaz and J.R. Huizenga, unpublished results

45) D. B. Thomson, Phys. Rev. 129 1649 (1963)

46) K. Tsukada, S. Tanaka, M. Maruyama and Y. Tomita Nucl. Phys. 78 369 (1966)

47) M. Maruyama, Nucl. Phys. A131 145 (1969)

48) O. A. Sal'nikov, G. N. Lovchikova, G. V. Kotel'nikova, V. S. Nesterenko, N. I. Fetisov, and A. M. Trufanov, Soviet Journal of Nucl. Phys. 12 620 (1971)

49) V.S.Stavinskii,Sov.J. of Nucl. Phys. 12 523 (1971)

50) C. H. Holbrow and H. H. Barschall, Nucl. Phys. 42
    264 (1963)
51) T. Ericson, Advances in Physics 9 425 (1960)
52) T. Ericson, Annals of Physics 23 390 (1963)
53) H. K.Vonach and J.R.Huizenga,Phys. Rev. 138B 1372(1965)
54) A. Richter, W. vonWitsch, P. vonBrentano, O. Häusser
    and T. Mayer-Kuckuk, Phys.Letters 14 121 (1965)
55) G. M. Braga-Marcazzan and L. Milazzo-Colli, Energia
    Nucleare, 15 186 (1968); Progress in Nucl. Phys.
    11 145 (1970)
56) M. Maruyama, K. Tsukada, K. Ozawa, F. Fujimoto, K.
    Komaki, M. Mannami, and T. Sakurai, Nucl. Phys.
    A145 581 (1970)
57) G. J. Clarke, J. M. Poate, E. Fuschini, C. Maroni,
    I. G. Massa, A. Uguzzoni, and E. Verondini,
    A.E.R.E.-R6756 (1971)
58) F. Brown, D. A. Marsden and R. D. Werner, Phys.
    Rev. Letters 20 1449 (1968)
59) W. M. Gibson and K. O. Nielsen, Proceedings of
    Second Symposium on the Physics and Chemistry of
    Fission, IAEA, Vienna (1969) P.861
60) S. A. Karamyan, Yu. Ta. Oganesyan and F.Normuratov,
    Dubna Preprint, P7-5512 (1970)
61) L.C.Vaz,C.C.Lu and J.R.Huizenga, unpublished results
62) C. Bloch, Phys. Rev. 93 1094 (1954)
63) H. Margenau, Phys. Rev. 59 627 (1941)
64) N. Rosenzweig, Phys. Rev. 108 817 (1957)
65) P.B. Kahn and N.Rosenzweig, Phys. Rev. 187 1193 (1969)
66) P. B. Kahn and N. Rosenzweig, Jour. of Math. Phys.
    10 707 (1969)
67) G. Yuen-Chan, F. C. Williams, Jr. and J. R. Huizenga,
    unpublished results
68) S. G. Nilsson, Mat. Fys. Medd. Dan. Vid. Selsk.
    29 No. 16 (1955)
69) F. C. Williams, Jr., Nucl. Phys. A133 33 (1969)
70) V. M. Strutinsky, Nuc.. Phys. A122 1 (1968)
71) A. V. Ignatyuk, V. S. Stavinskii and Yu. N.Shubin,
    Sov. Journ. of Nucl. Phys. 11 563 (1970)
72) G. Chenevert, I. Halpern, B. G. Harvey and D. L.
    Hendrie, Nucl. Phys. A122 481 (1968)
73) W. J. Nicholson, Jr., University of Washington
    Thesis (1960) unpublished
74) J. R. Huizenga, R. Chaudhry and R. Vandenbosch,
    Phys. Rev. 126 210 (1962)
75) D. S. Burnett, R. C. Gatti, F. Plasil, P. B. Price,
    W. J. Swiatecki and S. G. Thompson, Phys. Rev.
    134B 952 (1964)
76) A. Khodai-Joopari, UCRL 16489 (1966) unpublished
77) R. Chaudhry, R. Vandenbosch, and J. R. Huizenga,
    Phys. Rev. 126 220 (1962)

78) G. L. Bate, R. Chaudhry and J. R. Huizenga, Phys.
    Rev. 131 722 (1963)
79) R.F. Reising, G. L. Bate and J. R. Huizenga, Phys.
    Rev. 141 1161 (1966)
80) V. S. Ramamurthy, S. S. Kapoor, S. K. Kataria,
    Phys. Rev. Letters 25 386 (1970)
81) F. S. Chang, University of Rochester Thesis (1970)
    unpublished
82) A. P. Zuker, Private communication to J. B. French
    and F. S. Chang
83) F. S. Chang, Private communication
84) J. R. Huizenga and R. Vandenbosch, Phys. Rev. 120
    1305 (1960)
85) R. Vandenbosch and J. R. Huizenga, Phys. Rev. 120
    1313 (1960)
86) C. T. Bishop, J. R. Huizenga and J. P. Hummel,
    Phys. Rev. 135B 401 (1964)
87) H. Vonach, R. Vandenbosch and J. R. Huizenga, Nucl.
    Phys. 60 70 (1964)
88) R. Vandenbosch, L. Haskin and J. C. Norman, Phys.
    Rev. 137B 1134 (1965)
89) C. R. Keedy, L. Haskin, J. Wing and J. R. Huizenga,
    Nucl. Phys. 82 1 (1966)
90) H. K. Vonach and J. R. Huizenga, Phys. Rev. 149
    844 (1966)
91) C. Coceva, F. Corvi, P. Giacobbe and M. Stefanon,
    Contribution 6.12
92) J. M. Green and H. W. Hubbard, Contribution 6.10

DISCUSSION

MORETTO (Laboratorio di Radiochimica)
I would like to comment about your level density calculations
on Plutonium for different deformations. I do agree generally with
what you have said; however, I think it's a good thing to point out
that the pairing effects as a function of deformation change dramatically not because the pairing strength changes but because the
relative local density of the Fermi surface changes. Therefore,
you have a condensation energy which may vary a factor or two, or
more, over the range that you have considered. Therefore, the use
of a simple symbolic calculation without the addition of pairing
may lead to some inexact conclusions.

HUIZENGA
Yes, I should point out that these were very simple model
calculations that try to show you how things shift around when you
introduce realistic sets of single particle levels. Pairing has
not been included, and what Dr. Moretto says may or may not be true.
I't just hard to say how important those effects will be.

MORETTO
I have performed such calculations on the basis of the shell
model and BCS Hamiltonian. It turns out that: 1) pairing changes
dramatically as a function of deformation; 2) the condensation
energy is comparable with shell effects; 3) pairing and shell
effects are out of phase; namely, the former increases when the
latter decreases, thus producing a partial compensation.

NEWSTEAD (Saclay)
You gave an explanation in the case of a particular nucleus
for the discrepancy in the Fermi level density parameter as obtains
from neutron spacing and charged particle results. Could your
approach result in a unification of the body of level density data
from photo-nuclear, higher energy neutron scattering, and low
energy neutron spacing experiments? There has always been a strong
anomaly between the various values of the density parameters
derived from these different experiments.

HUIZENGA
I think a large part of the difficulty comes in the analysis
of the data and one will first have to go back through and
straighten that aspect of the problem out. But I think it's
fairly clear that there are certain nuclei for which you simply
cannot take the Fermi gas formula, put the parameters in and treat
them as energy independent parameters and hope to calculate a
realistic level density. That's the point I tried to emphasize;
and whether one is going to be able to bring all the information
together or not, I think rests on further corrections. As

Dr. Moretto has pointed out, one has to include the pairing and maybe get a better set of single particles levels; and then, last of all, I must defer that to my colleague, Dr. French. We have not included the interections, so whether this is a realistic approach one doesn't know, but this is sort-of the way an experimentalist goes about the business.

BERMAN (Livermore Radiation Laboratory)

With regard to the photo nuclear data, we've measured the level density parameter a and the delta values by comparing the $(\gamma,2n)$ to $(\gamma, total)$ cross-section ratio in twenty-nine medium and heavy nuclei. Typically, this covers an excitation energy of about 6-12 MeV. We've analyzed that with both the cases

$$\rho(U) \propto \exp [Z \sqrt{a(U-\Delta)}]$$

and $\rho(U) \propto \dfrac{1}{U^2} \exp [Z\sqrt{a(U-\Delta)}]$,

bracketing your preferred $\rho(U)\propto \dfrac{1}{U}\exp [Z\sqrt{a(U-\Delta)}]$.
We find, of course, as is obvious from your talk, that the a values change considerably when you use the different formulas, but the delta values do not. They always come out the same. Consequently, we think this might be a good measure of the combined shell and pairing effect of the delta value in these nuclei

KAPOOR (Bhabha Atomic Research Centre)

I would like to make a general comment regarding the intercept on the energy axis of the asymptotic behaviour of S^2 versus E_x plot which you showed. As you may know, in the work published by us about a year back, we found for Pb^{208} and Pu^{240} (spherical shape) the intercept is exactly equal to the shell correction obtained by Strutinsky's procedure. This could have been very well a coincidence, especially as the shell correction obtained by Strutinsky's method sometimes appears to depend on the parameters of the spreading function. In fact, it has been pointed out by others that Strutinsky's prescription for calculating shell effects may work only in some cases. I wish, therefore, to add that a better method of calculating the shell effects in the ground state is by finding the intercept of the asymptotic behaviour of S^2 versus E_x plot. If the pairing interactions are also included in the calculations this intercept will give shell plus pairing correction in the ground state.

HUIZENGA

I think time is just too short for any comment.

ROSENZWEIG(SUNYA)

A simple question. When you deduce the level density parameter "a" from various experiments, especially the energy dependence of

the level density, do you put in some values of sigma and moments
of inertia or are they also deduced from the experimental data?

HUIZENGA

I had planned to say something about this whole question of
sigma but due to shortage of time I did not. If you like the next
slide deals with that general subject. We normally put in values
which we determine experimentally,if that's a sufficient answer
to satisfy you at this point. But that's a very complicated subject, this whole question of sigma, and there are certain things
that I'm not sure I at all understand at this point--one of them
being the experimental measurements. I think this subject will be
discussed by one of the authors following my talk. If we get into
that subject I could show my slide at that point.

MEDICUS (R.P.I.)

In view of your calculations, how good a level density formula
is that of Gilbert and Cameron?

HUIZENGA

In philosophy you see I don't agree with the Gilbert-Cameron
formula because they want to use two energy independent parameters
"a" and delta and if you start with that approximation and then
use only pairing in the delta term, you can be off by orders of
magnitude. They've sort-of fit things to fit the neutron resonance
data; in other words, they've adjusted the parameters to fit that
data, and if you get off that energy range you can really go sour.
I think, however, one can maintain the Fermi gas formula if one is
willing to re-interpret the delta in terms of measuring the effect
of energy from some fictitious ground state--that ground state
being determined from pairing, structure, etc. I mean, you can't
even do that, you see, when you come to a shell. For certain local
regions in A then you're in great difficulty.

LE COUTEUR

I might just remark that the shell effect of lead really is a
dramatic one. The simple model known to me which gives the low
energy behavior and the high energy behavior is that of a gap at
the Fermi level in the single particle spectrum and the width of
that gap would be about 1 MeV. Then as the temperature increases
until it is large, compared to this, one finds the transition from
the appearance of the constant temperature behavior at low energy
to the pure Fermi gas at higher energy. I think this is the
simplest model that will give these results. Of course, more
detailed calculations with individual single particle levels as
Dr. Huizenga has done, numerically will do the same.

HUIZENGA

Right, essentially that's the calculation I've done. Simply
explained, it goes as you've indicated.

J-DEPENDENCE OF LEVEL DENSITY IN ^{180}Hf AND ^{178}Hf FROM STATISTICAL PROPERTIES OF SERIES OF RESONANCES WITH ASSIGNED SPIN

C. Coceva[*], F. Corvi[*], P. Giacobbe[*] and M. Stefanon[*]

CBNM EURATOM, Geel, Belgium, and

CNEN, Centro di Calcolo, Bologna, Italy

The spin cut-off parameter σ of the level-density formula may be determined directly from counting of highly excited states, without depending upon particular assumptions [1]) for the interpretation of the experimental results. Aim of this work was to evaluate σ from the densities of the two series of states with definite J excited by low-energy s-wave neutron capture in an odd-neutron nucleus. ^{177}Hf(I=7/2) and ^{179}Hf(I=9/2) were chosen because of the high spin and low p-wave strength; the probability that any of the observed resonances be p-wave was found to be negligible. The spin of the resonances was determined by the γ-cascade multiplicity method [2]), and for ^{179}Hf also by the low-level population method [3]). The experiment was performed at the electron linac of Geel, with the time-of-flight technique. Isotopically enriched samples were used. The σ value was deduced in each nucleus from the ratio R between the estimated mean spacings D_J of the two series of levels. To get a reliable estimate, either no level or the same percentage of levels must be lost in the two series. However, as the spins of the levels populated by s-wave capture differ by only one unit, the currently assumed statistical fluctuations of the level positions imply large errors also in the case in which no level is missed. For this reason we sought an experimental check of the more stringent conditions of the Dyson-Mehta theory [4]), in order to obtain estimates of σ affected by smaller errors. The influence of possible short-range anomalies of the level density was examined.

In ^{179}Hf, 93 resonances were observed below 430 eV, and the spin was assigned to 85 of them (see tab.I). The strength functions for the two spins, calculated from the data of ref. [5]), were found

---

[*] CNEN, Ispra, Italy.

Table I - Energies and spins of ^{179}Hf resonances.

E	J	E	J	E	J	E	J	E	J
5.680	5	101.2	5	188.5	5	270.0	(5)	381.7	4
17.65	4	103.7	5	189.8	4	273.5	5	385.6	-
19.13	5	107.8	4	192.9	4	276.3	5	389.0	4
23.66	5	117.2	4	197.9	5	284.7	4	390.6	5
26.54	4	120.1	4	202.6	5	288.7	4	395.3	-
27.40	5	121.9	4	204.1	4	292.4	5	401.1	5
31.14	4	122.6	5	205.8	4	300.6	4	403.8	-
36.50	5	129.9	4	213.1	(5)	306.3	4	408.8	-
40.12	5	137.2	5	213.1	(4)	313.6	5	411.4	4
42.29	4	138.1	4	224.0	4	322.5	4	413.6	-
50.77	4	144.2	5	224.5	5	327.0	5	423.4	5
51.11	5	147.0	4	227.7	5	332.9	4	428.9	5
54.79	4	152.3	4	229.7	4	338.2	5	430.9	4
69.03	4	156.3	5	241.6	4	346.1	4		
73.53	4	158.7	4	242.9	5	349.6	5		
76.63	5	165.7	5	245.3	4	356.1	4		
82.94	5	174.2	5	251.8	4	360.7	5		
85.42	4	174.9	4	253.3	5	364.4	5		
92.07	4	177.9	5	255.3	4	372.1	5		
92.67	5	182.6	4	262.9	4	375.2	4		

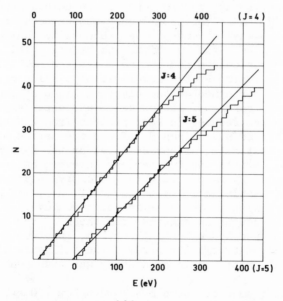

Fig. 1 - Cumulative graph of ^{180}Hf levels directly assigned to J=4 and J=5. Levels at 213.1 eV (possibly two) and 270.0 eV are not included. The straight lines are obtained by a least-square fit in both parameters in the range 0-210 eV.

Table II - Dyson-Mehta statistics for 25 J=4 and 22 J=5 ^{180}Hf levels. E<210 eV. The values are referred to the CM system.

| | J = 4 | | J = 5 | |
	Experiment	D.M. Model	Experiment	D.M. Model
Δ_3	0.35	0.32±0.11	0.39	0.31±0.11
Q	8.8	7.9 ±2.8	6.9	6.9 ±2.6
D_J	8.18±0.29		9.64±0.39	

Table III - Dyson-Mehta statistics for 37 J=4 and 32 J=5 ^{180}Hf levels. E<320 eV. The values are referred to the CM system.

| | J = 4 | | J = 5 | |
	Experiment	D.M. Model	Experiment	D.M. Model
Δ_3	0.42	0.36±0.11	0.41	0.34±0.11
Q	12.3	11.6±3.4	11.1	10.1±3.2
D_J	8.02±0.20		9.76±0.27	

to be coincident within the errors. Below 210 eV the spin was assigned to all observed resonances; on the basis of the behaviour of the staircases of fig.1 it was assumed that the two series of levels are complete up to this energy. In this case D_J may be estimated as the maximum likelihood value $D = [\pi/(4n) \sum_i^n D_i^2]^{1/2}$ of the Wigner distribution, the variance of which is var $D \approx 0.27\, D^2/n$, n being the number of spacings. In the range 210-400 eV, where the spin was assigned to 80% of the levels approximately, R was calculated as the ratio of the arithmetic means of the observed spacings, and the error was evaluated assuming a variance of D intermediate between those of Wigner and Poisson distributions. An overall estimate of R was obtained from a least-square composition of the two results; this is an acceptable procedure since in both intervals the relative errors are sufficiently small. From this estimate a value $\sigma = 3.51^{+0.86}_{-0.44}$ was obtained, the quoted errors being 68.3% confidence limits. In order to prove that the level positions are more correlated than implied by the Wigner spacing distribution, the "least-square" Δ_3 and the "energy" Q statistics of Dyson and Mehta [4] were calculated below 210 eV. The agreement between experimental and theoretical values shown in tab. II is at the same time a proof of the Dyson-Mehta theory and a check of the completeness of the two series. It is then justified to calculate D_J from the Dyson-Mehta optimum linear statistic (see tab.II). The resulting spin cut-off parameter is $\sigma = 3.70^{+0.32}_{-0.24}$. In the range 210-320 eV, only two peaks

in the (n,γ) yield could not be assigned: at 213.1 and 270.0 eV.
For the former, the experimental data are consistent with an overlapping of two resonances of different spin. Moreover, fig.1 suggests that one J=4 and two J=5 levels are missed in this range.
Following these indications we tentatively assume one J=4 resonance
at 213.1 eV and two J=5 resonances, at 213.1 and 270.0 eV respectively. Such assignments are supported by the calculated values of the
Dyson-Mehta statistics for the 69 levels up to 320 eV, which agree
with the theoretical values, as shown in tab.III. If the 270.0 eV
resonance is removed from the J=5 series, one gets a worse agreement
of the experimental value $\Delta_3=0.49$ with the theoretical one 0.34 ± 0.11.
From the spacings given in tab.III one obtains $\sigma=3.59^{+0.18}_{-0.16}$.

Table IV - Energies and spins of ^{177}Hf resonances.

E	J	E	J	E	J	E	J	E	J
1.100	3	49.60	4	103.1	3	156.0	3	223.4	4
2.381	4	54.78	4	104.2	–	160.1	4	224.8	3
5.909	3	56.37	3	111.2	3	163.1	3	226.7	4
6.596	4	57.06	4	111.9	4	167.4	3	229.2	3
8.889	4	59.28	3	114.6	–	171.1	3	232.5	4
10.95	3	62.19	3	115.1	3	174.4	4	236.3	3
13.68	4	63.50	4	121.3	3	176.2	3	238.8	(3)
13.97	3	66.76	3	122.6	–	177.0	4	240.9	4
21.97	4	70.03	4	123.8	3	181.1	4	249.0	3
22.28	3	71.38	4	126.2	4	184.7	4	255.1	3
23.43	4	72.28	3	131.7	3	186.9	–	259.7	4
25.63	3	75.60	3	134.1	4	192.8	4	264.7	3
27.02	3	76.05	4	136.3	3	194.2	3	267.8	4
31.59	3	82.30	4	137.6	4	199.3	4	272.3	4
32.82	4	84.67	4	141.2	4	201.9	3	273.4	3
36.09	3	85.31	3	143.2	3	208.8	4	283.3	–
36.96	4	86.79	4	143.8	4	210.2	3	284.9	3
43.06	4	88.56	3	145.6	3	212.2	3	288.2	4
45.14	4	93.18	3	148.6	3	217.2	4	294.5	3
46.23	4	97.10	4	151.1	–	219.7	3	298.7	4
48.82	3	99.04	4	152.9	4	222.2	3	299.7	3

In ^{177}Hf, 105 resonances were observed below 300 eV, and the
spin of 99 of them was determined (see tab.IV). The cumulative graphs
of fig.2 show an apparent loss of levels above 180 eV for the J=3
series and above 100 eV for J=4. The Dyson-Mehta statistics calculated below these limits, where the spin could be assigned to all observed levels, are in agreement with the theoretical model, as
shown in tab.V. At 300 eV, the J=3 staircase is 4 levels too low
with respect to the straight line (see fig.2), while the J=4 stair-

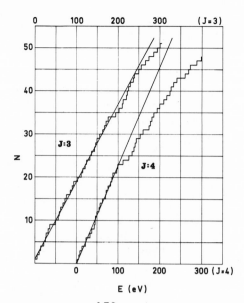

E (eV)

Fig. 2 - Cumulative graph of ^{178}Hf levels assigned to J=3 and J=4.
The straight lines are least-square fits below 180 and 100 eV respectively.

case is too low by 20 levels. The estimated values $2g\bar{\Gamma}_n^0 = 1.1^{+0.6}_{-0.3}$ for
J=3 and $2g\bar{\Gamma}_n^0 = 1.9^{+0.9}_{-0.5}$ for J=4 prevent from concluding that more J=4
than J=3 resonances are missed. Therefore the question arises wheth-
er the J=4 level density is actually changing over the observed energy range. If it is so, the value $\sigma=\infty$, with a lower 84% confidence
limit $\sigma_{min}=9.3$ obtained from the data of tab.V is meaningless. Moreover, the conclusion of a progressive density fall of the J=4 series
is substantiated by the fact that, if one skips the region between
100 and 125 eV, where there are three unassigned resonances, one obtains for the 19 levels between 125 and 250 eV, $\Delta_3=0.28$ in agreement

Table V - Dyson-Mehta statistics for 34 J=3 (E<180 eV) and 23 J=4
(E<100 eV) ^{178}Hf levels. The values are referred to the CM system.

	J = 3		J = 4	
	Experiment	D.M. Model	Experiment	D.M. Model
Δ_3	0.44	0.35±0.11	0.35	0.31±0.11
Q	11.0	11.0±3.3	8.1	7.2 ±2.7
D_J	5.60±0.14		4.35±0.17	

with the theoretical value 0.29±0.11. This indicates that very few
levels, possibly the only one at 151.1 eV, are missing in this region while, to have the same density as below 100 eV, nine levels
should be missing.

    Recently, Ideno and Ohkubo [6] observed that the neutron reso-
nances of the nucleus ^{177}Hf, among others, show a marked preference
for positions separated by constant intervals. In their analysis,
sequences of n equidistant windows with width $\Delta E << D$ and spacing x
are built starting from each level energy. In each sequence the number of windows containing a level is counted. Summing over all possible sequences, the value of a correlation function $A_n(x)$ is obtained. The mean value of $A_n(x)$ and the probability level of the deviations can easily be computed for uncorrelated series (sampling
from Wigner distribution would give the same results, except for
$x << D$). Ideno and Ohkubo found for the mixed set of J=3 and J=4 levels of ^{178}Hf a characteristic spacing of 4.4 eV. We repeated the
analysis on our series of J-assigned resonances and found that, as
shown in fig.3, this correlation is due exclusively to the J=3 se-

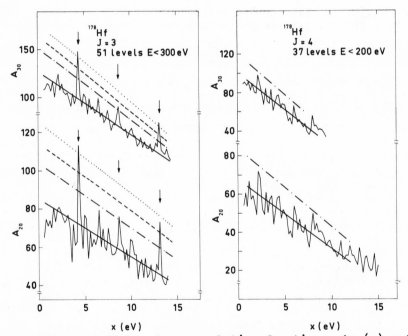

Fig. 3 - 20th and 30th order correlation functions, $A_{20}(x)$ and
$A_{30}(x)$, as a function of the step x for J=3 and J=4 ^{178}Hf levels.
The window width is $\Delta E=0.5$ eV. The solid curve gives the average
for uncorrelated series, curves -.-., --- and ..... are the
2.3×10^{-2}, 1.3×10^{-3} and 3.2×10^{-5} probability levels. The arrows indicate the peaks corresponding to multiples of x=4.4 eV.

ries, while the J=4 series is consistent with an uncorrelated distribution; the same lack of correlation was found for both series in ^{179}Hf. The attribution of the correlation found by Ideno and Ohkubo to a single series of levels with definite J suggests that this effect is of physical relevance.

In conclusion, very narrow confidence limits could be assigned to the estimate of σ in ^{180}Hf, as a consequence of the verification of the high degree of regularity of the level disposition foreseen by the Dyson–Mehta theory. This result shows clearly the importance of directing the efforts towards the detection of levels forming a complete, even if not numerous, sequence. For instance, when the level loss is such that the spacing distribution approaches the Poisson distribution, 400 levels are necessary to estimate D_J with a 5% error, while only 18 levels are needed in case of validity of the Dyson–Mehta theory. On the other side, the results of ^{177}Hf suggest that, in some cases, a σ value determined on such small energy intervals may be unreliable: even at an excitation as high as the neutron binding energy, the level density may not yet follow the smooth energy dependence required for its description in terms of statistical-model parameters, such as σ. It must be pointed out that, even in this case, it was possible to perform a significant test of the local validity of the Dyson–Mehta theory. The nature of the equi-spaced structure found in the J=3 series of resonances is not understood, but this effect may be considered as another symptom of a not yet fully statistical behaviour of the ^{178}Hf levels.

REFERENCES

1) J.R. Huizenga and R. Vandenbosch, Phys. Rev. 120 (1960) 1305 and Phys. Rev. 120 (1960) 1313.
2) C. Coceva, F. Corvi, P. Giacobbe and G. Carraro, Nucl. Phys. A 117 (1968) 586.
3) K.J. Wetzel and G.E. Thomas, Phys. Rev. C 1 (1970) 1501.
4) F.J. Dyson and M.L. Mehta, J. Math. Phys. 4 (1963) 701.
5) Neutron Cross Sections, BNL–325, 2nd ed. Suppl. 2 (1966).
6) K. Ideno and M. Ohkubo, J. Phys. Soc. Japan 30 (1971) 620.

DISCUSSION

GILAT (Soreg Nuclear Research Center)

This is really a comment and pertains both to Dr. Huizenga's talk and to the recent talk here. I would like to point out that the deviations from the simple Fermi gas model due to shell effects and discussed by Dr. Huizenga with respect to the $\underline{a}$ and Δ parameters are also reflected in the spin cutoff parameter σ. This is borne out, e.g., by the exact numerical solution of the combinatorial problem with realistic shell levels done by Hillman and Grover. Thus, the apparent disagreement with the predictions of the statistical model shown by Dr. Coceva is probably due to such fluctuations and shows the danger of deriving meaningful level density parameters from a measurement over a very limited range of energy and spins such as covered by the neutron capture resonance experiment described here. I would also like to point out that the normally used $\exp-[(J+1)^2/\sigma^2]$ spin dependence is just a first term in a series expansion, and when one calculates level densities at high energies, higher order terms ought to be included.

FLUCTUATION ANALYSIS OF TOTAL NEUTRON CROSS SECTIONS[*]

D. Kopsch and S. Cierjacks

Institut für Angewandte Kernphysik

Kernforschungszentrum Karlsruhe

1. INTRODUCTION

The cross section fluctuations occurring in the excitation functions of nuclear reactions have been investigated in several experimental and theoretical studies to determine average level densities and level widths. In addition, some experimental data have been the subject of a search for intermediate structure. However, many of these previous studies were restricted to narrow energy intervals in a few nuclei and gave partially inconclusive results.

The large set of neutron total cross sections measured with the Karlsruhe time-of-flight spectrometer was therefore considered as a favorable case for such study, because the data have high statistical accuracy and extend over the entire energy range in which significant fluctuations occur.

2. EXPERIMENTAL

The total neutron cross sections of the elements F, Na, Al, Si, S, K, Ca, V, Cr, Mn, Fe, Co and Ni have been measured during the last three years with the time-of-flight facility at the Karlsruhe isochronous cyclotron. This facility combines a 20 kHz pulsed neutron source of ~ 1,5 ns burst width, a 57 m flight path and a proton recoil counter as the neutron detector. Operational details of the facility have been described elsewhere [1].

The cross sections of the 13 elements were measured in the energy range 0.5 - 32 MeV by transmission experiments. Standard time-of-flight techniques were applied for the neutron energy determination. All transmission samples, except F, were in solid elemen-

tal form. For fluorine a sample of $(CF_2)_n$ was used. In this case
a carefully matched carbon sample was used for the sample-out
measurement.

Sample thicknesses generally were chosen to give approximately
40 % - 70 % transmission in most of the time channels. Data collection in the typically 8000 time channels was accomplished with a
digital time analyzer LABEN UC-KB and a CDC 3100 on-line computer.

The total neutron cross sections were calculated off-line by
combining sample-in, sample-out and background measurements.
Corrections for dead-time losses were applied using an analytical
equation which has been experimentally verified for the applied
conditions. The measurements were carried out with an energy resolution of typically 1 keV at 0.5 MeV increasing as $E^{3/2}$ to 70 keV
at 32 MeV. With the exception of the lowest and highest portions
of the excitation functions the measurements were performed with a
statistical uncertainty of ≤ 2 %. In fig. 1 the total neutron cross
sections of vanadium and chromium, which may serve as characteristic examples of our results, are shown on a double logarithmic
scale.

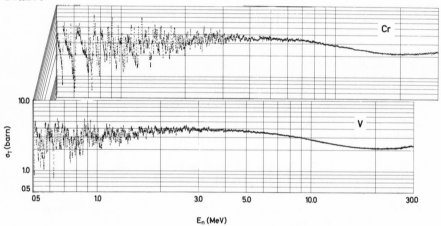

Fig. 1 Total neutron cross section of vanadium and chromium

3. ANALYSIS

Fluctuation analyses of the cross sections were carried out
in the energy region between 0.8 and 14 MeV. Above 14 MeV none of
the nuclei investigated here exhibited significant fluctuations.
The motivation for the analyses was threefold: (i) to determine
average level densities, (ii) to deduce mean level widths and (iii)
to search for intermediate structure.

Determination of average level densities: The determination of
average level densities was accomplished by the method adopted by
Carlson and Barshall [2]. Considering statistical fluctuations in

spacings as well as in widths of compound-nucleus levels, these
authors have shown that the variance of the compound-nucleus formation cross section can be expressed by

$$F=(\pi\lambda^2)^2 \cdot \frac{1}{\omega(E) \cdot \Delta_n} \sum_{J\pi} \frac{g^2(J)}{H(J\pi)} [k_W \sum_{l,s} (T^J_{l,s})^2 + k_n (\sum_{l,s} T^J_{l,s})^2] \qquad (1)$$

with the abreviations:

$$k_W = \frac{var[\Gamma_i(ls,J\pi)]}{\langle \Gamma_i(ls,J\pi)\rangle^2} \text{ and } k_n = \frac{var[N^{(n)}_{J\pi}]}{\langle N^{(n)}_{J\pi}\rangle} \qquad (2)$$

Here λ is the reduced wave length, $\omega(E) \cdot H(J,\pi) = N_{J\pi}$ is the
level density split into an energy dependent and a spin dependent
part, g is the spin weighting factor and T is the transmission
coefficient. The quantities k_W and k_n can be calculated from the
standard width and spacing distributions, respectively.

From the experiments we calculated the variances

$$F = \langle (\sigma_n - \overline{\sigma})^2 \rangle \qquad (3)$$

where σ_n is the average cross section in Δ_n and $\overline{\sigma}$ means the
average compound nucleus formation cross section. The latter was
obtained by subtracting the optical potential scattering cross
section from the measured total cross section. For $H(J,\pi)$ the
formula given by Gilbert and Cameron [3] and for T-values those
calculated by Mani, Melkanoff and Iori [4] were used in our
analysis.

For all 13 elements, analyses were carried out in 2 MeV wide
subintervals extending to high energies until no fluctuations were
observable. With the exception of F, Al, Si and S, reasonable
agreement was found between the deduced energy dependence of the
level density and the theoretical predictions.
Fig. 2 shows the results of Na, Ca and Co as typical examples. In

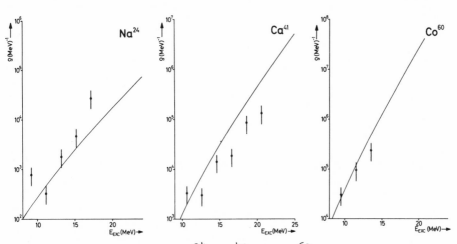

Fig. 2 Level densities of ^{24}Na, ^{41}Ca and ^{60}Co

this figure the solid circles are the results of the present
analysis, the solid lines are those of a calculation using the
Gilbert and Cameron formula.

Deduction of average level widths. For the deduction of average level widths the theory developed by Ericson [3] was applied.
In this theory it is assumed that all levels have the same total
width Γ and that the effect of fluctuations in level spacings may
be neglected. At energies at which a large number of levels overlap the following dependence of the self-correlation function on
the energy increment ε was deduced:

$$C(\varepsilon) = C(0) \cdot 1/\left[1 + (\varepsilon/\Gamma)^2\right] \qquad (4)$$

i.e. a Lorentzian form factor times the mean square deviation of
the average cross section $C(0)$. $\overline{\Gamma}$ therefore can be obtained from
the width (FWHM) of the autocorrelation function.

In the present analysis we calculated the self-correlation
function in the form modified by Pappalardo [6], as

$$C(\varepsilon, I) = \left\langle (\sigma(E) - \sigma_I(E))\, (\sigma(E+\varepsilon) - \sigma_I(E+\varepsilon)) \right\rangle \qquad (5)$$

Here the quantities $\sigma_I(E)$ and $\sigma_I(E+\varepsilon)$ mean average cross sections
in a sliding energy interval smaller than the total analyzing
interval. This is done to account for a slowly varying mean value.

Although the conditions for the Ericson theory may not be
satisfied for all measurements we have calculated the correlation
widths for all elements from 0.8 MeV to the highest energies at
which fluctuations occur. In all cases a nearly Lorentzian shape
of the self-correlation function was obtained.

Fig. 3 shows as a typical example the self-correlation function of vanadium calculated for the energy region between 2 and
4 MeV.

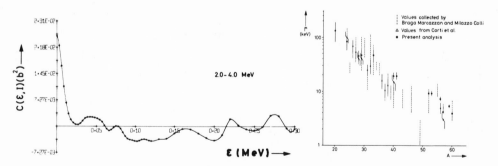

Fig. 3 Correlation function $C(\varepsilon, I_f)$ of   Fig. 4 Mass dependence
       vanadium   of average level
                                                           widths

The Γ values obtained from the analysis have been corrected for the effects of counting statistics and the energy resolution. The latter corrections were performed using the method proposed by Lang [7].

In Fig. 4 the mass dependence of the deduced average level widths is compared with recent results from (d,p) and (d,α) reaction measurements [8] at about 20 MeV excitation energy. In this figure our results are those for the highest energy subintervals. With the exception of the mass region near A = 40 there is agreement within the stated errors.

Search for intermediate structure: An investigation of the evidence of structure with intermediate widths, i.e. widths between approximately 50 and some hundred keV was performed following the procedure proposed by Pappalardo [9]. The occurrence of a second rise in the correlation function $C(0,I)$ given in eqn. 5 and the occurrence of two correlations with largely different widths in $C(\mathcal{E},I_f)$ (where f stands for "fixed") was taken as evidence that such intermediate structure exists. For all nuclei with the exception of Si, Mn and Co evidence was found in part of the investigated energy regions. No interpretation in terms of doorway states was tried. We hope to investigate this question by the inclusion of partial cross sections and a search for correlations in scattering angle as well as in reaction channels in a future study [10].

*Paper presented by G. Kirouac of Knolls Atomic Power Laboratory

References:

1. S. Cierjacks, B. Duelli, P. Forti, D. Kopsch, L. Kropp, M. Lösel, J. Nebe, H. Schweickert, H. Unseld, Rev. Sci. Instr. 39, (1968) 1279

2. A.D. Carlson, H.H. Barschall, Phys. Rev. 158, No. 4 (1967) 1142 and A.D. Carlson, Thesis, University of Wisconsin, 1967

3. A. Gilbert, A.G.W. Cameron, Can. J. Phys. 43, (1965) 1446

4. M. Mani, M. Melkanoff, I. Iori, Report CEA 2380, 1963

5. T. Ericson, Phys. Rev. Lett. 5, (1960) 430 and T. Ericson Ann. Phys. 23, (1963) 390

6. A. Agodi, G. Pappalardo, Nucl. Phys. 47, (1963) 129

7. D.W. Lang, Nucl. Phys. 72, (1965) 461

8. M. Corti, M.G. Marcazzan, L. Milazzo Colli, M. Milazzo, Energia Nucleare 13, (1966) 312 and M.G. Braga Marcazzan, L. Milazzo Colli, Progress in Nucl. Phys. Vol. II, 145, 1969

9. G. Pappalardo, Phys. Lett. 13, (1964) 320

10. F. Voß, Thesis, University of Karlsruhe, to be published

DISCUSSION

SCHRACK (National Bureau of Standards)
    We have measured total cross sections also at the same energy
region and for seven elements in which the level spacing and
frequency was adequate to make a correlation analysis. The conclusions that one has about intermediate structure are dependent,
I think, on what one wants to look for. I'm sorry he didn't show
the actual correlation analysis curves. That actual Papalardo
effect that you indicated on the top graph, the plateau effect, is
very seldom seen. I think a more realistic thing to look for is a
delayed rise. If you have no delayed rise, then there is no
second width. With that criteria you can see a distribution in
what you might call intermediate structure, strong cases, and weak
cases and the distribution of seeing intermediate structure is
approximately equivalent to a Monte Carlo mock-up that we've done;
so one can make any conclusion one wants to from this whether these
are purely statistical appearances or whether they have some real
door-way state meaning.

KIROUAC
    If I may, I'd like to make a couple of comments and then, if
I might turn things about, I'd like to ask you two questions.
First of all, when I say there was some evidence, that varied from
very strong in some cases to rather weak in others, and for three
materials there was absolutely no evidence whatsoever for intermediate structure. No interpretation was made obviously in terms
of doorway states. These are simply intermediate structures in
correlation functions. Now I did read your contribution and I have
two questions. First of all, I wonder if you could tell me what
your energy interval was and how high you did go.

SCHRACK
    The energy interval over which the correlation analysis was
made varied upon the structure change in the cross-section. We
did two types of analysis. Once we just went through all the data
and tried to pick out groupings of energy that had fairly similar
structure and did correlation analysis on those. Those varied in
widths from 1 MeV, 2 MeV to 10 MeV, depending on whether it was a
low energy or a high energy. Then we went back and we said that
we felt that the correlation analysis was really not very meaningful when the level overlap was high because there are two effects
coming in: one, the statistical fluctuation type effect; two, you
have many channels contributing and so really any significant
intermediate structure would have to lie in one spin channel so
that when you have many spin channels contributing, the theoretical
significance of the fluctuation is not clear. So we went back and
only did it from a half MeV to 1 MeV of where the levels could be
easily identified and we felt there was some significance.

KIROUAC
     Most of the intermediate structure observed in the present work
was observed above 1.5 MeV, usually above 2.0 MeV.

SCHRACK
     There was a particular case that I saw that was printed in an
earlier report at 2.7 MeV in aluminum. There is a very nice
plateau there. We examined that in some detail because we thought
it was so high up in excitation that we didn't think it could have
significance from the standard doorway state analysis. If you
look at it closely you find that this is actually based on an
interference type of shape in the fluctuation, and I'm not sure
what the significance is of it.

VONACH (Technische Universitat Munchen)
     From your slide on the gamma values, I noticed in the region
of high A 40-60 your gammas were quite a bit larger than those for
particle reactions. Might this be due to experimental resolution
or do you think this is a real effect.

KIROUAC
     I'm afraid I can't comment on that beyond the error bars that
have been put on the points.

SESSION VII

DECAY OF THE COMPOUND NUCLEUS

Chairman   - D. Sperber, Rensselaer Polytechnic
                              Institute
Scientific Secretary - H. Bakhru, State University of New York at
                              Albany

EVAPORATION SPECTRA*

I. Halpern

Department of Physics
University of Washington
Seattle, Washington

PROBLEMS IN THE ANALYSIS OF EVAPORATION SPECTRA

It has been known for many years that most particles emitted
in nuclear bombardments tend to come off with low energies, with
spectra resembling those of molecules evaporated from a hot body in
thermal equilibrium. We have not, however, learned so very much
about the properties of hot nuclei through the study of nuclear
evaporation spectra. The difficulties have been both experimental
and theoretical. To point out some of them, let me remind you that
in the simple Weisskopf[1] statistical theory for evaporation from
nuclei the spectra have the form

$$S(E) \sim \sigma(E)E\, \rho(U - B - E) \tag{1}$$

where $\rho(U - B - E)$ is the level density of the residual nucleus at
its excitation energy just after particle emission. U is the excitation energy just before particle emission, and B and E are the
binding and kinetic energies of the evaporated particle. The cross-
section $\sigma(E)$ is that presented by the excited residual for the
(time-reversed) emitted particle. In calculations of spectra, σ is
generally replaced by measured ground state cross-sections or those
inferred from such measurements.

To keep things simple ρ is usually characterized by a
single parameter. For example one can write it proportional to
$\exp(- E/T)$. To obtain a value of T from experiment for comparison
with available theoretical estimates one generally divides S by σE
and plots the log of this quantity against E. The slope of such a
plot should be $- 1/T$. It is best to limit oneself to the data to

465

the right of the spectrum peak (Fig. 1), but not too far to the
right because of the contributions of direct, non-evaporated particles to the spectrum.   If one wants to neglect say the top 10%
of the spectrum to avoid direct contributions, the range in E from
which T is determined extends over only a few MeV.   Many determinations of T which appear in the literature extend dangerously far to
high values of E.   Indeed there have been some measurements where
(because of high detector thresholds) the parameter T has been
determined from data which are almost entirely due to the direct
part of the spectrum.   Thus one major difficulty in spectrum analysis
is that there is generally a sizable direct component in the observed spectra and that there exists no generally valid prescription
for its removal.   (See, e.g. R.W. West[2] on proton spectra from α
bombardment for an attempt at systematic separation of direct from
statistical spectra.)

A second difficulty about the spectra is that, at higher energies, they are superpositions of the spectra emitted from the
original compound nucleus and spectra evaporated from later cooler
nuclei.   There is no way to measure the spectra associated with a

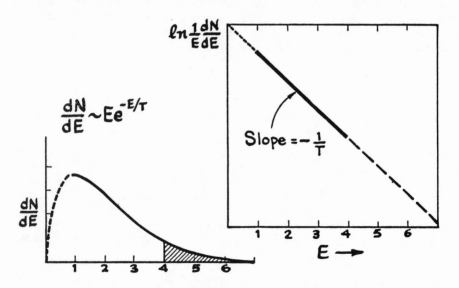

Fig. 1.   A hypothetical spectrum to show that if one ignores the
data below the peak and also the top 10% of the spectrum (to avoid
contamination from direct reactions)there is very little range left
from which to determine a precise value for the slope of the conventional log plot.

single stage in the decay all by itself.

One might therefore think that the easiest spectra to study and interpret would be those emitted by rather cold nuclei where there can be but one evaporation stage.[3-6] The main troubles here are that $\rho(E)$ is not expected to be a very simple function of energy or nuclear species at low energy, and that the direct and evaporated spectra tend to overlap badly in energy and angle.

In addition to these specific problems, there is a more general difficulty in obtaining significant information from observed evaporation spectra. These spectra are very smooth and amorphous. The theoretical expressions for ρ that one uses are, on the other hand, often fairly elaborate containing a number of parameters. Under such conditions, the empirical evaluation of the parameters from observed spectra can be an unhappy business.

CHARGED PARTICLE EVAPORATION SPECTRA AT HIGH EXCITATION IN HEAVY NUCLEI

As a result of the aforementioned difficulties it becomes desirable in the study of evaporation spectra to work first in domains where the information content of the observations is largest and the number of parameters in the theory is smallest. I believe that there exists a domain where the situation is particularly clean. This is for charged particle spectra emitted from moderately excited (U $\simeq$ 50 to 100 MeV) <u>heavy</u> nuclei.

The advantages here are the following:

(1) The direct components are not so troublesome as at lower energy since they are spread out in energy and overlap less seriously with the evaporated spectrum.

(2) The simple Fermi Formula[7] $\rho \sim \exp 2\sqrt{aU}$ (where we are ignoring a factor varying slowly with U) is likely to be more valid than it is at lower energies.

(3) In heavy nuclei those angular momentum complications which tend to make Eq. (1) invalid are minimized since the angular momentum per nucleon remains small, and finally

(4) The superposition problem is not too serious because most charged particle evaporation occurs in the first few stages of the evaporation cascade.

I should explain this last point since it may not be obvious. The probability that a charged particle will evaporate instead of (its main competition) a neutron, is proportional to $\exp\{-(E_c-E_n)/T\}$ where E_c is the energy cost of charged particle emission -- a little more than the Coulomb barrier height plus the particle binding energy. E_n is roughly the neutron binding energy. For example, $E_c - E_n$ is about 10 MeV for α particles evaporating from heavy nuclei. It follows that at low excitations ($T \leqslant 1$ MeV) α particle evaporation must be severely repressed. There is however an exponential increase in α emission probability with increasing excitation energy. From this it follows that α particles are evaporated with much greater chance early in a cascade than later.

Because alpha particle evaporation spectra from heavy nuclei at moderately high energy are clearer and stand out quite distinctly from the direct component, it is possible to characterize them in some detail experimentally. One can talk of their peak energy, peak cross-section and width. All of these features should be computable from the ratio of Γ_α to Γ_n where $\Gamma_\alpha \sim \sigma_\alpha(E)E \exp\sqrt{2aU_f}$ and where Γ_n is proportional to a similar expression.

It is instructive to pause here a moment to spell out how the peak energy and width are prescribed by Γ_α. It is easier for this purpose to write $\Gamma_\alpha \sim \sigma_\alpha E \exp(-E/T)$ than to use the Fermi form for $\rho(U)$ but the general idea is independent of how we characterize ρ. We can also ignore the factor E since it varies rather slowly near the spectrum peak. The behavior of $S(E_\alpha)$ is then given by

$$\ln S(E_\alpha) \sim \ln \sigma_\alpha - \frac{E_\alpha}{T} .$$

Clearly $S(E_\alpha)$ peaks where

$$\frac{d \ln \sigma_\alpha}{dE_\alpha} = \frac{1}{T} .$$

That is, if we happen to have an expansion of $\sigma_\alpha(E_\alpha)$ near the spectrum peak, say from an optical model, we must find where $d \ln \sigma/dE$ happens to equal $1/T$ and that is where the peak will be. This relation pins down the peak energy to a fraction of an MeV.

If now we expand $\ln \sigma_\alpha$ around the peak energy,

$$\ln S(E_\alpha) \sim \ln \sigma_\alpha\Big|_p + \frac{1}{2} \frac{d^2 \ln \sigma_\alpha}{dE_\alpha^2}\Big|_p \Delta E_\alpha^2 + \dots$$

From this it follows that the FWHM of the evaporation peak is

$$\Delta = \left[8 \ln 2/(-\frac{d^2 \ln \sigma}{dE_\alpha^2})_p \right]^{\frac{1}{2}} .$$

That is, an expansion of the optical model prescription for $\sigma_\alpha(E)$ is all one needs to determine the width of the evaporation spectrum. The width depends only on the curvature of $\ln \sigma_\alpha$ in the peak region.

I am giving these details to emphasize that in comparing theory and experiment, a number of observable features must simultaneously be accounted for with only one moderately free parameter, the so-called level density constant a of the Fermi formula. Actually the Fermi level density is generally written $U^{-n} \exp(2\sqrt{aU})$ so that n is a second parameter in the model. For reasonable, that is small, n the first factor here varies too slowly with U compared with the exponential to permit us to evaluate it from data. We ignore it.

Now let us look at some actual data. The spectra in Fig. 2 were obtained by Chenevert et al.[9] some years ago. The heavy lines are the observed spectra and the dashed lines are the results of calculations. In these calculations a single common value was used for the ratio of the level density parameter a to the atomic weight A, namely a = A/7. Actually a = A/8 gives a somewhat better overall fit. The sensitivity of the fits to the value of a/A permits one to determine a "best" value for this ratio with an accuracy of $\sim 15\%$.

The dark numbers are bombarding energies, so you see that in addition to everything else, the fit to the dependence on bombarding energy is also rather reasonable.

The agreement on the variation of cross-sections with A is not bad, but not perfect. Let me tell you what was done about the A dependence of the level density ρ in the calculations. It is

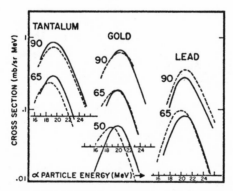

Fig. 2. A comparison of calculated (dashed) and observed[9] (solid) evaporation spectra for three targets at several bombarding energies. The calculated spectra were computed using a conventional formulation of the statistical theory with the level density constant a = A/7. The uncertainty of the "best" value for a/A is about 15%.

always a problem to know where to start measuring the excitation energy U. The actual location of the ground state depends on shell and pairing effects. What one requires for Γ_α/Γ_n is the <u>difference</u> of effective ground state energies after neutron emission and after α particle emission. This difference was obtained by subtracting from the original compound nuclear excitation energy, U, an <u>effective</u> binding energy for the emitted particle in place of its actual binding energy. This effective binding was calculated from the semi-empirical mass formula with shell and pairing corrections simply omitted. This procedure corresponds to the assumption that shell and pairing effects are completely healed out at the excitation energies (> 50 MeV) involved in these measurements. Had we used actual binding energies instead of the effective ones (a procedure which corresponds to the assumption that there is no healing at all from magicity in the observed range of excitations), the discrepancy between calculated and measured Ta/Pb yield ratios would have been more than an order of magnitude. Thus it appears that the assumption of full recovery from magicity by ∿50 MeV must lie close to the truth.

Should one expect the effects of magicity to be damped out by 50 MeV? Magic effects come about because the single particle level densities near the Fermi level are smaller than average for magic nuclei.[10] As a result there are fewer distinct configurations, i.e. fewer states that can be excited with a given excitation U in a magic nucleus than in a normal nucleus. As one raises the excitation energy, the band of single particle energies near the Fermi level which is relevant for the nuclear level density becomes wider ($\Delta \approx 2T$). One can certainly expect a washing out of magic effects at excitations so large (∿ 300 MeV in heavy nuclei) that Δ becomes comparable with the shell spacing $\hbar\omega_0$. The evidence suggests however that this washing out occurs at energies an order of magnitude smaller. The early onset of damping of magicity may be due in part to the effects of residual interactions between particles and holes. These interactions tend to disrupt the single particle level structure and the effective number of such interactions increases rapidly with temperature.

Returning to the main point of Chenevert's results (Fig. 2), it is that the <u>form</u> of the Weisskopf spectrum matches observed heights, widths, and locations of spectra over a wide range of excitation and nuclear species and that this match is achieved with one semi-free parameter. The overall fit in Fig. 2 would be conspicuously worse than it is had we taken the exponent in ρ to be proportional to U instead of to $\sqrt{U}$. Thus the general correctness of the Fermi Gas expression for nuclear level density is also verified here -- for higher excitations in large nuclei. It is encouraging to find this kind of agreement since there are no ready excuses available in this domain if the simple statistical theory were to fail to describe the data.

EVAPORATION SPECTRA AT LOWER EXCITATIONS

Because of the effects of residual interactions, one must not expect the lowest levels to distribute in the Fermi Gas form. These lowest levels must profit disproportionately from attractive residual interactions and one must consequently expect a stretching of the level diagram at its bottom end. The higher level densities (where effects of attractive and repulsive interactions are more nearly in balance) should be in better accord with the Fermi Gas character. This idea that the low levels are displaced downward from their "normal", independent-particle positions is supported by the empirical finding of Gilbert and Cameron[11] that the level density exponent goes more nearly as U at low energy than as $U^{\frac{1}{2}}$. Gilbert and Cameron have worked out a smooth joining recipe between the domains where these two dependences on U are assumed to hold. One reservation that one might raise about their joining procedure is that it does not explicitly conserve the total number of levels. As one turns on residual interactions in a Fermi Gas, the level density is expected to distort in a way that preserves the number of levels. This would require that in a real nucleus at higher excitations (where the level density has reached its asymptotic Fermi Gas form) the integrated number of levels which lie below any typical level would be the same as in the Fermi Gas.

It remains desirable to know better than we do just how the nuclear level density varies with U, J, A and Z in the lowest 10 or so MeV. Since $\rho(U)$ seems to follow the Fermi Gas form at higher excitations (at least for heavy nuclei), one is encouraged to see whether small modifications of this form can be found which provide an empirical fit to the observed spectra and to the other measures of level density which are available in this energy region. (See the review on level density determinations by Prof. Huizenga.) In making such fits on the basis of spectra one must be particularly careful to avoid contributions from direct reactions. At low bombarding energies direct contributions unfortunately tend to have energy and angular distributions that resemble those from statistical emissions more than they do at higher energies. One must also be careful in making fits to spectra to take into account the special effects associated with particle-photon competition and with nuclear distortion.

It is customary to describe empirically determined level densities with the Fermi Gas form $\rho(U) = U^{-n} \exp 2\sqrt{aU}$ where, of course, $\underline{a}$ must generally be regarded as some function of U instead of being $\underline{a}$ constant. For a Fermi Gas the integer n is equal to 2.[7] This value for n cannot be established experimentally from an examination of spectra at high energies (where the Fermi form with

constant a appears to be valid) because there the U dependence of
ρ(U) is very much dominated by that of the exponential factor. It
is also clear that the Fermi form with any positive n must be invalid at very low U. It has nevertheless been the general practice
to use the "full" Fermi form with n = 2 to describe observations.
In comparing a values measured by different authors it is important
to check that the same value for n was used in the analysis. It
is also important to check that translations from measured T's
(Fig. 1) to deduced a's were properly carried out in those regions
where a is still a function of U. In this case one has

$$\frac{1}{T} \equiv \frac{d}{dU} (2\sqrt{aU}) = \sqrt{\frac{a}{U}} (1 + \frac{d \ln a}{d \ln U}) \tag{2}$$

where the second term on the right can easily be comparable in size
to the first. Thus it is not always true that $U \simeq aT^2$. This is
valid only when a is constant. For example, where it is T that
stays constant as a function of U (i.e. where a goes as $\sqrt{U}$), the
correct relation is $U = 4aT^2$.

One final admonition. In comparing a values deduced by different authors it is important to make sure that one is measuring
U from the same effective ground state. Since $\rho \sim \exp 2\sqrt{aU}$ it is
clear that any fractional change ΔU/U will be compensated by an
equal and opposite Δa/a if one is holding ρ to match some observations.

The most pressing needs for improvement in our knowledge and
understanding of ρ(U) are those related to the lower excitation
energies. Although there remain some open questions at higher excitations (e.g., effects of higher order terms in the Fermi Gas
level density, effects of the energy dependence of the single particle level spacing including those associated with the taper of
the shell model potential,[12] etc.), these represent relatively
small refinements to the simple Fermi Gas form for ρ(U) and are
probably not accessible experimentally at the present time.

INVERSE CROSS-SECTIONS

I have been emphasizing the role of ρ(U) in the shaping of
particle spectra. The other interesting factor in Eq.(1) is the
inverse cross-section σ(E). We have seen that for α particles a
detailed knowledge of σ(E) from the optical model is needed to predict fundamental features of the spectra. The success of the predictions means that nuclei do not expand appreciably (as it had
sometimes been suggested they do) when they heat up to 50 or 100
MeV. Thus σ(E) for charged particles in the relevant energy range
has mainly to do with Coulomb barrier penetration and this does not
appear to depend significantly on U. The dependence of the inverse

cross-section on excitation energy may be more interesting for neutrons than it is for charged particles. For neutrons σ(E) is expected to reflect the location of the so-called size resonances.
For example the ratio of neutron reaction cross-sections of Ta to
Bi falls a factor of 1.6 between zero and 2 MeV because of such
effects. Indeed there already exists some evidence that the observed shapes of evaporated spectra show these size-resonance
effects.[13] With modern techniques it should be possible to measure
safely down to neutron energies of 1/4 or 1/2 MeV in order to see
the expected variations of σ(E) as a function of A more clearly.
An interesting question here is whether these variations persist
at high nuclear excitation. At 50 MeV in a heavy nucleus the Fermi
Sea is a foamy froth full of holes. Would such a nucleus exhibit
a significantly larger optical model W for 1 MeV neutron than a
ground state nucleus? Would the size resonances disappear? It
would be useful to measure carefully the low energy portions of
neutron spectra as a function of bombarding energy.

It would be appropriate to mention at this point a special
technique used recently by Cohen et al.[14] to measure the very lowest
parts of neutron spectra indirectly. One chooses a final residual
nucleus with a sizable gap between the ground and first excited
states and a reaction and a bombarding energy where one neutron can
be emitted (presumably evaporated) followed by a single proton.
One observes the spectrum of protons which feed the final ground
state. (They are numerous enough to appear distinctly in the observed proton spectra) and from this spectrum one deduces the population of the proton-emitting states and thus the spectrum of
neutrons which feeds these states. The preliminary results obtained
with this technique are somewhat puzzling. They call for further
study and extension to additional targets.

ANGULAR MOMENTUM EFFECTS ON EVAPORATION SPECTRA

I would like to comment only briefly about evaporation in
those situations where the angular momentum is large enough so that
Weisskopf's expression (Eq.(1)) for the emission width becomes in-
appropriate and one must consider the various possible angular
momentum changes, ΔJ, of the nucleus one at a time. The bombardments of Chenevert discussed earlier did not suffer from this difficulty. Indeed it had been shown even earlier[15] that in the heavy
mass region the Weisskopf formula works well for heavy ion bombardments in which the angular momentum inputs exceed those in
Chenevert's α particle work. The amount of angular momentum in a
nucleus for which the more complicated treatment is required is
roughly that where $J^2\hbar^2/2\mathcal{J}$ exceeds the equipartition kinetic
energy.[16-20] It can be shown that at this point the experimentalist
will begin to observe a noticeable anisotropy in the angular distribution of evaporated particles.

The main effect of larger amounts of angular momentum is that it influences the relative emission probabilities of various particles and the relative emission probabilities of particles vis-a-vis photons at the end of the evaporation cascade. Heavier particles tend to be emitted more easily from nuclei with large J since they are willing to remove angular momentum at a lower cost in emission energy; at low U, photons can be successfully emitted in competition with particles from high J states because of the unavailability of high J final states for the competing particle emissions. The latter effect has been extensively studied by Grover and co-workers,[21] the former by a number of workers, but not very systematically.

The angular momentum dependence of $\rho(U)$ has more easily observable consequences for the angular distribution of evaporation spectra than it does for their intensity or shape. The J dependence of the level density is generally described by a single parameter, the effective moment of inertia $\mathcal{J}$ of the nucleus.

$$\rho(U,J) = \rho(U,0)(2J + 1) \exp(\frac{J(J + 1)\hbar^2}{2\mathcal{J}T}). \tag{3}$$

For non-interacting particles in a Fermi Gas, this parameter is a constant independent of U and J. If one uses (3) to describe level densities for real nuclei one must regard $\mathcal{J}$ as being energy dependent, and perhaps even J dependent. An average value of $\mathcal{J}T/\hbar^2$ (often written σ^2) for an evaporation cascade can be determined from angular distributions of evaporated particles and a rough insight into its energy dependence can be gotten from studies of angular correlations between pairs of coincident evaporated particles.[22] At higher excitations one finds in this way that the effective moment of inertia is consistent with the rigid body value, but this determination has not been made with high precision. At lower excitation the effective moment of inertia seems sometimes to be smaller than the rigid value. In some nuclei -- for example in strongly deformed ones -- it may not be reasonable to characterize the low-lying spin dependence of level density with the form (3) and the single parameter $\mathcal{J}$.

Some suggestion that this may be the case comes from the study of the photon spectra emitted in bombardments of deformed nuclei. Most of such photons represent the further cooling of the nucleus after all energetically possible particle emission has taken place. In α particle or heavy ion bombardments the photon transitions de-excite the nucleus from states many of which have high J, and in so doing, they illuminate the high J, low lying level structure in the residual nuclei. Some of the more interesting recent developments in evaporation studies have been provided by measurements of these statistically emitted photons.[23,24] For example, in a report to this conference, Dr. Ejiri provides new evidence in support of the idea that in addition to J and parity, the K quantum number is important in steering the course of statistical emissions through the lower lying states of deformed nuclei.

THE DIRECT COMPONENT IN REACTION SPECTRA

Every student of evaporation spectra must be concerned with the direct spectra because they present him with a background which he must find a way to subtract. These subtractions generally go best at backward angles where the ratio of direct to evaporation components is smallest. When this ratio is small subtractions seem to be safe enough, but it is often not small and then one becomes aware of the fact that no reliable universal procedure exists for such subtractions.

I would like to call attention to an insufficiently emphasized point about the direct reactions. The particle that tends to come out in direct reactions is the incident projectile. That is, the inelastic direct cross-section is the largest -- and it can be very large indeed.

Thus the integrated direct α,α' cross-section in Chenevert's work is three to four hundred mb for heavy targets at $E_\alpha = 90$ MeV. The integrated evaporation cross-section is only about one tenth of that. (Fortunately most of the direct cross-section is forward and does not interfere with our perception of the evaporation component at back angles.) Similarly Chenevert's direct ^{3}He,^{3}He' cross-sections run about 200 to 250 mb. The ^{3}He evaporation cross-section is negligible. The direct transfer cross-sections are much smaller than the inelastic scattering cross-sections. Thus the ^{3}He,α cross-section is about 1/3 that for ^{3}He,^{3}He' and the direct $\alpha,^3$He cross-section is about 10% of that for direct α,α'. It appears that although ^{3}He's change into α particles more readily than α particles change into ^{3}He's, both particles prefer to remain intact i.e., to keep their integrity as they pass through the nucleus.

The integrated direct inelastic cross-sections are large also in p,p' reactions. Thus the direct p,p' cross-section in heavy targets is 250 mb at 29 MeV[25] and 500 mb at 62 MeV (Fig. 3). It has about the same value for lighter targets at this energy. The sum of the direct cross-sections for d,t, ^{3}He and ^{4}He production add up to a little more than 100 mb or 1/5 of the inelastic cross-section. Thus for protons or alpha particles or ^{3}He's the integrated inelastic cross-sections at moderate energies are quite comparable being about 1/5 of the reaction cross-section. I understand that cross-sections of similarly large magnitude have been seen for deuterons. It does not seem that the binding energy or compactness of the projectile matters too much. Perhaps even carbon and other heavy ions would exhibit large integrated inelastic cross-sections.

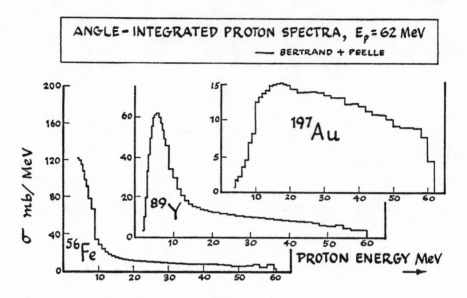

Fig. 3. Integrated inelastic cross-sections for protons. For all
targets the evaporation components are most clearly separated from
the direct ones at backward angles. In the lighter targets the
ratio of evaporated to direct components is large enough so that
this separation is possible even in the integrated spectra.

        The reason that the inelastic reaction dominates is probably
connected with the fact that scatterings tend to be forward. When
the incident projectile scatters on a nuclear component it tends to
retain rather than to share its energy and momentum. The optical
model imaginary potential W describes the rate of removal of the
projectile from the entrance channel -- either by absorption and
decomposition or by scatterings which leave the projectile intact.
The large inelastic cross-sections require that a good fraction of
the encounters responsible for W are scattering rather than absorptive encounters. In the α,α' work the mean free paths for absorption
and for scattering in nuclear matter appear to be comparable.

        Returning to the evaporation problem -- one consequence of the
large direct inelastic cross-sections is that it makes somewhat
uncertain the distribution of angular momentum depositions in the
compound-nucleus-producing fraction of the reactions. The direct
reactions presumably tend to use the larger impact parameters but
one must rely on specific models to assign impact parameters to
the direct and compound statistical parts of the cross-section.
These uncertainties can play a role in the studies of the angular
momentum history of compound nuclear decay like those I mentioned
earlier in connection with photon emission.

A second possible implication of the large inelastic cross-sections for evaporation studies is that one must be careful to subtract all direct components in experiments where favored re-emission of the incident particle is taken as evidence for isospin conservation in statistical reactions. (See the comprehensive discussion of this problem by Prof. Miller.)

Let me conclude by just mentioning a recent development that relates direct and evaporation spectra and angular distributions. You will notice (Fig. 3) that at moderate energies the direct spectra are quite smooth with energy. This suggests that they might be simply described -- indeed that some statistical considerations might apply. Griffin, Blann and others[26-28] have been trying to look at the direct spectra from this point of view. As the projectile makes its way into a nucleus it produces a "shower" of nuclear collisions. Some of the particles embroiled in the shower manage to escape from the nucleus. The chance for such escape is largest at early stages where the number of particles participating is still small and the energy per involved particle is large. To the extent that the early portions of the collision cascade contribute most to the direct spectra, the individual collision matrix elements and the detailed geometry of the collisions must be expected to matter. It is therefore hard to see how one could hope for a statistical approach which would have universal applicability. This problem is reflected in the present formulations of the theory of "pre-equilibrium" emissions by the introduction of a pair of parameters that are to be fitted empirically to the observed spectra. The values of these parameters vary from reaction to reaction and it is too soon to say whether a simple systematics exists in the assignments of the parameters which will permit one to correlate various observed direct spectra in a useful way.

REFERENCES

*   Work supported in part by the U.S. Atomic Energy Commission.
1.   V.F. Weisskopf, Phys. Rev. 52, 295 (1937).
2.   R.W. West, Phys. Rev. 141, 1033 (1966).
3.   N.O. Lassen and V.A. Sidorov, Nucl. Phys. 19, 579 (1960).
4.   C.H. Holbrow and H.H. Barschall, Nucl. Phys. 42, 264 (1963);
       R.M. Wood, R.R. Borchers, and H.H. Barschall, Nucl. Phys. 71,
       529 (1965).
5.   A. Alevra et al., Nucl. Phys. 58, 108 (1964).
6.   V.A. Sidorov, Nucl. Phys. 35, 253 (1962).
7.   T. Ericson, Advan. Phys. 9, 425 (1960);
       D. Bodansky, Ann. Rev. Nucl. Sci. 12, 79 (1962);
       A. Bohr and B.R. Mottelson, Nuclear Structure, Benjamin, N.Y.
       (1969);
       M. Böhning, Int. Conf. Nucl. React. (Heavy Ions), N. Holland
       Pub. N.Y. (1970).
8.   J.R. Huizenga and G.J. Igo, Nucl. Phys. 29, 462 (1962).
9.   G. Chenevert, I. Halpern, B. Harvey, and D. Hendrie, Nucl. Phys.
       A122, 481 (1968);
       G. Chenevert, Ph.D. Thesis, University of Washington (1969).
10.   T. Ericson, Nucl. Phys. 8, 265 (1958);
       E. Gadioli, I. Iori, N. Molho, and L. Zetta, Nucl. Phys. A138,
       321 (1969).
11.   A. Gilbert and A.G.W. Cameron, Can. J. Phys. 43, 1446 (1965).
12.   D.B. Beard and A. McLellan, Phys. Rev. 131, 2664 (1963).
13.   K.K. Seth, R.M. Wilenzick, and T.A. Griffy, Phys. Lett. 11,
       308 (1964).
14.   B.L. Cohen et al., Phys. Rev. Lett. 26, 23 (1971).
15.   H.C. Britt and A.R. Quinton, Phys. Rev. 124, 877 (1961).
16.   I. Halpern, B.J. Shepherd and C.F. Williamson, Phys. Rev. 169,
       805 (1968).
17.   T.D. Thomas, Ann. Rev. Nucl. Sci. 18, 343 (1968);
       T.D. Thomas, Nucl. Phys. 53, 558 (1964).
18.   D. Sperber, Phys. Rev. 141, 927 (1966); 142, 578 (1966).
19.   C. Brun, B. Gatty, M. Lefort, and X. Tarrago, Nucl. Phys. 116,
       117 (1968).
20.   A.C. Douglas and N. MacDonald, Nucl. Phys. 13, 382 (1959).
21.   J.R. Grover and J. Gilat, Phys. Rev. 157, 802,814,832 (1967);
       J.R. Grover, Phys. Rev. 127, 2142 (1962).
22.   C.R. Gruhn, Ph.D. Thesis, University of Washington (1961).
23.   S.M. Ferguson, Ph.D. Thesis, University of Washington (1969).
24.   J.O. Newton et al., Nucl. Phys. A141, 631 (1970).
25.   F. Bertrand and R. Peele, Oak Ridge Nat. Lab. Rep. 4450,55,56,
       60 and 69.
26.   J.J. Griffin, Phys. Rev. Lett. 17, 478 (1966); Phys. Lett. 24B,
       5 (1967).
27.   M. Blann, Phys. Rev. Lett. 21, 1357 (1968).
28.   F.C. Williams, Jr., Phys. Lett. 31B, 184 (1970).

DISCUSSION

MORETTO   (Berkeley)

I am very much interested in your comments about the washing out of the shell effects. Some calculations that I have performed on the basis of the shell model and pairing hamiltonian, show a very rapid washing out of the shell effects; this can be seen for instance in the disappearance of the deformation with excitation energy. Typically, there is no deformed nucleus that can stay deformed at an excitation energy of the order of 35 to 40 MeV. At first it seems that indeed the washing out of shell effects should occur when $T \sim hw_o$ or $T \sim 7MeV$. An accurate analysis of the problem shows that the washing out of the shell effect goes asymptotically as $\exp(- \pi^2 T/hw_o)$. There is this remarkable π^2 term that multiplies the temperature that makes the thing different. So, a 0.7 MeV termperature makes the shell effects decrease by a factor of e in agreement with my calculations.

HALPERN

I'd like to hear about those.

HILLMAN   (B.N.L.)

We've also observed the washing out of the shell effect, and I want to make a comment about another subject, the dependence of the inverse crossection on the volume, with the volume increasing with excitation energy. It may very well be that you don't observe any such effects because the level density is also dependent on the volume; the volume increases with excitation energy and the level density goes down with increasing volume.

HALPERN

I think it would be a great accident for that to happen over that entire range, that is,both of these are rapidly varying functions that we're talking about and it may be so. One has to be always suspicious.

EJIRI   (Osaka)

I'd like to ask you about the dependence of level density on the angular momentum, and some typical angular momentum transfer in the inelastic alpha scattering and also about the angular distribution of the α'.

HALPERN

You're talking about the direct component or the evaporation component?

EJIRI

Mostly the evaporation component.

HALPERN

The issue, when you have to worry about angular momentum is roughly the following: You worry about it when the amount of angular momentum you have in the nucleus is large compared to the equipartition amount. May be that doesn't help experimentally. What the experimentalist needs to know is that he has to stop using this but start using the more complicated formula J x J. That's when he begins to see some antisotropy. It turns out that in this particular case the angular distribution of the evaporation component measured over range of angles is quite isotropic. The other thing I could say is that the very same region was explored with heavy ions by Quintin over ten years ago. They too, found the Wiskopff formula worked very nicely. Things were not happening in a very dramatic way with angles. The reason really is that you don't have that much angular momentum per nucleon.

COHEN (Pittsburgh)

On this business of subtracting a direct reaction, I think probably at these high energies, it's not so dangerous, but at the 20 MeV region, where a lot of work has been done, we think we have a method that really separates direct reactions from compound nucleus. We have found that conventional methods for (p,p') are very dangerous. For example, direct reactions give many low energy protons, and their angular distributions are rather isotropic.

NEWSTEAD

Do you have any evidence for semi-direct interactions?

HALPERN

You mean for the intermediate structure in inelastic alpha particle spectrum, or what?

NEWSTEAD

What I really have in mind is an example where an alpha particle would be ejected by a secondary neutron, the alpha parti-cle perhaps having pre-existed in the nuclear surface. In modern terms we would probably call it a quartet.

HALPERN

You're talking really then, about the direct component. Isn't that right? I would be looking for effects like that in the direct component.

NEWSTEAD

Yes, it would be in the direct component but it would also be a semi-indirect process.

HALPERN

No. The direct component doesn't have all that much structure in it. What structure there is we've been having some difficulty

interpreting.   I'd say no, there's no evidence for those two-step
processes or anything like that.

NEWSTEAD
     One further comment: How does the alpha particle excitation
function follow, say the neutron emission excitation function?
Is there a constant ratio between the two?

HALPERN
     No.   What happens is you don't get much alpha particle emission
until you really heat up the nucleus, quite a bit as far as evaporation
is concerned.   Of course, in these heavy nuclei, what you don't get
in the way of charged particles has to come out in neutrons.   So
everything that isn't alpha particle actually and is part of the
reaction process, ends up coming out as neutrons.

YIELDS OF K ISOMER, BETA AND GROUND

BANDS IN Yb $(\alpha,\text{xn } \gamma)$ ^{174}Hf REACTIONS

H.EJIRI[+][*], G.B.HAGEMANN[*], and T.HAMMER[*]

+ Dept. of Physics, Osaka-Univ, Toyonaka, Osaka, JAPAN

*The Niels Bohr Institute, Univ. Copenhagen, Copenhagen

The purpose of the present work is to study a role of K-quantum number (projection of spin I on the symmetry axis) in de-excitation of high spin compound nuclei in deformed region. If K-quantum number may still be a moderately good quantum number in the excitation region above the energy gap $(2\Delta \approx 1.5$ MeV), gamma decays may be affected due to possible K-selection rules. Consequently populations of low-lying levels in $(\alpha,\text{xn } \gamma)$ reactions on deformed nuclei may be somewhat different from what one would expect on a conventional statistical model, where K selection rule is not explicitly taken into account. In fact, a role of K-selection rules in the $(\alpha,\text{xn } \gamma)$ reactions has been suggested to explain anomalously large damping of the angular momentum in the gamma de-excitations.[1~6]

We investigated the populations of the levels with same spin and parity, but with different K, in the reactions ^{171}Yb$(\alpha,\text{n})^{174}$Hf and ^{172}Yb$(\alpha,2\text{n})^{174}$Hf at Eα=20~28 MeV. The ^{174}Hf is very suitable for this purpose since there are three low-lying rotational bands based on the K-isomer ($K^{\pi}=6^{+}$), β vibration $(K^{\pi}=0^{+})$ and ground $(K^{\pi}=0^{+})$ states.[7] The beta and the K isomer bands are located at about the same excitation energy, but their K-quantum numbers are quite different (K=0 and 6).

The α beam was provided from the super FN Tandem Accelerator at the Niels Bohr Institute. The gamma rays following the (α,xn) reactions were observed by a 40cc Ge(Li) detector at θ_{lab}=55 deg, where P$_2$ $(\cos\theta)$=0.

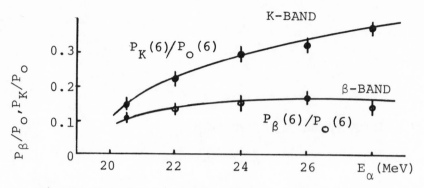

Fig.1. Relative populations of the K isomer (K=6)
and beta (K=0) bands to those of the K=0 (β and the
ground) bands

Targets used were self supporting metalic foils of about
1.5 mg/cm^2. Gamma lines from the levels in the K, β
and ground bands were well observed. We obtained from
these gamma intensities the populations of the levels in
question. Here we introduce the following notations,

  $P(I^\lambda)$: Population of the level with spin I in the λ
       band (λ=K,β,g), excluding the contributions
       from other levels in λ=K, β and ground bands
  $P_\lambda(I) = \sum_{I' \geq I} P(I^\lambda)$: Population of the levels with $I' \geq I$ in
       the λ band (the population of the λ band).
  $P_o(I) = P_\beta(I) + P_g(I)$: Population of the K=0 bands

The ratios of the observed populations, $P_K(6)$ $(P_\beta(6))$,
for the K (β) band to those of the K=0(β and ground)
bands, $P_o(6)$, (all above I=6) are shown in Fig.1.

i) K=0 BANDS: The relative populations of the β-band
increase very slowly with increase of E_α. For large E_α,
they are nearly constant, namely $P_\beta(6)/P_0(6) \approx 0.15$. The
value for the 20 MeV (α,n) reaction is about the same
as those for the 24~28 MeV (α,2n) reactions. Here the
mean excitation energies E_x of the residual nuclei after
neutron evaporation in both the (α n) and (α 2n) reactions are about the same. This means that the relative
populations of the β band to the ground band increase
slowly with increase of the mean excitation energy E_x
at lower E_x (≤3 MeV) and become constant at higher E_x.
This weak energy dependence may be due to the fact that
the β band lies somewhat higher than the ground band (E^β
$-E^g \approx 0.7$ MeV). The value $P_\beta/(P_\beta + P_g) \approx 0.15$ suggests that
the β and ground bands are populated statistically from
levels lying higher than the β band by $\delta \approx 1$ or 1.5 MeV,
assuming dipole or quadrupole transitions.

ii) K-ISOMER BAND: The relative populations, P_K (6), of
the K-band increase much more rapidly than the K=0 bands as the
E_α increases. The ratio P_K/P_O for the 20 MeV (α,n)
reaction   with   higher   excitation of the residual nucleus is smaller than those for the 24~26 MeV (α 2n) reactions with larger angular momentum input. Therefore
the K band populations increase with increase of the input angular momentum. The absolute values of P_K/P_0 are
much larger than those of P_β/P_0 , although the K(K=6) and
β (K=0) bands lie at about the same excitation energy.

These outstanding behavioursof the K-isomer band in
comparison with the K=0 bands may be qualitatively explained in terms of the K effects. The populations of
the levels with large K increase with increasing input
angular momentum since the spin I≥K. The levels with
large K then feed preferentially levels with large K,
finally populating the levels in the K-isomer bands, instead of decaying down to the low-lying levels with
small K (ground, β,γ bands,etc.).

We now try to understand quantitatively the K populations by taking explicitly into account the K quantum
number and the K selection rules. For simplicity we
write the populations as
$$P(I^\lambda) = P_i(I)\ B_\lambda(I)\ f_\lambda(E_x),$$
where $P_i(I)$ is the spin distribution at the initial stage
of the gamma de-excitation after neutron evaporation,
and $B_\lambda(I)$ is the effective branching ratio of gamma decays populating finally the levels in the λ band. (note
$B_K(I)+B_\beta(I)+B_g(I)=B_K(I)+B_0(I)=1$) The weak dependence on
the residual excitation energy E_x after neutron evaporation is given by $f_\lambda(E_x)$.

The initial spin distributions $P_i(I)$ are obtained by
applying an optical model for the formation of the compound nuclei and a statistical model for the neutron
evaporation[8]. The ratios f_β/f_g were obtained so as to
reproduce the β-ground ratio $P_\beta(6)/P_g(6)$. The value
$f_\lambda(E_x)$ is almost unity for higher excitation E_x ($E_x>3$ MeV),
and the values $f_K(E_x)$ and $f_\beta(E_x)$ become a little smaller
than the value $f_g(E_x)$ for lower E_x since the K and β
band lie higher by ≈0.8 MeV than the ground band. For
the K-isomer band, one may assume $f_K=f_\beta$ because the K
and β bands lie at about the same excitation region.
Then we can get the effective branching ratio of the K-
isomer band, b_K, from the observed ratio P_K/P_0. This is
$$b_K \equiv (\sum_{I\geq K} P_i(I)B_K(I))/\sum_{I\geq K} P_i(I) = P_K(P_K+P_0 f_K/f_g)^{-1}$$
The values b_K are shown as a function of the mean

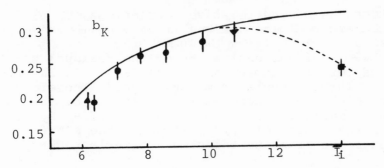

Fig.2. The branching ratios for the K-isomer band. Experiments: ▲αn, ♦α2n, ♦α3 n[5] ▼ O[18] 4n[10] The solid line:pre-
sent calculation. The dotted line is calculated assuming $B_K(I)=0$ for $I \geqslant 17$.

angular momentum $\bar{I}_i = \Sigma I P_i(I)/\Sigma P_i(I)$ at the initial stage
of the gamma de-excitation. The K band branching ratios,
b_K, really increase as the initial $\bar{I}_i$ increase.

     Now let us calculate the K-branching by taking ex-
plicitly into account the K-selection rules in the two
quasi-particle region $(2\Delta<E<4\Delta)$. We assume the following
de-excitation process.[4,5]

   i) Levels with spin I lying more than 2.6 MeV above
the ground band level I^{δ} feed uniformly (statistically)
lower levels in two quasi-particle bands. (Note this
energy $E(I)+2.6$ MeV ($2.9 \sim 3.2$ MeV for $I=4 \sim 6$) corresponds
to the boundaries of the two and four quasi particle regions.) Therefore the K populations in this two quasi-
particle region are simply proportional to the number of
levels with given K (K distribution of the levels in two
quasi-particle region). They are obtained from the single particle energies ε_i given in the Nillson diagram and
the energy gap $2\Delta=1.5$ MeV. Actually the K distribution
is rather uniform and insensitive to the assumed energy
boundary.
   ii) The levels with $K \geqslant 6$ in the two quasi-particle region populate finally the levels in the K-isomer band.
On the other hand the levels with $K \leqslant 5$ contribute to the
β and ground (K=0) band populations.

     The calculated values for b_K agree well with the
experimental b_K as shown by the solid line in the Fig.2.
The b_K increase as increase of $\bar{I}_i$ in the region $\bar{I}_i \lesssim 11$
since $P_i(I)$ with $I \geqslant K=6$ increase there. They become constant in the higher spins $I_i \geqslant 11$ since the K in the two-
quasi-particle region is limited to $K \leqslant 11$.

In short the populations of the K-isomer band in the (α, xn) reactions indicate an important role of some non-statistical de-excitation process, which depends on the K quantum number. They are well explained by introducing the K selection rules in the two quasi-particle region, supporting the basic idea of the previously proposed model.[4,5]

The K quantum number is considered to be moderately good in the two-quasi-particle bands with the present spin region $I \lesssim 16$, since the K mixing interaction $<H_c>$ (Coriolis force) is still smaller than the energy separation of the two interacting levels. However K will become indefinite for higher spin $I \gtrsim 17$ since $|<H_c>|^2 \propto (I^2 - K^2)$. Actually the K branching ratios b_K obtained from the 41 MeV $(\alpha, xn \gamma)$ reactions[5] with $\bar{I}_i \simeq 14$ are smaller than the calculation based on the present K selections $(B_K(I))$, and no strong populations for K_9-isomers have been found in the heavy ion xn reactions[9] with $\bar{I}_i = 30 \sim 50$.

The K selection rules make the high spin populations in the ground band much smaller than what they would be on a conventional statistical decay.[1~6] This is so because high spin levels with large K do not feed high spin levels in the ground band. They decay down along the bands with large K and finally feed the low spin levels $(4_g, 6_g)$ in the ground band through the isomer.

We have found similar results on the populations of the K, β, and ground bands in ^{176}Hf, too. Recently isomer ratios[11] in ^{182}Os have been analyzed on the similar method like the present one[4,5].

1. B.J.Shepherd, et. al. Phys. Rev. Letters 17, 806.
2. C.W.Williamson, et. al. Phys. Rev. 174, (1968) 1544.
3. I.Halpern, et. al. Phys. Rev. 169 (1968) 805.
4. H.Ejiri and I.Halpern, Bull. Am. Phys. Soc. 13 (1968) 700; Univ. Wash., Annual Report (1968).
5. I.Halpern and H.Ejiri, Joutsa, Symposium, Finland (1970); S.M.Ferguson, Thesis, Univ. of Wash. (1969).
6. C.D.Kavaloski, et. al, Nucl. Phys. A124 (1969) 401.
7. H.Ejiri and G.B.Hagemann, Nucl. Phys. A161 (1971) 449; J.Borgreen et. al, Nucl. Phys. A96 (1967) 561, H.Ryde, et. al, private communication (1970).
8. T.Ericson and V.Strutinski, Nucl. Phys. 8 (1958) 284; J.H.Jett and D.A.Lind, Nucl. Phys. A155 (1970) 182.
9. J.O.Newton, et. al, Nucl. Phys. A141, 631 (1970).
10. G.B.Hagemann, Private communication (1971).
11. M.Ishihara et. al, private communication (1971).

PRODUCTS OF THE REACTION ^{12}C + ^{209}Bi

J.C. Bell, I.S. Grant, K. Gregory and R.J. Williams

University of Manchester, U.K.

ABSTRACT

Excitation functions for the production of $^{114-118}$Ac in the ^{209}Bi (^{12}C,xn) reaction have been measured at bombarding energies from the Coulomb barrier to 115 MeV. Calculations taking fission and grazing reactions into account give cross-sections for the spallation products in satisfactory agreement with experiment. Mass distributions of Nb, Cd and Sb fission fragments in this reaction support the UCD hypothesis. Measurements of the total energy carried away from fission fragments suggest that the initial spin of a typical fragment varies appreciably with bombarding energy.

When heavy ion beams are used to cause fission in compound nuclei with very large angular momenta, the fissioning nuclei are at sufficiently high excitation that we need not worry about shell structure influencing the fission process. Any effects which are observed as the angular momentum and excitation change can be regarded as characteristic of nuclear matter. Unfortunately in systems of very high angular momentum there are always other processes in competition with fission. The nature of the nuclei which do actually undergo fission, and their angular momentum distributions, are usually rather uncertain. This paper describes measurements on several different reactions in the same system, namely (^{12}C + ^{209}Bi). Because the reacting system is so complex, the measurements are very far from complete. Nevertheless they constrain the parameters used to describe the reactions, and so lead us to have more confidence in estimations of the properties of the fissioning species.

Spallation Reactions

The system ($^{12}C + ^{209}Bi$) was chosen because even at 115 MeV, the highest energy of the carbon beam from the Manchester University HILAC, there is no appreciable fission following incomplete momentum transfer [1]. (This was checked by measuring the angular correlation of fission fragments in the reaction plane). At this bombarding energy, the compound nucleus ^{221}Ac can emit as many as eight neutrons, and fission is multichance, since it may occur after the emission of one or more neutrons. As a marker sensitive to the fission/neutron competition, we chose to measure cross-sections for the production of successive actinium isotopes along the neutron evaporation chain. These isotopes are all α-active and all have short lifetimes. Since the α-energies are rather close together, the only way to differentiate the Ac activities is to stop the recoil compound nuclei in a thin catcher foil after they have left the target, and to observe the foil in situ. Accurate cross-sections cannot be derived directly from observed count-rates, because the proportion of recoil nuclei stopped in the foil depends on their poorly known energy loss near the end of the range, and their angular spread on emerging from the target. However, the 19 min activity of ^{212}Fr is produced over the whole range of bombarding energies from 115 MeV down to the Coulomb barrier. After irradiation of a target placed just in front of a catcher foil thick enough to stop all recoils, the yield of this long-lived activity can be found by removing the catcher foil and counting in 2π geometry.

The excitation function for ^{212}Fr is shown in figure 1, with the cross-section for the parent ^{216}Ac which has been measured relative to ^{212}Fr. At a bombarding energy of 80 MeV, ^{216}Ac is the most abundant actinium isotope, and at this energy its yield is almost the same as that of the daughter ^{212}Fr. At higher energies where there is no ^{216}Ac activity, ^{212}Fr is formed either by α-particle evaporation or by an incomplete momentum transfer reaction in which an α-particle fails to fuse with the target. The other Ac-Fr parent-daughter pairs have similar excitation functions, except for ^{218}Ac, which at its peak appears to have a yield about 25% higher than the daughter ^{214}Fr. This may indicate that there is an isomer of ^{218}Ac decaying to the high-spin ^{214m}Fr, but we cannot be sure of this from the excitation functions alone, since ^{214m}Fr is also produced from direct processes not passing through ^{218}Ac.

Another hint that ^{218}Ac possesses an isomer comes from the lifetime. After chopping the beam in 1μ sec bursts, the 9.208 MeV ^{218}Ac activity decays out of beam with a half-life of 1.96 ± 0.20 μ sec (with poor statistics because of the drastic reduction in the duty cycle of the pulsed machine). The isotope ^{218}Ac has not previously been observed as a product of neutron evaporation, but Borggreen et al [2] recently attributed a 9.205 MeV activity to the ^{218}Ac daughter of ^{222}Pa, and measured a half-life of 0.47 ± 0.04

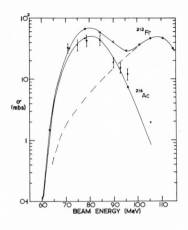

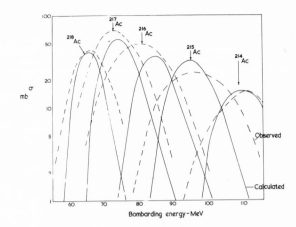

Fig. 1: ^{216}Ac and ^{212}Fr   Fig. 2: Comparison of observed and
yields   calculated yields

μsec. For the 9.60 MeV ^{217}Ac activity, another isotope which has
not previously been observed except as a daughter product following
the decay of ^{221}Pa, our lifetime of 0.44 ± 0.06 μsec is in good
agreement with the measurement of Borggreen et al.

Theoretical estimates of the spallation cross-sections have
been made using the method of Sikkeland[3]. Penetration factors
in the ingoing channel are found for each ℓ-value in the WKB approximation, taking a parabolic representation of the top of the
barrier presented by the Woods-Saxon potential which gives the best
fit to carbon-bismuth elastic scattering at 115 MeV. A fraction .03$A_I$of
the compound nuclear cross-section is assigned to grazing reactions,
which are then separated from compound nucleus formation with a
sharp angular momentum cut off. At 115 MeV compound nucleus for-
mation is cut-off at ℓ= 56, whereas $|\eta_\ell|$ is one-half at ℓ = 61.
The compound nucleus is assumed to decay by emission of neutrons at
a fixed temperature, with fission competition for each angular momentum evaluated at all stages of the spallation using Vandenbosch
and Huizenza's formula[4] for $\Gamma n/\Gamma f$. The calculated and observed
Ac cross-sections are compared in figure 2 using the value 0.45 for
the ratio of moments of inertia of the saddle-point and ground state
and with a neutron temperature rising from 1 to 1½ MeV over the
range of bombarding energies. In the figure the calculated values
have been multiplied by a factor 0.21; a discrepancy in magnitude
is expected, because the calculations have taken no account of proton and α -particle evaporation from the compound nucleus. Apart
from the normalizing factor, the excitation functions are quite
well reproduced over the whole energy range, suggesting that the
fission/neutron competition has been satisfactorily described.

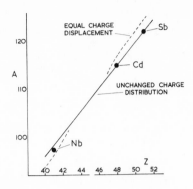

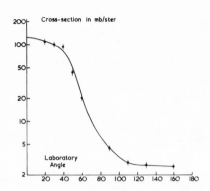

Fig. 3: α-particle cross-
         sections at 115 MeV

Fig. 4: Most probable masses
         of Nb, Cd and Sb

Cross-sections for Fission and Particle Emission

Fission, α-particle and proton cross-sections were measured at
115 MeV, identifying the particles by time-of-flight. The measured
fission cross-section is 1.64 ± 0.10 barns, in fortuitously good
agreement with the value of 1.642 barns calculated from Sikkeland's
recipe. The angular distribution of α-particles (figure 3) has an
isotropic component as well as the forward-oriented direct reaction
component. The total cross-section for α-particle evaporation is
40 mb - assuming the backward cross-section to be entirely due to
evaporation   leaving a 360 mb direct reaction cross-section. The
proton angular distribution varies all the way to backward angles,
and cannot be separated into direct and evaporation components: the
total proton cross-section is 220 ± 20 mb. The calculated grazing
cross-section is 557 mb, about the same as the sum (580 mb) of proton and direct α-cross-sections.

Neutron time-of-flight spectra have been observed, both singles
and in coincidence with fission fragments perpendicular to the beam.
In the reaction plane the average energy of the coincident neutrons
is 6.2 MeV: the kinematics is very complicated and the number and
temperature of post-fission neutrons has not yet been evaluated.

Mass Distributions of Separated Fragments

The mass distribution of the fragments has been found at a bombarding energy of 115 MeV from energy and flight-time measurements
using a flight path of 75 cm. The mean fragment mass is 106.5,
and the FWHM of the mass distribution is 31 mass units. Relative
mass yields have been found from chemically separated Nb, Cd and Sb

isotopes by γ-spectroscopy. (The antimony work was done in collaboration with A. Ahmed and V. Robinson of the Manchester University Chemistry Department). These elements were chosen because they
span the centre of the fragment distribution, and because each has
several isotopes for which an independent yield can be obtained.
The mean masses are plotted in figure 4 for the three elements, together with the predictions given by assuming either an unchanged
charge distribution (UCD) or equal charge displacement from stability (ECD) for the most probable atomic number of the primary fragments. To derive the predicted values one must know the fissioning
nucleus and the number of post-fission neutrons. According to the
calculations described above fission is important throughout the
neutron evaporation chain, but about half the fission occurs in the
compound nucleus ^{221}Ac and on average only 1.6 neutrons are emitted
before fission. The predicted values of most probable mass are not
very sensitive to whether the neutrons are emitted before or after
fission, and the lines in figure 4 are calculated assuming that all
fission occurs in ^{119}Ac. It has also been assumed that the number
of neutrons emitted by the fragments is proportional to the fragment mass. We have no direct information about this but have estimated that total energy carried away by evaporated particles by
measuring the angular correlation of fragments out of the reaction
plane. The evaporated particle energy is nearly independent of
mass ratio, though a high energy is associated with very asymmetric
fragments with a large kinetic energy. This is in agreement with
the work of Plasil and co-workers [5,6], who have measured pre- and
post-neutron emission mass distributions in (p,f) and (α,f) reactions at high energies. They find that the neutron emission tends
to be roughly proportional to mass at high excitation. Assuming
that this is also true in the (C + Bi) reaction, figure 4 shows that
the mass distributions are in agreement with the UCD hypothesis. It
is hard to imagine that the Cd and Sb isotopes emit more neutrons
than we have assumed, but if they should emit less - not unreasonable
for near closed-shell fragments - the disagreement with the ECD
hypothesis becomes even more marked. Yaffe [7] also favoured the
UCD hypothesis in energetic fission initiated by protons.

Prompt γ-radiation

In an attempt to learn something about the de-excitation of
the fragments after neutron emission, we have observed prompt
γ-spectra in coincidence with fission. A surface barrier fission
detector and a NaI (Tl) counter were set up perpendicular to each
other and to the beam. The fragment energy, the γ-pulse height
(above a discriminator level set at 200 KeV) and the time separation
were recorded for coincident events. Time spectra taken with the
γ-counter at different distances from the target indicated that
less than 5% of the events were caused by neutrons; the low background is obtained because of the peaking of energetic neutrons

along the fission fragment axis. No peaks corresponding to individual transitions were observed, and the smooth spectra were stripped assuming that the detector had a δ-function photopeak and a rectangular Compton background. The photopeak efficiencies and peak-to-Compton ratios were measured with a series of standard sources placed at the target position. The γ-energy per fission and the number of quanta per fission are given in the table for three different bombarding energies. There are several sources of uncertainty in these numbers. As well as the inadequacy of the stripping procedure, errors are introduced because of the uncertain neutron background, because of anisotropy of the radiation, and because almost 20% of the γ-energy is delayed by more than 5 nsec, and is emitted by fragments which have already reached the detector or the walls of the beam tube. The 10% accuracy quoted in the table is simply a reasonable guess at the result of combining these errors. At 115 MeV the energy from transitions above 200 KeV is 14 MeV,

Beam Energy	Observed spectrum		After correction for low energy γ's		Fragment spin
	Energy	Number	Energy	Number	
115 MeV	14.0 ± 1.4	11.3 ± 1.5	15.0	19.3	19
84	13.1 ± 1.3	10.9 ± 1.5	14.1	18.9	18
78	10.9 ± 1.1	9.1 ± 1.2	11.9	17.1	15

well above the prompt γ-energy of 6 ± 1 MeV per fission[8] observed in thermal neutron fission of ^{235}U. This indicates that the fragments initially have a high angular momentum, since neutron emission continues until the fragment is a few M_eV above the yrast line.

Grover[9] has calculated the energies of yrast levels in a number of nuclei of which ^{111}In is the one most nearly representative of fragments from (C + Bi) fission. To illustrate the magnitude of spins required to explain the observed γ-energies let us assume that a typical fragment has a $J(J \pm 1)$ yrast line which smoothly fits the ^{111}In energies, and that after neutron evaporation it is left with spin J at an excitation 3 MeV above the yrast level of spin J. After decaying to a yrast level near J, the fragment then cascades down to its ground state in a series of quadruple transitions. The last three or four of these transitions in each fragment are below the detector bias: suppose 8 transitions carrying a total energy of 1 MeV are lost. The total energy and number of quanta are then as given in columns 4 and 5 of the table. At 115 MeV, of the 7.5 MeV per fragment, 3 MeV is the excitation above the yrast line, implying a yrast level at 4.5 MeV. The smooth yrast line has this energy at J = 19, requiring a cascade of about 9 quanta from each fragment, in reasonable accord with the observed number. The fragment angular momenta estimated in the same way are given in the table for the other bombarding energies.

About 20% of the observed radiation is delayed by more than 5 nsec, and the average energy of the radiation arriving from 15 to 65 nsec after fission is 0.52 MeV at a beam energy of 115 MeV. For E2 transitions this corresponds to a hindrance factor of about 100 by comparison with the single particle rate. This is not unreasonable for the slow transitions, but does not rule out the possibility that they are dipole (hindrance about 10^8).

To sum up, the γ-spectrum emitted in (C + Bi) fission is consistent with the view that the fragments have a rather high initial spin and that they de-excite in a predominantly quadrupole cascade. We are currently checking this conclusion by measuring isomer ratios in antimony isotopes both in fission and in $Sn(\alpha,pn)$ reactions. The latter reactions can yield antimony nuclei in the angular momentum range up to $20\hbar$.

REFERENCES

1. T. Sikkeland and V.W. Viola, Proceedings of the Third Conference on Reactions between Complex Nuclei, University of California Press (1963).
2. J. Borggreen, K. Valli, and E.K. Hyde, Phys. Rev. 2C, 1841, (1970).
3. T. Sikkeland, Arkiv för Fysik 36 539 (1966).
4. J.R. Huizenza and R. Vandenbosch, Nuclear Reactions (eds. P.M. Endt and P.B. Smith), North-Holland (1962).
5. F. Plasil, R.L. Ferguson and H.W. Schmitt, IAEA Symposium on Physics and Chemistry of Fission, IAEA (1969).
6. S.C. Burnett, R.L. Ferguson, F. Plasil and H.W. Schmitt, Phys. Rev. 3C 2034 (1971).
7. L. Yaffe, IAEA Symposium on Physics and Chemistry of Fission, IAEA, (1969).
8. H. Maier Leibnitz, P. Armbruster and H.J. Specht, IAEA Symposium on Physics and Chemistry of Fission, IAEA (1965).
9. J.R. Grover, Phys. Rev. 157 832 (1967).

YRAST TRAPPING IN HEAVY ION COMPOUND NUCLEUS REACTIONS[†] **

Richard A. Gough* and Ronald D. Macfarlane

Department of Chemistry and Cyclotron Institute

Texas A&M University, College Station, Texas 77843

I. Introduction

A distinguishing feature of heavy-ion reactions is that large
ℓ-wave transfers have a significant probability because of the
large center-of-mass of the interacting system. When a compound
nucleus is formed, states of very high J are populated. In the
de-excitation sequences, nucleon emission proceeds as a statistical
process until the phase space is reduced by energy and J-dependent
effects. At this stage of de-excitation, α-particle and γ-ray
emission may favorably compete with nucleon evaporation at energies
greater than nucleon binding energies. The yrast curve, the locus
of J-states of lowest energy as a function of excitation energy,
represents the limit of the phase space. The slope of the yrast
curve can reflect specific properties of the ground state of the
nucleus such as a predominance of high-J quasi-particle configurations. Nuclei in the region of N=126 possess this property. It
is the purpose of this study to determine whether compound nucleus
reactions in this region are sensitive to the predicted high density of large J-states at low excitation energy.

** The paper was given by D. MacFarlane.

[†]Work supported by the U.S. Atomic Energy Commission and
Robert A. Welch Foundation.

*Present Address: Lawrence Radiation Laboratory, Berkeley,
California.

II. Experimental Details

The compound nucleus, ^{221}Ac, was studied via the reactions:
^{209}Bi + ^{12}C, ^{207}Pb + ^{14}N, and ^{205}Tl + ^{16}O. The Yale heavy-ion
accelerator was used in this work. Targets with thicknesses of
100-300 μg/cm^2 were bombarded with heavy ions ranging in energy
from 5-10 MeV/nucleon. Short-lived α-emitting nuclear reaction
products were detected by their characteristic α-decay properties.
The complexity of products is indicated by the α-spectrum shown
in Fig. 1. Absolute excitation functions were obtained using
targets containing known amounts of rare earth isotopes which
produced, simultaneously, ^{150}Dy α-activity which served as a
standard for obtaining absolute cross sections.[1] A representative
set of excitation function data for the reaction ^{209}Bi(^{12}C, xn)
221-xAc with x=3-8 is shown in Fig. 2.

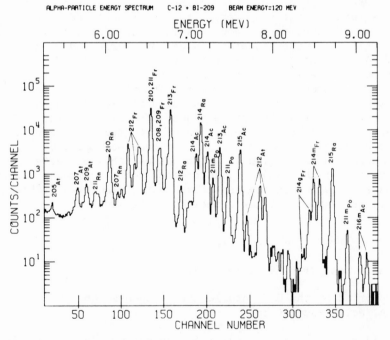

Figure 1 - Alpha spectrum of products from the reaction
^{209}Bi + ^{12}C at an energy of 120 MeV.

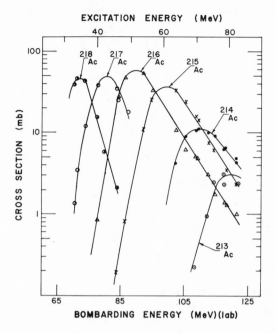

Figure 2 - Excitation functions for the reactions $^{209}Bi(^{12}C,$ xn) $^{221-x}Ac$.

III. Results and Discussion

For nuclei containing neutrons in excess of the N-126 closed shell, the $j_{15/2}$ and $i_{11/2}$ neutron states, and $i_{13/2}$ proton states make significant contributions to the density of states at moderate excitation energy(0-30 MeV). Yrast curves calculated using the Grover-Gilat program clearly show this effect.[2] Their calculation assumes that the neutrons and protons are non-interacting fermions(except for pairing) in an infinite spherical well. The BCS theory is used to calculate the pairing energy. The lowest energy configuration for a particular J is obtained and the results are used to calculate the yrast curve. The nature of the calculation automatically takes account of contributions from high-j valence nucleons such as exist in the region around N=126. Figure 3 shows a set of yrast curves calculated for $^{211-221}Ac$. The nuclei, ^{211}Ac, and ^{216}Ac, on the neutron-deficient side of N=126, have yrast curves similar to that observed for the rare earth region. On the neutron excess side, the shape of the yrast curves changes dramatically due to the effect of the $j_{15/2}$ and $i_{11/2}$ neutron states. Angular momentum states of 50ℏ are predicted at energies as low as 7 MeV for ^{221}Ac while the J=50ℏ yrast level is predicted to occur near 16 MeV for ^{211}Ac.

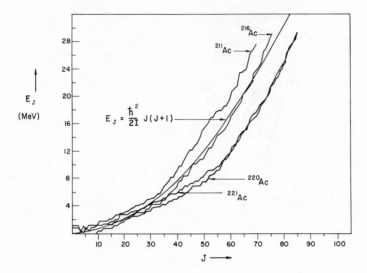

Figure 3 - Calculated yrast curves for Actinium nuclei in the
region of N=126.

An increase in the slope of the yrast curve as neutron evaporation proceeds from the compound nucleus has the effect of introducing an additional phase space restriction for neutron emission,
and can give rise to "yrast traps." This can result in increased
gamma-ray emission and can also give rise to fluctuations in the
widths of excitation functions. Both of these effects have been
observed in this work.

Figure 4 shows a plot of the average energy per evaporated
neutron $\bar{\epsilon}$ taken from the excitation function data. This number
can give an indication of the contribution of gamma-ray emission
to the de-excitation. Previous results have shown that the
kinetic energy spectrum of the neutrons emitted in heavy-ion
compound nucleus reactions in the heavy elements is approximately
that expected for a statistical decay process, giving rise to an
$\bar{\epsilon}_n \sim 2.7$ MeV. An excess may represent energy appearing as electromagnetic radiation. As shown in Figure 4, the limiting value of
2.7 MeV is approached at high initial excitation energy when up
to 9 neutrons are emitted. At these energies, most of the compound nuclei, particularly those with high-J, undergo fission so
that these results essentially represent the behavior of low-J
compound nuclei which are relatively insensitive to yrast level
effects. At lower excitation energies where 3 to 4 neutrons are
evaporated and the final state is on the neutron excess side of
N=126, $\bar{\epsilon}$-values increase to as high as 6 MeV. By contrast, results
obtained by Bimbot et al. in the same region but on the neutron-

deficient side of N=126 give consistent $\bar{\varepsilon}$ values of $\sim$ 3.2 MeV over the same range of excitation energies.[3] It is clear that there is an enhancement in the amount of γ-ray emission from compound nuclei in the region around N=126 and the effect of the neutron-dependent yrast level schemes seems to provide a logical explanation.

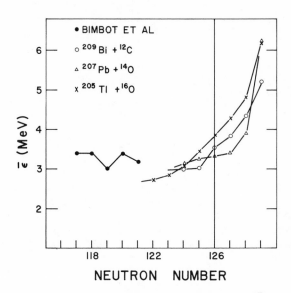

Fig. 4.   Average energy per evaporated neutron ($\bar{\varepsilon}$) as a function of neutron number for the region near the N=126 closed shell.

The other effect of the neutron-dependent yrast levels results in fluctuations in the widths of the excitation functions. Statistical model calculations were performed first assuming no effect of fission competition, and a constant yrast level locus at each stage of the evaporation. The width(FWHM) was found to be a smoothly-varying function of excitation energy, increasing in value at larger excitation energies where the J-distribution of compound nuclear states is the broadest. The effect of fission competition is to generally reduce the width because of the fractionation of high-J states. Incorporating the neutron-dependent yrast levels introduces fluctuates in the curve which are very similar to the experimental observations. These results are shown in Figure 5. The fluctuations may also have some contribution from odd-even effects in the level density. The extremes of the excitation functions reflect in a sensitive way, the competition between neutron and gamma ray emission in the final stages of de-excitation.

We are grateful to the staff of the Yale heavy-ion accelerator for their excellent support.

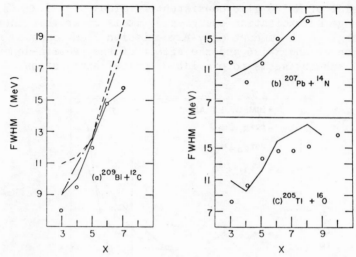

Fig. 5. Variation of excitation function widths with neutron emission
         multiplicity. Statistical model calculations for no fission
         competition, constant yrast line(---), fission competition,
         constant yrast line (—·—·) and fission competition, neutron-de-
         pendent yrast line (———) are included for comparison.

References

1. J. M. Alexander and G. N. Simonoff, Phys. Rev. 130, 2383(1963)

2. J. M. Grover and J. Gilat, Phys. Rev. 157, 802(1967)

3. R. Bimbot, M. Lefort, and A. Simon, J. Physique 29, 563(1968)

DISCUSSION

HILLMAN

I'm not sure that your energy is high enough but angular momentum can be carried off by emission of other particles besides γ -rays and fission reaction. At above 100 MeV evaporation of particles such as He^5 becomes important, even more important than He^4 because of the high spin of the ground state of the former.

GILAT (Israel)

With reference to the comment of Dr. Hillman, first of all I would like to say that probably just plain alpha emission from high angular momentum states plays a much more important role. Some of it is mentioned in the paper we have in the proceedings here that will be forthcoming shortly. As a matter of fact, at sufficiently high energies and spins, alpha particle emission replaces gamma emission as the most important mode of removing excess angular momentum. There may be evidence for this in the excitation functions, shown by Dr. Macfarlane, especially that of Ac^{213}. In addition to that, I would like to make a very general comment refering to two previous papers. We can now fit these excitation functions with statistical calculations much better than the fits shown by Dr. Grant; as a matter of fact, our calculated curves are usually within a factor of 2 and often much closer to the experimental ones

MORETTO (Berkeley)

I would like to make a comment. You indicated that the region where a nucleus is paired extends up to 2 MeV, usually calculations indicate something quite different, namely, it may happen that for nuclei very close to magic number, pairing does not extend much above 2 MeV, but for ordinary nuclei of that size, pairing correlation extends from 4 to 10 MeV. My second point regards the yrast line; again, for a specific case the nucleus will remain spherical along the yrast line, but in general one ought to make sure that the information is not changing with angular momentum.

WIGNER

I have a very down to earth question. The neutrons from the fission process have a much lower energy (about 1 MeV) than the neutrons emitted in the reactions you discussed. Could you tell us what the principal reason is for the difference.

MACFARLANE

The E values aren't necessarily the kinetic energy of the neutrons, but they, since all we have is information from the excitation function data, also include energy involved in electromagnetic radiation. The average kinetic energy for heavy-ion reactions is about 2 or 2.5 MeV.

WIGNER

    Why is that so much higher than in the case of this?

HUIZENGA   (Rochester)

    That's a very complex question. The dependence of the energy of fission neutrons on angular momentum is weak. However, the average energy of fission gamma rays is less than the average energy of neutron capture gamma rays.

THE ROLE OF ANGULAR MOMENTUM AND ISOBARIC SPIN IN TESTS OF THE INDEPENDENCE HYPOTHESIS[*]

J. M. Miller

Department of Chemistry, Columbia University

INTRODUCTION

This paper will be concerned with compound-nucleus reactions that lead to residual nuclei excited into the continuum. Thus, not only will direct processes be ignored, but scant attention will be given to compound-nucleus reactions leading directly to discrete states of residual nuclei.

Any attempt to analyze a nuclear reaction within the context of the compound-nucleus model clearly requires assurance that one is indeed dealing with a compound-nucleus reaction. Such assurance may be obtained from four different kinds of more or less satisfactory criteria:
1) Do the measured cross sections conform to the quantitative predictions of the compound-nucleus model? In the past, this criterion has occasionally amounted to a self-fulfilling (or, from time to time a self-unfulfilling) prophecy.
2) Is the angular distribution of the emitted particles symmetric about a plane that is perpendicular to the incident beam?
3) Does the excitation function exhibit fluctucations of the type described by Ericson?
4) Is the probability that the reaction populates a given exit channel the same for <u>different</u> entrance channels that give compound systems with the <u>same</u> constants of the motion? In other words, do the reactions conform to the "Independence Hypothesis" of Bohr?

In this paper, attention will mainly be focused on the
first and fourth of these criteria.

STATISTICAL-MODEL CALCULATIONS

The heart of the compound-nucleus model entails the
idea that the reaction may be divided into two independent steps: the formation of the compound nucleus with
a given set of values for the constants of the motion
followed by the decay of the compound nucleus in a manner determined only by that set of the constants of the
motion.   Thus, the cross section for the emission of a
particle b with kinetic energy ε at an angle Θ in a compound-nucleus reaction from entrance channel i may be
expressed in compact form as:

$$d^2\sigma(i,b)/d\varepsilon d\Omega = \sum_{J,M,T} \sigma_i(U,J,M,T) \; W(b,\varepsilon,\Theta|U,J,M,T) \tag{1}$$

The first factor in each term of the summation is the
cross section for forming the compound nucleus with excitation energy U, spin J, Z-component of spin M, and
isobaric spin T.   The isobaric spin is introduced among
the constants of the motion in anticipation of matters
that will be discussed later on; parity will be suppressed because it will be assumed that at the excitation
energies considered here, there are essentially equal
numbers of states of each parity.   The second factor in
each term is the probability that a compound nucleus
with the given set of quantum numbers decays into the
exit channel of interest.   The explicit formalism for
evaluating equation (1),effectively under the assumption
that T is not a constant of the motion, has been described and discussed in detail by Douglas and MacDonald[1]
and by Thomas[2].   Recently, Liggett and Sperber[3] have
described a method for the evaluation of a counterpart
equation 1 that is particularly useful when more than
one particle is emitted.

While the two independent steps of the compound-
nucleus model are explicitly seen in the two factors in
each term in the summation, it is clear that the spectrum of values of J, M, and T as well as the dependence
of the decay probability upon these quantities make it,
in general, impossible to factor equation (1) into two
terms: a total cross section for compound-nucleus formation multiplied by a decay probability.

If equation (1) is integrated over angle, the resulting expression must be independent of any polarization of the compound nucleus and thus does not depend
upon M:

$$d\sigma(i,b)/d\varepsilon = \sum_{J,T} \sigma_i(U,J,T) \ W(b,\varepsilon \ |U,J,T) \qquad (2)$$

Although $W(b,\varepsilon \ |U,J,T)$ is considerably less sensitive to J than is $W(b,\varepsilon,\Theta|U,J,M,T)$, it is still, in general, a dubious approximation to factor equation (2) into two terms. Fig. 1a taken from Reedy et al[4] indicates how the calculated average kinetic energy depends on the spin of the compound nucleus. It can be seen from this figure that the main J-dependence occurs for compound nuclei with large J and that the dependence is complex: the average energy of emitted nucleons tends to decrease while that for alpha particles tends to increase with increasing J. This behavior stems from a complicated interplay between the two quantities that dominate $W(b,\varepsilon,| \ U,J,T)$: the transmission coefficients for the emitted particles which depend on their kinetic energy and orbital angular-momentum, and the spin-and-excitation energy dependence of the level density of the residual excited nuclei.

If Equation (2) is integrated over ε, there remains an expression giving the cross section for the emission of particle b:

$$o(i,b) = \sum_{J,T} \sigma_i(U,J,T) \ W(b|U,J,T) \qquad (3)$$

Fig. 1b also taken from Reedy et al[4] illustrates the significant dependence of the probability for the emission of a particular type of particle on the spin of the compound nucleus. While the behavior is again seen to be rather complex; in general, the fact that the centrifugal barrier diminishes with increasing mass of the emitted particle coupled with the spin dependence of the level density results in the enhancement of the emission of heavy particles with increasing spin of the compound nucleus. As may be seen from the study by Grover and Gilat[5], as exhibited in Fig. 2, the behavior can become quite dramatic as the spin of the compound nucleus approaches the highest spin that is available at that excitation energy - the yrast level. The rapid decline in the probability of alpha-particle emission after the abrupt increase reflects the decisive importance of gamma-emission when angular momentum conservation severly restricts the number of channels open to particle emission.

Since nuclear reactions induced by nucleons and alpha particles do not impart very large angular momenta to the compound nuclei that are formed, it is often, but not always, a reasonable approximation for these re-

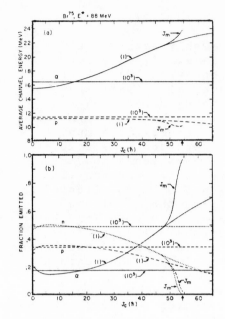

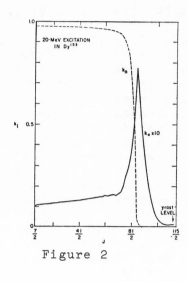

Figure 2

Figure 1

1. (a) Average channel energies of emitted protons and
   alpha particles, and (b) probabilities for the emission of neutrons, protons, and alpha particles, calculated for the first particle emitted from a Br^{75}
   compound nucleus at 88 MeV as a function of spin of
   the compound nucleus. The curves marked (1) took the
   moment of inertia as that of a rigid body; those
   marked (10^3) took it 10^3 times larger. The branches
   marked J_m include the effects of yrast levels. Figure
   from reference 4.

2. The probability of emitting neutrons (dashed curve)
   and alpha particles (solid curve) from the compound
   nucleus Dy^{153} excited to 20 MeV as a function of
   angular momentum. Figure from reference 5.

----

actions to take an average value for the slowly varying
$W(b|U,J,T)$ which, if we also suppress the effect of T,
leads to the classical expression:

$$\sigma(i,b) = <W(b|U,J)> \sum_J \sigma_i(U,J) = \sigma_i(U)W(b|U) \qquad (4)$$

Since heavy-ion accelerators were relatively rare until
recently, studies of compound-nucleus reactions usually
entailed projectiles no heavier than alpha particles
and thus Equation 4 was usually employed. For the same

reason, analyses of energy spectra usually involved a
version of Equation 2 in which the J-dependence of the
emission-probability term was ignored with, it should
be added, occasional unsettling consequences.

The foregoing discussion has been concerned with
the emission of a single particle from a compound nucleus at excitation energy U; in many experiments the excitation energy is high enough so that several particles
are emitted sequentially in the de-excitation of the
compound nucleus and thus the appropriate one of
Equation 1-4 must be employed at each step in the sequence. This can become a formidable task when it is
realized that compound nuclei that are characterized by
the same set of quantum numbers will, after the emission
of the first particle, generate intermediate compound
nuclei with a spectrum of quantum numbers and that each
member of the spectrum will in turn generate a new spectrum after the emission of the second particle, etc.
This complexity has encouraged efforts to reduce the dimensionality of the problem by ignoring angular-momentum
effects largely by effectively assigning the compound
nucleus a very large moment of inertia as indicated by
the curves labelled (10^3) in Fig. 1. In addition, the
model carries the implication that the emission of each
particle in the sequential decay of a compound nucleus
is independent of the emission of each preceding particle except insofar as they alter the constants of
motion of the initial compound nucleus. This property
of the de-excitation process immediately suggests the
usefulness of the Monte Carlo technique as a means of
calculating the consequences of the sequential decay.

An extensive calculation employing the Monte Carlo
technique and suppressing the effects of angular-momentum was carried out by Dostrovsky, Fraenkel and Friedlander[6] on a fast computer in which they computed excitation functions for the formation of various product
nuclei in essentially all of the nuclear reactions that
had been studied up to that time. The decisive quantity
in a calculation of this type which suppresses angular
momentum effects is the probability per unit time and
energy that the compound nucleus at excitation energy U
emits particle b with kinetic energy ε

$$I(\varepsilon,b) \sim \varepsilon\sigma_{inv}(\varepsilon,b) \; \rho_B(U-\varepsilon-S_b) \qquad\qquad (5)$$

where $\sigma_{inv}(\sigma,b)$ is the cross section for the inverse
of the emission process and $\rho_B(U-\varepsilon-S_b)$ is the density of
energy states in the residual nucleus as S_b is the
binding energy of particle b. The state density was expressed in the usual approximate form

$$\rho_B(U-\varepsilon-S_b) \sim e^{2a^{1/2}(U-\varepsilon-S_b-\delta)^{1/2}} \tag{6}$$

where $\underline{a}$ is the standard level-density parameter that is usually taken as proportional to the mass number of the nucleus and δ is related to the "pairing energy" and adjusts the level density for the odd or even character of the neutron and proton number as well as other effects which tend to make the ground state of the nucleus in question unusually stable. The parameter $\underline{a}$ in this expression is decisive in determining the shape of the energy spectrum of the emitted particles while the relative probabilities for the emission of various types of particles are most sensitive to $\underline{\delta}$. The latter quantity is of great importance and its role merits some explication.

The Coulomb barrier, manifesting itself through the inverse cross section in Equation 5, tends to inhibit the emission of charged particles in compound-nucleus reactions. Accordingly, it became a commonplace that neutron emission should predominate in these reactions and any significant emission of charged particles was taken as evidence for non-compound processes. This view held sway despite the fact that the work of S.N. Ghoshal[7], which will be discussed in some detail later, had graphically shown at the beginning of the 1950's that it was certainly not universally correct. In particular, the observation that the cross section for the (p,pn) reaction with targets of mass number in the vicinity of forty-five to seventy was often as large as or even larger than that for the (p,2n) reaction was taken as evidence for either the direct ejection of the neutron by the incident proton or the direct inelastic scattering of the proton followed by the evaporation of a neutron. The significance of the fact that in all instances of unusually large cross sections for the (p,pn) reaction the product formed was an odd-odd nucleus, was illustrated, for example, in the observation by Houck and Miller[8] that the cross section for the $Fe^{54}(\alpha,pn)Co^{56}$ reaction was some sixty times larger than that for the $Fe^{54}(\alpha,2n)Ni^{56}$ reaction at the peak of their excitation functions. This result had been ascribed by them to the relatively low density of energy levels in the even-even and doubly magic product Ni^{56} as compared to the odd-odd Co^{56} rather than to non-compound processes. This, and many other such sets of data, could be encompassed within the compound-nucleus model[6] by the assingment of a $\underline{\delta}$ to each even neutron and proton number.

On the whole, fairly good agreement between observed

and calculated excitation functions were achieved with
the use of Equations 5 and 6.   There was also, however,
a serious and systematic disagreement that often appeared:
excitation functions for a particular reaction did not
diminish rapidly enough with increasing energy beyond
that at the peak.   An example of this divergence may be
seen in Figure 3 where a comparison is shown between a
calculated and observed[9] excitation function for the
La^{139} (α,n) Pr^{142} reaction.   An implication of this
discrepancy is that particles are emitted with greater
kinetic energy than that expected from this simple
version of the statistical model thereby more often
leaving residual nuclei in bound states.   This explanation immediately implies that these higher-energy particles are produced in some non-compound process.   As
has been seen in Fig. 1a, though, the proper incorporation of angular-momentum conservation in the statistical
model can also have a significant effect on the kinetic
energy of the emitted particles.   In addition, Fig. 2
illustrates the point that particle emission, although
energetically allowed, can become negligible compared
to gamma decay for excited nuclei with spins approaching
that of the yrast level and this too could give rise to
the observed discrepancies.   Accordingly, the limitations
of Equation 4 and the quantitative effects of angular-
momentum conservation required investigation.
     The discussion thus far has emphasized the first of

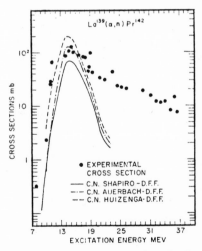

Figure 3.   Comparison between calculated and experimental excitation functions for the La^{139}(α,n) Pr^{142}
reactions.   Figure from reference 9.

the four criteria: comparison between calculated and
experimental cross sections. For the investigation of
the role of angular-momentum conservation, and, to
anticipate, that of isobaric spin, it is useful to
sketch briefly the mode of experimental tests of the
Independence Hypothesis.

THE INDEPENDENCE HYPOTHESIS

The Independence Hypothesis for compound nucleus
reactions in the continuum reflects the hope that an
expression such as Equation 1 is a useful one and, in
particular, that $W(b,\varepsilon,\Theta|U,J,M,T)$ does not explicitly
depend on the entrance channel i. For low-energy reactions that proceed through isolated resonances, there
is no problem. For higher-energy reactions that entail
a large number of overlapping states, the Independence
Hypothesis rests on the consequences of random phases
for these many states in a manner than has received considerable theoretical investigation[10]. We shall not
here pursue the theoretical questions, but instead, shall
look into the experimental investigation of the Independence Hypothesis.

What is required, of course, is to prepare identical
compound systems from at least two different entrance
channels, i and i', then to see if the identical compound systems formed in these two different ways give
the same relative probability for decaying into the
various open exit channels. The compound system is
meant to be specified only by the constants of the
motion Z,N,U,J,M, and T. There is little difficulty in
finding at least two different entrance channels that
will give compound systems with the same Z,N,U, and T.
There is in general, however, no way of preparing compound systems with a given J and M (except for bombarding energies low enough so that s-waves are the only
ones involved) and thus the only hope is to make compound systems from different entrance channels with the
same Z,N,U, and T, and with the same distribution of J
and M. As a further simplification, if the exit channel
does not depend upon the observation of an angle, the
distribution of M may be suppressed. What this means,
then, is that if there are two distinguishable entrance
channels, i and i', such that
$$\sigma_i(U,J,M,T) = Const.\sigma_i'(U,J,M,T) \tag{7}$$
for all values of J,M and T at a given value of U, and
if b and b' are two distinguishable exit channels, the
Independence Hypothesis requires that

$$\sigma(i,b)/\sigma(i,b') = \sigma(i',b/\sigma(i',b') \qquad (8)$$

The first investigation of this sort was carried out by S.N. Ghoshal[7] in which, however, the distribution of J and T were not the same for the two entrance channels that were employed. In particular, Ghoshal investigated the behavior of the Zn^{64} compound system by measuring the excitation functions for the (x,n), (x,2n) and (x,pn) reactions where x was a proton with a Cu^{63} target and an alpha particle with a Ni^{60} target. A verification of the Independence Hypothesis that may be derived from his results is shown in Fig. 4 where it is seen that the relative probability of decay of the compound nucleus into these two exit channels is essentially the same for the two entrance channels despite the differing J and T distributions.

Not all subsequent experiments showed such clear-cut results. An example is shown in Figure 5 taken

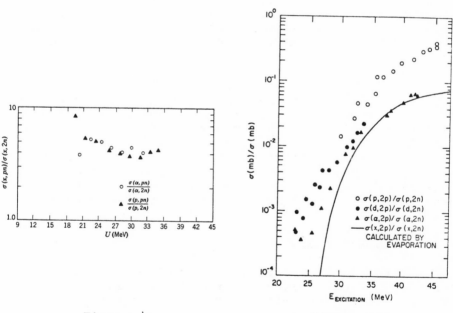

Figure 4   Figure 5

Figure 4. Comparison of the behavior of an excited Zn^{64} compound nucleus made in two different ways: p + Cu^{63} and α + Ni^{60}. Data are from Reference 7.

Figure 5. Ratio of cross sections for (x,2p), and (z,2n) reactions as a function of excitation energy where x = p,d, and α, for Ni^{62}, Ni^{61}, and Co^{59} targets, respectively. Figure from Reference 11.

from the work of Stearns and Moorhead[11] in which is
plotted the ratio of the cross sections for the (x,2p)
and x,2n) reactions as a function of the excitation
energy of the Cu^{63} compound systems where x is a proton
with a Ni^{62} target, a deuteron with Ni^{61}, and an alpha
particle with Co^{59}. The curve in the figure was calculated by the method of reference 6 and thus effectively
ignored the effects of angular-momentum conservation on
these reactions despite the fact that the spin distribution in the three different entrance channels were
quite different. From the figure it is clear that in
this instance the mode of decay of the compound system
is not independent from its mode of formation; what is
less clear is whether this is a consequence of non-compound processes or of the divergence among the distributions of T and J in the three entrance channels.

ROLE OF ANGULAR-MOMENTUM CONSERVATION

The effect of not satisfying the condition expressed
in Equation 7 on the requirement of Equation 8 was carefully investigated by Grover and Nagle[12]. These authors
measured the excitation functions for the (x,n) and
(x,2n) reactions in the vicinity of the threshold of the
latter reaction where x was a proton with a Bi^{209} target
and an alpha particle with a Pb^{206} target. The result of
this study, as presented in Fig. 6, shows a clear divergence from Equation 8. The important point, though, is
that through the sequential use of equations similar to
(2) and (3), these authors showed that in this instance
the divergence could be entirely explained by the difference in the angular-momentum distribution in the two
entrance channels and thus reinstated the Independence
Hypothesis.

A recent experiment by D'Auria et al[13] more directly examined the validity of Equation 8 by employing
entrance channels that essentially conformed to the condition in Euqation 7 (except for M). They measured the
energy and angular distribution of alpha particles and
protons emitted from the Br^{75} compound system excited to
49 MeV and formed by $C^{12}+Cu^{63}$ and $O^{16}+CO^{59}$ entrance
channels. Figures 7,8, and 9 show the calculated angular-momentum distributions in the two entrance channels,
and the channel-energy distributions of emitted protons
and alpha particles, respectively. The factor of 1.16
by which the cross sections for reactions from the O^{16}
entrance-channel are multiplied represents the ratio of
the C^{12} to O^{16} reaction cross sections and is to be com-

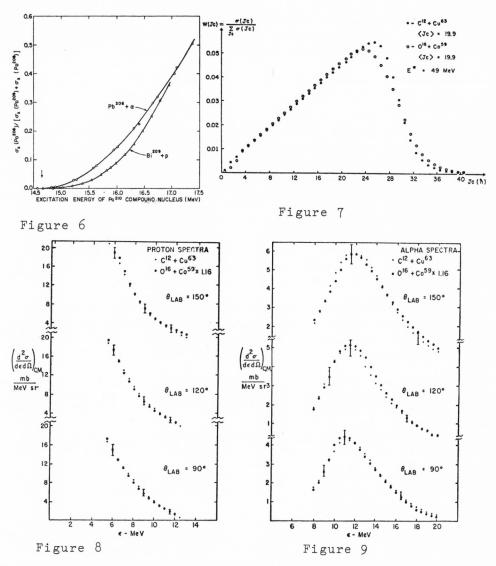

Figure 7

Figure 6

Figure 8

Figure 9

Figure 6. Experimentally measured values of $\sigma(Po^{208})/$ $\{\sigma(Po^{208})+ \sigma(Po^{209})\}$ as a function of the excitation energy of the Po^{210} compound nucleus prepared in two ways: p + Bi^{209} and α+ Pb^{206}. Figure from Ref. 12

Figure 7. The normalized angular-momentum distribution of Br^{75} compound systems at an excitation of 49 MeV. Figure from Ref. 13.

Figure 8. Energy spectra of emitted protons. Cross sections from $O^{16}+Co^{59}$ multiplied by 1.16. Fig. from Ref. 13.

Figure 9. Energy spectra of emitted alpha particles. Cross sections from $O^{16}+Co^{59}$ multiplied by 1.16. Fig. from ref 13.

pared with a ratio of 1.2 given by the optical-model
calculation that generated the angular-momentum distributions. A glance at figures 8 and 9 shows verification
in this instance of the Independence Hypothesis. If the
results are integrated over angle and energy, the verification may be cast in the form of Equation 8:

$$\sigma(C^{12},\alpha)_{10-20\,MeV}/\sigma(C^{12},p)_{6-12\,MeV} = 0.61 \pm 0.02$$
$$\sigma(O^{16},\alpha)_{10-20\,MeV}/\sigma(O^{16},p)_{6-12\,MeV} = 0.61 \pm 0.03$$

In a subsequent experiment[14], the Br^{75} compound nucleus
again at an excitation energy of 49 MeV was produced in
yet another entrance channel, $B^{11} + Zn^{64}$, in which, however, the angular-momentum distribution was different
from that for $C^{12} + Cu^{63}$ and $O^{16} + Co^{59}$ channels. The
corresponding ratio was found to be 0.50 ± 0.03 as compared to the 0.61. The diminution of the probability
for alpha emission is qualitatively consistent with the
diminution in the angular momentum carried in by the
B^{11}.

Further evidence[14] for the effect of angular-
momentum conservation was provided by an experiment
which again prepared Br^{75} compound-systems by the same
three target-projectile combinations that were just en-
umerated, but at an excitation energy of 88 MeV and, as
is illustrated in Figure 10, with differing angular-
momentum distributions. Integration of the results
over angle and energy showed that the ratios of the cross
sections for the emission of alpha particles with channel
energies between 10 and 22 MeV to that for protons between 6 and 14 MeV are 0.66, 0.76, and 0.96 for $B^{11} + Zn^{64}$,

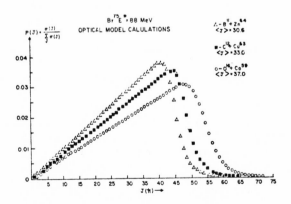

Figure 10. The normalized angular-momentum distribution
of Br^{75} compound systems at an excitation of 88 MeV.
Figure from Ref. 14.

$C^{12} + Cu^{63}$, and $O^{16} + Co^{59}$, respectively. Thus, when the angular momentum distributions were not matched, the mode of decay was not independent of the mode of formation.

From these data it is possible to determine the decay behavior of compound-nuclei within a rather narrow range of spins by a method of analysis that is described in Reference 14. Utilization of this method showed that the ratio of alpha particle to proton emission for the energy regions given above was 1.3 and 1.9 for angular-momentum regions about $10\hbar$ wide at half maximum and with average values of $48\hbar$ and $54\hbar$, respectively. While the method of analysis is quite general and does not depend upon the calculation of either transmission coefficients or any other quantity, the assignment of spins to the regions of spin that are isolated does depend upon an estimation of the transmission coefficients for compound-nucleus formation. Accordingly, the values of $48\hbar$ and $54\hbar$ are probably just upper limits since it was assumed in their estimation, as is certainly not true, that all of the reactions were compound-nucleus reactions.

In an extensive and detailed calculation, Gilat and Grover[15] showed that the difference between the decay of Br^{75} compound nuclei at 49 MeV excitation formed by $B^{11} + Zn^{64}$ on the one hand, and $C^{12} + Cu^{63}$ or $O^{16} + Co^{59}$ on the other, could, indeed, be quantitatively encompassed within the statistical model if proper account were taken of the angular-momentum distributions in the entrance channels.

In sum, then, the validity of the Independence Hypothesis has been directly verified for different entrance channels which are characterized by the same constants of the motion. Some of the _apparent_ violations of the Independence Hypothesis have been shown to be caused by differing angular-momentum distributions in the corresponding entrance-channels and the divergent behavior of the compound nuclei that are formed have been shown to conform to the predictions of the compound-nucleus model when special care is taken to formulate carefully the level densities of the relevant residual nuclei and to conserve angular momentum.

ROLE OF ISOBARIC SPIN CONSERVATION

In the previous sections of this report it has been implicity suggested that isobaric spin may be a conserved quantity for compound-nucleus reactions proceeding

through, and leading to, continuum states. It is not
obvious that this is correct; indeed, there exists good
reason to believe that isobaric spin is not strictly
a constant of the motion in these reactions[16]. Nevertheless, as will be described below, there have been
experimental observations that indicate that isobaric
spin is at least partially a conserved quantity in
these compound-nucleus reactions in the continuum.

It might be best to start with a qualtitative consideration of the consequences of isobaric-spin conservation for the type of reactions that are being discussed
here. First, it is only when the target and projectile
have opposite signs for the third component of isobaric
spin that there will be compound nuclei formed in more
than one isobaric-spin state. In practice, then, this
means bombardment by either protons or He^3 particles
which can lead to compound nuclei with isospin of either
$T_O + 1/2$ or $T_O - 1/2$ where T_O is the isospin of the target.
Formation of compound nuclei with the same Z and N by
any entrance channel in which the third components do
not have opposite signs lead only to the $T_O - 1/2$ states.
As is usual, the $T_O + 1/2$ state will be referred to as
$T_>$ while $T_<$ will be used for the $T_O - 1/2$ states. If it
is assumed that the optical potential does not depend
on the isospin of the entrance channel, then the $T_>$
states will be produced with a statistical weight of
$1/2T_O + 1$, while that for the $T_<$ states is $2T_O/2T_O + 1$.
The situation for the particular instance of forming
the Cu^{63} compound nucleus via two different channels,
$p + Ni^{62}$ and $\alpha + Co^{59}$, is illustrated in Fig. 11.

Second, the question arises of whether there is any
difference in the decay characteristics of the $T_>$ and $T_<$
compound-nuclei. There are, as mentioned earlier, two
decisive factors involved: the level densities of the
residual nuclei and the cross sections for the inverse
of the decay processes. To speak of the level density
first, it may be seen in Fig. 11 that neutron, proton,
and alpha emission from the $T_<$ states (the T = 5/2 states
of Cu^{63}) can populate any and all states of the respective residual nuclei consistent with the conservation of
energy and angular momentum. From the $T_>$ states, on the
other hand, while any and all states subject to the above
constraint may be populated by proton emission; neutron
and alpha emission would only populate states with isobaric spin at least one unit greater than that of the
ground state. From Fig. 11 is is seen that this, effectively, increases the separation energy of the alpha
particle by 9.5 MeV and of the neutron by 4.5 MeV there-

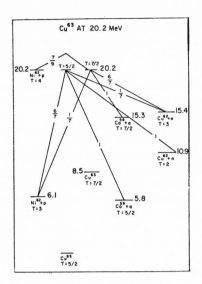

Figure 11. Ground-state level scheme of the $T_<$ and $T_>$ states relevant to the formation and decay of Cu^{63} compound nuclei at an excitation of 20.2 MeV. The number beside each level is the energy in MeV of that level above the ground state of Cu^{63}. The number in the lines connecting levels is the statistical factor for the isospin change in that transition. Figure from Ref. 17.

----

by drastically reducing the probability that these particles are emitted from the $T_>$ state as compared to the $T_<$ state. The second factor, the cross sections for the inverse of each decay process, must be weighted by the statistical factors for the isobaric-spin change in that mode of decay. These factors are given by the numbers on the lines connecting levels in Fig. 11 if it is assumed that the optical-model parameters for formation of the $T_>$ and $T_<$ states in Cu^{63} are the same. Calculations[17] have shown that in this instance the level-density effect is the decisive one and thus, despite the unfavorable statistical-weight of 1/7 for proton emission from the $T_>$ state, the $T_>$ state decays essentially completely by proton emission. The net result, then, is that since the proton entrance-channel can form the $T_>$ state while the alpha entrance-channel cannot, there should be an enhancement of the cross section for the (p,p') reaction with, however, the enhanced proton emission showing an angular and energy distribution that is characteric of a compound-nucleus reaction.

There have, of course, been many observations of the preferential emission of protons in proton-induced reactions. These have usually been discussed in terms of some sort of non-compound process which, indeed, is appropriate when either or both the energy and angular distributions of the emitted protons diverge from expectations of the compound-nucleus model. In 1966, however, Dudey et al[18] reported on the energy and angular distributions of the alpha particles and protons emitted from the Cu^{63} compound system at 20.2 MeV excitation energy prepared via two different entrance channels: $\alpha + Co^{59}$ and $p + Ni^{62}$. They found that although the energy and angular distributions of the emitted particles conformed to the requirements of the compound nucleus model, the probability for proton emission was greater from the incident-proton channel than it was from the incident-alpha channel. The authors pointed out that the distributions of angular momentum and of isotopic spin were not the same in the two entrance-channels but that calculations showed that the observed enhancement of proton emission did not appear to be a consequence of the angular-momentum differences. A more complete description and an extension of their work, taken from reference 17, is summarized in Table I where there are also presented similar results for the behavior of a Ni^{60}

	Cu^{63} system	
	$\dfrac{\sigma Co^{59}(\alpha,p)Ni^{62}(5\text{-}8\text{ MeV})}{\sigma Co^{59}(\alpha,\alpha')Co^{59}(7\text{-}12\text{ MeV})}$	$\dfrac{\sigma Ni^{62}(p,p')Ni^{62}(5\text{-}8\text{ MeV})}{\sigma Ni^{62}(p,\alpha)Co^{59}(7\text{-}12\text{ MeV})}$
Experimental	(1.25±0.06)	(3.4±0.3)
Calculated (J)	1.57	1.44
Calculated (J,T)	1.35	2.66

	Ni^{60} system	
	$\dfrac{\sigma Fe^{56}(\alpha,p)Co^{59}(5\text{-}9\text{ MeV})}{\sigma Fe^{56}(\alpha,\alpha')Fe^{56}(7\text{-}12\text{ MeV})}$	$\dfrac{\sigma Co^{59}(p,p')Co^{59}(5\text{-}9\text{ MeV})}{\sigma Co^{59}(p,\alpha)Fe^{56}(7\text{-}12\text{ MeV})}$
Experimental	(1.9±0.1)	(4.2±0.4)
Calculated (J)	2.15	2.37
Calculated (J,T)	1.79	3.86

Table I. Comparisons among experimental values for the ratios of cross sections with those calculated including angular-momentum conservation [Calculated (J)] as well as those calculated including both angular-momentum and isospin conservation [Calculated (J,T)]. Table from Ref. 14.

compound nucleus excited to 23.6 MeV from two different
entrance-channels: p + Co^{59} and α+ Fe^{56}. It may be
seen from the table that the proton emission from proton
entrance-channels is enchanced[19] by a factor of 2.7
in the Cu^{63} system and 2.2 in the Ni^{60} system. The rows
labelled "Calculated (J)" are the result of a statistical-model calculation assuming rigid-body moments of
inertia and complete mixing of all isospin states; those
labelled "Calculation (J,T)" assume isospin conservation.
The latter calculation includes the isospin statistical
factors in the transmission coefficients for the inverse
processes and takes the level densities of the T$_>$ states
that are involved as being equal to that of their parent
isobaric analogues at the same excitation energy above
the lowest state of the isospin in question. Grimes et
al[20] have recently decribed a similar method of calculation. It is evident from Table I that the calculations that include isospin conservation give better
agreement with the experimental results for these two
systems. It should be pointed out, however, that it is
probably not impossible to reproduce the experimental
results essentially as well with a calculation that
assumes complete isospin mixing but leaves as free parameters the moments of inertia of the several residual
nuclei. In particular, reduction of the moment of inertia of the Ni^{62} residual nucleus to 50% or less of
the rigid-body moment moves the calculated results closer
to the experimental. As pointed out, however, by Lu et
al[21] a reduction of this magnitude for the moment of
inertia appears to be inconsistent with the observed angular distributions of the emitted particles.

     Although the possibility that the observed enhance-
ment of proton emission was merely a consequence of some
non-compound process was obviated by the energy and angular distributions of the emitted protons, Lu et al[21]
and Jaffe and Fluss[22] have closely examined this point
by studying the same compound systems at lower excitation
energies where the possibility of interference from se-
cond-chance proton emission at the lower end of the energy spectrum is diminished and thus attention can be
focused on lower-energy protons where non-compound processes are a priori expected to be less important. Lu
et al find an enhancement of 2.1 for protons between 2.2
and 8.7 MeV for Cu^{63} at 18.9 MeV, and 1.6 for protons
between 2.0 and 9.0 MeV from Ni^{60} at 22.3 MeV. Jaffe
and Fluss find an enhancement of 2.2 for protons between
2.0 and 7.0 MeV from Cu^{63} at 18.1 MeV.

     In sum, there is strong circumstantial evidence that

isobaric spin can be a "good" quantum number for these
compound-nucleus reactions leading to states in the continuum. That this should be so depends on the lifetime
of the compound nucleus as compared to the mixing time
for different isobaric-spin states. On this basis, Lu
et al[21] conclude, for example, that the Coulomb matrix element that would mix states in Cu^{63} must be less
than about 5 keV. The possibility, however, that the
enhancement of proton emission that has been observed
may result at least partly from the differing spin distributions in the two entrance channels merits further
attention.

SUMMARY

The range of validity of the compound-nucleus model
for the description of nuclear reactions proceeding
through continuum states and leading to continuum states
is extended if proper account is taken of all of the
constants of the motion and of the detailed variation
of level density from nucleus to nucleus. Adjustment
of the level-density formula for those factors that
make the ground state of the nucleus unusually stable
can have dramatic effects on the probability for the
emission of various types of particles. The conservation of angular momentum along with careful evaluation
of the spin dependence of level densities can have large
effects on the shapes of energy and angular distributions
as well as the probabilities for the emission of the
various types of particles. Finally, isobaric-spin conservation can increase the probability for proton emission in proton-induced reactions. This is particularly
true when there is a very small probability for proton
emission from the $T_<$ states of the compound nucleus.

*Work supported by the U.S. Atomic Energy Commission

References

1. A.C. Douglas and M. MacDonald, Nucl. Phys. 13, 382 (1959).
2. T.D. Thomas, Ann. Rev. Nucl. Sci., 18, 343 (1968).
3. G. Ligett and D. Sperber, Phys. Rev. C3, 167 (1971).
4. R. Reedy, M. Fluss, G. Herzon, L. Kowalski, and J.M. Miller, Phys. Rev. 188, 1771 (1969).
5. J.R. Grover and J. Gilat, Phys. Rev. 157, 802 (1967).
6. I. Dostrovsky, Z. Fraenkel, and G. Friedlander, Phys. Rev., 116, 683 (1959).
7. S.N. Ghoshal, Phys. Rev., 80, 939 (1950).
8. F.S. Houck and J.M. Miller, Phys. Rev. 123, 231 (1961).
9. E.V. Verdieck and J.M. Miller, Phys. Rev. 153, 1253 (1967).
10. See, for example: Theodore J. Krieger, Ann. Rev. Phys. (N.Y.) 31, 88 (1965) and the references quoted therein.
11. Figure taken from E.G. Moorhead, Nuclear Reactions of 10-20 MeV Deuteron with 61-Nickel, Ph.D. Thesis, Columbia University (1964).
12. J. Robb Grover and Richard J. Nagle, Phys. Rev. 134, B1248 (1964).
13. J.M. D'Auria and M.J. Fluss, G. Herzog, L. Kowalski, J.M. Miller, and R.C. Reedy, Phys. Rev., 174, 1409 (1968).
14. R.C. Reedy, M.J. Fluss, G.F. Herzog, L. Kowalski and J.M. Miller, Phys. Rev., 188, 1771 (1969).
15. Jacob Gilat and J. Robb Grover, Phys. Rev. C3, 734 (1971).
16. See for example: A.M. Lane and R.G. Thomas, Rev. Mod. Phys. 30, 344 (1958).
17. M.J. Fluss, J.M. Miller, J.M. D'Auria, N. Dudey, Bruce M. Foreman, Jr., L. Kowalski and R.C. Reedy, Phys. Rev., 187, 1449 (1969).
18. N. Dudey, M. Fluss, B. Foreman, L. Kowalski and J. Miller, International Nuclear Physics Conference, pp. 803-806, Academic Press, New York (1967).
19. Enhancement here is defined as $\sigma(p,p')\sigma(\alpha,\alpha')/\sigma(p,\alpha)\,\sigma(\alpha,p')$.
20. S.M. Grimes, J.D. Anderson, A.K. Kerman, and C. Wong, UCRL-72759 (1970).
21. C.C. Lu, J.R. Huizenga, C.J. Stephan, and A.J. Gorski, Nucl. Phys., A164, 225 (1971).
22. G. Jaffe and M. Fluss, personal communication.

DISCUSSION

VONACH

I'd like to make a comment on Ghosal's experiment. It has
been checked by several authors. As those excitation curves which
we are dividing to form the ratio are very energy dependent, they
have the usual rising and falling, this change of the energy scales
produces a considerable change of the ratios.

MILLER

The point of that first slide was that the ratio must be
properly chosen. You have to take the $\sigma(i,b)$ to $\sigma(i,b')$. In that
curve the ratio over a large range of excitation is independent of
the excitations. I could have moved the alpha points 2 or 3 MeV
either way you want.

VONACH

The constant ratio was produced by dividing two curves of the
same form but with strong energy dependence. Then of course, you
get a constant ratio. If you displace these two curves against
each other, then the ratio changes.

MILLER

The two curves were for the (α,pn) and the (α, 2n) reactions,
so if you're going to displace them, you have to displace them both
at the same time and then only the ratios remain unchanged.

COHEN

I feel that Dr. Miller is to be complimented in finding this
very interesting isospin effect, and I agree with his method of
applying it. However, we have recently applied our method of
separating direct reactions from compound nucleus reactions, and
we have found a lot of direct reaction in $Ni^{62}(p,p')$ at back angles.
It's something to worry about.

MILLER

I'm concerned about this too, Bernie. There's a way of doing
this, maybe. To get this isobaric spin effect, what you have to
have is a neutron excess target and a neutron deficient projectile.
Say, you take He^3 process. Conversely, if the He^3 particle comes
out, it too should be enhanced by isobaric spin. Perhaps another
handle on this that might make you happier is if instead of seeing
an enhancement of the (p,p') reaction if one were to see an enhancement of the (p,He^3) reaction. That might be quite nice.

BERMAN (Livermore)

Surely the recently performed, but as yet unanalyzed, experiments
of J. Baglin and Barney Cook on photon-induced reactions through both
$T_<$ and $T_>$ states in Zn^{64} will throw light on this problem.

MILLER
    Is this for the giant resonance?

BERMAN
    Yes-- many excitation energies, but the limit of the machine
is 28 MeV, so we didn't go above that.

HUIZENGA   (Rochester)
    I think that Cohen's comment is well taken. One certainly
has to worry about the direct reactions, but I think he's being a
little over-optimistic about the magnitude of the direct reaction
effect. I think one can see this very quickly if one looks at two
isotopes of the same element. For example, we studied the two
systems of Zn^{64} and Zn^{66}, and there's a very strong effect associated
with the fact that in one system proton emission is fairly intense,
whereas in the other system it is not. What I am saying is that
when neutron emission predominates, the isospin effect is very
strong, and as far as the direct reactions go, these are two nuclei
that are quite similar, and there doesn't seem to be any reason to
believe that one nucleus will have a higher percentage of direct
reactions than the other. In fact, the spectra look just like
evaporation type, and it's a little hard for me to believe that this
sort of a spectrum represents a lot of direct reaction.

MILLER
    John, we seem to be faced with direct reactions which are
isotropic and have evaporation spectra.

COHEN
    I really think we have a clean-cut way of seeing what is this
direct reaction. We saw how much was direct reaction, and there's
plenty.

HUIZENGA
    I think the burden of proof may be on Cohen at this point.
If he thinks there is a lot of direct reaction and says it's easy
to separate the two,and one has exactly the same sort of spectra,
the same kind of angular distribution, then I would say the
separation is not easy, and I think the technique he describes is
absolutely not applicable to systems that Dr. Miller has described.
I would like to see you apply the technique you're describing to
these systems. I think it would be very interesting to the
conference for you to show those pictures.

COMPOUND NUCLEUS REACTIONS BETWEEN OXYGEN IONS AND ^{18}O, ^{12}C, AND ^{14}C TARGETS

Y. Eyal, I. Dostrovsky and Z. Fraenkel

The Weizmann Institute of Science

INTRODUCTION

The basic assumption underlying the analysis of compound nucleus reactions is the Bohr independence hypothesis which supposes the decay of the compound nucleus to be independent of its mode of formation. The applicability of the independence hypothesis to heavy ion reaction has previously been validated for the kinetic energy spectrum of the particles emitted from the compound nucleus ^{75}Br formed by the reactions ^{16}O + ^{59}Co and ^{12}C + ^{63}Cu at one excitation energy[1]. We have tested the validity of the independence hypothesis for heavy ion reactions by comparing the cross sections for the residual nuclei ^{28}Al and ^{29}Al from the reactions ^{18}O + ^{12}C and ^{16}O + ^{14}C over the excitation energy range of the compound nucleus ^{30}Si of 30 - 40 MeV. We have also investigated the importance of the angular momentum in compound nucleus reactions of heavy ions by comparing the experimental formation cross sections of several residual nuclei from the above reactions and from the ^{18}O + ^{18}O and ^{18}O + ^{14}C reactions with statistical model calculations which neglect angular momentum effects and with calculations which take into account in a detailed fashion the effects of the angular momentum on the decay of the compound nucleus.

EXPERIMENTAL DETAILS

The experiments were performed at the EN Tandem Van de Graaff accelerator at E.T.H. Zurich with ^{16}O and ^{18}O beams of 13 - 41 MeV. The ^{18}O targets were made of a thin layer (75 μgr/cm^2 oxygen) of 98% ^{18}O enriched Ta$_2$O$_5$ on thick (21 mg/cm^2) tantalum backing. The ^{12}C targets were 140-280 μgr/cm^2 natural carbon foils on 245 mg/cm^2 gold backing. The ^{14}C targets were made of 125 μgr/cm^2 enriched

(83% ^{14}C) polyacetylene sintered on 245 mg/cm^2 gold backing. The thick backings of the targets stopped all reactions products. The products of interest in this work were radioactive and their formation cross sections were measured by direct counting of the target activity after the beam was turned off. The gamma ray activity was measured with a 3" x 3" NaI crystal. Special care was taken to prevent build-up of natural carbon on the ^{14}C targets. The various counting uncertainties led to a combined 10% error in the measurements in addition to errors of 0.5 - 15% due to counting statistics.

EXPERIMENTAL RESULTS

We show in Fig. 1 the excitation functions for ^{28}Al and ^{29}Al produced in the reactions ^{18}O + ^{12}C and ^{16}O + ^{14}C. Natural carbon contains 1.1% of ^{13}C and therefore the cross section for the

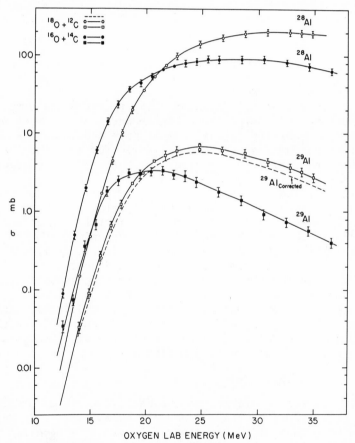

Fig. 1. Experimental excitation functions of the residual nuclei ^{28}Al and ^{29}Al produced in the compound nucleus reactions ^{18}O + ^{12}C and ^{16}O + ^{14}C.

$^{18}O(^{12}C,p)^{29}Al$ must be corrected for the contribution of the ^{13}C impurity. We have calculated this contribution on the basis of the optical model (formation cross section of the compound nucleus ^{31}Si) and the statistical model (decay of the compound nucleus to yield ^{29}Al). The ^{29}Al excitation function corrected in this manner for the ^{13}C contribution is also shown in Fig. 1. Similar calculations showed that the contribution of the ^{13}C in the ^{12}C and ^{14}C targets to formation of ^{28}Al entirely negligible (it obviously does not contribute to the ^{29}Al cross section from the reaction $^{16}O + ^{14}C$).

The solid lines in Fig. 2 show the experimental excitation functions for the formation of ^{30}Al, ^{29}Al and ^{27}Mg from the reaction $^{18}O + ^{14}C$. The ^{29}Al cross section was corrected for the fraction of ^{12}C and ^{13}C in the ^{14}C target. The correction of ^{12}C was based on the experimental $^{18}O(^{12}C,p)^{29}Al$ cross section measured in the present experiment whereas the correction for ^{13}C was based on calculations. The ^{29}Al cross section was also corrected for the additional ^{12}C

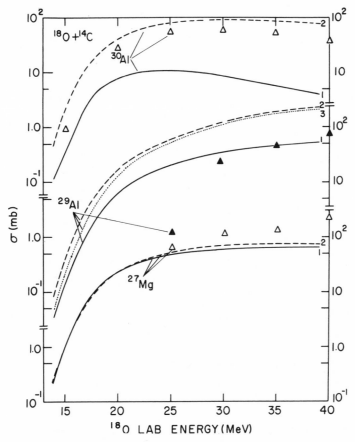

Fig. 2.   The excitation functions of ^{27}Mg, ^{29}Al and ^{30}Al produced in the reaction $^{18}O + ^{14}C$. Curve 1 (solid lines) - exptl. results. Curves 2, 3 and triangles - calculated results.

which was deposited on the target. Its amount was estimated from
the increase in the ^{28}Al activity during the irradiation. The total
corrections for the ^{29}Al formation cross section changed from 30%
at the lowest bombarding energy to 2% at the highest energy. Calculations showed that the corrections of the ^{30}Al and ^{27}Mg cross sections
due to the ^{12}C and ^{13}C contents in the target could be neglected.

The solid lines in Fig. 3 show the experimental excitation
functions for ^{34}P, ^{30}Al and ^{27}Mg from the reaction ^{18}O + ^{18}O. None
of these products were seen in the reaction ^{18}O + ^{16}O and therefore
the cross sections shown in Fig. 3 did not have to be corrected for
the ^{16}O content of the Ta$_2$ ^{18}O$_5$ target.

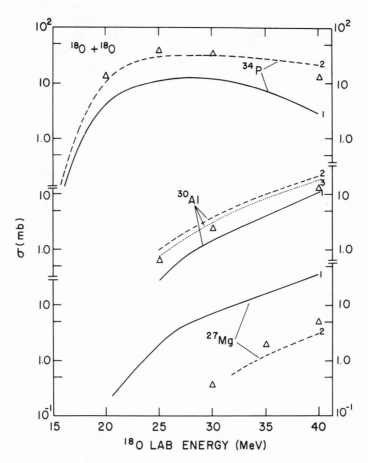

Fig. 3.   The excitation functions of ^{27}Mg, ^{30}Al and ^{34}P produced
         in the reaction ^{18}O + ^{18}O. Curve 1 (solid lines) -
         experimental results. Curves 2, 3 and triangles -
         calculated results.

DISCUSSION

The reactions $^{18}O + ^{12}C$ and $^{16}O + ^{14}C$ form the same compound nucleus ^{30}Si. These reactions can therefore be used to study the validity of the Bohr independence hypothesis in heavy ion reactions. In view of the fact that the compound nucleus formation cross sections may be different for the two reactions the validity of the independence hypothesis cannot be tested by comparing the cross sections for the same residual nucleus formed by the two reactions. However, such a test is provided by the comparison of the ratios of cross sections of two residual nuclei (such as $\sigma(^{28}Al)/\sigma(^{29}Al)$) for the two reactions. It can be readily shown that for a given excitation energy of the compound nucleus ^{30}Si the relation

$$\frac{\sigma(^{18}O(^{12}C,pn)^{28}Al)}{\sigma(^{18}O(^{12}C,p)\ ^{29}Al)} = \frac{\sigma(^{16}O(^{14}C,pn)^{28}Al)}{\sigma(^{16}O(^{14}C,p)\ ^{29}Al)} \tag{1}$$

is a test of the validity of the independence hypothesis provided (1) the angular momentum distribution of the compound nucleus ^{30}Si is identical for the two reaction paths $^{18}O + ^{12}C$ and $^{16}O + ^{14}C$, or (2) the decay of the compound nucleus ^{30}Si is independent of its angular momentum, or that both conditions (1) and (2) hold. Conversely, the experimental verification of Eq. 1 does not only prove the validity of the independence hypothesis for these reactions but also shows that at least one of the conditions (1) and (2) is fulfilled.

We show in Fig. 4 the cross-section ratio $\sigma(^{28}Al)/\sigma(^{29}Al)$ of the reaction $^{18}O + ^{12}C$ and $^{16}O + ^{14}C$ for the ^{30}Si excitation energy range 29 - 39 MeV. It is seen that the ratios are identical within the experimental error although they vary by more than an order of magnitude over this energy range. This result may thus be taken as a proof of the validity of the independence hypothesis for these reactions. Fig. 4 also shows the calculated cross section ratio based on the statistical model (DFF evaporation code[2] with level density parameter a=A/16). The calculated curve agrees very well with the experimental ratios.

We have also calculated the absolute excitation functions of ^{28}Al and ^{29}Al for the reactions $^{18}O + ^{12}C$ and $^{16}O + ^{14}C$. This was done by calculating the total reaction cross section on the basis of the optical model and assuming it to be approximately equal to the compound nucleus formation cross section. The decay of the compound nucleus was calculated on the basis of the statistical model. We used the ABACUS-2 code[3] for the optical model calculations. The parameters of Shaw et al.[4] were used for $^{18}O + ^{12}C$ whereas for the reaction $^{16}O + ^{14}C$ we used the parameters of Maher et al[5]. The decay of the compound nucleus was calculated with the DFF code[2] which does not take into account effects of angular momentum. In order to examine the effect of the angular momentum on the decay of the compound nucleus we repeated this part of the calculation with the GROGI-II code[6] which takes into account in detailed fashion

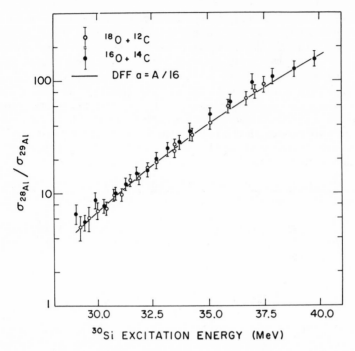

Fig. 4. The experimental cross section ratio $\sigma(^{28}\text{Al})/\sigma(^{29}\text{Al})$.
For the reactions $^{18}\text{O} + ^{12}\text{C}$ and $^{16}\text{O} + ^{14}\text{C}$ as a function
of the excitation energy of the compound nucleus ^{30}Si.
The curve shows the calculated ratio.

the effect of the angular momentum on the evaporation cascade.
Good agreement with the experimental results was obtained for the
$^{18}\text{O} + ^{12}\text{C}$ reaction products when the GROGI-II was used for the
evaporation stage. Distinctly worse agreement was obtained for
the $^{16}\text{O} + ^{14}\text{C}$ reaction. Since the evaporation calculation is the
same for the two reactions these results seem to indicate that the
optical model parameters of Maher et al do not describe well the
reaction $^{16}\text{O} + ^{14}\text{C}$. The results of the DFF evaporation code are
slightly worse than those of the GROGI-II calculation.

We have repeated the above calculations for the products of
the reactions $^{18}\text{O} + ^{14}\text{C}$ and $^{18}\text{O} + ^{18}\text{O}$. The results are also shown
in Figs. 2 and 3. In these figures the lines 2 and 3 denote the
results of the DFF calculation whereas the triangles are the
results of the GROGI-II calculation. For the optical model calculation we used the parameters of Shaw et al (curves 2 and triangles).
We have also repeated some of the calculations with the optical
model parameters of Voos et al[7] (curves 3). It is seen that
agreement with the experimental results is mostly very poor and

that the inclusion of angular momentum effects (triangles) does not improve the agreement substantially. Moreover the results of the GROGI-II and DFF calculations are mostly quite similar. (The very large discrepancy between the GROGI-II results and both the experimental results and the DFF calculation for the excitation function of ^{29}Al from ^{18}O + ^{14}C at low energies (see Fig. 2) is due to the neglect of deuteron emission in the GROGI-II calculation). We conclude that the angular momentum has little effect on the decay of the compound nucleus in these reactions, i.e. that condition (2) (see above) is approximately satisfied. The discrepancy between the experimental and calculated results must be partly blamed on the optical model calculations which give poor results for the compound nucleus formation cross section, at least for the reaction ^{16}O + ^{14}C. As seen from curves 3 in Figs. 2 and 3, the use of quite different optical model parameters[7] has little effect on the calculated results. The deficiencies in evaporation calculations are probably due to the inadequate description of the level density in these calculations. A more detailed report of this work will be published elsewhere.

ACKNOWLEDGEMENT

We wish to thank Dr. J. Gilat for his help with the GROGI-II calculations. We are deeply grateful to Prof. P. Marmier for his help and hospitality which made these experiments possible.

REFERENCES

1. J.M. d'Auria, M.J. Fluss, G.F. Herzog, L. Kowalski, J.M. Miller, and R.C. Reedy, Phys. Rev. 174, 1409 (1978).
2. I. Dostrovsky, Z. Fraenkel and G. Friedlander, Phys. Rev. 116, 683 (1959).
3. E.H. Auerbach, BNL Report 6562, 1962 (unpublished).
4. R.W. Shaw, Jr., R. Vandenbosch and M.K. Mehta, Phys. Rev. Letters 25, 457 (1970).
5. J.V. Maher, M.W. Sachs, R.H. Siemssen, A. Weidinger and D.A. Bromley, Phys. Rev. 188, 1665 (1969).
6. J.R. Grover and J. Gilat, Phys. Rev. 157, 802 (1967); and J. Gilat and J.R. Grover, Phys. Rev. C3, 734 (1971).
7. U.C. Voos, W. von Oertzen and R. Bock, Nucl. Phys. A135, 207 (1969).

DISCUSSION

GILAT (Israel)

The problem you experience in fitting some reaction channels with an evaporation model are probably not due to level density effects but rather to problems associated wtih compound nucleus formation crossections. For reactions involving evaporation of several charged particles there may be transfer channels (such as Be^8 transfer, for example) that give rise to the same products you attribute to purely compound processes.

FRAENKEL

This may be so, but one should mention at this point that the over-simplified version of the very naive model has advantage that you can calculate essentially the evaporation of any old particle involved and still these calculations give just as pure results as the calculations which calculate the emission of a few particles in very sophisticated way.

COMPOUND NUCLEAR LIFETIMES FROM BLOCKING EXPERIMENTS IN Ge AND Mo

G.J. CLARK[+], E. FUSCHINI[*], C. MARONI[*], I.G. MASSA[*],
F. MALAGUTI[*], A. UGUZZONI[*], E. VERONDINI[*] and D.WILMORE[+]

+A.E.R.E., Harwell, Berks, U.K. and *Istituto Nationale
Di Fisica Nucleare, Sezione Di Bologna and Istituto Di
Fisica Dell' Universita, Bologna, Italia

INTRODUCTION

Recent developments (1,2,3) have made the use of the blocking effect of charged particles in a single crystal a viable technique with which to measure compound nuclear lifetimes in the range 10^{-16}-10^{-18} seconds. This is a lifetime range difficult to study by other techniques but one of considerable interest as the fine structure of medium mass nuclei has lifetimes of this order.

The method is based on the fact that when a positively charged particle is emitted from a normal lattice position in a crystal it is almost completely inhibited from moving in directions parallel to major axes or planes. Consequently the angular distribution of such particles emerging from the crystal shows a minimum along the axial or planar direction. If, however, a displacement into a planar or axial channel arises during a nuclear reaction it is possible for a subsequently emitted particle to move in those directions. These particles will cause a filling of the blocking-dip along the axial or planar direction. From the amount of filling-in of the dip it is possible to determine the mean displacement from the equilibrium lattice sites and consequently, knowing the recoil velocity, the reaction time.

We have utilized $\langle 110 \rangle$ axial blocking to make measurements for the reactions $^{70}Ge(p,p')^{70}Ge^{*}$ (1.04 MeV, 2[+]) and $^{72}Ge(p,p')^{72}Ge^{*}$ (0.835 MeV, 2[+]), the particles being detected in a two dimensional position sensitive detector of the "Chequer Board" type.

A computer program which simulates the passage of particles through a crystal lattice was used to predict the effects of compound

535

nuclear lifetimes on axial blocking dips. Experimental blocking
data and the computer results were then compared to assign values
to the average lifetimes of compound nuclear levels in ^{71}As and
^{73}As at various excitation energies. Using a natural Mo target
and the reaction Mo(p,p')Mo($\sim$1.5 MeV, 2^+) we have similarly been
able to assign a value to the average lifetime of compound nuclear
levels in ^{93}Tc.

These results are average compound nuclear lifetimes, as the
level width is much smaller than the energy loss in the crystal,
over which the results are averaged. From these measurements of
the average compound nuclear lifetime it is possible to obtain an
estimate of the level density parameters.

METHOD AND RESULTS

The experimental techniques are described elsewhere.[3] A proton
beam from the Harwell tandem Van de Graaff was used. The 2-D
chequer board counter was positioned 1m from the crystal at an
angle of 88° to the beam. Behind the chequer board a large area
surface barrier counter was placed to stop any protons that were
not stopped in the totally depleted chequer board counter. The
energy resolution of the system was 70 keV and the angular
resolution of each square on the chequer board, 0.14°.

i)   REACTIONS ^{70}Ge(p,p')^{70}Ge*(1.04 MeV, 2^+) and ^{72}Ge(p,p')^{72}Ge*
      (0.835 MeV,2^+)

Measurements were made at 5.0, 5.4 and 6.0 MeV incident proton
energy. All the counts in the concentric frames of the chequer
board are summed and a one dimensional projection of the dip can
be constructed as shown in fig. 1 where the filling of the dip is
clearly evident. The elastic dips were measured at incident
energies where the elastically scattered groups had the same energy
as the inelastic groups of interest. This procedure was adopted
because the dips are dependent on the energy of the scattered
protons and it is therefore necessary to measure the reference
zero-lifetime blocking dip at the same scattered energy as the
inelastic group.

A useful parameter[1] for defining the blocking dip is R =
(S-D)/S where S and D are the counts in the shoulder and dip
respectively summed over five frames. The parameter R, therefore
is a measure of the extent of the blocking dip. Table 1 lists the
ratios R_I/R_E of the inelastic and elastic R's for three incident
energies. A ratio of unity means a zero lifetime effect whereas
less than unity indicates a lifetime effect. The errors quoted
are statistical.

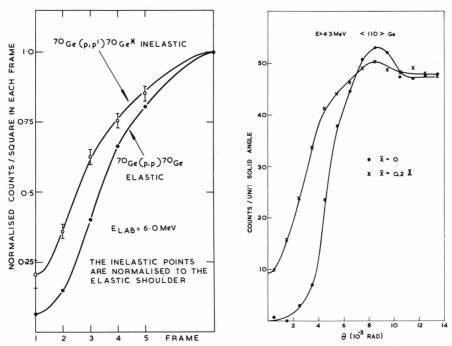

Fig. 1. A comparison of elastic and inelastic blocking dips. The dips on the left are experimental and on the right computer simulated. Each frame on the absicissa corresponds to 2.4 x 10^{-3} radians.

ii) REACTION Mo(p,p')Mo* ($\sim$ 1.5 MeV, 2^+)

Measurements were made at 8.0 MeV incident proton energy. Figure 2 is a one dimensional projection of the blocking dip illustrating the filling in of the dip. As all the Mo isotopes have 2^+ or $3/2^+$ states in the energy region of 1.55 ± 0.08 MeV it was not possible to distinguish between inelastic proton groups going to excited states at these energies in the various isotopes. Hence the R_I/R_E value estimated, 0.720 ± 0.002, is an average result for all the Mo isotopes.

ANALYSIS OF BLOCKING DIPS

A Monte Carlo type computer program[3] was used to simulate the blocking process and subsequently examine the dip shapes for various compound nuclear lifetimes. The program calculated blocking dips (Figure 1) for various values of the mean recoil distance, $\bar{x}$, of the excited nucleus from the lattice site. $\bar{x} = v\tau$ where v is the recoil velocity and τ is the mean compound nuclear lifetime. In this way theoretical R_I/R_E values were constructed for various lifetimes. If there is a sharp cut-off

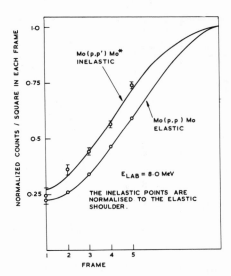

Fig. 2   Comparison of elastic and inelastic blocking in a
         natural molybdenum crystal.

distance $\bar{x} = r_c$ beyond which no blocking is evident, the ratio
should have the form

$$R_I/R_E = 1 - \exp - (r_c/v\tau) \qquad (1)$$

This simple model gives an excellent fit to the computed values.
It is of interest that the best fit value of r_c ($r_c = 0.123$Å for
the Ge case) is very close to the Thomas Fermi screening distance
($a_{TF} = 0.141$Å for the Ge case).

     It is anticipated in this analysis that there is not a
significant difference between the multiple scattering contribution to the elastic and inelastic dips.

DISCUSSION

     The experimental observation of a reduction in the blocking
effect for inelastically scattered protons compared to that for
protons elastically scattered at the same beam energy indicates
that we are seeing an effect resulting from the finite reaction
lifetime.

i)   REACTIONS $^{70}Ge(p,p')^{70}Ge*(1.04$ MeV, $2^+)$ and $^{72}Ge(p,p')^{72}Ge*$
      $(0.835$ MeV, $2^+)$

     The comparison of the measured R_I/R_E with the computer calcul-
ations enabled average lifetimes to be assigned to the compound
nuclear states in ^{71}As and ^{73}As at various excitation energies
between 9.5 and 11.5 MeV. The lifetimes are quoted in table 1.

E_{LAB} (MeV)	^{70}Ge(1.04 MeV, 2^+)			^{72}Ge(0.835 MeV,2^+)		
	R_I/R_E	τ(sec)x10^{-17}	Γ (eV)	R_I/R_E	τ(sec)x10^{-17}	Γ (eV)
5.0	0.71+0.04	(2.2+0.4)	28	0.60+0.04	(3.1+0.5)	20
5.4	0.67+0.03	(2.4+0.4)	27	0.51+0.04	(3.9+0.6)	16
6.0	0.74+0.03	(1.9+0.3)	34	0.93+0.03	(0.9+0.2)	72

Table 1. Measured filling-in parameters and derived lifetimes.
Errors in R_I/R_E are statistical. Errors in τ are derived from
a combination of experimental and theoretical errors giving values
of $\sim 20\%$. If the errors in τ are derived from the experimental
spread in R_I/R_E then the errors are typically 10%. This is the
more meaningful error when comparing lifetimes as a function of
bombarding energy.

All the lifetimes quoted in table 1 are experimental averages
over an energy interval corresponding to the energy lost by the
incident protons in transversing the crystal i.e. ~ 200 keV. It
should be emphasised that reaction times from zero to infinity all
contribute to these averages. In the computer model it is assumed
that the probability of particle emission from the recoiling
nucleus is exponential with displacement from the string. This
assumption is satisfactory here as calculations show that the $5/2^+$,
$3/2^+$, $3/2^-$ compound states have 80% of weight with lifetimes varying
by a factor of two only.

Calculations[3] predict that for the ^{70}Ge and a 6.0 MeV incident beam the compound elastic cross section at 90^0 is approximately
3% the shape elastic cross section. At other incident energies and
for the ^{72}Ge measurement the corresponding proportion of shape
elastic is even smaller. Calculations of the direct inelastic
cross section indicated that at all incident energies and for both
^{70}Ge and ^{72}Ge this was always less than 2% of the compound inelastic
cross section. Hence these calculations show the zero life-time
normalization from the elastic group to be satisfactory and the
inelastic reaction to be dominantly compound. The mean level
widths $\langle \Gamma \rangle$ calculated from the uncertainty principle ($\langle \Gamma \rangle . \tau = \hbar$)
are also given in table 1.

At a beam energy of 5.21 MeV the ^{70}Ge(p,n) channel opens and
becomes the dominant reaction channel as these proton energies
are below the Coulomb barrier for Ge. The opening of this
dominant channel has the effect of reducing the compound nuclear
lifetime as illustrated in table 1 for the ^{72}Ge(p,p')$^{72}Ge^*$ experimental results at 6.0 MeV. The lifetime is a factor of 3 smaller

at 6.0 MeV than at 5.0 MeV. No such result is observed for the $^{70}Ge(p,p')^{70}Ge*$ measurement as the neutron channel does not open until 7.39 MeV. A drop in the ratios of the differential cross section for $^{72}Ge*(p,p')$ and $^{70}Ge*(p,p')$ was also observed experimentally and predicted accurately.

The filling in parameter R_I/R_E may be analysed in terms of a level density formalism[3,4]. Broadly the method is to assume that R_I/R_E can be considered as the weighted average of the $(R_I/R_E)J\pi$ due to states of the same $J\pi$ weighted by the total observed cross-sections which can be calculated using a Hauser-Feshbach formalism. As the quantity $(R_I/R_E)J\pi$ can be expressed as a function of the level density parameter, a, using the level density formula of Gilbert and Cameron[5] the R_I/R_E can subsequently be calculated as a function of a. From the experimental values for R_I/R_E and the theoretical curves relating R_I/R_E and a, we deduced values for the level density parameter a. The results are shown in Table 2. Also shown are the level spacings, $D_{\frac{1}{2}}$, of the $J = \frac{1}{2}$ levels calculated from the level density formula of Gilbert and Cameron using our derived a values. As shown in Table 2 the results are in agreement with those deduced from the semi-empirical formula given by Gilbert and Cameron[5]. A semi-empirical fit to fast neutron data[6] gives, for the nuclei of interest here, values for the level density parameter, a, in agreement with our result as shown in Table 2.

^{71}As			^{73}As		
Excit. (MeV)	a (MeV^{-1})	$D_{\frac{1}{2}}$(eV)	Excit. (MeV)	a (MeV^{-1})	$D_{\frac{1}{2}}$(eV)
9.2-9.4	10.50+0.23	480+80	10.1-10.3	10.32+0.23	210+35
9.6-9.8	11.02+0.20	225+60	10.5-10.7	10.88+0.23	100+25
10.2-11.4	10.98+0.20	135+40	11.1-11.3		
Ref 6)	11.17			11.75	
Ref 7)	10.1			10.7	

Table 2. Values of level density parameter 'a' deduced from the experiment. The results are compared with a values deduced from semi-empirical formulae[5] and fast neutron data.[6]

ii) REACTION $Mo(p,p')Mo*$ ($\sim$ 1.5 MeV, 2^+)

As a natural Mo target was used, a comparison of the measured R_I/R_E with the R_I/R_E calculated using a cut-off value of $r_c = 0.112\text{Å}$ in equation (1) enabled an average lifetime of $\tau = (3.5 \mp 0.6) \times 10^{-17}$

sec to be assigned to the compound nuclear states ^{93,95,96,97,98,99,101}Tc.
The weight of the various nuclei in determining the lifetime is
dependent on the relative isotopic abundances and (p,p') differential cross sections. However, as this measurement was made at
8.0 MeV i.e. at an energy below the high ^{92}Mo(p,n) threshold,
Q = -8.84 MeV, and well above the (p,n) thresholds for the other
Mo isotopes it was expected that the ^{92}Mo(p,p') cross section
would be much larger than the (p,p') cross sections to the states
at ~1.5 MeV in the other isotopes. Measurements we made of the
^{92,94,95,96,97,98,100}Mo(p,p') (~1.5 MeV) differential cross
sections at 88° and 8.0 MeV incident energy showed the ^{92}Mo(p,p')
(1.54 MeV) differential cross section to be the dominant
contributor to the total Mo(p,p')(~1.5 MeV) differential cross
section.

Calculations similar to those carried out for Ge, show the
zero lifetime normalization from the elastic group to be satisfactory and the ^{92}Mo(p,p') inelastic reaction to be dominantly
compound.

We thus feel justified in assigning a lifetime $\Upsilon = (3.5 \pm 0.6) \times 10^{-17}$ to the compound nuclear states in ^{93}Tc at an excitation
energy of 11.5 - 11.8 MeV, the results being averaged over
approximately 300 keV. This lifetime corresponds to a mean
level width of $\langle \Gamma \rangle$ = 20 eV.

REFERENCES

1) M. Maruyama, K. Tsukada, K. Ozawra, F. Fujimoto, K. Komaki,
   M. Mannami and T. Sakurai. Nuclear Physics A145 (1970) 581.

2) W.M. Gibson and K.O. Nielsen, Phys. Rev. Lett. 24 (1970) 114
   and Proceedings of the Second International Symposium on
   the Physics and Chemistry of Fission, Vienna, IAEA (1969) 861.

3) G.J. Clark, J.M. Poate, E. Fuschini, C. Maroni, I.G. Massa,
   A. Ugguzzoni and E. Verondini. AERE Report R-6756 and
   Nuclear Physics (to be published).

4) F. Malaguti, A. Uguzzoni and E. Verondini, Nuovo Cimento
   (to be published).

5) A. Gilbert and A.G.W. Cameron, Can. Jour. Phys. 43, (1965)
   1446.

6) V. Benzi and G. Reffo, ENEA Neutron Data Compilation Centre,
   Newsletter Bulletin 10, (1969) 6.

We would like to acknowledge and thank John Porte for his
collaboration in the first part of this work.

DISCUSSION

BERMAN (L.R.L.)

Just a small experimental point: Did you take data far enough off axis so that the "shoulder effect" (resulting from conservation of particles) did not introduce an error in your normalizations?

CLARK

That's an experimental problem when you work with very thin crystals. We're working here with crystals something like 10 microns thick, and the shoulder does not appear because of multiple scattering effects. However we definitely did go out far enough to be beyond the shoulder. On our computer simulations which are for thinner crystals, you could actually see the shoulder.

TEMMER   (Rutgers)

I would like to announce with some pride--some recent preliminary results from our Rutgers' Bell Lab collaboration, using extremely thin Ge crystals ($\sim$1 micron), using the crystal-Blocking life-time technique just described by Dr. Clark. We were able to demonstrate that the average life-time of fine structure (unresolved, of course) within an analogue resonance ($\Delta E \sim$30 keV) is significantly shorter than the corresponding life-time away from such a resonance; in other words we can see the Robson enhancement of time structure widths. [See Gibson, Maruyama, Mugay, Sellschop, Temmer, and VanBree (Bull Am. Phys. Soc. 16,557 (1971)]

SHELDON   (Lowell)

Dr. Clark's abstract whetted my appetite by mentioning some results on $Mo^{92}(p,p')$ blocking experiments. Do you have further information on the outcome of the analysis?

CLARK

Yes. With 8 MeV proton energy, we've extracted a life-time of something like $(2.0 \pm 0.6) \times 10^{-17}$ seconds. That corresponds to a width of something like 33 eV, a result which was slightly surprising to us. At the moment, we are starting to extend these measurements.

SESSION VIII

LIMITATIONS OF THE COMPOUND NUCLEUS

Chairman   - N. Francis, Knolls Atomic Power Laboratory
Scientific Secretary - C. Lubitz, Knolls Atomic Power Laboratory

INTERMEDIATE STRUCTURE

C. Mahaux

University of Liège at Sart Tilman

Liège, Belgium

1. INTRODUCTION

It is not only hopeless but also of little physical interest to reproduce the detailed energy dependence of the nuclear cross sections from dynamical models, except perhaps in very light nuclei. Rather, one attempts to describe the main features of the experimental data in terms of statistical distributions of resonance parameters. The standard statistical assumptions are remarkably successful in reproducing the gross structure of the data. Nevertheless, systematic deviations from the statistical assumptions have been detected. In this review, we shall mainly discuss those deviations which are restricted to fairly narrow energy ranges, of size intermediate between the average spacing between the compound states on the one hand and the energy dependence characteristic of the optical model on the other hand. Hence the name "intermediate structure" coined for this phenomenon. These energy intervals range from several eV (neutron induced subthreshold fission) to several MeV (giant photonuclear resonances). In the doorway state model, one proposes that the standard statistical assumptions apply to these regions as well, after one has singled out and treated explicitly one particular configuration, the "doorway state". Phrased in these terms, the associated concepts of intermediate structure and doorway state appear to leave little room for controversy in the interpretation of the experimental data. Nevertheless, the existence of intermediate structure due to doorway states has long been subject to discussions and this interpretation remains doubtful in many cases. The main reason is that the experimental data are in general insufficient for deciding wether or not the standard statistical assumptions apply once the doorway configuration has been singled out. Too few compound resonances parameters are available for a reliable statistical analysis. Quite often, only the average cross sections are measured.

In order to avoid too much overlap with the contents of other sessions of this Conference, we shall discuss only one aspect of the doorway state concept, namely the intermediate structure phenomenon. Accordingly, we shall barely mention applications like the deviation of the strength function from the predictions of the optical model (1 and session 5), the interpretation of the high energy tail in the neutron evaporation spectra (2, session 7), the correlations between partial widths of different channels (3, session 4), the statistics of radiative capture widths (4, session 3), the intermediate structure in the mixing between bound states (5), etc. This list is not exhaustive but suffices to demonstrate how fruitful the concept of doorway states has been and will remain, even if intermediate structure resonances due to doorway states do not occur as frequently as was once hoped.

In the following, we make no effort to be complete concerning the early work on the subject. The concepts of intermediate structure (6) and of doorway states (7) already existed in the mid-fifties, but it is only in the last decade that these phenomena were systematically and eagerly investigated, mainly under the impetus of the MIT group (1), (8). We refer to refs. (8-10) for a description of the state of the problem around 1968.

The main purpose of the present paper is to discuss to what extent and in what sense one can check experimentally whether a doorway state phenomenon is operative and to review sqme recent experimental data. We shall first recall a few basic equations and definitions (section 2). In section 3, we discuss the problem of the identification of a doorway state phenomenon and give a few examples. Section 4 is devoted to a discussion of recent progress related to three well-established cases of intermediate structure due to doorway states. The conclusions are drawn in section 5.

2. DOORWAY STATES

We mentioned before that a doorway state is some configuration that one singles out before making the usual statistical assumptions. Since configurations are specified in terms of nuclear models, it is convenient and in my opinion even necessary in the present context to write the full Hamiltonian H in the form

$$H = H_o + V \qquad , \qquad (1)$$

where H_o is a model Hamiltonian and V the corresponding residual interaction. In the dynamical approach to nuclear reactions, H_o has two kinds of eigenstates, those belonging to the continuous spectrum, which we shall denote by χ_E , and the quasi-bound states corresponding to the discrete spectrum. We assume that H_o is chosen in such a way that

$$<\chi_E|H|\chi_{E'}> = E\delta(E-E') \qquad , \qquad (2)$$

i.e. that its eigenstates χ_E give a good description of the actual channels (10). The role of the residual interaction V is twofold. Firstly, it couples the quasi-bound states θ_j together, yielding the true compound states. Secondly, it couples the latter states to the continuum, giving thereby rise to resonances.

As an example, let us choose a case where the shell-model description is appropriate, specifically the elastic scattering of a nucleon by a doubly closed shell nucleus. The quasi-bound states θ_j are (m+1) particle-m hole ((m+1)p-mh) states. If the residual interaction V is a sum of two-body forces, most matrix elements $<\chi_E|H|\theta_j>$ vanish identically since only the 2p-1h states are directly coupled to the continuum χ_E corresponding to the target in its ground state. The other, more complicated, states are coupled to this continuum indirectly, via the chain (m+1)p-mh → mp-(m-1)h → ... → 2p-1h → χ_E . Thus, the 2p-1h states may be called doorway states, since they constitute a necessary intermediate step in the decay of a (m+1)p-mh state.

A linear combination of doorway states is itself a doorway state. The states Φ_m obtained by diagonalizing H among the 2p-1h states have more direct physical meaning than the unperturbed configurations. We call ϵ_m the associated energies. We also diagonalize H among the more complicated (m+1)p-mh states, call ψ_j the resulting configurations, e_j their eigenvalues. At high excitation energy in light nuclei and already right above threshold in heavy nuclei, the average distance d between neighbouring energies e_i is much smaller than the average energy distance D between neighbouring doorway states. In certain cases, some features of the cross section in an energy range smaller than D could therefore be dominated by the doorway state whose energy falls in this energy range. In order to discuss what happens then, we take the extreme model where only one channel exists, the M complicated states ψ_j being coupled to it via only <u>one</u> doorway state, Φ_o . This model is characterized by the following dynamical equations, together with eq. (1),

$$\langle \Phi_o | H | \psi_j \rangle = v_j \quad , \quad \langle \psi_j | H | \chi_E \rangle = 0 \quad , \tag{3}$$

$$2\pi \langle \Phi_o | H | \chi_E \rangle^2 = \Gamma^\dagger \quad . \tag{4}$$

For simplicity, we assume that the <u>escape width</u> $\Gamma^\dagger$ defined by eq. (4) is independent of energy. The bound states are normalized to unity and χ_E to a δ-function of the energy.

We must now investigate wether the existence of only one doorway state leads to specific features in the cross section. The scattering function of the model is given by the following formula

$$S(E) = \exp(2i\delta) \; \frac{E - \varepsilon_o - \frac{1}{2}i\Gamma^\dagger - \sum\limits_{j=1}^{M} \dfrac{v_j^2}{E - e_j}}{E - \varepsilon_o + \frac{1}{2}i\Gamma^\dagger - \sum\limits_{j=1}^{M} \dfrac{v_j^2}{E - e_j}} \quad , \tag{5}$$

where δ is the smooth potential scattering phase shift associated with χ_E . Equation (5) can <u>identically</u> be written in the form

$$S(E) = \exp(2i\delta) \; \frac{1 - i \sum\limits_{\lambda=1}^{M+1} \dfrac{\gamma_\lambda^2}{E - E_\lambda}}{1 + i \sum\limits_{\lambda=1}^{M+1} \dfrac{\gamma_\lambda^2}{E - E_\lambda}} \quad . \tag{6}$$

Disregarding the trivial quantity δ , the scattering function is characterized by the $2M+2$ parameters $(\varepsilon_o, \Gamma^\dagger, v_j^2, e_j)$ or $(\gamma_\lambda^2, E_\lambda)$. The similarity between expression (6) and the familiar R-matrix parametrization implies that it is always possible to obtain a good fit of any experimental data displaying at most $M+1$ resonances with expression (6) or (5). We conclude that the one doorway state model implies no characteristic feature for the cross section. The question whether the cross section is compatible with the existence of a single doorway state can therefore always be answered in the affirmative. In other words, it is always possible to find a separation of H between a model Hamiltonian H_o and a residual interaction in such a way that only one bound eigenstate of H_o is coupled to the continuum (11). This conclusion drawn in the one-channel case holds very generally.

3. INTERPRETATION OF THE EXPERIMENTAL DATA

We have just seen that the one-doorway state model implies nothing specific about the shape of the cross section. Therefore, it is not meaningful to check experimentally whether a doorway state model mechanism is operative, unless some additional assumptions are added to the model. Since the two main ingredients of the model are the doorway state Φ_o on the one hand and the complicated states ψ_i on the other hand, these additional assumptions are essentially of two types, the nature of the doorway state and the statistical properties of the complicated states.

3.a. Nature of the Doorway State

One can check whether the data are compatible with a specific assumption concerning the nature of the doorway state Φ_o . In the one-channel case, the fit of the cross section with eq. (5) yields the quantities ε_o and $\Gamma^{\uparrow}$, which can then be compared with the values derived from a model calculation. In order to perform this analysis, one must be able to resolve the individual resonance peaks, which is only rarely possible. Moreover, the test mainly concerns the energy ε_o , because the value of $\Gamma^{\uparrow}$ is fairly sensitive to the choice of the scattering wave function χ_E .

One series of examples of an interpretation of this type is provided by some isobaric analogue resonances in light nuclei (A $\lesssim$ 40, ref. (12)). The only other case of which we are aware is presented at this conference by Beres and Divadeenam (16) and will be briefly described here, as an illustrative example. A 58 keV wide $1/2^{+}$ resonance is observed at about 500 keV incident energy in the total cross section of neutrons on ^{208}Pb. In the reaction $^{206}Pb(n,n)$, the neutron partial widths are enhanced near the same bombarding energy (13). The analysis of the eleven resonances participating in the enhancement with expression (5) yields the values ϵ_o = 470 keV , $\Gamma^\uparrow$ = 60 keV (13, 14). The striking analogy between the two sets of parameters for the targets ^{208}Pb and ^{206}Pb gives a hint for the nature of the doorway state involved in ^{207}Pb. Several years ago, Lande and Brown (15) emphasized that the best candidates as prominent doorway states, i.e. doorway states with large values of $\Gamma^\uparrow$, are those where a single particle state is coupled with a vibrational excited state of the core. Beres and Divadeenam (16, contribution to this conference) show that in both reactions, the values of ϵ_o and $\Gamma^\uparrow$ are compatible with the assumption that Φ_o results from the coupling of the $2g_{9/2}$ neutron orbital to the 4^+ vibrational excited state of ^{206}Pb and ^{208}Pb, respectively. The agreement with experiment is rather convincing in view of the fact that all parameters are obtained from other sources and are not fitted to this piece of data. An additional check on the validity of the identification of Beres and Divadeenam would be provided by the radiative width of Φ_o . This will be discussed in section 3.d.

3.b. Statistical Assumptions

Prominent doorway states have by definition a large overlap with the state $V|\chi_E>$. Hence, they have a simple configuration if expressed in terms of the channel state χ_E . On the contrary, the states ψ_j are complicated and one may expect that statistical assumptions can be valid for the corresponding parameters. More specifically, one can assume that the eigenvalues e_j and matrix elements v_j^2 follow the Wigner and Porter-Thomas distributions, respectively. In practice, it is very difficult to check these assumptions experimentally. Indeed, the number of fine structure peaks which can be resolved and analyzed is usually so small that no reliable statistical analysis can be made. This is so because the number of enhanced fine structure peaks is either really small or on the contrary so large that only the average cross section can be measured. Care must also be taken because the resonances associated with states with small values of v_j^2 can be missed so that the measured distributions are incomplete. For these reasons, one can in practice test only less specific statistical assumptions.

In most analyses, the assumption which is tested is that the quantity

$$R(E) = \sum_{j=1}^{M} \frac{v_j^2}{e_j - E} \tag{7}$$

possesses the following property, characteristic of the statistical R-functions (17) in the limit of large M

$$R(E+iI) \simeq -i \Gamma^{\downarrow} \quad , \tag{8}$$

where the spreading width $\Gamma^{\downarrow}$ is independent of E and of I , provided that $d \ll I \ll e_M - e_1$. Using eqs. (7) and (8), we find

$$\Gamma^{\downarrow} = I \sum_{j=1}^{M} \frac{v_j^2}{(E-e_j)^2 + I^2} = \frac{2\pi}{d} v^2 \quad , \tag{9}$$

where v^2 is by definition the average of the quantities v^2 in an interval of length 2I around E . The validity of approximation (8), i.e. of the independence of $\Gamma^{\downarrow}$ upon E , has been checked in ref. (18) in the case of an intermediate structure found in $^{56}Fe(n,n)$. This analysis was corrected by Jeukenne and Mahaux (11) and applied to the case $^{206}Pb(n,n)$ described above. They find that the imaginary part of the quantity R(E+iI) varies between -140 and -190 keV in the range 200-600 keV.

We note that assumption (8) is equivalent to two other relations which we now give, in the sense that any one of these implies the two other ones. Using relation (8) and identifying the expression of S(E+iI) obtained from eqs. (5) and (6), we find the distribution of the partial widths γ_λ^2 . The average γ^2 , in the sense of eq. (9), is given by

$$\gamma^2 = \frac{d}{4\pi} \frac{\Gamma^{\uparrow} \Gamma^{\downarrow}}{(E-\varepsilon_o)^2 + \frac{1}{4}(\Gamma^{\downarrow}+2I)^2} \quad . \tag{10}$$

Hence, the strength function γ^2/d displays an enhancement in the vicinity of the energy of the doorway state, if assumption (8) holds. This characteristic is often taken as a signature of the doorway state model (plus the statistical assumption (8)). As will be discussed later on, a similar local concentration of strength can, however, also be due to random fluctuations.

Inserting relation (8) in the expression S(E+iI) of the average scattering function, we find

$$<S(E)>_I = S(E+iI) = \exp(2i\delta)(1-i\frac{\Gamma^{\uparrow}}{E-\varepsilon_o+\frac{1}{2}i(\Gamma^{\uparrow}+\Gamma^{\downarrow}+2I)}) \tag{11}$$

Hence, the measurement of the _average_ total cross section yields the quantity $\Gamma^{\downarrow}$ and can verify whether it is independent of energy. This check does not imply a measurement of the fine structure but is less sensitive than the use of eq.(8) because of two reasons. Firstly, the quantity $\Gamma^{\downarrow}$ in eq.(10) appears in the combination $\Gamma^{\uparrow}+\Gamma^{\downarrow}+2I$ and must be determined from a difference between measured quantities of comparable magnitude. Secondly the average cross section may include contributions due to different angular momentum or parity states. They can usually be disentangled if the fine structure is measured.

In practice, relation (10) or its generalization to the many-channel case is the only one which is comparable with experimental data because, as we said above, either too few fine structure peaks lie within the intermediate structure or more often because too many exist to be resolved. The latter situation is the most frequent one. Fortunately, it is also the one where intermediate structure due to random statistical fluctuations is less likely, as we discuss below.

3.c. Intermediate Structure from Statistical Fluctuations

The signature of a doorway state mechanism if accompanied by the statistical property (8) is that the average cross section displays a resonance. Such bumps of intermediate width appear rather frequently in measured cross sections. This leads at first to the belief that many isolated doorway states were observed. It became, however, soon clear that the occurrence of such bumps should be expected from the random fluctuations in level density and width. This was explicitly exhibited by Singh et al.(19) and Moldauer (20). In a contribution to this conference, Schrack extends the study of ref.(19) where a code generated total cross section data from random distributions in width and spacing. His results and those of ref.(18) concern the region of Ericson fluctuations, i.e. $<\Gamma>/d \gg 1$.

A similar study for the case $<\Gamma>/d \ll 1$ has recently been published by the Livermore group (21) who also present part of their results at this conference. The lower part of figure 1 shows a typical set of resonance widths and energies, generated by a random choice of widths and spacings, with Porter-Thomas and Wigner distributions, respectively. The full curve in the upper part of the figure is the average cross section ($I = 8d$),

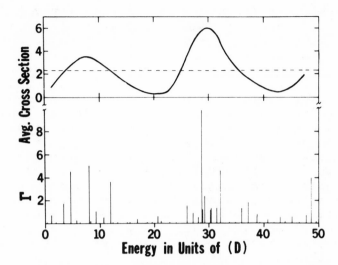

Fig. 1. A typical set of widths and energies generated by
a random choice is shown on the lower plot. The solid
curve in the upper plot shows the corresponding average
cross section (I=8d) and the dashed line the mean cross
section (from ref.(21)).

the dashed line the mean cross section. The energy unit
in the abscissa is d , the average level spacing. We see
that intermediate structure, i.e. a local concentration
of large widths or bumps of intermediate width in the average cross section, must inevitably appear within the
standard statistical assumption. The main question is of
course with what frequency this"statistical intermediate
structure" occurs. Several indications are provided by
ref.(21) where, however, I did not find the value of the
ratio Γ/d , where Γ is the average width. It has probably been chosen appropriate for the $^{206}Pb(n,n)$ case,
i.e. about 1/10. We emphasize that the statistical intermediate structure discussed here is not due to Ericson
fluctuations, which are discussed below. Baglan, Bowman
and Berman (21) counted the number of times that a statistical intermediate structure occurs per 1000 resonances.
They define a statistical intermediate structure as a bump
(in the average cross section) at least 1.5 times higher
than the mean cross section. The intermediate resonance
as defined in ref.(21) is also characterized by the minimum number n of individual resonances contributing to
the structure. If n = 14 , 10 structures exist for 1000
resonances. Since the average level spacing in ^{207}Pb is
d = 40 keV and 14 fine structure peaks exist, a statis-

tical intermediate structure occurs once every 4 MeV.
While this is a valid argument in favour of the doorway
interpretation of the structure, in this case, one should
always keep in mind that statistics impose to unlikely
events to happen sometimes. An interesting result of ref.
(21) is that the number of statistical intermediate structures decreases rapidly when the number n of contributing levels increases, i.e. when the width of the structure increases. If n = 10 , fifteen statistical structures are found per 1000 resonances. When n = 20 , only
two statistical structures appear per 1000 resonances.
This clearly shows that the giant dipole resonances and
isobaric analogue resonances in medium-weight and heavy
nuclei are not due to statistical intermediate structures,
independently of any other theoretical evidence. In their
cases, however, one is usually in a situation where Γ/d
$>> 1$, so that Ericson fluctuations occur. We now discuss
this situation.

     Singh et al.(19) showed both numerically and in the
frame of analytical approximations that bumps of intermediate width occur rather frequently in the region $\Gamma/d >>$
1 . In the case $\Gamma/d = 33$, for instance, they find a
30% chance of finding a structure of width 6Γ wide with
a cross section at the peak at least equal to 1.4 times
the mean cross section. The situation is thus similar to
the case $\Gamma/d << 1$. Experimentally, intermediate structures were found by a number of authors in neutron total
cross sections in the MeV region (22). Careful measurements
by Carlson and Barschall (23) confirm and extend these results. These authors, however, demonstrate that in all cases where bumps were found, i.e. Mg, Al, Si, S, Ti, Fe,
^{206}Pb and Pb, these structures can be accounted for by the
expected fluctuations in widths and spacings of compound
nucleus levels. Their method of analysis is similar to
that proposed by Agodi and Pappalardo (24). In the case
of overlapping levels, it therefore appears necessary to
check whether the structure is present in a given angular
momentum and parity state before suggesting the possible
existence of a doorway state phenomenon.

             3.d. Correlation between Partial Widths

     For simplicity, we dealt up to now with a one-channel
case. The model can, however, be extended to a many-channel situation (10). The inclusion of photon channels is
particularly trivial. We distinguish between prominent
(or strong) doorway states, which may produce an observable intermediate structure phenomenon, and weak ones,

which provide a kind of background strength function (25).
In general, the doorway states are different from channel
to channel and only rarely will a doorway state be prominent in several channels. This can happen, however. For
instance, Lane (3, session 4) proposes that a pygmy resonance may be a prominent common doorway for the neutron
and photon channels, in a number of nuclei with A ≤ 150.
In this case, however, the energy ε_o is located far
below threshold and no modulation of the neutron strength
function is observed, but rather a correlation between
photon and neutron partial widths. If the common prominent
doorway is located above threshold, both the neutron and
photon widths should be enhanced and they should, in addition, be correlated. Baglan, Bowman and Berman (21 and
contribution to this conference) studied the reaction
$^{207}Pb(\gamma,n)$, searching whether the doorway detected in the
reaction $^{206}Pb(n,n)$ is also prominent in the photon channel.
This may be expected from the model of ref.(16). Baglan
et al.(21) find an enhancement of the radiative widths at
the appropriate energy. The authors argue that two
statistical structures overlap within an energy interval
of 200 keV only once per 40 MeV. Moreover, they find that
a correlation exists between photon and neutron partial
widths with 96% probability and that the degree of correlation, measured by some quantity r is $r = 0.45 \pm 0.1$,
the values $r = 0$ corresponding to no correlation and
$r = 1$ to full correlation. Using this value for r , the
authors determine the photon escape width of the doorway
state to be equal to 38 eV. It would be of interest to
compare this number with the model of Beres and Divadeenam (16).

Much care should be taken before concluding to the
existence of a common doorway state. The Livermore group
(21) studied several cases, for instance $^{57}Fe(\gamma,n)$. They
find an enhancement of the photon widths at about the same
energy as the one observed for neutron widths (11,18,26).
The enhancement involves about eleven resonances, with
d ≃ 40 keV. According to ref.(21), two enhancements in
different channels should coincide only once per 40 MeV,
if both enhancements have statistical origin. However,
no significant correlation is observed in the present case
(r = 0.2) between the partial widths (21). This should
remind us again that even unlikely events do occur.

We conclude this paragraph by recalling that while
a correlation between partial widths may indicate the
existence of a doorway state, the absence of correlations
does not disprove it, since a doorway state may very well
be prominent in only one channel. Moreover, other mecha-

nisms, like the interference with direct reactions or the correlations between partial widths corresponding to the states of a common rotational band may provide correlations (session 4).

3.e. Discussion

Besides the data related to ^{207}Pb discussed above and the well-established cases reviewed in the following section, no convincing example has been found of an intermediate structure due to an isolated doorway state. Why is this so ? In order for a doorway state to produce a bump in the average cross section, two conditions must be fulfilled. Firstly, the doorway state must be isolated, which implies

$$\Gamma^\uparrow + \Gamma^\downarrow \lesssim D \qquad\qquad , \qquad (12)$$

where D is the distance to the next prominent doorway state. Secondly, the doorway state must be prominent which requires

$$\Gamma^\downarrow \lesssim 2 \, \Gamma^\uparrow \qquad . \qquad (13)$$

Payne (27) estimated the values of D , $\Gamma^\uparrow$ and $\Gamma^\downarrow$ for a number of spherical target nuclei and typical three quasi-particle states. He finds that conditions (12) and (13) can only be fulfilled for light and doubly magic nuclei, for incident neutron energy less than a few MeV. In medium-weight and heavy nuclei (A $\gtrsim$ 60) with partially filled shells, the value of $\Gamma^\downarrow$ is so large that condition (13) is not fulfilled. Hence, we expect isolated doorway states to show up in excitation curves only if they are coherent, so that their escape width is particularly large (15).

4. RECENT PROGRESS IN WELL-ESTABLISHED DOORWAYS

4.a. Giant Photonuclear Resonances

The giant dipole resonances were the first examples of intermediate structure resonances to be detected and properly analyzed in terms of doorway states (6). The giant resonances observed by Barschall (28) in total neutron cross sections at low energy should not be associated with a doorway state phenomenon (29). In recent years, a number of calculations of the giant dipole resonance have been performed which explicitly include several channels and 1p-1h excitations. In the cases of ^{12}C (30), ^{16}O (31, 32,33), ^{28}Si (34), ^{40}Ca (30) and ^{208}Pb (35) the gross

structure of the resonance was fairly well reproduced.
The location of the dipole resonance obtained in ref.(35)
is in agreement with experiment, while it was found too
low in bound state calculations (36). It would be interesting to investigate whether this is due to the use of
a Woods-Saxon basis or of a suitable residual interaction,
in particular whether the latter also yields correctly
the low lying collective states in ^{208}Pb in ref.(35). Few
attempts exist to include some complicated states in the
calculations. Gillet, Melkanoff and Raynal (37) proposed
that the fine structure in ^{16}O is mainly due to 2p-2h
states but recently Shakin and Wang (38) obtained an impressive fit to the experimental data by assuming that the
complicated states are 3p-3h excitations.

In heavier nuclei, the spreading width of the giant
dipole resonance was estimated by Danos and Greiner (39)
to be about a few MeV and recently by Dover, Lemmer and
Hahne (40). The latter authors use the Green function
approach and obtain a width ranging from 2 to 5 MeV in
^{208}Pb (the experimental value is about 4 MeV), depending
upon the choice of optical potential parameters.

In recent years, a number of theoretical and experimental efforts have been devoted to the isospin splitting
of the dipole resonance, to giant resonances corresponding
to M1 and E2 transitions, to pygmy resonances, etc.
We refer to sessions 4 and 8 for a review of these problems. In a contribution to the present session, Fubini,
Oliva and Prosperi give indication that they have observed
the lower member of the quadrupole giant resonance.

4.b. Isobaric Analogue Resonances

The first detailed theoretical interpretation of the
isobaric analogue resonances is due to Robson (41). A very
complete theoretical and numerical study was recently completed by Auerbach, Hüfner, Kerman and Shakin (42). The
agreement between the experimental values of ε^- and $\Gamma^\uparrow$
and the theoretical calculations is excellent. The spread-
ing width $\Gamma^\downarrow$ cannot be computed very accurately. It
appears that the theoretical values of $\Gamma^\downarrow$ are slightly
too small.

Isobaric analogue states produce intermediate structure resonances because their spreading width is quite
small. Indeed, the isobaric analogue state has (almost)
pure isospin T+1 , while the complicated states have
(almost) pure isospin T . In many cases, the internal

coupling between Φ_o and ψ_j via the proton channels

$$< \Phi_o | V | \chi_E > < \chi_E | V | \psi_j > \qquad , \qquad (14)$$

since the nuclear residual interaction is operative here.
This special feature is responsible for the Robson asymmetry which states that in the case of pure external mixing the enhancement of the fine structure widths in a proton channel c is described by

$$\gamma^2_{\lambda c} = \gamma^2_{jc} \frac{(E - \varepsilon_o)^2}{(E-E_o)^2 + \frac{1}{4}(\Gamma^\downarrow)^2} \qquad . \qquad (15)$$

Here, ε_o is the energy of the doorway state, $\Gamma^\downarrow$ the spreading width, E_o the resonance energy (usually $E_o < \varepsilon_o$), γ^2_{jc} is the background width, i.e. the width that level j would have in the absence of the isobaric analogue state. The reduced widths in the neutron channels n are not enhanced in this external coupling model. Nevertheless, the isobaric analogue resonance shows up in (p,n) reactions. This can be understood from the Hauser-Feshbach formula describing the fluctuating part of the total (p,n) cross section. The transmission factor T_p has the same energy dependence as expression (12) and the same holds for σ_{pn} provided that $\Sigma T_n \gg T_p$.

In a contribution to this conference, Grosse, Richter and Tepel have detected an isobaric analogue resonance (IAR) in the reaction $^{89}Y(\alpha,n)^{92}Nb$ where both the entrance and exit channels have isospin $T_<$. This IAR is well-known from $^{92}Zr(p,p)$ and $^{92}Zr(p,n)$. The latter cross section is in nice agreement with the external coupling model. The energy dependence of the fluctuating part of the (α,n) cross section as given by the Hauser-Feshbach formula arises from T_p only, in the external coupling model. This corresponds to a dip followed by a peak. This shape appears at variance with data reported by the Heidelberg group. If their preliminary results are confirmed by more accurate measurements, they probably indicate that internal mixing is significant and perhaps that the direct part of the cross section is not negligible in this case.

Temmer et al. report their recent experimental data (43) at this conference. They observe a number of narrow substructures within an isolated $1/2^+$ IAR in ^{71}As. These substructures have $\ell = 0$ signature, are correlated at several angles, in different channels and in this region of excitation $\Gamma/d \simeq 1$. Hence, they are not Ericson fluctuations. Since each of the substructures encompasses about 50 resonances, an interpretation in terms of statistical intermediate structures is excluded. Two inter-

pretations appear plausible. The first one, proposed by
the Rutgers group (43) is that the simple states responsible for the substructures are hallway states, i.e. states of next degree of complexity. It would be interesting
to see from density of levels arguments whether this is
a reasonable assumption. We believe that the density of
hallway states is actually much larger than the observed
substructures would indicate. We can, however, assume that
there exist "prominent" hallway states which are strongly
coupled to the isobaric analogue state (IAS). This would
give them an appreciable escape width thereby allowing
them to become in turn prominent doorways for the surrounding hallway states which are weakly coupled to the
IAS. If this interpretation is correct, the branching ra-
tios and the ratio $\Gamma^\uparrow/\Gamma^\downarrow$ of each substructure should be
identical to those of the IAR. The second possible inter-
pretation is specific to an IAR. Let us assume that each
substructure is due to a weak doorway state. Normally,
these would not give rise to an observable intermediate
structure because their escape width is too small compared
to their spreading width. In the external mixing model,
the escape widths of the weak doorway states are multiplied by the enhancement factor (eq.(15)). Thus, their
escape width increases and they may produce an observable
intermediate structure bump. If this interpretation holds,
the branching ratios should not be equal to those of the
IAR, but rather to those characteristic of each weak doorway. Thus, it is in principle possible to distinguish experimentally between the two possible interpretations,
which correspond to internal and external mixing, respectively.

4.c. Subthreshold Neutron Induced Fission

Striking intermediate structures were discovered in
neutron induced fission around 1967 (44,45). They exhibit
the existence of prominent doorway states in the fission
channel, which are interpreted as quasi-bound states in
the second minimum of a double-humped fission barrier
(46-51). There exists some evidence (46,52) that in the
present case $\Gamma^\uparrow \gg \Gamma^\downarrow$, in contrast with the examples
reviewed above. More details concerning neutron induced
fission can be found in session 3 of this conference.

5. CONCLUSIONS

Whenever an intermediate structure resonance can be
ascribed a given spin and parity and contains twenty or

more compound states, a statistical origin is very unlikely and a doorway state interpretation appears suitable.
When less than about ten compound states exist, the interpretation in terms of doorway states should be attempted
only when a number of favourable circumstances are gather-
ed, for instance a knowledge of the doorway configuration
(like in the isobaric analogue resonances), or correlations between individual partial widths in several channels, or a large frequency of appearance (like in the experiment of the Rutgers group (43)). In most other cases,
it appears more natural to ascribe the existence of an
intermediate structure resonance to random statistical
fluctuation.

REFERENCES

(1) B. Block and H. Feshbach, Ann. Phys. (N.Y.) 23 (1963)
    47 ; C. Shakin, Ann. Phys. (N.Y.) 22 (1963) 373
(2) J.J. Griffin, Phys. Rev. Letters 17 (1966) 478
    M. Blann, Phys. Rev. Letters 21 (1968) 1357
(3) A.M. Lane, Ann. Phys. (N.Y.) 63 (1971) 171
(4) P. Axel, K.K. Min and D.C. Sutton, Phys. Rev. 2C
    (1970) 689
(5) C. Bloch, N. Cindro and S. Harar, Progress in Nucl.
    Phys. (D.M. Brink and J.H. Mulley, eds) 10 (1969)
(6) A.M. Lane, R.G. Thomas and E.P. Wigner, Phys. Rev.
    98 (1955) 693
(7) K.A. Brueckner, R.J. Eden and N.C. Francis, Phys.
    Rev. 100 (1955) 891
(8) V.F. Weisskopf, Phys. Today 14, 7 (1961) 18
    A.K. Kerman, L. Rodberg and J.E. Young, Phys. Rev.
    Letters 11 (1963) 422 ; A. Lande and B. Block, Phys.
    Rev. Letters 12 (1964) 334 ; H. Feshbach, A.K. Kerman
    and R.H. Lemmer, Ann. Phys. (N.Y.) 41 (1967) 230
(9) H.P. Kennedy and R. Schrils, eds, Intermediate Struc-
    ture in Nuclear Reactions (University of Kentucky
    Press, Lexington, 1968)
(10) C. Mahaux and H.A. Weidenmüller, Shell-Model Approach
    to Nuclear Reactions (North-Holland Publ. C°, Am-
    sterdam, 1969)
(11) J.P. Jeukenne and C. Mahaux, Nucl. Phys. A136 (1969)
    49
(12) A.M .Lane, in Isospin in Nuclear Physics, D.H. Wil-
    kinson, ed. (North-Holland Publ. C°, Amsterdam 1969)
(13) J.A. Farrell, G.C. Kyker Jr., E.G. Bilpuch and H.W.
    Newson, Phys. Letters 17 (1965) 286
(14) A. Lejeune and C. Mahaux, Z. Physik 207 (1967) 35
(15) A. Lande and G.E. Brown, Nucl. Phys. 75 (1966) 344
(16) W.P. Beres and M. Divadeenam, Phys. Rev. Letters

25 (1970) 596

(17) A.M. Lane and R.G. Thomas, Revs Mod. Phys. 30 (1958) 257

(18) A.J. Elwyn and J.E. Monahan, Nucl. Phys. A123 (1969) 33

(19) P.P. Singh, P. Hoffman-Pinther and D.W. Lang, Phys. Letters 23 (1966) 255

(20) P.A. Moldauer, Phys. Rev. Letters 18 (1967) 249

(21) R.J. Baglan, C.D. Bowman and B.L. Berman, Phys. Rev. 3C (1971) 2475

(22) K.K. Seth, Phys. Letters 16 (1965) 306
F. Manero, Nucl. Phys. 65 (1965) 419

(23) A.D. Carlson and H.H. Barschall, Phys. Rev. 158 (1967) 1142

(24) A. Agodi and G. Pappalardo, Nucl. Phys. 47 (1963) 129 ; G. Pappalardo, Phys. Letters 13 (1964) 320

(25) F. Iachello, Ann. Phys. (N.Y.) 52 (1969) 16

(26) C.D. Bowman, E.G. Bilpuch and H.W. Newson, Ann. Phys. (N.Y.) 17 (1962) 319

(27) G.L. Payne, Phys. Rev. 174 (1968) 1227

(28) H.H. Barschall, Phys. Rev. 86 (1952) 431

(29) K.W. McVoy, Phys. Letters 17 (1965) 42

(30) M. Marangoni and A.M. Saruis, Nucl. Phys. A132 (1969) 649

(31) B. Buck and A.D. Hill, Nucl. Phys. A 95 (1967) 271

(32) J. Raynal, M.A. Melkanoff and T. Sawada, Nucl. Phys. A101 (1967) 369

(33) J.D. Perez and W.M. MacDonald, Phys. Rev. 182 (1969) 1066

(34) M. Marangoni and A.M. Saruis, Nucl. Phys. A166 (1971) 397

(35) R.F. Barrett and P.P. Delsanto, Phys. Letters 34B (1971) 110

(36) T.T.S. Kuo, J. Blomqvist and G.E. Brown, Phys. Rev. Letters 31B (1970) 93

(37) V. Gillet, M.A. Melkanoff and J. Raynal, Nucl. Phys. A97 (1967) 631

(38) C.M. Shakin and W.L. Wang, Phys. Rev. Letters 26 (1971) 902

(39) M. Danos and W. Greiner, Phys. Rev. 138 (1965) B876

(40) C.B. Dover, R.H. Lemmer and F.J.W. Hahne, preprint TH71-15 (Orsay, June 1971)

(41) D. Robson, Phys. Rev. 137 (1965) B535

(42) N. Auerbach, J. Hüfner, A.K. Kerman and C.M. Shakin, (to be published in Revs Mod. Phys.)

(43) G.M. Temmer, M. Maruyama, D.W. Mingay, M. Petrascu and R. Van Bree, Phys. Rev. Letters 26 (1971) 1341

(44) D. Paya, J. Blons, H. Derrien, A. Fubini, A. Michaudon and P. Ribon, J. de Physique 29 (1968), suppl.1

159 ;
A. Fubini, J. Blons, A. Michaudon and D. Paya,
Phys. Rev. Letters 20 (1968) 1373

(45) E. Migneco and J.P. Theobald, Nucl. Phys. A112
(1968) 603

(46) J.E. Lynn, Nuclear Structure, Dubna Symposium 1968
(I.A.E.A., Vienna, 1968) p. 463

(47) J.E. Lynn, The Theory of Neutron Resonance Reactions
(Clarendon Press, Oxford 1968)

(48) H. Weigmann, Z. Physik 214 (1968) 7

(49) V.M. Strutinsky, Nucl. Phys. A95 (1967) 420 ;
ibid A122 (1968) 1

(50) S.G. Nilsson et al., Nucl. Phys. A115 (1968) 545 ;
ibid A131 (1969) 1, A140 (1970) 289

(51) D.K. Sood and N. Sarma, Nucl. Phys. A151 (1970) 532

(52) H. Weigmann, G. Rohr and J. Winter, Phys. Letters
30B (1969) 624.

DISCUSSION

WIGNER

You mentioned a correlation between neutron widths and gamma
rays, in the case of a single doorway state. Could you say a little
bit more about it? Barschall told me that it is almost impossible
to see this. If you have a background amplitude one tenth of the
doorway amplitude then, because of the way the amplitudes combine,
it can be constructive or destructive. The ratio of the resulting
intensities is four. (See discussion following H.E. Jackson's
talk) It's fantastic, but apparently true. It would be good to
know. Could you say a little more about that case? How strong is
the statistical correlation?

MAHAUX

Yes. I skipped over this case, because Dr. Berman is going to
give a contributed paper on exactly that problem. I just wanted
to show it in order to introduce what he is going to say in the
frame of the general discussion.

EVIDENCE FOR INTERMEDIATE STRUCTURE NEAR AN ISOBARIC ANALOGUE RESONANCE[*]

G. M. Temmer, M. Maruyama[†], D. W. Mingay[‡]

M. Petraşcu[§] and R. Van Bree

Department of Physics, Rutgers University
New Brunswick, New Jersey 08903, U.S.A.

INTRODUCTION

In an investigation of the elastic and inelastic scattering of protons on separated isotopes of ^{70}Ge and ^{72}Ge for the purpose of locating isobaric analogue states of ^{71}Ge and ^{73}Ge in ^{71}As and ^{73}As, respectively, we found one $\ell = 0$ resonance in each isotope, displaying pronounced structure within their "widths". The purpose of this Contribution is to present evidence that this structure is intermediate between the gross analogue resonance structure on the one hand, and the ultimate compound-nucleus fine structure on the other hand. It will also be argued that these structures are not Ericson fluctuations.

THE CASE OF ^{70}Ge + p

The $\ell = 0$ analogue resonance of interest here lies at 9.66-MeV excitation in ^{71}As (5.06-MeV bombarding energy), and was shown to correspond to the $J = 1/2^+$ state in ^{71}Ge at 2.22-MeV excitation. This correspondence is established by the matching of about a dozen $\ell = 0$ and $\ell = 2$ resonances (to within ~ 30 keV) to corresponding known states in the parent nucleus ^{71}Ge.[1] A special (d,p) experiment was performed on our broad-range spectrograph[2] to insure that only one $\ell = 0$ state exists in this region of excitation of ^{71}Ge; moreover, the nearest-neighbor $\ell = 0$ resonances (over 200 keV away) were

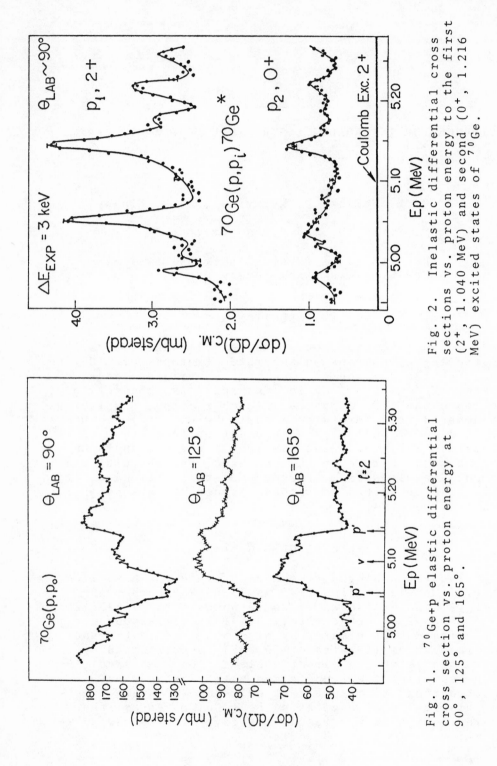

Fig. 2. Inelastic differential cross sections vs. proton energy to the first (2^+, 1.040 MeV) and second (0^+, 1.216 MeV) excited states of ^{70}Ge.

Fig. 1. ^{70}Ge+p elastic differential cross section vs. proton energy at 90°, 125° and 165°.

located in our excitation curves. These two facts rule
out the possibilities that either additional parent
states, or anomalously large Coulomb-energy shifts
might account for the cluster of subresonances seen
within the analogue resonance. The elastic and inelastic excitation curves (to the first 2^+ and second
0^+ excited states of ^{70}Ge) over the region of interest
are shown in Figs. 1 and 2, respectively. Angular distributions of protons to the first five excited states
were found to be isotropic, consistent with the $1/2^+$
assignment of the resonance. It is to be noted that
the substructures have widths between 10 and 20 keV,
whereas the overall experimental resolution is about
3 keV.

Let us now examine the question whether the observed
substructure might not in fact reflect the ultimate fine
structure, i.e. the structure resolved in the Duke experiments[3], and represent the lowest rank of the
hierarchy in compound-nucleus formation; in other words,
whether individual states of 10-20 keV widths could
exist in this region of excitation. First of all,
Maruyama et al.[4], in their pioneering blocking lifetime
measurements with the same target nucleus, at about the
same excitation energy, determined average widths of
ultimate fine structure of about 20 eV; this is also
very close to a theoretical estimate, based on a detailed
Hauser-Feshbach calculation[5]. Our very recent extension
of this work[6] confirms this result and shows that fine-
structure widths, either off or on analogue resonances,
certainly lie between 10 and 200 eV. One would be hard
put to understand widths of about 100 times these values
in terms of individual fine-structure resonances. Moreover, the question of level spacing must be examined. A
reliable estimate of the density of $J=1/2^+$ states at
9.6-MeV excitation, based on a simple extrapolation of
the measured s-wave neutron resonance density at the
neutron binding energy of 7.4 MeV, yields the value 3
per keV[5]. Once again, one is hard put to explain
structure of the order of 10 keV width on this basis;
about 30 to 60 fine-structure states must be contained
within each structure.

Finally we note that the structures are correlated
at different angles and in different outgoing channels;
the ratio of width to spacing of the fine-structure
levels lies in the range 0.1-1.0; these three facts together rule out Ericson fluctuations as the origin of
the observed structure.

Inelastic scattering can, in principle, yield the key to the interpretation of these "doorway" states in terms of their particle-hole hierarchy, once we know something concerning the structure of the target states involved.

THE CASE OF ^{72}Ge + p

Considerable structure, 10-20 keV wide, was found within the analogue resonance in ^{73}As at an excitation energy of 12.0 MeV (6.36 MeV bombarding energy) corresponding to a unique $\ell=0$ parent state in ^{73}Ge at 2.96 MeV excitation[7]. As in the case of ^{70}Ge, we are sure that only one $\ell=0$ state exists in that region[7], and that we account for all neighboring s-wave resonances. 10 analogue states are found to line up to better than 30 keV. The elastic excitation curve in the region of interest is shown in Fig. 3.

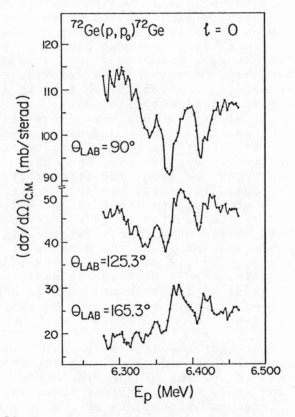

Fig. 3. ^{72}Ge+p elastic differential cross section vs. proton energy at 90°, 125.3° and 165.3°.

Now there is one important difference between the
^{71}As and ^{73}As cases: in the former, the resonance lies
well below the opening of the neutron channel at 7.12
MeV, while in the latter, we find ourselves above the
neutron threshold at 5.21 MeV. This would account for
considerably wider fine-structure resonances in the
latter case, and was, in fact, beautifully demonstrated
by the crystal-blocking measurements of Maruyama et al.[4]
Once again, Hauser-Feshbach estimates yield widths of
about 2700 eV[5]. Hence, fine-structure widths of the
order of our energy resolution ($\sim$3 keV) are reasonable.
However, the level-density estimate militates against
the interpretation of the substrucure as individual,
ultimate fine-structure states: for the case of ^{73}As,
an extrapolation similar to the one carried out for ^{71}As[5],
leads to a level density of J=1/2$^+$ states of about 15
per keV, or at least 150 levels per substructure; about
50 levels contribute within our energy resolution interval. Unfortunately, the inelastic excitation curves in
this case (Fig. 4) are not as revealing as in the previous isotope.

The ratio of width to spacing for the fine structure
in this nucleus is seen to be about 50, and hence this
condition for Ericson fluctuations is fulfilled. Since
all the enhanced fine-structure states within the con-
fines of an analogue resonance are of the same spin and
parity (1/2$^+$), correlations could exist in the substructure at different angles, and we cannot rule out
Ericson fluctuations as being the cause of the observed
phenonenon in ^{73}As.

CONCLUSION

The substructure we find in analogue resonances in
compound nuclei of arsenic are definitely not due to
individual fine-structure states from what we know
concerning the widths and level densities of the latter.
These substructures encompass of the order of 50 or more
ultimate fine-structure states, and have widths about
an order of magnitude greater than the experimental
resolution.

Since the structures in question are more pronounced
at the level of $\sim$20 keV than at the level of the experimental resolution near 3 keV, we believe that the
observed fluctuations can also not be blamed on ordinary finite-sampling effects in the (Porter-Thomas)
level-width distribution.

   We are probably dealing with a modulation associated with "doorway" phenomena, at least in the case of
^{71}As, where Ericson fluctuations are ruled out.

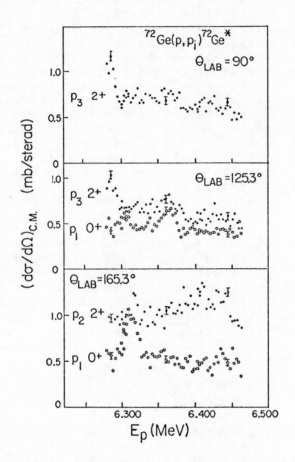

Fig. 4. Inelastic differential cross sections vs.
proton energy to the first (0$^+$, 0.693 MeV), second
(2$^+$, 0.837 MeV), and third (2$^+$, 1.486 MeV) excited
states of ^{72}Ge at three different angles.

REFERENCES

*Supported in part by the National Science Foundation.

†On leave from Japan Atomic Energy Research Inst., Tokai, Japan. Partly supported by Bell Telephone Labs.

‡University of the Witwatersrand, Johannesburg, R.S.A. Partly supported by Bell Telephone Labs.

§Institute for Atomic Physics, Bucharest, Rumania.

[1] L. H. Goldman, Phys. Rev. 165, 1203 (1968); G. M. Temmer et al., Phys. Rev. Lett. 26, 1341 (1971).

[2] W. Darcey, J. Fenton, T. H. Kruse and M. E. Williams, unpublished.

[3] See e.g., J. C. Browne et al., Nucl. Phys. A153, 481 (1970).

[4] M. Maruyama et al., Nucl. Phys. A145, 581 (1970).

[5] J. R. Huizenga, private communication. We are greatly indebted to Dr. Huizenga for making these estimates for us.

[6] W. M. Gibson et al., Bull. Amer. Phys. Soc. 16, 557 (1971), and to be published.

[7] G. Heymann et al., Z. Physik 218, 137 (1969); however, a number of the ℓ_n-value assignments given therein were found to be in error: W. Darcey, private communication of unpublished Aldermaston results.

DISCUSSION

NEWSON
    Could you say a little more about the hierarchy of doorway
states you mentioned.

TEMMER
    Well, let's start at the top. The broadest thing you have are
size resonances or giant dipole resonances or something like that.
That's the biggest envelope. If you go within that, one example
of what you can find is an isobaric analogue resonance, mainly two-
particle, one hole, at least near closed shells. All right?

NEWSON
    I thought the isobaric analogue was a single particle state.

TEMMER
    No. If you T-minus the parent, you get mostly two-particle,
one-hole states. In our case, it's six-sevenths of that type and
only one-seventh of the other. Within that type of structure, you
find what we have been talking about, what you so beautifully find
every day of the week, 2500 of them. I claim that that's it. In
this particular case, there isn't anything in between, and that's
another mystery. What is the hierarchy? The isobaric analogue
state in this case by the way, is not two-particle, one-hole.
It's possibly four-particle, three-hole.

NEWSON
    What about the next step?

TEMMER
    The next step below? Whatever the isobaric analogue state
happens to be. This is an experimental paper. I'm not trying to
put an interpretation on it.

FALLIEROS
    I wanted to ask you about the spectroscopic factors of the
parents. In the first case that you showed, the number was very
small, .13 or something like that. Can you say more about the
other cases?

TEMMER
    Not much, except to say yes. It so happens that in Ge, which
isn't a nucleus that theorists are raving mad about, for obvious
reasons---it's far from closed shells. It's not particularly
exciting. The spectroscopic factors are small, .1 or .2 in the
parent, and presumably also in the daughter, although we haven't
done a very good analysis of it. We're far from the center of
strength of the s-wave.

FALLIEROS
    In the second case that you showed, it was considerably higher
in energy, wasn't it? Can you remind me how high?

TEMMER
    12 MeV excitation

FALLIEROS
    And the first one was...

TEMMER
    9.6 MeV

FALLIEROS
    I would have guessed that you moved closer to the single
particle.

TEMMER
    That's true, but let me put it this way--the data are fairly
recent, and I cannot tell you what the spectroscopic factors are
there.

RICHTER   (Heidelberg and Bochum)
    I have a question related to Professor Fallieros' question.
If I remember one of your slides right, in Ge^{70} (d,p) you had 3 or
4 $\ell=0$ states. Do you see for these other cases also,fine structure
in the resonances? Did you look?

TEMMER
    Yes. We saw much less or none at all. We are only showing
typical cases here.

LANE (Harwell)
    I am concerned with the identity of these doorway states into
which the analogue fragments. Let's say they are discrete, inner
doorway states, possibly 4-particle, 3-hole or 5-particle, 4-hole
states, something fairly high up in the hierarchy. Apparently,
they are separated from each other, if we accept Professor Temmer's
interpretation, but this is very curious. States lower down in the
hierarchy, two-particle, one-hole are, of course, notoriously
difficult to find experimentally. One would conclude that apparently
they overlap, except maybe near a closed shell. When you go higher
up, they don't. I've never heard of a theory that predicted that.
Rather, just the opposite. There's a Gerry Brown argument going
back to '58, which said that the lower levels of doorways in the
hierarchy might be isolated but eventually they'll begin to overlap,
and then its "curtains" until, of course, you achieve the ultimate fine
structure. But here, apparently we have just the opposite. I know
of no theoretical explanation for this phenomenon.

TEMMER
      I would like to make one remark on an important technical
point. I know that theorists would very much like (if there be
such structure) that it be only near the analogue resonances, and
not away from them. The data seem to indicate that this is true,
but a word of caution is in order. As an experimentalist, I must
say that we have no way of proving it. When you're away from the
analogue resonance, you excite, as I said before, many different
spins and parities, and we have no way of pulling them apart.
The group at Duke has enough patience and fortitude (that with
luck) and some day they might actually get enough cases together to
do it. To use an optical analogy, you get sort of white light of
all wave lengths. When you get in the analogue, you suddenly have
a focusing effect that pulls up the spin 1/2+ states (in this case)
and then you see them. The theorists say of course, this must
happen but experimentally you cannot really support it.

MAHAUX
      In the vicinity of any strong intermediate structure, one
can hope to see "secondary" intermediate structure phenomena
because of the selective enhancement of states of given spin and
parity--as emphasized by Dr. Temmer and also because the ratios
$\frac{\Gamma\uparrow}{\Gamma\downarrow}$ of these "smaller" doorways are enhanced. The intriguing

problem raised by Dr. Temmer's results concerns the smallness of
their spreading width of the secondary peaks, because I don't see
a selection rule operating. But I agree that he has detected
intermediate structure, i.e. that his data does not reflect random
fluctuation.

INTERMEDIATE STRUCTURES IN THE NEUTRON
SCATTERING CROSS SECTIONS OF IRON *

Y. Tomita, K. Tsukada and M. Maruyama

Japan Atomic Energy Research Institute

Tokai-mura, Ibaraki-ken, Japan

Intermediate structures at incident neutron energies 0.36 and 0.7 MeV observed in the neutron scattering cross sections of iron have been shown to be consistent with the doorway-state hypothesis by Elwyn and Monahan[1]. The present authers also observed[2] two structures at about 1.6 and 2.05 MeV in the energy dependence of angular distributions of elastic and inelastic scattering of neutrons by iron. Whether these structures may be regarded as doorway-state resonances or not is interesting.

We present here an analysis of these structures based on the doorway-state picture. Since the observed structures are not so pronounced, we do not take the pure doorway-state mechanism but allow also for direct coupling between the "compound" states and the open channels. The situation is quite similar to that considered by de Toledo Piza and Kerman[3] for isobaric analog resonances. Scattering matrices averaged over an appropriate "intermediate" interval I can be expressed as follows:

$$\langle S^{J\pi}_{n\ell j; n'\ell'j'}\rangle = S^{opt}_{n\ell j}\, \delta_{nn'}\delta_{\ell\ell'}\delta_{jj'} - i\sum_d \frac{e^{i(\Delta^d_{n\ell j}+\Delta^d_{n'\ell'j'})}\sqrt{\Gamma^{\uparrow}_{d;n\ell j}\,\Gamma^{\uparrow}_{d;n'\ell'j'}}}{E_d - E + \frac{1}{2}\Gamma_d} \quad , \quad (1)$$

$$\Delta^d_{n\ell j} = \delta_{n\ell j} + \varphi_{d;n\ell j} \qquad\qquad , \quad (2)$$

$$\Gamma_d = \sum_{n\ell j}\Gamma^{\uparrow}_{d;n\ell j} + \overline{\Gamma}^{\downarrow}_d + I \qquad\qquad , \quad (3)$$

where $J\pi$ is the total angular momentum and parity of the system and, n and d denote target states and doorway-states (in the present case n = 0 and 1 represent the groung state and the first excited state). $S^{opt}_{n\ell j}$ and $\delta_{n\ell j}$ are the S-matrix and the real part of

phase shift of a complex optical potential model, respectively.
The imaginary part of the potential and the additional phase $\mathscr{G}_{d;n\ell j}$ are
due to the direct coupling between the "compound" states and the
open channels. The quantities $\Gamma_{d;n\ell j}^{\uparrow}$ and $\Gamma_{d}^{\downarrow}$ are the decay
width and the damping width. The summation is taken over the
doorway-states with spin and parity $J\pi$. Cross sections are sum
of the "shape and resonance" cross sections given by the averaged
S-matrix (1) and the compound cross sections, which are calculated
by means of Moldauer's theory with the largest possible values[4] of
the parameter Q_c. Transmission codfficients for the compound
process are given by

$$T_{n\ell j}^{J\pi} = 1 - \sum_{n'\ell'j'} \left| \langle S_{n\ell j;\, n'\ell'j'}^{J\pi} \rangle \right|^2 . \tag{4}$$

For the first excited state channel, several partial waves couple
simultaneously to a doorway-state. Since experimental information
is not sufficient to determine $\mathscr{G}_{d;i\ell j}$, we neglect $\mathscr{G}_{d;i\ell j}$ and do
not take into account the interferences between different partial
waves in the first level channel. In order further to reduce the
number of parameters we employ an approximation regarding the decay
width. The decay width is expressed as

$$\Gamma_{d;n\ell j}^{\uparrow} = 2\pi \left| \int dr\, d\xi\, \Psi_d^{\dagger}(\xi, r)\, \widetilde{H}_{dp}\, \Phi_n(\xi)\, \psi_{n\ell j}^{(+)}(r) \right|^2 , \tag{5}$$

where $\Psi_d(\xi)$ is the doorway-state wave function, $\widetilde{H}_{dp}$ is the interaction
between the open channels and the doorway-states including the
virtual transitions to the open channels and the "compound" state,
$\psi_{n\ell j}^{(+)}(r)$ is the neutron radial wave function of the optical model,
and $\Phi_n(\xi)$ is the wave function in channel (n 1 j) of other degrees
of freedom ξ than r. We have no knowledge about Ψ_d . But the
quantity

$$h(r) = \int d\xi\, \Psi_d^{\dagger}(\xi, r)\, \widetilde{H}_{dp}\, \Phi_n(\xi) \tag{6}$$

is expected to be significant in the surface region. Therefore we
adopt the form factor $g(r)$ of the imaginary potential as $h(r)$
in the first excited state channel, and $\Gamma_{d;n\ell j}^{\uparrow}$ is expressed as
follows:

$$\Gamma_{d;1\ell j}^{\uparrow} \propto \Gamma_{d;1}^{\uparrow} \left| \int dr\, g(r)\, \psi_{1\ell j}^{(+)}(r) \right|^2 ,$$

$$\Gamma_{d;n}^{\uparrow} = \sum_{\ell j} \Gamma_{d;n\ell j}^{\uparrow} . \tag{7}$$

The potential we use is of the form

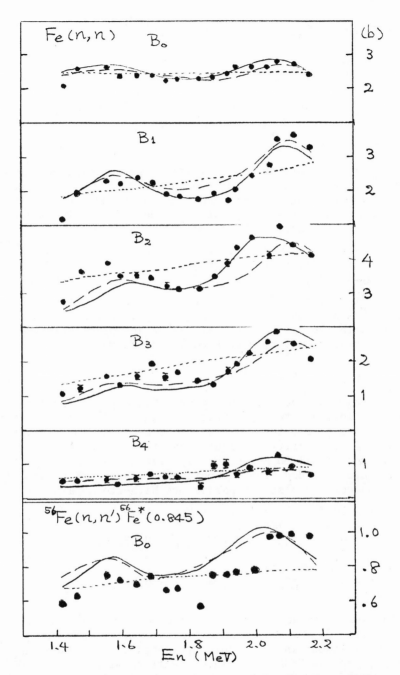

Differential cross sections expressed in the form (10). Points are the measured values. Dotted lines are calculated with the parameters (9) without including doorway resonances. Dashed lines and solid lines are the results of two and three resonance fit. respectively.

$$-V f(r) - W g(r) + V_{so} \lambda_\pi^2 (\sigma \cdot \ell) \frac{1}{r} \frac{df(r)}{dr}$$

$$f(r) = \frac{1}{1 + \exp \frac{r - r_0 A^{1/3}}{a}} \tag{8}$$

$$g(r) = \frac{4 \exp \frac{r - r_s A^{1/3}}{b}}{\left(1 + \exp \frac{r - r_s A^{1/3}}{b}\right)^2}$$

In order to determine the geometrical parameters, measured cross sections are averaged over the measured energy range, and parameters reproducing the averaged cross sections are searched for at the averaged incident energy of 1.812 MeV without including doorway-state resonances. The resultant parameters are

$$\begin{array}{lll}
V = 46.4, & W = 15.1, & V_{sc} = 14.6, \\
r_0 = 1.241, & r_s = 1.264, & \\
a = 0.631, & b = 0.255. &
\end{array} \tag{9}$$

These parameters are considered to give an overall fit to the measured cross sections in this energy range. Cross sections calculated with these parameters are shown in the figure by dotted lines. Cross sections are expressed in the form

$$\frac{d\sigma}{d\Omega} = \frac{1}{4\pi} \sum_\ell B_\ell P_\ell (\cos\theta) \tag{10}$$

Fixing r_0 , r_s , a and b to the values (9), depth parameters V, W and V_{so} and doorway-state parameters $\mathcal{E}_d$, Γ_d , $\Gamma_{d;n}^\uparrow$ and φ_d are varied in order to fit the measured cross sections for various choice of spin and parity $J\pi$ of doorway-states. Values of $J\pi$ greater than or equal to 5/2 generally give too large structures in B_4, and are excluded from the parameter search. For two doorway-state resonances the best fit parameters hitherto obtained are

$$\begin{array}{lll}
V = 50.3, & V = 25.7, & V_{so} = 10.0, \tag{11} \\
\mathcal{E}_1 = 1.5, & J\pi = 3/2-, & \\
\Gamma_1 = 0.2, & \Gamma_{1;0}^\uparrow = 0.03, & \Gamma_{1;1}^\uparrow = 0.03, \\
\varphi_1 = 0.1, & & \\
\mathcal{E}_2 = 2.0, & J\pi = 3/2-, & \tag{12} \\
\Gamma_2 = 0.2, & \Gamma_{2;0}^\uparrow = 0.06, & \Gamma_{2;1}^\uparrow = 0.08, \\
\varphi_2 = 0.2, & &
\end{array}$$

though the search is not complete. Here energies and widths are given in units of MeV. Cross sections calculated with these parameters are shown in the figure by dashed lines. Adopting as I the energy spread FWHM of incident neutrons of 0.05 MeV, $\Gamma_1^\downarrow$ = 0.09 and $\Gamma_2^\downarrow$ = 0.01 are obtained as the damping widths. Though these values have large uncertainties, $\Gamma_2^\downarrow$ is too small compared with the value $\Gamma^\downarrow$ = 0.08 obtained by Elwyn and Monahan[1] at lower neutron energies. Large discrepancy is seen in B_2 near

E_n = 2.0 MeV. These two facts suggest the existence of another resonance, if we persist to the doorway-state mechanism. Therefore an aditional resonance with $J\pi$ = 3/2+ is assumed near E_n = 2.0 MeV. Fixing the depth parameters to the values (11), a preliminary fit is obtained with parameters

$$\varepsilon_1 = 1.55, \qquad J\pi = 3/2-, \qquad \Gamma_1 = 0.2,$$
$$\Gamma_{1;0}^{\uparrow} = 0.04, \qquad \Gamma_{1;1}^{\uparrow} = 0.03, \qquad \varphi_1 = 0.2,$$
$$\varepsilon_2 = 1.95, \qquad J\pi = 3/2+, \qquad \Gamma_2 = 0.2,$$
$$\Gamma_{2;0}^{\uparrow} = 0.05, \qquad \Gamma_{2;1}^{\uparrow} = 0.035, \qquad \varphi_2 = 0, \qquad (13)$$
$$\varepsilon_3 = 2.05, \qquad J\pi = 3/2-, \qquad \Gamma_3 = 0.2,$$
$$\Gamma_{3;0}^{\uparrow} = 0.05, \qquad \Gamma_{3;1}^{\uparrow} = 0.035, \qquad = 0.3,$$

and calculated cross sections are shown in the figure by solid lines. These parameters give $\Gamma_1^{\downarrow}$ = 0.08 and $\Gamma_2^{\downarrow}$ = $\Gamma_3^{\downarrow}$ = 0.065. Though these values are not unreasonable, the quality of the fit is not yet impressive.

We used here a surface absorption potential. It is likely that doorway-states are responsible for the surface absorption, and that the use of a volume absorption potential is preferred when doorway-states are treated explicitly. Calculation using a volume obsorption potential will be tried next to see whether or not this choice will give a better fit and result in a smaller value of W.

Further calculations have been carried out since the manuscript was written. Better fits have been obtained with depth parameters

$V\cong 47, \quad W\cong 14, \quad V_{so}\cong 24,$

and with nearly the same resonance parameters as in the text.

The best fit hitherto obtained is with the following parameters:

$V=46.55, \quad W=14.82, \quad V_{so}=19.47,$

$\varepsilon_1 =1.598, \quad J\pi=3/2-, \quad \Gamma_1 =.149,$

$\Gamma_{1;0}^{\uparrow} =.013, \quad \Gamma_{1;1}^{\uparrow}=.061, \quad \varphi_1 =.940,$

$\varepsilon_2=1.991, \quad J\pi=3/2+, \quad \Gamma_2 =.194,$

$\Gamma_{2;0}^{\uparrow} =.055, \quad \Gamma_{2;1}^{\uparrow}=.058, \quad \varphi_2=-.248.$

1) A.J. Elwyn and E.J. Monahan, Nucl.Phys. A123,(1969) 33.
2) Y. Tomita, K. Tsukada, M. Maruyama and S. Tanaka,"Nuclear Data For Reactors" (IAEA, Vienna, 1970)Vol. 11, p. 301.
3)A.F.R. de Toledo Piza and A.K. Kerman, Ann.Phys. (N.Y.) 303,43,1967.
4)T. Tomita, JAERI, 1191 (1969)
* This paper was given by K. Tsukada.

DISCUSSION

NEWSTEAD
    Your optical model parameters seemed quite interesting, in
that you had a very high spin-orbit term, approaching almost 20
MeV, while the diffusemess of the absorptive term was about half
of what one would normally expect. Could you comment on that and
in general on the inclusion of doorways in the optical potential?

TSUKADA
    No, except to say that these parameters fit a rather wide
range of experimental data in this energy region.

FRANCIS
    It is known that if you include explicit doorway states in a
coupled-channel calculation,   the imaginary potential in the
ground state channel decreases. Although one might hope that the
spin-orbit term and diffuseness would stay the same, once one
does different calculations like this, they might change too. Has
anybody in the audience had experience with nonconventional optical
model parameters in doorway state interpretation?

    (No response)

$1/2^+$ PARTICLE-VIBRATION DOORWAYS IN COMPOUND NUCLEI Pb^{209} AND Pb^{207}*

M. Divadeenam, North Carolina Central University and TUNL,
Durham, North Carolina 27706 and W. P. Beres, Wayne State
University, Detroit, Michigan 48202

The doorway state concept has become important in recent years. Such states are composed of simple excitations and might be observed as single resonances or as several narrow resonances. The well known[1] Pb^{209} $1/2^+$ state at 500 keV neutron energy with a width of 58 keV is an example of the former type. Some similar doorways are found in neutron scattering experiments on target nuclei Y^{89}, and Ca^{48}[2], while fragmented doorways have been identified in compound nuclei, Sr^{89}, Pb^{207} and Pb^{208}. The dissolution of a doorway into narrow resonances implies a width for spreading into more complex states. We have attempted to study doorway states in terms of two different models: 1) 2p-1h configurations and 2) particle-vibration configurations. Reference 2 treats the former approach for compound nuclei Ca^{49}, Sr^{89}, and Zr^{91}. However the latter mode of excitation seems to be preferable in the case of Pb isotopes. In this paper we demonstrate the applicability of the particle-vibration model for interpreting the neutron resonances observed in Pb^{208} and Pb^{206} neutron scattering experiments.

The coupling of an odd particle to a nuclear vibration is a very general and important problem in nuclear physics. The particle-vibration weak-coupling model[3,4] has been successfully applied to study nuclear level structure. In particular the almost pure single particle nature of pertinent neutron and proton states and the collective behavior of the excited states of Pb isotopes facilitate the application of this model for Pb and Bi isotopes. Usually the particle-vibration** states considered are located in the bound region.

---

* Work supported in part by the U. S. Atomic Energy Commission.
** Both the terms "particle-vibration" and "core-particle" are used in this paper.

However, some of these states could be above particle threshold and observable as resonances. (Incidentally Motelson[3] has pointed out the need for
the evaluation of radial matrix elements when the single particle state is at a
known energy in the continuum.) We will refer to such resonances as particle-
vibration doorways.

As pointed out earlier a very wide $1/2^+$ resonance in Pb^{209} was observed in a neutron scattering experiment at $E_n = 500$ keV with a width of
58 keV. The experimental cross sections for the Pb^{207} and Pb^{208} compound nuclei exhibit much fine structure ($J^\pi = 1/2^+$) in the vicinity of 500 keV neutron energy. The individual sum of the Pb^{207} and Pb^{208} $1/2^+$ neutron reduced
widths is approximately equal to the Pb^{209} $1/2^+$ reduced width[1,5]. This fact
lead Newson and his collaborators to conjecture that a common doorway (in
Pb^{209}) is responsible for the fine structure observed in the other two lead isotopes.

Arguments were given in ref. (6) as to why the 2p-1h picture could
not account for the 500 keV Pb^{209} $1/2^+$ resonance. Similar conclusions may
be drawn from the results of Vergados[7]. Following a suggestion* by Sha-
kin[8] we have shown in an earlier paper[6] that the $\{4^+ \otimes 2g_{9/2}\}_{1/2^+}$ state located nearly at the experimental energy has a width of about 1 to 2 times the
observed resonance width. Here we report an extension of our particle-vibration model calculation to the other possible $1/2^+$ doorway states in Pb^{209} and
Pb^{207}. The conjecture that the Pb^{209} and Pb^{207} doorways have similar structure
is also verified. An extension of the present model for the Pb^{208} (Pb^{207} + n)
case is expected to give similar results. The Hamiltonian in the weak-coupling model is

$$H = H_{core} + H_{part.} + V_c \quad . \tag{1}$$

Here H_{core} and $H_{part.}$ are the vibrational core and single particle parts of
the total Hamiltonian and V_c is the coupling interaction chosen to be of the
following form: [3]

$$V_c = k(r) (2\lambda+1)^{1/2} (a_\lambda Y_\lambda)_0 \quad . \tag{2}$$

Where λ is the vibrational quantum number and $k(r)$ a form factor given by

$$k(r) = -r \frac{dV(r)}{dr} \quad . \tag{3}$$

The potential $V(r)$ is taken as a Woods-Saxon well. The magnitude of the
vibrational amplitude a_λ can be determined from the cross section for excitation of one vibrational quantum. The matrix element of the coupling interaction V_c is completely determined by the properties of the single particle
motion and the vibration, and is of the form [3]

---

* A similar suggestion was made by G. E. Brown at the time of publication
of the paper quoted in ref. 1.

$$\langle J_c \otimes j_p | V_c | O^+ \otimes j_p' \rangle = \langle j_p | k(r) | j_p' \rangle \langle j_p \| Y_\lambda \| j_p' \rangle \frac{1}{\sqrt{2}} (\hbar \omega \sqrt{2 C_\lambda})^{1/2} . \quad (4)$$

Here the last factor represents the vibrational amplitude. For the present case the above relation may be expressed as

$$\langle (J_c \otimes j_p)_{1/2} + | V_c | O^+ \otimes s_{1/2}^{confin} \rangle = \langle j_p | K(r) | u(s_{1/2}, Kr) / r \rangle \langle j_p \| Y_\lambda \| \phi(s_{1/2}) \rangle \frac{1}{\sqrt{2}} (\frac{\hbar \omega_\lambda}{2 C_\lambda})^{1/2} , \quad (5)$$

where K is the wave number of the neutron at E_n = energy of the particle-plus-core state. The wave functions $u(s_{1/2}, Kr)$ and $\phi(s_{1/2})$ represent respectively the radial and spin angular wave function of the continuum s-wave neutron. Since V_c is a scaler, only the $\ell = 0$, $j = 1/2$ partial wave need be considered in calculating the neutron escape width, which is given[6] by

$$\Gamma_d^{\uparrow} = \frac{4m}{K\hbar^2} |\langle (J_c \otimes j_p)_{1/2} + | V_c | O_{\circ}^{\uparrow} u(s_{1/2}, kr) / r \phi(s_{1/2}) \rangle|^2 . \quad (6)$$

The details of the calculation of $\Gamma_d^{\uparrow}$ are given below.

The Pb^{208} and Pb^{206} core excited states are located at the same energies with similar vibrational amplitudes. In the present context we consider the triplet of positive parity vibrational states that are above 4 MeV excitation in the target nuclei; these are 2^+, 4^+, and 6^+. The enhanced B(Eλ) experimental values for the $Pb^{208}(Pb^{206})$ 2^+, 4^+, and 6^+ states respectively are 8(6.2), 15(15), and 5.5(6.2) single particle units.[9,6,10] For the Pb^{206} case the two $3p_{1/2}^{-1}$ states merely act as spectators, but as pointed out later they could conceivably play some role in spreading the doorways. The 8^+ state does not contribute for angular momentum and parity reasons. The microscopic structure of these positive parity states has not been studied so far.[11] They could possibly contain 2p-2h in addition to 1p-1h components, and the states formed by coupling a single neutron to such vibrations can decay by neutron emission to the 0^+ ground state via the ground state (2p-2h) correlations, which are non-negligible.[11] In the lead region the available positive parity single neutron states are $2g_{9/2}$, $1i_{11/2}$, $3d_{5/2}$, $4s_{1/2}$, $1g_{7/2}$, and $3d_{3/2}$. The odd parity single neutron state $1j_{15/2}$ is not considered because of the unavailability[11] of a collective 7^- state in the target nucleus. Furthermore the position of the $1j_{15/2}$ neutron state is not well established.[12]

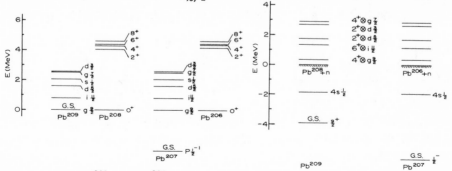

Fig. 1. Left. Pb^{208} and Pb^{206} core excited states and the available single particle states in the neighboring nuclei. Right. Pb^{209} and Pb^{207} doorway states.

On the left-hand side of fig. 1 the relative positions of the single particle states and the core excited states considered are shown. Note that the energies of the Pb^{208} and Pb^{206} 2^+, 4^+, and 6^+ states are almost the same. In addition the relative positions of the various particle states with respect to the $2g_{9/2}$ are similar. As a result, the position of a doorway corresponding to a particular particle-vibration configuration is almost the same in both the compound nuclei. All the particle-vibration doorway states with their configurations and the $4s_{1/2}$ neutron state are shown on the right-hand side of figure 1. While it is possible that these states could mix with each other, such mixings are neglected because of the previous success of the weak-coupling model in the Pb region.[3,9] The lowest $1/2^+$ state in both nuclei is essentially pure $4s_{1/2}$ single particle in character. For example according to ref. 9 in Pb^{209} the admixture of the particle + core states into the lowest 1.97 MeV $1/2^+$ state (Spectroscopic factor = .92) is $\leq 4\%$. It is conceivable that all the doorways under investigation contain both 2p-1h and 3p-2h components in their microscopic forms. Various types of correction terms to matrix elements discussed by Mottelson[3] are not considered here. However one of the corrections due to the antisymmetrization of the microscopic core ingredients and the extra particle in the correct particle + core wave function are not large in the Pb region.[7]

To calculate the neutron escape width (Eq. 6) of each doorway Hamamato's Woods-Saxon potential parameters are used for generating the radial wave functions of both the bound single neutron and continuum s-wave neutron states. In addition Perey and Buck's local equivalent parameters are

Table 1a

Configuration[b]	Pb^{209} Theory			Pb^{209} Expt.[1,14]		Pb^{207} Theory			Pb^{207} Expt.[1]	
	E_d	$_c\gamma_d^2$	$_d$	E_{res}	γ_n^2	E_d	$_c\gamma_d^2$	$_d$	E_{res}	$\Sigma\gamma_n^2$
$4^+\otimes2g_{9/2}$	0.365	37.8	22.2	0.500	22.5	0.361	37.1	23.6	~.43	23
$6^+\otimes1i_{11/2}$	1.265	3.2	2.4	1.314	1.7	1.159	3.7	2.8	no data	
$2^+\otimes3d_{5/2}$	1.700	11.8	6.6	{1.735}{1.87}	{12.5}{2.6}	1.691	9.1	5.7	avail-	
$2^+\otimes3d_{3/2}$	2.670	5.5	3.1			2.661	4.2	2.7	able	
$4^+\otimes2g_{7/2}$	2.865	13.0	8.9			2.863	12.9	9.4		

a) Predicted doorway and experimental resonance energies are in MeV, while the reduced widths are in keV. b) The symbol $\otimes$ represents the coupling of two angular momentum states. c) Widths calculated using Hamamoto's parameters for the continuum neutron. d) Buck and Perey's parameters used for the continuum neutron.

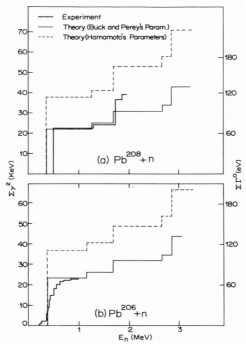

Fig. 2. Histogram plots of predicted and observed reduced width sums $\Sigma\gamma^2$ for Pb^{209} (Fig. 2a) and for Pb^{207} (Fig. 2b) .

used for the latter. The criteria for choosing a particular set of parameters to generate each continuum s-wave neutron radial wave function have been discussed earlier[6]. The computer code ABACUS[13] was used to compute the radial matrix elements (Eq. 5).

The predicted doorway escape widths $\Gamma_d^{\uparrow}$ for the Pb^{209} and Pb^{207} particle-vibration doorways are given in table 1 and compared to the experimental results. Figures 2a and 2b show a histogram plot of $\Sigma\gamma_d^2$ versus neutron energy. Here γ_d^2 are reduced widths obtained from the predicted escape widths. Experimental resonance reduced widths γ_n^2 are also shown and a comparison of theory and experiment is discussed below.

$\underline{Pb^{209}}$ (Pb^{208} + n). In fig. 2a the dashed and solid thin-line histograms correspond respectively to theoretical predictions which make use of Hamamoto's parameters and Buck and Perey's parameters for the continuum neutron state. The solid thick-line histogram represents the experimental situation. The first experimental resonance[1] at 500 keV has been referred to earlier, while the rest are taken from Fowler's[14] work. Note that the three lowest predicted doorway energies are in good agreement with the experimental resonance energy positions (See also table 1). Since there is no fine structure observed in the vicinity of the first three predicted doorways, the observed resonances may be interpreted as doorways of the particle-vibration type. This interpretation suggests a very small spreading width ($\Gamma_d^\downarrow \sim 0$) for the doorways. However the 1.87 MeV observed resonance (spin doubtful $1/2^\pm$) could conceivably be considered to be a fragmented resonance of the 1.7 MeV (third) predicted doorway. Alternatively, this resonance may correspond to the $\{(7^- \otimes j_{15/2})\}_{1/2+}$ configuration if a 7^- level is identified in Pb^{208}. Of the two sets of predicted particle-vibration doorway reduced widths the lower set corresponding to Buck and Perey's parameters is in excellent agreement with experiment and the upper is almost as good.

Pb^{207} ($Pb^{206} + n$). Histograms similar to fig. 2a are given in fig. 2b for Pb^{207}. The experimentally analyzed high resolution data does not exist beyond about 1 MeV neutron energy. Measured neutron cross section data beyond 1 MeV are not suitable for extracting resonance parameters. The experimental $1/2^+$ resonance reduced width sum $\Sigma\gamma_n^2$ (shown as a solid thick-line histogram) should be compared to the predicted .36 MeV particle-vibration doorway reduced width. There is observed fine structure[1] in the vicinity of the lowest predicted doorway and the sum $\Sigma\gamma_n^2 = 23$ keV exhausts the entire strength (lower plot) of the theoretical doorway state ($\gamma_d^2 = 23.6$ keV). In other words the particle-vibration doorway appears to be fragmented into several narrow resonances, and the spreading width $\Gamma_d^\downarrow \gg 0$. From the inspection of the figure a spreading width of about 200 keV is in fair agreement with the value obtained in ref. (15). Again as in the case of Pb^{209}, the lower predicted histogram is in better agreement with experiment.

Comparing the predicted widths and positions for Pb^{209} and Pb^{207} $1/2^+$ doorways, we can conclude that the observed Pb^{209} and Pb^{207} doorways have similar structure, except for the spreading widths of the lowest doorways. The sharp contrast between the spreading widths in Pb^{209} ($\Gamma_d^\downarrow \sim 0$) and Pb^{207} ($\Gamma_d^\downarrow \sim 200$ keV) may be explained qualitatively in the following manner.

According to the 2p-1h doorway interpretation the spreading width is related to a matrix element of the type $< (2p-1h) | V | (2p-1h) + p-h >$. However, in the present case the spreading may arise from matrix elements of the type $< (core + p) | V | (core + p) + p-h >$. The p-h state (assuming to be not a part of core ingredients) can be considered either in the target or in the compound nucleus. Since we are studying $1/2^+$ states, the p-h states could have $J^\pi = 0^+$, 1^+. In the case of Pb^{208} there are two 1^+ p-h states around 6 MeV above the Pb^{208} ground state. In other words the corresponding $|(c+p)+p-h>$ states are separated by about 6 MeV from the lowest predicted doorway, and the above mentioned matrix element may be expected to be very small; i.e., $\Gamma_d^\downarrow \sim 0$. However in the case of the Pb^{206} target, two neutrons are lacking from closed shells considering the $3p_{1/2}$ shell as empty the $\{3p_{1/2}, 3p_{3/2}^{-1}\}_{1+}$ state is at 1.78 MeV above the Pb^{206} ground state.[16] In addition two states[16] (based on a filled $3p_{1/2}$ shell) at 1.314 MeV ($\{f_{5/2}^{-1}\}_{0+}^2$) and 2.17 MeV

($\{f_{5/2}^{-1}, p_{3/2}^{-1}\}_{1+}$) might play some role in addition to the $\{p_{1/2}, p_{3/2}^{-1}\}_{1+}$ p-h state in spreading the .36 MeV predicted Pb^{207} particle-vibration doorway.

In conclusion, the particle-vibration model accounts quantitatively for the observed neutron scattering data of Pb^{208} and Pb^{206} for the $1/2^+$ case. We emphasize that no free parameters have been used in the present calculations. The present model has been extended to Pb^{209} p- and d-wave doorways and the results will be discussed elsewhere. The alternate model (2p-1h configuration) is much less successful for the Pb^{209} $1/2^+$ case (with a limited basis) for it does not predict the right doorway energy and escape width but is successful in explaining the observed resonances in Sr^{89} and Ca^{49}.

We wish to thank Professor Henry Newson for reading the manuscript and for helpful suggestions.

References

1. J. A. Farrell, G. C. Kyker, Jr., E. G. Bilpuch and H. W. Newson, Phys. Letters 17, 286 (1965).
2. M. Divadeenam, W. P. Beres and H. W. Newson, Ann. of Phys. (in press).
3. B. R. Mottelson, J. Phys. Soc. Jap. 24, 87 (1968).
4. J. Hamamoto, Nucl. Phys. A126, 545 (1969) and A135, 576 (1969).
5. F. Seibel, E. G. Bilpuch and H. W. Newson, Ann. of Phys. (in press).
6. W. P. Beres and M. Divadeenam, Phys. Rev. Letters 25, 596 (1970).
7. J. D. Vergados, Phys. Letters 34B, 121 (1971).
8. C. Shakin, Private communication.
9. N. Auerbach and N. Stein, Phys. Letters 25B, 628 (1969).
10. G. Vallios, J. Saudinos and O. Beer, Phys. Letters 24B, 512 (1967).
11. N. Stein, Proc. Int. Conf. on Properties of Nuclear States, Montreal (1969) p. 337.
12. D. G. Kovar, Yale Univ. Thesis (unpublished, 1971).
13. E. H. Auerbach, BNL Report no. 6562 (unpublished).
14. J. L. Fowler, Phys. Rev. 147, 870 (1966).
15. A. Lejeune and C. Mahaux, Zeitschrift fur Physik 207, 35 (1967).
16. W. W. True, Phys. Rev. 168, 1388 (1968).

ADDENDUM

A qualitative explanation was given (in the text of our paper) for the large spreading width for the $Pb^{207}\{4^+ \otimes g_{9/2}\}_{1/2^+}$ doorway. Here we give an alternative explanation based on the density of the higher configuration states. Writing

$$\Gamma d^{\downarrow} \sim \frac{|<\lambda \otimes P|V|(\lambda \otimes \lambda') \otimes P>|^2/D}{|<\lambda P|V|\lambda\lambda'P>|^2/D}$$

Here $|\lambda \otimes P>$ (or $|\lambda P>$) and $|(\lambda \otimes \lambda') \otimes P>$ (or $|\lambda\lambda'P>$) are respectively the particle + core and two-phonon + particle states. The latter state is of higher configuration type. D represents the average spacing of two-phonon + particle states. In the case of Pb^{209} there are no $1/2^+$ higher configuration states, while there is a high density of such states (10 in 1 MeV range) near the vicinity of the doorway in Pb^{207}. We can write the ratio of the Pb^{207} and Pb^{209} spreading widths:

$$\frac{\Gamma d^{\downarrow}(207)}{\Gamma d^{\downarrow}(209)} = \frac{D(209)}{D(207)} \cdot \frac{|<\lambda P|V|\lambda\lambda'P>|^2_{207}}{|<\lambda P|V|\lambda\lambda'P>|^2_{209}}$$

Now we make a drastic assumption regarding the matrix elements involved in the above equation:

$$|<\lambda P|V|\lambda\lambda'P>|^2_{207} = |<\lambda P|V|\lambda\lambda'P>|^2_{209}$$

Since $D_{209} \sim 9$ MeV and $D_{207} = \frac{0.72}{14}$ (14,1/2 resonances observed in 0.72 MeV range). We have:

$$\Gamma d^{\downarrow}(207) \stackrel{<}{\sim} 175 \; \Gamma d^{\downarrow}(209) = 175 \text{ keV (assuming } \Gamma d^{\downarrow}(209) \stackrel{<}{\sim} 1 \text{ keV)}.$$

Our rough estimate of the Pb^{207} doorway spreading width agrees with Dr. Newson's empirical estimate and with that of Dr. Mahaux.

DISCUSSION

FRANCIS
     Since you can calculate the positions and widths of these
states, is there any prediction that you would like to make for
experimentalists to look for states at particular energies?

DIVADEENAM
     We have not done that, but it can be done. So far we have
taken only cases where there is already evidence for a doorway
phenomenon, and then we go back to the model and calculate and
compare. However, one could do the predictions very easily.

FRANCIS
     Are there any general rules concerning what to look for e.g.,
certain nuclei or certain energies?

DIVADEENAM
     Yes. You have a target nucleus, and if you know the energies
of the single-particle states in the neighboring odd nucleus, with
respect to the ground state of the compound nucleus, you add those
energies to the hole energies of the target nucleus. This energy
will be the energy in the compound nucleus. If this energy is
above the particle threshold, you take the difference between this
energy and the separation energy. That will give the energy at
which you might expect a resonance.

NEWSON
     I would like to remark that some of the work reported in my
paper was undertaken in response to the calculations of Divadeenam
and Beres, particularly of strontium 88. We had a little scratchy
data before, which we repeated and extended, and on the whole, I
think the check was pretty good.

INTERMEDIATE STRUCTURE IN PHOTON INDUCED REACTIONS

S. Fallieros[†]

Brown University

Providence, Rhode Island

We will discuss, in this report, the sequence of events that follow the absorption of a photon by a nucleus. We are interested in the distribution of the absorption strength as a function of the excitation energy and in the dependence of the strength, total or localized, on the particle number of the nuclear system.[1] Some features of the transition densities, the functions whose moments determine the intensity of a radiative transition, will also be considered.

The photon carries a certain amount of energy and momentum, zero charge, and a mixture of isospins (zero or one). It does not change the particle number (particles minus holes) but tends to change the quasiparticle number (particles plus holes) by two i.e., to create particle-hole pairs. The interactions of the particles and the holes with each other and with the other nucleons determine the locations i.e., the energies around which radiative strength is concentrated, while their coupling to excitations of increasing complexity (two-particle-two-hole, etc.) characterizes the degree of localization i.e., the spreading width. We will find it helpful to refer to the amplitude

$$T_{AB}(\varepsilon) = \sum_n \frac{A_{on} B_{no}}{\varepsilon^+ - \varepsilon_n}$$

$$= \langle 0 | A \frac{1}{\varepsilon^+ - H} B | 0 \rangle \tag{1}$$

[†]

Supported in part by the U.S. Atomic Energy Commission

which is closely related to the response function of the nucleus
(a term with a plus sign in the denominator is not of interest here
and is omitted). The states $|n\rangle$ are the stationary states of the
Hamiltonian H and ε_n are the eigenvalues. A and B are electromagnetic transition operators for photons of energy ε and they are
allowed to be functions of q the momentum of the photon. In general
we let $q \neq \varepsilon/\hbar c$ i.e., virtual photons are considered. For $q = \varepsilon/\hbar c$
we set $A = B = M$ the latter being the operator for the appropriate
radiative transition. In the second line of Eq.(1) the states $|n\rangle$
have been eliminated and we can insert instead, non-stationary, doorway states[2] i.e., configurations that can be reached directly through
the action of A and B on the ground state $|0\rangle$. It is also useful to
consider the function

$$R_{AB}(\varepsilon) = \frac{1}{\pi} \operatorname{Im} T_{AB}(\varepsilon)$$
$$= \sum_n A_{on} B_{no}\ \delta(\varepsilon - \varepsilon_n) \tag{2}$$

and the standard sum rules

$$\int_0^\infty R_{AB}(\varepsilon)\ d\varepsilon = \langle 0|A\ B|0\rangle \tag{3}$$

and

$$\int_0^\infty R_{AB}(\varepsilon)\ \varepsilon\ d\varepsilon = \frac{1}{2} \langle 0|[A,\ [H,\ B]]|0\rangle$$
$$= \sum_n \varepsilon_n A_{on} B_{no}\ . \tag{4}$$

Average values are obtained by the conventional rule

$$\langle T(\varepsilon)\rangle = \int T(\varepsilon') \frac{I/\pi}{(\varepsilon - \varepsilon')^2 + I^2}\ d\varepsilon'$$

$$= T(\varepsilon + iI) \tag{5}$$

where I is the energy interval over which the average is taken and,
if D_n is the level spacing

$$\langle R_{AB}(\varepsilon)\rangle = \sum_n A_{on} B_{no} \frac{I/\pi}{(\varepsilon - \varepsilon_n)^2 + I^2}$$
$$= \left\langle \frac{A_{on} B_{no}}{D_n} \right\rangle\ . \tag{6}$$

In particular

$$\langle R_{MM}(\varepsilon)\rangle = \left\langle \frac{|M_{on}|^2}{D_n} \right\rangle \tag{7}$$

is the conventional strength function[1] for the radiative strength.

Models can now be introduced by the insertion of appropriately selected states into Eq.(1). (The radiative case will be considered here; we will return to the more general expressions in the discussion of inelastic electron scattering). We will discuss first what we will call the single-doorway approximation where we will consider giant collective modes, total radiative strengths and strength functions. After that we will introduce a multi-doorway representation and discuss particle-hole states and finer structures, isovector excitations and analog states and finally sum rules and isoscalar electroexcitations.

The single-doorway description involves the introduction of a single (non-stationary) state $|s>$ whose presence in a level $|n>$ is responsible for the radiative strength of this level i.e.,

$$|n> = c_n^s |s> + \sum_{i(\neq s)} c_n^i |i>$$

$$M_{on} = M_{os} c_n^s , \quad M_{oi} = 0$$

and

$$\sum_n |M_{on}|^2 = |M_{os}|^2 \sum_n |c_n^s|^2 = |M_{os}|^2 . \tag{8}$$

The amplitude $T(\varepsilon)$ can be written as

$$T_{MM}(\varepsilon) = \frac{|M_{os}|^2}{\varepsilon^+ - \varepsilon_s - \Sigma_s(\varepsilon)} \tag{9}$$

where ε_s is the average energy and $\Sigma_s(\varepsilon)$ the self-energy of $|s>$, a rapidly varying function of ε which includes all the effects due to the coupling with all other excitations. This expression (Eq.(9)) is still exact with the doorway state providing only a representation. An approximation is introduced when the average value of T is taken if the assumption is made that the average Σ i.e., $\Sigma(\varepsilon + iI)$ is essentially a constant independent of energy i.e.,

$$<T_{MM}(\varepsilon)> = \frac{|M_{os}|^2}{\varepsilon - \varepsilon_s + i\Gamma_s/2} \tag{10}$$

where the real part of $<\Sigma(\varepsilon)>$ has been included in ε_s and

$$\Gamma_s = - 2 \operatorname{Im} <\Sigma(\varepsilon)> . \tag{11}$$

The strength function (for a given multipolarity) finally becomes

$$\left\langle \frac{|M_{on}|^2}{D_n} \right\rangle = \frac{|M_{os}|^2}{\pi} \frac{\Gamma_s/2}{(\varepsilon - \varepsilon_s)^2 + \Gamma_s^2/4} . \tag{12}$$

Similar expressions have been considered by a number of authors.[3,4,5]
The excitation energy for each mode is believed to be inversely proportional to the nuclear radius i.e.,

$$\varepsilon_s \propto A^{-1/3} \quad . \tag{13}$$

This follows easily either from hydrodynamic considerations[6,7] or
from the shell model where

$$\varepsilon_s = \hbar\omega + n \, \Delta$$

with

$$\hbar\omega \propto A^{-1/3}, \; \Delta \propto A^{-1}, \; n \propto A^{2/3} \quad . \tag{14}$$

The symbols $\hbar\omega$, Δ and n represent respectively the average values
of the single particle energy, the particle-hole-interaction matrix
element and the number of particle-hole pairs. A rough idea of the
mass dependence of the width is obtained from

$$\Gamma_s \underset{\sim}{\sim} 2\pi <|H'|^2> \rho(\varepsilon_s) \tag{15}$$

where H' represents the coupling between one-particle-one-hole and
the two-particle-two-hole excitations and ρ is the level density of
the latter. If we assume the matrix element to be inversely proportional to the nuclear volume and let[8]

$$\rho(\varepsilon_s) \propto A^{7/3} \, \varepsilon_s \tag{16}$$

we find (keeping Eq.13 in mind) no systematic dependence of Γ_s on A.
This crude result does not of course exclude variations of Γ_s related to shell closure etc.

We can finally discuss the mass dependence of the total radia-
tive strength $|M_{os}|^2$. For electric dipole radiation we have

$$|M_{os}|^2 \propto A^{4/3}$$

which is to be contrasted with the $A^{2/3}$ dependence of the single-
particle estimate. The additional $A^{2/3}$ factor represents the
number of particle-hole states (the factor n in Eq.(14)). More
generally, from sum rules of the form

$$\varepsilon_s |M_{os}|^2_{E\lambda} = \Sigma_n \; \varepsilon_n \; |M_{on}|^2_{E\lambda}$$

$$\propto \; A <r^{2\lambda - 2}> \tag{17}$$

and with Eq.13 kept in mind, we find

$$|M_{os}|^2_{E\lambda} \propto A^{2/3} \, A^{2\lambda/3} \quad . \tag{18}$$

This completes the discussion of the simplest features of the gross structure. As mentioned earlier $\Sigma(\varepsilon + iI)$ is not really constant unless one averages over unreasonably large intervals $I \gg \Gamma_s$. While the variation of $\Sigma(\varepsilon)$ is related to the complete fine structure of the compound nucleus, the variation of $\Sigma(\varepsilon + iI)$ should give rise to intermediate structures of various degrees of complexity. The simplest structure we can consider is related to isospin. The electromagnetic operators M have isoscalar components leading to $T \to T$ transitions (T is the ground-state isospin) and isovector components leading to both $T \to T$ and $T \to T + 1$ transitions. The $T + 1$ excitations are clearly analog states. For E1 transitions there is no isoscalar components and we obtain

$$<T_{MM}(\varepsilon)> = \frac{|M_{os}^T|^2}{\varepsilon - \varepsilon_s^T + i\Gamma_s^T/2} + e^{i\phi} \frac{|M_{os}^{T+1}|^2}{\varepsilon - \varepsilon_s^{T+1} + i\Gamma_s^{T+1}/2} \quad . \tag{19}$$

The phase factor in the second term is related to the common interaction that the states T and T+1 have with the various open channels. It leads to an expression for the strength function which is asymmetric around the T+1 peak i.e.,

$$<R_{MM}(\varepsilon)> = \frac{|M_{os}^T|}{2\pi} \frac{\Gamma_s}{(\varepsilon - \varepsilon_s^T)^2 + (\Gamma_s^T/2)^2}$$
$$+ \frac{|M_{os}^{T+1}|^2}{2\pi} \frac{\Gamma_s^{T+1}\cos\phi - 2(\varepsilon - \varepsilon_s^{T+1})\sin\phi}{(\varepsilon - \varepsilon_s^{T+1})^2 + (\Gamma_s^{T+1}/2)^2} \quad . \tag{20}$$

The energy ε_s^T can be identified with the position of the giant dipole resonance (for $T \geq 1$) i.e.

$$\varepsilon_s^T \sim 80 \text{ MeV } A^{-1/3} \quad . \tag{21}$$

Recent estimates[9] have given the results

$$\varepsilon_s^{T+1} - \varepsilon_s^T \sim \frac{T+1}{A} 60 \text{ MeV}$$

$$|M_{os}^{T+1}|^2 = |M_{os}|^2 \frac{1}{T+1} (1 - \frac{2T}{A} \xi)$$

$$|M_{os}^T|^2 = |M_{os}|^2 \frac{T}{T+1} (1 + \frac{2}{A} \xi)$$

$$\xi \sim \frac{3}{4} A^{1/3} \quad . \tag{22}$$

Similar results can be written down for other multipolarities but

our knowledge in most cases is not as detailed. It should be mentioned that the above expression for the T+1 strength is obtained
from rather simple shell-model considerations and does not include
renormalizations due to the presence of background states of isospin
T or to the open-channel continuum. The comparison of the estimates
of Eq.(22) with experiments[10] is quite satisfactory. A rough estimate of the relative widths of the T and T+1 excitations can be
written down. The result is

$$\Gamma_s^{T+1} = \Gamma_s^{T}(1 - 0.5 \frac{A}{100})^2$$

which although obtained by very crude approximations seems to reproduce qualitatively the observed trends. In addition to other
simplifying assumptions we should mention that terms such as T/A and
$Z/A^{1/3}$ have been set proportional to $A^{2/3}$ in deriving the above
result.

We can now consider the case where all possible doorway excitations are inserted between the operators A and B in Eq.(1) and the
intermediate-size variations of $\Sigma_s(\epsilon + iI)$ are taken into account.
We then obtain

$$T_{AB}(\epsilon) = \sum_d \frac{A_{od} B_{do}}{\epsilon^+ - \epsilon_d - \Sigma_d(\epsilon)} \tag{23}$$

and

$$<T_{AB}(\epsilon)> = \sum_d \frac{A_{od} B_{do}}{\epsilon - \epsilon_d + i\Gamma d/2} \tag{24}$$

where "d" denotes the various doorway states. These can include
particle-hole excitations, not necessarily of "giant" character, and
other excitations resulting from e.g. particle-vibration or dipole-
quadrupole etc. couplings. They are expected to appear in e.g. the
low-energy side of the giant resonance or as structure on top of the
giant resonance. A theoretical spectrum[11] showing the giant resonance (isospin T), analog states (isospin T+1) and other weaker states
(isospin T) is presented in Fig. 1. Some levels were too weak to be
shown individually on the left side of the figure but were included
in the mathematical averaging represented by the solid curve. Although relatively weak, the analog states in the giant-resonance
region can (and have) been observed[10] because of their narrowness.
The general comment we should make in connection with Eq.(24) (and
equivalent expressions for the strength function) is that one can
seldom expect the picture of a single isolated doorway resonance to
be perfectly valid. The tails of other (both nearby and faraway)
doorways should be quite important and the resulting interference
can have a significant effect on e.g. the correlations between widths

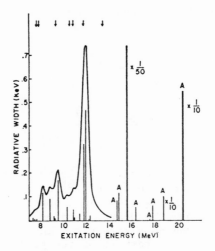

Fig. 1. Theoretical spectrum of dipole
excitations in ^{88}Sr.

in different channels.

We will avoid repeating at this point the general discussions
or the illustrations of individual cases presented in this conference by several authors.[12] We will consider instead the process of
inelastic electron scattering and discuss the possible significance
of doorway-state dominance in this case. For this purpose we will
allow the matrix elements of A and B in Eq.(1) to represent form
factors describing the electroexcitation of nuclear levels at some
momentum transfer q. For simplicity we will limit our discussion to
isoscalar monopole electroexcitations and present results for other
cases later on. For B we then chose

$$B = \frac{1}{2Z} \sum_i j_o(qr_i) = F(q) \tag{25}$$

where $j_o(qr)$ is a Bessel function. For A we can use the same expression of a similar one evaluated at a different momentum transfer q'.
We will prefer however, to consider the low-momentum-transfer limit
and set

$$A = \frac{1}{2} \sum_i r_i^2 = M \tag{26}$$

i.e. the (isoscalar part of the) pair-formation operator. Clearly

$$M_{on} = -6Z \frac{d}{dq^2} F_{on}(q) \Big|_{q=0} \tag{27}$$

the amplitude T, which now depends on the momentum transfer, then
becomes

$$T_{MF}(\varepsilon;q) = \sum_n \frac{M_{on} F_{no}(q)}{\varepsilon^+ - \varepsilon_n}$$

$$= \sum_d \frac{M_{od} F_{do}(q)}{\varepsilon^+ - \varepsilon_d - \Sigma_d(\varepsilon)} \quad . \tag{28}$$

We will now introduce the assumption of doorway dominance and approximate the sum in the last line of Eq.(28) by a single term. The
particular selection for A, Eq.(26), is important at this point:
we are assuming that among the doorway states in the sum, one tends
to contain most of the monopole strength. This would be a collective
monopole excitation, the counterpart of the giant resonance in the
dipole case. We thus let

$$T_{MF}(\varepsilon;q) = \frac{M_{oc} F_{co}(q)}{\varepsilon^+ - \varepsilon_c - \Sigma_c(\varepsilon)} \tag{29}$$

where c denotes the collective state. We note that the collective-
dominance assumption does not imply that the spreading width

$$\Gamma_c = - 2 \text{ Im } <\Sigma_c(\varepsilon)>$$

is small compared e.g. with the spacing of the various doorway states.
We are simply assuming that one of the M_{od} namely M_{oc} is much
larger than the others. Eq.(29) brings us back in a way to the
single doorway description discussed earlier. In fact for small
values of q Eq.(29) becomes identical to Eq.(9) which as we noted
is a representation rather than an approximation. As q increases,
Eq.(29) becomes an approximation and we can expect that after a critical value of q it will no longer be valid. Our interest is in
investigating the momentum-transfer region in which this assumption
is both reasonably valid and non-trivial. From Eqs.(28) and (29) we
immediately obtain

$$M_{on} = M_{oc} c_n^c \; ; \quad F_{no}(q) = F_{co}(q) c_n^c \tag{30}$$

i.e., we find that all form factors are the same (apart from the
scale factor c_n^c) and their shape is determined by the collective-
state form factor. Clearly this is a very restrictive assumption
and can at best be expected to be valid only for a limitted range
of q. The idea is to investigate this range first and then see to
what extent one can go beyond this by introducing additional doorways from the right-hand-side of Eq.(28). Interfering doorways will
then introduce differences in the individual form factors F_{on}.

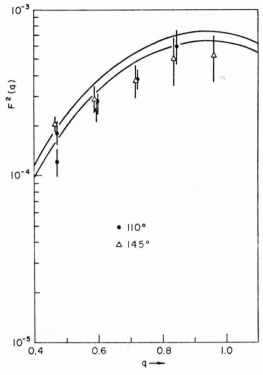

Fig. 2.   Theoretical (solid lines) and
           experimental form factor of the
           first-excited state of ^{16}0. The
           two solid lines correspond to
           experimental uncertainties in the
           value of M_{on}.

    In order to compare with experiment we have to obtain first an
expression for the collective-state form factor. This can be done
with the help of the sum rule of Eq. (4) which now reads

$$\sum_n \epsilon_n M_{on} F_{no}(q) = \frac{1}{2} <0|[M, [H, F(q)]]|0> . \tag{31}$$

The double commutator is easy to evaluate since the isoscalar
nature of the commutator makes it possible to ignore charge exchange
terms. If we also insert on the left-hand side of Eq. (31) the results of Eq. (30) we find

$$M_{oc} F_{co}(q) \sum_n \epsilon_n |c_n^c|^2 = \frac{\hbar^2}{m} q^2 \frac{d}{dq^2} F_{el.}(q) \tag{32}$$

where $F_{el.}(q)$ denotes the elastic-scattering form factor. After some minor algebraic manipulations we finally obtain

$$F_{no}(q) = \frac{M_{no}}{Z<r^2>} q^2 \frac{d}{dq^2} F_{el.}(q) \qquad (33)$$

where $<r^2>$ is the mean-square radius. This expression determines the inelastic-scattering form factor if the quantities on the right are known from experiment. A comparison[13] of a form factor obtained from Eq.(33) with experiment[14] is shown in Fig. 2 for the case of the 6.052 MeV first-excited state ($J^{\pi} = 0^+$, $T = 0$) of ^{16}O. Similar results, showing again a satisfactory comparison with experiment, have been obtained[15] for a number of other transitions both monopole and isoscalar quatrupole or dipole. In some cases (e.g. the first 2^+ state in ^{16}O) discrepancies start appearing after $q \simeq 0.8$ fm^{-1}. Work, attempting to improve the behavior of the individual form factors at higher momentum transfers by introducing additional doorways is still in progress.

References

1.   For an excellent discussion of questions related to those presented here see Peter Axel's report: "Simple Nuclear Excitations Distributed Among Closely Spaced Levels"; Proc. of the International Symposium on Nuclear Structure, Dubna July 4-11, 1968.

2.   H. Feshbach, A. K. Kerman and R. H. Lemmer, Annals of Physics, 41, 230 (1967).

3.   P. Axel, Phys. Rev. 126, 671 (1962).

4.   V. Shevchenko and N. Yudin, Atomic Energy Rev. 3, 3 (1965).

5.   J. E. Lynn; "The Theory of Neutron Resonance Reactions" Clarendon Press, Oxford (1968).

6.   A. Bohr and B. R. Mottelson, Nuclear Structure, Benjamin (1969).

7.   H. Arenhovel and W. Greiner; Progress in Nucl. Phys. Vol. 10, p. 167 (1968).

8.   T. A. Hughes, S. Fallieros and B. Goulard, to be published.

9.   See e.g. S. Fallieros and B. Goulard, Nucl. Phys. A147, 593 (1970); R. O. Akyuz and S. Fallieros, preprint (1971) and references mentioned therein.

10.   See e.g. S. S. Hanna in Isospin in Nuclear Physics, North-
        Holland (1969); P. Paul, J. F. Amann and K. A. Snover,
        (preprint) and references mentioned therein.

11.   B. Goulard, T. A. Hughes and S. Fallieros; Phys. Rev. 176,
        1345 (1968).

12.   See for instances the reports by H. Feshbach, A. M. Lane,
        C. Mahaux, H. E. Jackson and B. L. Berman in this volume.

13.   E. I. Kao and S. Fallieros; Phys. Rev. Letters 25, 827 (1970).

14.   J. C. Bergstrom, W. Bertozzi, S. Kowalski, X. K. Maruyama,
        J. W. Lightbody, S. F. Firozinsky and S. Penner; Phys. Rev.
        Letters 24, 152 (1970).

15.   T. J. Deal and S. Fallieros, to be published.

DISCUSSION

FESHBACH

A question of information: In the prediction of the q-dependence that you showed at the end, were there arbitrary constants, or not?

FALLIEROS

In some cases, there was no arbitrary constant. We knew the lifetime from experiments, and we put it in. In some cases, a couple of them, the lifetime was not known, so we fitted our curves to one point.

GROUND-STATE RADIATION WIDTHS FOR RESONANT STATES IN ^{57}Fe[*]

H. E. Jackson and E. N. Strait

Argonne National Laboratory, Argonne, Illinois 60439

Because the radiative properties of a highly excited nuclear state in the continuum are described by its partial widths for radiative decay, a substantial experimental effort has been focused on their study. Neutron capture data have provided most of the information available for radiative transitions in the 7—11-MeV region. Recently, high-resolution studies of photoneutron spectra at threshold[1,2] have become a promising source of information. Nuclear states of interest are excited by photon absorption from a bremsstrahlung beam of very precisely determined maximum energy and are observed through their decay by neutron emission. With this photon-energy definition, excitation can be limited to a band of excited levels just above the neutron binding energy; and these levels can decay by neutron emission only to the ground state of the daughter nucleus. Because the total yield for each level is given by $2\pi^2 \lambdabar^2 g \Gamma_{\gamma 0} \Gamma_n/\Gamma$ and $\Gamma_n/\Gamma \approx 1$, $\Gamma_{\gamma 0}$ can be determined from the resonance areas. In the work presented here, we have determined the ground-state radiation width $\Gamma_{\gamma 0}$ for neutron resonances in the range 20—700 keV in ^{57}Fe by measuring the photoneutron cross section for ^{57}Fe(γ,n) near threshold with high resolution. Our analysis of the data emphasizes problems which, in the past, have been studied in resonance neutron capture— namely, the determination of resonance spins and parities, radiative strength functions, distributions of ground-state radiation widths, and testing for possible correlations between neutron and radiative widths.

The measurements were performed at the ANL threshold photo-neutron facility shown schematically in Fig. 1. An energy-analyzed

---

[*]Under auspices of U. S. Atomic Energy Commission.

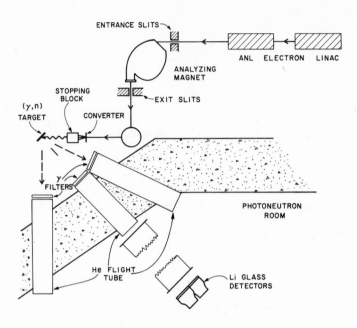

Fig. 1.   Threshold photoneutron facility.

pulsed electron beam with a pulse width of 6 nsec and a repetition
rate of 800 pps from the ANL LINAC irradiates a 40-mil Ag converter.
The resulting bremsstrahlen pass through a block of aluminum
(which stops the electrons) and irradiates the target in which the
(γ,n) reaction occurs. The photoneutrons are observed along flight
tubes placed at 90°, 135°, and 155° to the direction of the
bremsstrahlung beam. Neutrons are detected in a ^{6}Li glass detector
system or in a proton-recoil detector system (the choice depending
upon the excitation region under study), and their energies are
determined by their flight times.

    The photoneutron time-of-flight spectrum covering the low-
energy region, 20—300 keV, is shown in Fig. 2. The data result
from irradiation of a 40-g target of enriched ^{57}Fe with
bremsstrahlen whose end-point energy was 8.45 MeV. From the
extensive data available on s-wave resonances in the total cross
section of ^{56}Fe, a complete list of expected s-wave ($J=\frac{1}{2}^+$) levels
can be obtained. Where the resonance structure was resolved,
i.e., below 225 keV, a measurable yield in the (γ,n) spectrum was
observed for all but one of the known s-wave levels. One of the
striking features of the results for ^{57}Fe as well as other nuclei
studied in this region is the strength of resonances with $\ell > 0$.
The integrated strength of the higher angular momentum component
is greater than that of the s-wave component.

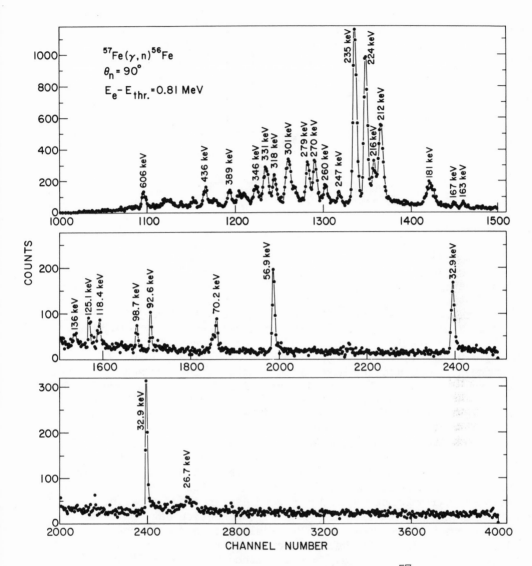

Fig. 2.   Photoneutron time-of-flight spectrum for $^{57}Fe(\gamma,n)$.
Points in the center section are sums of two adjacent
channels; in the lower section they are sums of four
adjacent channels. The data were taken with ^6Li
glass detectors.

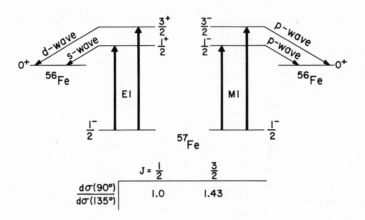

Fig. 3. Decay scheme for ^{57}Fe(γ,n).

To explore this aspect of the data further, the spins of strong neutron groups were determined by measuring the angular distribution of the photoneutrons. The possible modes of excitation and decay for ^{57}Fe are illustrated in Fig. 3 for absorption of dipole radiation. For the spin sequence $\frac{1}{2} \rightarrow \frac{1}{2} \rightarrow 0$, the photoneutron angular distribution will be isotropic; for $\frac{1}{2} \rightarrow \frac{3}{2} \rightarrow 0$, the ratio $d\sigma(90°)/d\sigma(135°)$ will be 1.43. Spin assignments were made by normalizing the data for 90° and 135° so that the relative yields gave isotropy for the strong s-wave level at 212 keV and calculating the corresponding ratio for the other resonances. Table I lists the assignments for all levels observed with sufficient intensity to allow unambiguous calculation of the 90°/135° ratio. Also shown are weak resonances known (from total-cross-section measurements in ^{56}Fe) to be s-wave. The intensity ratios observed are consistent with the assumption that only dipole photon absorption is important in the excitation process. The parities of resonances with j = $\frac{1}{2}$ were assigned by comparing the photoneutron results with the total cross section data on ^{56}Fe. Those already known from the ^{56}Fe results to be j = $\frac{1}{2}^{+}$ were identified in the photoneutron spectra and the remainder were assigned a spin and parity j = $\frac{1}{2}^{-}$. Although both j = $\frac{3}{2}^{+}$ and j = $\frac{3}{2}^{-}$ levels can in principle be excited by dipole absorption, our results for neighboring nuclei, ^{53}Cr and ^{61}Ni indicate that in this excitation region d-wave emission is not significant. Presumably the yield of d-wave resonances is inhibited by small values Γ_{n}/Γ, and the probability of error is small in assigning all j = $\frac{3}{2}$ resonances a negative parity corresponding to p-wave emission. We have followed this convention and interpreted all of the resonance structure with $\ell > 0$ as a strong p-wave component excited by M1 transitions.

TABLE I. Angular-momentum assignments for resonances in the reaction ^{57}Fe(γ,n).

E_n (keV)	$\dfrac{d\sigma(90°)/d\Omega}{d\sigma(135°)/d\Omega}$	J^π	E_n (keV)	$\dfrac{d\sigma(90°)/d\Omega}{d\sigma(135°)/d\Omega}$	J^π
26.7	s-wave	$\frac{1}{2}^+$	181.	1.05 ± 0.07	$\frac{1}{2}^+$
32.9	1.10 ± 0.08	$\frac{1}{2}^-$	212	1.00 ± 0.03	$\frac{1}{2}^+$
56.9	1.08 ± 0.09	$\frac{1}{2}^-$	216	0.90 ± 0.10	$\frac{1}{2}^-$
70.2	s-wave, other	$\frac{1}{2}^+$, ?	224	1.32 ± 0.04	$\frac{3}{2}^-$
118.4	s-wave	$\frac{1}{2}^+$	235	1.43 ± 0.04	$\frac{3}{2}^-$
125.1	s-wave	$\frac{1}{2}^+$	270	1.36 ± 0.09	$\frac{3}{2}^-$
136	s-wave	$\frac{1}{2}^+$	278	0.99 ± 0.07	$\frac{1}{2}^-$
163	s-wave	$\frac{1}{2}^+$			

The individual resonance yields were analyzed to determine the values of $g\Gamma_{\gamma 0}\Gamma_n/\Gamma$. For all s-wave resonances and for p-wave levels above 100 keV, $\Gamma_n/\Gamma \approx 1$ so that $\Gamma_{\gamma 0}$ can be obtained from the yields and spin assignments. The s-wave and p-wave resonances give electric-dipole and magnetic-dipole partial radiation widths, respectively. Such data for ^{57}Fe were analyzed to obtain individual and average ground-state widths and the corresponding reduced widths $\overline{k}_{E1}$ and $\overline{k}_{M1}$. For E1 transitions $\overline{k}_{E1} \times 10^3 = 0.9^{+0.8}_{-0.3}$, which is much smaller than the usual single-particle estimate $\overline{k}_{E1} \times 10^3 = 3.2$ but is consistent with the estimate obtained by extrapolation of the giant dipole resonance to low energies, namely, $\overline{k}_{E1} \times 10^3 = 1.1$. For M1 transitions our result is $\overline{k}_{M1} \times 10^3 = 10^{+10}_{-5}$. Bollinger[3] has observed that the data on nonmagic nuclei between A = 80 and 250 are consistent with $\overline{k}_{M1} \times 10^3 \approx 20$. Our data indicate that the region of validity of this observation extends down to A $\approx$ 50; furthermore, the prominence of the p-wave resonances in the (γ,n) spectra does not arise from an enhancement of the M1 radiative strength. Rather it results from a relatively low mean intensity of electric-dipole transitions in the lighter nuclei, which can be interpreted as a consequence of an $A^{8/3}$ mass dependence in the giant-dipole-resonance estimate.

The data were tested for a correlation between the reduced neutron widths Γ_n^0 for the s-wave resonances and the corresponding values of $\Gamma_{\gamma 0}$. The strength of such a correlation can be interpreted as a measure of the importance of direct (i.e., valence-nucleon) types of transitions in explaining the data for radiative widths. Recently, Block and coworkers[4] have reported evidence for a strong correlation between Γ_n^0 and the total radiation width Γ_γ

for 12 resonances in nuclei near ^{57}Fe. Because the capture
spectrum is dominated by strong high-energy transitions, the
observed effect suggests the existence of correlations between
Γ_n^0 and Γ_γ^0. We find the value of the correlation coefficient for
9 s-wave levels to be 0.106—which is consistent with no
correlation and suggests that if a special reaction mechanism is
present, its contribution to the ground-state radiation width is
small.

Although the resonance structure was completely resolved only
below a photoneutron energy of 250 keV, the local distribution of
radiative strength was studied over the full 20—700 keV. The
measurement covering the 0—300 keV interval was extended to 700
keV in the data shown in Fig. 4. This photoneutron time-of-flight
spectrum for neutron energies from 200 to 700 keV was observed
with a recoil-proton detector placed at 90° to the photon beam.
The analysis is not complete, but the data for p-wave levels
indicate two anomalous concentrations of M1 strength. Below
300 keV only three resonances with $J^\pi = \frac{3}{2}^-$ are observed, including
a very intense pair at 224 and 235 keV. Other data indicate that
a much larger number of p-wave resonances were undetected and the
results for s-wave levels would imply that about 30 resonances
with $J^\pi = \frac{3}{2}^-$ should be found below 300 keV under the usual
statistical assumption that the density of levels is proportional
to 2J + 1. The mean value of $g\Gamma_{\gamma_0}$ resulting from the doublet
alone is sufficiently high to guarantee that a substantially
larger number of resonances should have been observed if they were

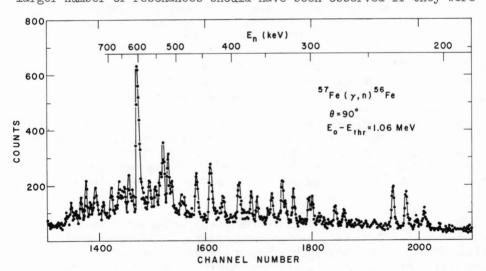

Fig. 4.   Photoneutron time-of-flight spectrum for ^{57}Fe(γ,n)
          as observed with a proton-recoil neutron detector.

drawn from a Porter-Thomas distribution containing 30 levels. We estimate less than 10^{-4} for the probability that the observed concentration of strength might result from a Porter-Thomas population of 30 resonances whose mean width is chosen to give the observed integrated radiative strength. Among the $j = \frac{1}{2}^-$ resonances, an anomalously strong resonance is observed at 606 keV. The spin and parity assignment of this level is based upon an isotropic angular distribution and the lack of correlation of the resonance energy with the s-wave resonances observed in the ^{56}Fe total neutron cross section. Our preliminary estimate of $\Gamma_{\gamma 0}$ for this level is 1.5—3.0 eV. Although this is perhaps not as dramatic a case as the intense doublet, we estimate in the same manner as above that the probability of obtaining a level whose width is greater than that of the 606-keV resonance from a Porter-Thomas population of the appropriate mean value and spacing is less 0.03.

Thus the photoneutron data suggest the presence of intermediate structure in the cross section for radiative excitation of p-wave levels in ^{57}Fe, and therefore in the M1 radiative strength function. We have considered the possibility that this structure could be attributed to the presence of a strong two-quasiparticle doorway excitation of the proper character. For both protons and neutrons, ^{57}Fe is close to the N = 28 shell closure of the $f^{7/2}$ level; calculations of energy eigenvalues above the Fermi energy in a Woods-Saxon potential show that near A = 50 the energy difference between the filled $1f_{7/2}$ level and the empty $1f_{5/2}$ orbital is about 8 MeV, quite close to the excitations studied in this experiment. A spin-flip transition in which a $f_{7/2}$ nucleon is transferred to the $f_{5/2}$ orbital would generate a doorway excitation of the appropriate energy, spin, and parity. The observed intermediate structure can be attributed to a doorway consisting of a $(f_{5/2})(f_{7/2})^{-1}$ particle-hole pair coupled to the ^{57}Fe ground state. Such a doorway would be expected to affect both the $j^{\pi} = \frac{1}{2}^-$ and $j^{\pi} = \frac{3}{2}^-$ channels since the particle-hole pair can couple to give either spin value. This configuration should also generate two $\frac{3}{2}^-$ doorways corresponding to two possible ways in which the particle-hole and core angular momenta could couple to $j^{\pi} = \frac{3}{2}^-$. Analysis of the higher energy data is continuing.

References

[1] W. Bertozzi, et al., Phys. Rev. Letters 6, 108 (1963).

[2] R. J. Baglan, et al., Phys. Rev. C 3, 679 (1971).

[3] L. M. Bollinger, in Nuclear Structure, Dubna Symposium, 1968 (International Atomic Energy Agency, Vienna, Austria, 1968) p. 317.

[4] R. G. Stieglitz, et al., Nucl. Phys. A163, 592 (1971).

DISCUSSION

LANE   (Harwell)

Fe^{57} was a nucleus mentioned 12 or 15 years ago very commonly, but these comments one used to hear about it just aren't heard anymore. Perhaps it's in order to remind ourselves of them. First of all, it's the only nucleus in the mass range from 40 to 65 which shows no anomalous capture whatsoever. It's very, very special. All of the others show a tremendous amount of capture. On the theoretical side, it is the only nucleus in the whole periodic table for which the main parentage is supposed to come from an excited state of the core, Fe^{56} rather than the ground state. This of course, is very relevent to the absence of correlations with the neutron channel, which is essentially a ground state channel. On the other hand, Cr^{53} is more normal in this respect, and I've no reason to think any such effects should apply there at all.

WIGNER

I have two questions. The first one concerns your statement that there should be 30 J=3/2 levels. Did that take into account the spin cutoff factor?

JACKSON

No.

WIGNER

Then it isn't perhaps so surprising that there are few higher spin levels.

JACKSON

I think that the normally accepted values of the spin cutoff parameter, sigma, wouldn't diminish the density that much. I think this is much lower than any reasonable value of sigma would lead you to believe.

WIGNER

Well, what would one expect?

JACKSON

I would guess a 10 or 20% effect at most. This is a relatively low spin. We're talking about missing 27 levels.

WIGNER

In other words, you don't believe that, that is an explanation. My second question is, if you add up the gamma ray transition probabilities, how far are you from a single-particle width?

JACKSON

The group at 600 keV is half a Weisskopf unit.

WIGNER

Then it's very large. Barschall has explained to me, and I
think correctly, that it's very difficult to observe correlations,
because what is added and subtracted are amplitudes.

A large value of the sum of these transition probabilities is
necessary to produce even the small correlation coefficient which
was obtained. A background amplitude which would give only 10%
of the amplitude due to the doorway state is only $\sqrt{0.1}=0.32$ times
smaller. The ratio of the intensities of two lines in one of which
the two amplitudes reinforce,in the other of which they weaken,
each other is $(1+0.32)^2/(1-0.32)^2$ or almost 4.

FRANCIS

Could you explain why you believe that the correlation should
be largest when you consider the ground state as one of the states?

JACKSON

Mainly because the ground state has such a large spectroscopic
factor, and as I understand these valence calculations, the correlated
crosssection is proportional to the spectroscopic factor, which is
a measure of the single-particle strength in that state.

FRANCIS

This wouldn't be true if it were a doorway state, though?

JACKSON

No.

FESHBACH

Just a question of information. You said that the width of
this intermediate structure that we were observing was too narrow.
I don't know if you ever quoted a number for its width. If you did,
I missed it.

JACKSON

The width doesn't mean very much when there are only 2 resonances,
but it would have to be around 20 keV.

NEWSON

You made a point about seeing very few p-levels. Isn't that
connected with the slide of Bob Bloch's that I showed in which he
measured some p-wave strength functions in this region of Fe, and
found them to be practically zero?

JACKSON

You have to be very careful about this point, because what
you're saying essentially is that for a small p-wave strength
function, you worry about the decay being very small. The only
thing that's important is that the neutron widths be large compared
to the radiative widths. We feel that that's sufficiently true

in this region because we have enough sensitivity to determine it.

NEWSON

　　　Were you considering that the p-wave strength function was
about 1/10 of the usual optical model value?

JACKSON

　　　We actually used Bob's values for the strength function in
our estimate.

THRESHOLD PHOTONEUTRONS AND PHOTON DOORWAYS[*]

B. L. Berman, C. D. Bowman, and R. J. Baglan[**]

Lawrence Livermore Laboratory

University of California

It has long been recognized that the measurement of photo-
nuclear cross sections with an energy resolution as fine as that
obtained heretofore only for neutron cross sections would be a
powerful probe for investigating the properties of nuclear states.
This would bring to bear the electromagnetic selection rules as a
spectroscopic tool, and open for investigation all the stable
nuclei not accessible to neutron-induced reactions. This goal now
has been accomplished, for the first MeV or so above the (γ,n)
threshold, by the threshold photoneutron technique.[1] In this experiment, bremsstrahlung from a pulsed, nearly monoenergetic
electron beam is directed at the sample under study. The neutrons
ejected in the (γ,n) reaction are detected and their energy is
measured by the neutron time-of-flight technique. The energy of
the electron beam is adjusted so that the tip of the bremsstrahlung
spectrum barely exceeds the (γ,n) threshold of the sample. Then
neutron transitions from levels of the compound nucleus to excited
states of the residual nucleus cannot take place -- only ground
state transitions are energetically possible. Thus a measure of
the neutron energy uniquely determines the energy of the photon
which induced the reaction. The energy of the photon typically is
100 times larger than the energy of the neutron, but the absolute
uncertainties in the neutron and photon energies are the same.
Therefore, a neutron energy resolution of 1%, which is obtained
easily, results in a photon energy resolution of 0 01%. Also, it
should be noted that the high-intensity pulsed electron beams
from linear accelerators can be utilized directly in this experiment. This allows the counting rates for these measurements to
compare favorably with those for neutron-induced reactions, and
moreover permits neutrons with high ℓ-values to be observed with

relative ease. The technique is illustrated in Figs. 1 and 2.
Further experimental details are given in Refs. 2.

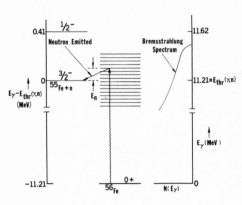

Fig. 1 Schematic diagram for
the reaction $^{56}Fe(\gamma,n)^{55}Fe$.
If ^{56}Fe is irradiated with
bremsstrahlung having a maximum
energy between 11.21 and 11.62
MeV, all the resonances seen in
the neutron energy spectrum can
be identified unambiguously as
representing levels in ^{56}Fe having excitation energy equal to
the neutron binding energy plus
the resonance energy. If the
bremsstrahlung end-point energy
now is raised to 12 MeV, the
resonances in the neutron energy
spectrum above 400 keV must cor-
respond to ground-state transtions; only those peaks below 400 keV
which were not present in the low-energy run can represent
excited-state transitions.

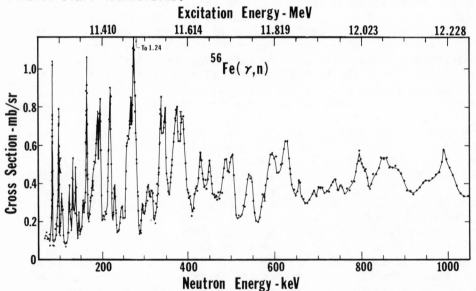

Fig. 2 The 135° differential $^{56}Fe(\gamma,n)$ cross section. At the
highest energies, the deterioration in the energy resolution resulting from the finite resolving time of the experimental appara-
tus is evident. From these data (Ref. 3), values for $g\Gamma_{\gamma 0}\Gamma_n/\Gamma$
have been extracted for 28 resonances, and all but 5 have been
identified with ground-state transitions.

The threshold photoneutron technique has been used to study many nuclei for the presence of intermediate structure in the photon channel. The fine-structure resonances in the (γ,n) reaction measured with high resolution by this technique were examined for clusters of ground-state γ-ray transition strength associated with excited states all having the same J^π.. Evidence for doorway states in the form of envelopes of γ-ray widths was observed for ^{207}Pb near 125 and 400 keV, ^{57}Fe near 50 and 250 keV, and ^{208}Pb near 500 keV above the (γ,n) threshold.[4]

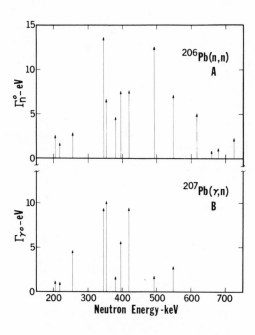

Fig. 3 Reduced neutron widths (A) and ground-state γ-ray widths (B) for $1/2^+$ resonances in ^{207}Pb.

Probably the best example of a doorway state detected by neutron-induced reactions was discovered in ^{207}Pb by Farrell et al[5] in a measurement of the neutron total cross section for ^{206}Pb. They found a number of $J^\pi = 1/2^+$ states in the energy region near 500 keV. The values for the reduced neutron widths Γ_n^o show an envelope, Fig. 3A, whose width is $\simeq$ 250 keV. This envelope was taken to represent a 2p-1h doorway state, with fine structure representing 3p-2h states. If so, it might be expected to appear as a common doorway in the photon channel as well. We therefore attempted to find this doorway using the threshold photoneutron technique.

The 135° differential ^{207}Pb(γ,n) cross section, from 200 to 700 keV above threshold, is shown in Fig. 4. Below 600 keV, ten $1/2^+$ resonances could be assigned from the ^{206}Pb(n,n) data;[5] these are indicated by the unlabeled arrows above the data. The ground-state γ-ray widths $\Gamma_{\gamma o}$ for these states are shown in Fig. 3B. The distribution of these values for Γ shows an envelope very similar to that for Γ_n^o; this fact alone is very strong evidence for the existence of a common doorway.

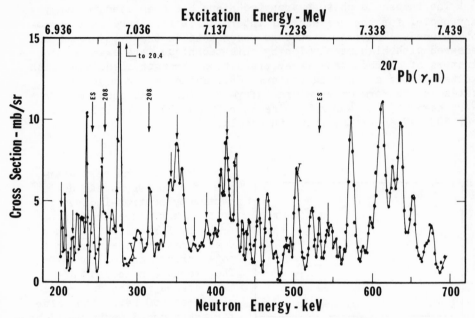

Fig. 4   Threshold photoneutron cross section for 207Pb.   The
arrows labeled "208" and "ES" indicate contaminant peaks from
208Pb and excited-state transitions, respectively.   The unlabeled
arrows indicate resonances with $J^\pi = 1/2^+$ (see text).

        The neutron and γ-ray widths for the fine-structure reson-
ances of the doorway state should be correlated if indeed the same
doorway state is seen in both channels.   The rank correlation coefficient, which tests the existence of correlation, was computed
for the two sets of widths.   The results give a probability of
0.96 that some correlation exists; this is further evidence for
the presence of a doorway state.   The product-moment correlation
coefficient, which measures the degree of correlation, also was
computed; it was used to determine the ground-state γ-ray width of
the doorway state $\Gamma_{\gamma 0}^D$ to be 36.5 eV.   Since the total ground-state
γ-ray width of the ten $1/2^+$ resonances was measured to be 48.6 eV,
about 75% of the γ-ray strength is derived from the doorway state.
The value 36.5 eV is about an order of magnitude smaller than the
single-particle width calculated for an E1 transition in 207Pb.
Thus, most of the γ-ray strength of the single-particle state
apparently has been transferred into the giant resonance; the residual strength which is measured here perhaps is typical of the
many single-particle states possible in 207Pb.

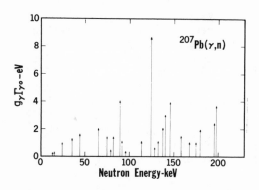

Fig. 5   Values for $g_\gamma\Gamma_{\gamma 0}$ for low-energy resonances in ^{207}Pb.

Figure 5 shows the values for $g_\gamma\Gamma_{\gamma 0}$ versus neutron energy for all the resonances seen in the ^{207}Pb (γ,n) cross section below 200 keV. Only one of these (at 66 keV) has J^π=1/2$^+$. Therefore, most of these resonances probably represent 3/2$^+$ states which are excited by El photons from the ground state of ^{207}Pb and which decay by d-wave neutrons. The averaged (γ,n) cross section contains a peak centered at 125 keV. The large number of resonances (25) tends to rule out statistical

fluctuations as the cause of the structure. Therefore, this structure might represent a 3/2$^+$ doorway state, whose total width is 120 keV and whose ground-state γ-ray width is 16.3 eV. Such a state is expected simply by recoupling the 1/2$^-$ neutron hole and the 1$^-$ p-h or collective state to give a 3/2$^+$ doorway addition to the 1/2$^+$ doorway described above.

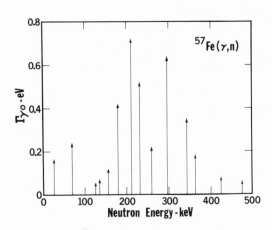

Fig. 6   Ground-state γ-ray widths for 1/2$^+$ resonances in ^{57}Fe.

Figure 6 shows the values for $\Gamma_{\gamma 0}$ obtained for 1/2$^+$ resonances in ^{57}Fe. The J^π assignments come from ^{56}Fe(n,n) measurements.[6] These values for $\Gamma_{\gamma 0}$ form an envelope centered near 250 keV, and thus appear to indicate the presence of a doorway state. However, the correlation with the Γ_n^0 values of Ref. 6 is not significantly greater than zero. This structure, then, either represents a 1/2$^+$ doorway state with γ-ray strength but very little neutron strength, or else it results merely from statistical fluctuations in the γ-ray widths.

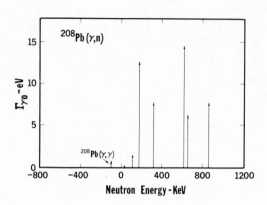

Figure 7 shows the values for $\Gamma_{\gamma 0}$ for 1^+ resonances in ^{208}Pb, which deexcite by p-wave neutrons.[7] The values for J^π for these resonances were obtained from the ratio of the (γ,n) cross sections obtained at two angles ($90°$ and $135°$), combined with ^{207}Pb(n,n) data.[8] The M1 strength in ^{208}Pb arises from spin-flip transitions from the $i_{13/2}$ neutron shell and the $h_{11/2}$ proton shell; these p-h states form the doorway state through which the compound-nucleus states observed here are reached. The total γ-ray strength for these states is 50.8 eV, more than 5 Weisskopf units, and constitutes at least half and perhaps all of the total M1 strength calculated for this nucleus.[9] This, then, is the long-sought M1 giant resonance, a collective doorway, in contrast to the 2p-1h case for ^{207}Pb.

To date, we have measured the threshold photoneutron cross sections for 16 nuclei, and over 250 resonances have been identified and in most cases parameterized. Through this work we also have obtained information on several other phenomena of current interest in nuclear physics, including E1 strength functions,[3] analog states and their isospin purity,[10] nonresonant neutron capture,[11] the curious threshold resonance in ^{9}Be,[12] and the source of neutrons in stars.[13] Moreover, the technique shows promise of further refinement, and no doubt will be a rich source of additional interesting nuclear physics information.

References

* Work performed under the auspices of the US Atomic Energy Commission. A more detailed account has appeared as B. L. Berman, UCRL-73003, 1971 (unpublished).

** Present address: Vanderbilt University, Nashville, Tenn.

1. W. Bertozzi, C. P. Sargent, and W. Turchinetz, Phys. Letters 6,108 (1963);
    B. L. Berman, G. S. Sidhu, and C. D. Bowman, Phys. Rev. Letters 17, 761 (1966);
    C. D. Bowman, G. S. Sidhu, and B. L. Berman, Phys. Rev. 163, 941 (1967);
    C. D. Bowman, B. L. Berman, and H. E. Jackson, Phys. Rev. 178, 1827 (1969).
2. R. L. Van Hemert, C. D. Bowman, R. J. Baglan, and B. L. Berman, Nucl. Instrum. Meth. 89, 263 (1970);
    R. L. Van Hemert, Ph.D. Thesis, UCRL-50501, 1968 (unpublished);
    R. J. Baglan, Ph.D. Thesis, UCRL-50902, 1970 (unpublished).
3. R. J. Baglan, C. D. Bowman, and B. L. Berman, Phys. Rev. C 3, 672 (1971).
4. R. J. Baglan, C. D. Bowman, and B. L. Berman, Phys. Rev. C 3, 2475 (1971).
5. J. A. Farrell, G. C. Kyker, Jr., E. G. Bilpuch, and H. W. Newson, Phys. Letters 17, 286 (1965).
6. C. D. Bowman, E. G. Bilpuch, and H. W. Newson, Ann. Phys. (N.Y.) 17, 319 (1962).
7. C. D. Bowman, R. J. Baglan, B. L. Berman, and T. W. Phillips, Phys. Rev. Letters 25, 1302 (1970).
8. E. G. Bilpuch, private communication.
9. M. S. Weiss, private communication.
10. B. L. Berman, R. J. Baglan, and C. D. Bowman, Phys. Rev. Letters 24, 319 (1970).
11. C. D. Bowman, R. J. Baglan, and B. L. Berman, Phys. Rev. Letters 23, 796 (1969).
12. B. L. Berman, R. L. Van Hemert, and C. D. Bowman, Phys. Rev. 163, 958 (1967).
13. B. L. Berman, R. L. Van Hemert, and C. D. Bowman, Phys. Rev. Letters 23, 386 (1969).

SESSION IX

CONCLUSION OF THE CONFERENCE
Round Table Discussion

CHAIRMAN - E. Wigner (Princeton University)
PANELISTS-G. A. Bartholomew (Chalk River Laboratories)
        L. Bollinger (Argonne National Laboratory)
        B. Cohen (University of Pittsburg)
        H. Feshbach (M.I.T.)
        K. J. LeCouteur (Australian National University)
        J. Rainwater (Columbia University)

WIGNER(Princeton University)

    We decided on all of us first making a four minute statement
on the subject about which we have learned most during the
conference, and on the subject which we would wish to work most
when we get home. After that we will ask the audience to make
remarks or to pose questions either to a particular member of the
panel or to the entire panel as such. Later we may ask questions
each other and try to answer them also. Answering questions is
often a little more difficult, than asking them. But before making
my own four minute statement, I would like to ask Dr. Bartholomew
to say something.

BARTHOLOMEW(Chalk River Nuclear Laboratory)

    Thank you Mr. Chairman, I would first like to move a sincere
vote of thanks to Dr. Garg and his committee for the excellent
organization and the arrangements that they made for this conference.
I think that we all appreciate what they have done for us all.
They were always willing to help us with all arrangements and I
think that they have done a very superb job.

WIGNER

    Dr. LeCouteur would also like to say something.

LeCOUTEUR(Australian National University)

    I would like to associate myself with this motion of the
initiative of Dr. Garg. His work on this conference has been a
very valuable one. It has stimulated many ideas and apart from
that the organization has been outstandingly good.

WIGNER

    Thank you very much, both of you. Before making my own four
minute statement I was asked to recall to all of you that the
proceedings of this conference will be published as a memorial
volume to C.E. Porter and R.G. Thomas and I am very happy to tell
you about this. Now I will make my own four minute statement and
I hope that it will be a little bit shorter. What impressed me most
and what was most novel to me was the wealth of evidence for
structure in what we consider the Statistical Region. We can call
this structure, giant resonance, as I think was first done by
Goldhaber and Teller, or evidence for doorway states. It doesn't
really matter. It is a structure so clear and pronounced that I
would be inclined to modify the statistical theory as a total theory
in such a way so that it describes the heavy fluctuations causing
accumulations of levels or more accurately of transition probabilities
to restricted regions. The existence and manifold of these accumulations
of transition probabilities to groups of reasonably closely spaced
levels is what impressed me most, so that as I said, I am inclined
to suggest that we develop a theory or statistics of levels and
particularly transition probability accumulations rather than
confining ourselves to the statistics of individual levels and

transition probabilities as we practiced it so far. As far as levels
are concerned, this has been done to a considerable extent already
by Mehta, Gaudin and Dyson on the spacing of levels and they have
also developed a theory which accounts for the magnitude of correlations between the adjoining spacings and second and higher order
spacings. The correlation of adjoining spacings was negative.
Apparently the correlation between adjoining transition probabilities
is positive but we do not know it. I do hope that it might be
possible and this is of course a sort of dreaming to develop a
theory of such accumulations of transition probabilities. This is
what impressed me most and this is what I take home most prominently
as a new element, at least for me a new element from this conference.
Well Herman it is your turn.

FESHBACH(Massachusetts Institute of Technology)
     Well I must apologize because the ground rules for this
discussion were set while I was away and originally I was planning
to make some comments about doorway states and intermediate structures. But that sounds a little, how shall I say it, improper in
view of the fact, that now I am supposed to talk about what
impressed me most. In any event, to keep within that domain, I
think that we agree with Professor Wigner that the evidence for
intermediate structure is quite there to see in many cases and I am
impressed by the evidence which is quite subtle and requires very
careful study and we can be grateful that there are a few very
simple cases that we can study and see what the structure is. We
have doorway state entrance and exit channels, we have overlapping
doorways to different channels we have many overlapping doorways;
and in fact Anderson and Grimes at Livermore Laboratories have
applied statistical concepts to doorways, so that you can discuss
random phase approximations and things of that kind until you get
an idea as to the nature of the spectrum of neutrons, let's say as
an example that are produced in the alpha neutron reactions. When
you get away from the domain in which temperature concepts seem to
apply in so called equilibrium region. We have secondary doorways
which we have seen in the photons, originally photon induced
reactions and now discovered in isobars. I think that it is
encouraging because it indicates that even in this domain of high
excitation there are many many levels and much **going** on, that
simple structures, simple parts of the wave functions can be seen.
It's not at all universal in my opinion and you have to look in
special domain in order to really see it that visibly. In a single
doorway state situation it's just not that way all of the time.
But near closed shells when nuclear states can be described by a
weak coupling of the valence nucleon to a core like in the core
excitation model, or, which is particularly important when the core
has collective states, when symmetries act to reduce the spreading
width like the example of the isobaric analogue. In those cases
these things become visible. One might turn that around and say
that perhaps in looking for intermediate structure, there may be a

way for finding new dynamical symmetry. Symmetries that we don't
suspect, symmetries that wouldn't appear simply because they are
dynamical rather than geometrical. One has perhaps a couple of
cases of that kind. The heavy ion experiments with alpha particle
transfers done at Argonne and at Saclay are examples of that kind
of dynamical symmetry that has to do with the special structure of
the alpha particles. In a way the most exciting thing to me, is the
prospect that one might discover new fundamental things about nuclear
physics from looking at these states.

WIGNER
    Thank you very much. Now Dr. Rainwater.

RAINWATER(Columbia University)
    I also find that the rules have been changed since I last
heard what they were. I find that the conference has been useful
to obtain an exposure to very beautiful new data that has been
accumulating in many areas in the last year of so and in particular
there were presentations of the results on doorway state, intermediate structure from the experimental side rather than just
having to see the same theoretical talk over and over, which is fine
but one would like to see that this actually occurs in reality and
I believe now, that the theories are being backed up by reality.
I found that particularly interesting, the enlargement in the
experimental evidence. The main topic that I have personally been
involved with was the one that Professor Wigner pioneered back
around 1956 and on the business of the level spacing distribution.
This is a subject which was only in a sense possible as a combination of the experimental and theoretical situations, the great
orders of magnitude improvements in accelerators, detectors, on
line computers which make the data sufficiently reliable so that
one can see what happened. It was before 1956, essentially nothing
was known what one would expect for reduced width distribution or
for nearest neighborhood spacings or other distributions. I believe
that the random matrix theory, as developed by Professor Wigner
was done somewhat more in the sense of pure mathematics, but at
least it seemed to apply and this was an effect which had the
unfortunate feature of being somewhat unphysical when you take
random matricies, that the joint distribution is the product of all
possible differences between the levels, the absolute values of the
spacings but then you have to have some convergence feature,
which when you put in the guassian ensemble was the semicircular
distribution, a law for the density about the center. This is a
very non-physical effect, certainly if you are going to compare it
with the way data are where the densities are reasonably uniform
it isn't needed. Dyson's circular ensemble was one somewhat obvious
way of correcting it but to say obvious and to say that you can do
it in the way that the mathematicians are happy, these are two different things. At least it had the feature of the new developments
that you can have, something that is more realistic and you can

doctor it up now that it appears to get any slow exponential
growth whatever you consider appropriate. The experimental side
has been somewhat unfortunate I believe in the past, in that there
has been a tendency to look **at** thorium and uranium data as being
particularly pertinent. These elements, the thorium 232 and
uranium 238 happen to lie in a mass region where the S wave
strength function isn't that much stronger than the P wave strength
function and the nature of the Porter Thomas Distribution is such
that you have 8% of the S wave reduced widths below one percent
of the mean, 3% below the tenth of a percent and so on. So it
really goes down low and it's very easy for the P wave population
to get mixed up in an impossible manner. Earlier comparisons in
fact were such that when one was looking at long range effects in
the statistic, it seemed utterly hopeless until very recently
to get anything near a small value the theory predicts. Dyson has
developed a theory for the Delta statistic, to remind you, that if
you plot the staircase of the number of levels against energy, on
the average you have to have 20,000 levels between the mean square
deviation from the staircase in and the straight line is up to
unity, and of course, this is a fantastic regularity for long range
order. Recently, we have a large collection of rare earth data
which, I believe, the theorists were not really aware of, most of
them, before they came to the conference, and the evidence now
seems to be somewhat different from that stated in some of the
speeches--it seems to favor for the Kth distribution, and so on,
the random matrix theory, and not the form of the two body interaction force, as treated by Bohigas and Flores, and French and
Wong. We have copies of these papers, not here, that we can send,
for those of you who are interested in seeing this new collection
of experimental data. That's all.

WIGNER
        Thank you very much. I probably did not give the right
introduction but that doesn't matter. Bollinger is next.

BOLLINGER(Argonne National Laboratory)
        Thank you. I wish to spend my four minutes, outlining for you
the statistics of the neutron gamma experiments, about which you've
been hearing, and their relationship to the statistical model.
Until this time, much of the effort has been spent in testing the
statistical model, and in attempting to find cases where it can be
demonstrated quite conclusively that there are departures from the
model. We are now at the stage, where there appears to be some
such departures that have been identified, and so the effort now
must be devoted to attempting to map out ranges of energy and
nuclei, where such effects are observed. One hopes, that within
the foreseeable future this can be accomplished, and what I mean
by this is that, it appears that with existing technology one
should be able to now obtain a fairly refined description of the
behavior of the gamma radiation produced by neutron capture. Now

back to the question of our present status of the data. I'll
divide the data into three categories: those having to do with
correlations, distributions of widths, and absolute values of
widths. First as to correlations: as you have heard both the
valency model and some of the models having to do with doorway
states, predict firmly that there should be correlations between
the radiation widths and the neutron widths. Yet, in my opinion,
it seems that there are no cases where it is yet demonstrated with
absolute certainty. There are cases where it is suggested but much
needs to be done yet. One of the more convincing cases, which was
capture in thulium 169 was essentially disintegrated at this
conference. There is also expected to be correlation between the
radiation widths and the (d,p) spectroscopic factors having to do
with the single particle character of the final state. Here again,
for heavy deformed nuclei, there is not, in my opinion, any convincing
example of such a correlation. On the other hand, for thermal capture,
in the light nuclei, those in the range of Iron say, there are great
many cases where the correlation is absolutely clear cut and where
it can be interpreted in terms of the existing models. Similarly,
there's now beginning to be a good body of evidence from resonance
capture in nuclei in the region of molybdenum, zirconium and
lighter where there again seems to be clear cut evidence for such
correlations, and this is the kind of thing I say that we'll
presumedly be rapidly be getting more of during the next few years.
Now going on to the distributions of widths. There have been
several nuclei for which it was claimed that there was a clear cut
departure from the expected Porter-Thomas Distribution. Perhaps
the more spectacular of these has been withdrawn-these claims have
been withdrawn during the last year--and in my opinion all of the
others are doubtful at this time. The problem is, that there are
various ways in which one can bias the data in by failing to resolve
either the initial or the final state--one can get a wrong result
and I think there is reason to have suspicion that this has
happened. Going on to the final major topic: absolute values.
Relatively little attention has been paid to the absolute values.
Perhaps because in the past the absolute values were not very
reliably known. This is unfortunate because often adding the
information about absolute values to the more qualitative information
having to do with the spectra where some lines appear to be strong,
abnormally strong, adding this information about absolute values
can shed a new light on the matter. For example, we've been hearing
for ten years, or more now, about the anomalously strong radiative
transitions to low lying states in the lightest nuclei in the
neighborhood of iron. There can be no doubt at all that these
spectra are peculiar in the sense that most of the strength, or a
large part of the strength, does appear to go to these low lying
states. But nevertheless, if one looks at the absolute values,
they are not abnormally strong as to what one would infer by
extrapolating from other nearby regions of the periodic table. So
perhaps at least it is as meaningful a question to ask as to why

are these transitions abnormally strong as it is to ask why are
the others abnormally weak. Thank you.

WIGNER
    Thank you very much. Dr. LeCouteur you are next on our program.

LeCOUTEUR(Australian National University)
    Well,Mr. Chairman, I'm impressed by the burst of activity in
this field. It is a tribute to the vision of Weisskopf and Bethe,
that their ideas are still being useful to this conference after
more than thirty years. Now it seems to me that one cannot
separate the discussion of level densities from some consideration
of how they can be determined; and the uses to which they can
properly be put. So for that reason, I admired the sophistication
of Miller's attempt to test the Bohr's hypothesis. Because the
direct test is very difficult, at another level, one can get a
feeling for the consistency of our ideas by seeing where the
different types of experiment yield consistent values for the
parameters in plausible theories. At first, people just concentrated on determining the parameter "a" in the exponent of the
Fermi gas level density formula and there is evidence from this
meeting that one's ideas on that are converging in a fairly
satisfactory way. In the 1950's one thought that little "a" was
equal to big "A" over 8. In 1968, the Italian groups gave little
"a" equal to about .127A or sometimes little "a" equal to big
"A" divided by seven and a half and this conference helped in presenting data which suggests that the experimental little "a" should
be equal to big "A" divided by 8. So that hasn't changed it very
much but one must be clear that this is an average value and that
there are very striking deviations at closed shells, notably near
Pb^{208}. Now there is some very important work of Rosenzweig which
has also been extended by Rosenzweig and Kahn and Böhning, which
shows that the level density of a system with discrete shell
structure of the single particle levels can be represented by the
continuum Fermi gas formula if an additive shift of the base level
from which one counts discrete energies before correction, is made
to the excitation energy. That shift being a measure of the
departure from a continuous distribution of single particle levels.
The systematic experiments of Huizenga and his group provides a
determination of this shift and also of the moment of inertia
which is the other important parameter of the theory. I was
pleased to see that although the shift is large in the region of
closed shells, that in between the shells the perfect base level
seems to be close to the ground state of even nuclei. Which is
where you would expect it on the basis of the pairing model.
I am also glad to see that a contributed paper (Number 57) says
that Vonach and Böhning are making a systematic survey of the data
in terms of an exponential parameter a, and an energy shift parameter. This is not too easy to do because level counts are
confirmed by selection rules and reaction data by direct processes.

Now, coming to the question of models, we have had excellent surveys from Bloch and French. It is clear that the useful model is that of particles in Nilson's single particle levels with a pairing interaction between the particles. It is however worthwhile noticing that if you use the BCS theory and then one finds that there is a transition temperature and above which the gap in the quasi particle spectrum vanishes and this transition usually takes place below the neutron binding energy so that the simple minded theories are not too bad about that level so, one has the choice of working with effective parameters and shifts or with the detailed levels and this depends on how much computation you wish to do -- the results of the one process can be represented in terms of the other. We've also had discussion from Bloch and French on the more important questions of how far these levels are displaced by interaction between particles. Well, clearly the levels must be spread, on the other hand, if you use a statistical model, the mathematics in any case will average the level density of the range of excitation energies of the order of magnitude of the nuclear temperature so that if the spreading is not more than that, then the results will not be affected very much. I think, it remains for someone to calculate the spreading rather more carefully, it's a very difficult job to do but I think the work with random matrices and random residual interactions will be of great help in clarifying the matter.

WIGNER
    Thank you very much. I think Dr. Cohen you are next.

COHEN (Pittsburgh University)
    I was asked, I thought quite specifically, to talk about direct reactions, so I prepared a talk on that. I may not have been the right person to be asked here, because someone accused me last night, in a friendly way of course, of being partial to direct reactions, and it is true that I've made my living with them for about twelve years or so, but actually in the last two years I've been spending more of my time on compound nucleus. Well anyhow, for the next three and a half minutes, or whatever is left I'll talk about direct reactions.
    First of all, how do you recognize a direct reaction? Back in the old days, we used to say that if the angular distribution is forward peaked, the forward peaked component is direct and the back angle part gives the compound nucleus contribution which is assumed to be isotropic. I remember giving a talk at the Amsterdam Conference in 1956 on the analysis of data in this way, but that work was never published because about that time we started to recognize direct reactions by the levels they excite. And then we came to understand their angular distributions first empirically and then with DWBA. It was then clear that direct reactions are not necessarily forward peaked. For example, there was an important piece of work from MIT on Bi(d,p) in which all angular distributions

increased monotonically with angle, reaching a peak at 180°. So
the old method doesn't work.

Thus when we decided, about a year or two ago, to study (p,p')
direct reactions exciting states in the continuum, our first problem was to find a method for separating compound nucleus (CN) from
direct reactions (DR). One of our methods is illustrated in Fig.1.
Here

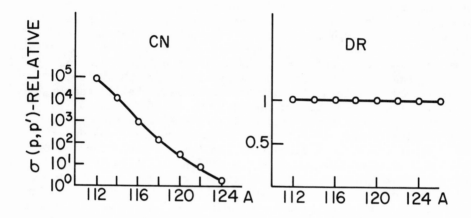

FIGURE 1

we plot the theoretical cross section for (p,p') reactions in the
isotopes of Sn on the assumption of CN and DR. In the CN mechanism,
$\sigma(p,p')$ is very sensitive to neutron-proton competition and hence
to the 0-value for the (p,n) reaction. This varies rapidly with
A in the Sn isotopes, whence $\sigma(p,p')$ varies by a factor of 10^5 between
Sn^{112} and Sn^{124}. With DR, there is no question of competition;
$\sigma(p,p')$ depends only on nuclear structure, which is very similar
among all the even isotopes of Sn their collective 2^+ and 3^- states
are at the same energy and are excited with the same $B(E-\ell)$, etc.
Hence we expect no appreciable A-dependence for $\sigma(p,p')$.

The experimental $\sigma(p,p')$ are shown by the circles in Fig.2.
In the heavier isotopes, DR is clearly predominant, and since we
expect DR to vary slowly with A, we may estimate its contribution
to $\sigma(p,p')$ to be the dashed line in Fig. 2. On this assumption,
the CN contribution is obtained by subtracting the DR from the
experimental values, giving the crosses in Fig. 2. It turns out
that these crosses agree with the theoretical CN $\sigma(p,p')$ with
surprising accuracy, which gives credence to the whole method.

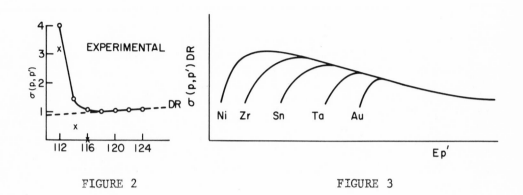

FIGURE 2   FIGURE 3

You may note that we are operationally defining DR as those
reactions in which competition with neutron emission is not a
factor. Unfortunately, that is also the signature of Jack Miller's
process -- he says that if the compound nucleus is in a $T_>$ state,
neutron emission is impossible. There are reasonably strong object-
ions to using it to explain the Sn data here: Miller's theory
predicts $\sigma(p,p') = (2T+1)^{-1} \sigma_R$ (the reaction cross section) which
is 40 mb in Sn^{124} as compared to our experimental value of 100 mb.
It predicts that $\sigma(p,p')$ should decrease by 30% from Sn^{118} to Sn^{124},
whereas it actually seems to increase by about 10%. It predicts the
same spectral shape in Sn^{112} and Sn^{124}, whereas experimentally they
are somewhat different.

But Jack Miller is not talking about Sn, but about Ni, and
there the situation is not so clear. Even in the heaviest isotope,
Ni^{64}, CN is not negligible, so one must estimate its contribution
theoretically. Moreover, $\sigma(p,p')$ in Ni^{64} is not larger than
$(2T+1)^{-1} \sigma_R$, and the spectral shape differs little from that of
Ni^{60}. The relative importance of Miller's mechanism and DR in
Ni^{64} is therefore open to question.

Let me mention some other evidence, though it is somewhat
circumstantial. If we believe our method outlined above for
separating CN and DR, the DR contribution $\sigma(p,p')$ is as shown in
Fig. 3 for various nuclei. It appears that DR $\sigma(p,p')$ follows a
universal curve for all nuclei except for the cut-off by the
coulomb barrier. Since our method is very probably right for
heavy elements, adopting the alternate explanation in Ni^{64} would
ruin this regularity.

Another method we have been using for telling when we have

DR in (p,p') depends on the fact that (d,d') is almost certainly
DR. We have shown by use of ^{DWBA} that

$$\sigma(p,p')_{DR} = k\ (\theta,A,E')\sigma(d,d')$$

where k is a slowly varying function of θ, A, and E' (except for
Coulomb barrier effects) which is calculable from DWBA. If we
assume that $\sigma(p,p')$ in Ni^{64} is nearly all DR, we obtain the same
value of k in Ni as in Sn which seems to support our assumption.
Unfortunately, however, DWBA predicts k $\tilde{\ }$ 3 in both Ni and Sn,
whereas experimentally we find k $\tilde{\ }$ 5 in both cases. We've had other
problems with DWBA in reactions leading to the continuum, so I don't
trust it quantitatively. I only hope we can trust it qualitatively
enough to believe that k is about the same for Ni and Sn.
        I thought I might mention one other method we have been using
to separate CN and DR in (p,p') even though we don't yet have any
results applicable here. We have developed a method for measuring
the gamma ray multiplicity, the number of prompt gamma rays emitted
following a reaction. It seems that there are fewer gamma rays
following DR than following CN. We hope to establish and use this
method to distinguish between the two processes.

WIGNER
        Thank you very much.

COHEN
        I think the other people should have a chance to reply as
soon as possible.

WIGNER
        Yes, we have one more speaker Dr. Bartholomew.

BARTHOLOMEW(CHALK RIVER)
        First, I would like to say, as an experimentalist, several
things have stood out in experimental methods discussed at this
conference. Firstly, we've all been impressed, I'm sure, by the
demonstrations by what can be done with a well-stabilized high
resolution Van-de-graaff beams and the work from Duke University.
This work has been in progress for some time but it has brought
out to me here, the fact that the field of statistical properties
of resonances is not the sole preserve of the neutron physicist
but equally good work can be done with proton scattering and
capture reactions, at least in some elements of the iron region.
Remembering that this was done with a three Mev Van-de-Graaff,
one wonders how much further it could be extended with higher
energy machines and with further refinements and stability and
resolution. A second technique, also not so very new, but which
was, has produced already much beautiful work in the field previously reserved for neutron techniques, is the threshold photo-

nuclear method used at both the Lawrence Radiation Lab. and at Argonne, and described so ably this morning. The tracking down of
the M1 giant resonance in Lead 208, for example, seems to be a most
important contribution from Dr. Berman's work. This is the same
technique we heard from Jackson about a strong M1 excitation in
Iron 57 and from both speakers about the facility provided by this
technique of determining resonance parameters. And of course, we
have also seen, somewhat earlier in the conference, the gain and
capability for resolving resonances in the kilivolt region
provided by new Linacs. These were three impressive techniques
which set a tone of high quality in data taking at this conference.
Of course there were other techniques described which are also
very important such as the Blocking technique, and the bombardment
of alpha particles of heavy ions, but I haven't time to comment on
these. One thing that struck me as very important and here I repeat things said, emphasized by our chairman and by various
speakers is of course the intermediate structure effects and departures from the extreme statistical model. I need say very
little about this, there is much evidence for this particularly in
the talks by Chrien and Mughabghab, and the photo-nuclear work.
This was covered very ably from the theoretical viewpoint by
Dr. Lane and by Dr. Mahaux. These effects include the evidence
for various types of correlation and direct valency capture
effects. Dr. Bollinger has covered the neutron gamma correlation
question so I might turn, for a moment, to the work I know somewhat better, which is namely the existence of non-statistical
gamma ray spectral distributions in some elements just below neutron closed shells. This seems to be explanable as Doorway
capture, where one can think, in the most general terms, of a
single particle excitation, coupled to an excited state of the
target. The excited particle undergoes a transition of dipole
type leaving the product nucleus in an excited state. This seems
to be a quite generally accepted mechanism at this conference,
although problems remain, and remain to explain how the El strength
remains at the unperturbed particle energy. Similarly, the
applicability of the valency captured model has come to some surprise that such a simple picture should work so well. Dr.
Mughabghab gave some very good examples of this near the p-wave
strength function maximum. It's almost trivial or obvious to
make the observation, but all of these effects are important because they mean that the resonances are not always the faceless
members of a great ensemble, but manage to preserve interesting
personalities and family connections, in spite of their great
inherent complexity. This was brought out not only by the effect
just mentioned but of course by the increasing body of evidence
from Doorway states as shown by Dr. Newson and by the isobaric
analog work described by Dr. Temmer. Now there was another region
where non uniform strength functions are appearing, and that is
in electron capture decay to highly excited states and heavy elements. The beta decay strength function is sensitive to shell

structure in the initial and final states. This effect was only
brought out in discussion by Dr. Hanson whose contribution on
this subject to this conference, I understand was somehow waylaid.
The last important point in my estimation has to do with non-
statistical properties of groups of levels in the resonance region,
following excitation of the region by different reactions. And
Bernie Cohen has said something about this and that levels
excited by different reactions behave differently, as is of course
well known at low excitations , but what hasn't been generally
appreciated particularly outside the charged particle field, I
think, is that such effects persist into the resonance region.
Charged particle physicist are, seem to be quite widely aware of
this, and this brought out yesterday by Dr. Miller's talk, but
I'm somewhat embarrassed that I chose that time during his talk to
play hookie and so I didn't know in detail what he said. But I
would like to tell the conference, we've recently completed an
experiment at Chalk River, somewhat similar to Miller's and to
some of the things that Bernard Cohen was saying. And I hoped it
wouldn't be out of place to take a moment to discuss this. We
chose platinum 195 and chose to excite the region just above
threshold by the (dp) and (pp') reactions. We adjusted the
energies as best we could to give about the same distribution of
population of the J states, in this region. Then we measured on
one hand the cross-section ratio of the (d, pn) reaction to the
(d,pγ) reaction, and on the other hand the cross section ratio
for the (p, p'n) to (p, p'γ). When we found that these were
very different, by about a factor of thirty, for states about
five-hundred kilovolts above threshold. And this would be rather
hard to put down to special structure effect because we varied
the angle and also the excitation energy. And going along with
this, we found that the gamma ray spectra to be very different.
The gamma rays from the (dp) reactions showing the 5.5 Mev
anomoly, where the (pp') gamma rays did not. So perhaps this will
be an interesting area that is opening up in the future. Now I
have given a demonstration of a well known recognized effect, I
think, what I have gotten out of the conference so far is mostly
some aspects of things I had some acquaintance with before. But
I hope to learn more from the proceedings.

WIGNER
     Thank you very much. Well, I think it is now up to the
audience to make comments, ask questions, either from a single
person or from the panel (which will be more difficult to answer).
Is there anybody who has some ideas?

BERMAN (Livermore)
     I don't have an idea, I would just like to hear Dr. Feshbach
elaborate a bit on what he means by new dynamical symmetries that
one might discover from Doorway states and their distributions.

FESHBACH
        That's what I get for saying something like that. Well I
gave one case and I will go back to that,...I might just parenthetically say that perhaps George Temmer's very narrow and unexplained second order doorway states might be an example of some
new principle which is working which we don't get enough. But
another example which I did mention, was in the case of alpha
particle transfer in the collision of heavy ions. It is found
way up, not way up, but fairly far up in the continuum, that the
alpha particle suddently finds a window, actually transfer very
easily, at very narrow levels sticking up in a place where there
are thousands or hundreds (I forget the exact numbers of levels
there) just clean as a whistle. Actually, some of that was discovered years ago at Argonne by Singh in an alpha particle collision with Mg^{24}- a very similar phenomena was found except not that
clearly exposed. Now you might ask: Why is it that alpha particles, which all optical model calculators say, has a very large
absorption, large imaginary parts of the optical potential, could
have this very, how shall I say it, well a Doorway I guess is still
a good word, a very easy way, a very easy route or easy way of
getting into another nucleus. Why is it that you get, in the alpha
particle, from magnesium, very clean resonances--where does it
come from? Now, what might be going on is that there is a
symmetry principle operating, perhaps connected with Professor
Wigner's SU_4 or some other dynamical principle which says--which
like in the isobaric analog says these states are very clean, they
don't couple with all the fine structure underneath because the
fine structure underneath belongs to an opposite part of the
symmetry space. So it's really at this present moment, just
wishful thinking, that this is going to occur--it serves as an
example.

BERMAN
        There's probably no connection, but that reminds me of the
nuclear Josepheson effect we heard about for a while. Where you
take tin 118 and tin 122 and come out with two tin 120's, and they
go up in back and you get different yields as a function of angle
up because how many times, how long do they have to exchange pairs
of neutrons? Is this business with alpha particles--could that be
treated on an exchange mechanism?

FESHBACH
        I really don't know. The only theory that's presently avail-
able is a theory due to Gilet and Arima which says that--it has to
do with a quartet structure and I think--I don't want to elaborate
on that--I just don't know.

WIGNER
        It's a very difficult question to discover a new symmetry,
certainly the SU-4 worked out only moderately well. And the

other question or comment...Dr. Garg.

GARG(SUNYA)
    As an experimentalist, I am certainly interested in knowing,
how you can analyize this beautiful resonance data taken in the
neutron total cross section especially for the light nuclei and
medium weight nuclei where we know that the scattering is the
most predominant process. The capture widths are usually small.
Normally we have been using this R-matrix formula in its reduced
form, sort of, formulas where we take the radiative channel as a
sort of perturbation to the R-matrix and I would like to pose this
question to the panelists, if that formula is in fact, meaningful,
to extract the resonance parameters for the S-wave, P-wave and D
wave resonances which are observed in this mass region.

WIGNER
    What about you? You would be the next person to be heard
from.

RAINWATER
    Well, I've worked with Dr. Garg, I think he wants it from
one of the theorists. (Laughter)

WIGNER
    Well, I think that about the present calculation to which you
refer. I like that, now that doesn't mean that everybody else will
like it, but I like it -- perhaps somebody in the audience does not
feel it very strongly. What I don't like is the expansion of the S
matrix in the general case as a sum of $\gamma^2/E-E_o-i\gamma$ I think ...
Dr. Moldauer has pointed out that such an expansion as a rule is
not unitary. And if we want to maintain something on the property
of the collision matrix, we should try to make it unitary. I mention this, because in the particular case which you point out, even
though the collision matrix which you obtain is not unitary, it does
not matter, because the deviation from unitary nature is not significant and I think there is good reason to believe that the perturbation calculation is reasonable. Perhaps somebody else in the
audience has a different view.

FESHBACH
    As usual I almost agree with Professor Wigner, but not
totally almost...I think in my own particular way of doing things
you can formulate a theory in which the gamma ray is not treated
as a perturbation. Generally speaking, it is not important, except I would say to do it exactly, to do it with precision, so
that you preserve particularly, unitarity. However, there is an

energy domain in which the neutron width is smaller than the
gamma ray width. And in that region I think perturbation calculations would not be valid. (Chairman: "Of course.") And, therefore, one has to rely on--one has to be careful in that domain .

WIGNER
     I am glad to report that in this case, Dr. Feshbach and I
agree completely. Dr. Garg...you wanted to say something?

GARG
     Thank you very much. I think I will accept that point of
view myself. One thing is that of course there are people who
have tried to analyize some of high resolution data with other
formulas, I have been told that they sometimes obtain the peak
cross sections at the maximum of the resonance which is larger
than given by the usual $4\pi \chi^2 g$. And that has been somewhat disturbing to me because we usually utilize this as a method for
determining the spins of the resonances. And it has been somewhat of a disturbance as to how much reliance one can place on
this kind of formula to determine the proper spins of the
resonances in fitting the experimental data.

FESHBACH
     But there are, after all, formalisms. I'm not refering to
mine, this time...there are after all, forms which can be used to
fit the data which do fit unitarity, which are not just on the
matrix theory. You have to put in the conditions that Professor
Humblet wrote down on the board. You can have an expansion in
partial fractions if you at the same time say that all the constants have constraints. Now you can do either one but you've
got to do it right, and I agree to that.

WIGNER
     Dr. Humblet's idea, I did not appreciate fully when he spoke
about it. And it is of course true that his comments are complex.
And that will not subject to the usual analysis as we have. But
perhaps there is somebody else who came to appreciate it more
since he heard about it. Dr. Lane do you have any comment on this?
(Dr. Lane: "No, I don't...how about changing the subject?)

TEMMER
     I'd like to address a question to Professor Wigner: I
wonder if you could give us an inkling, or a hint, or a dream...
an elaboration on your dream on how you might modify statistics
in such a way that concentrations of strength would occur that are
essentially nonstatistical.

WIGNER
     Well, I can tell you what I am learning to do and of course
at almost every brand misfires and I fully expect this to misfire

also, but I will change it then. I think it may be a useful
thing if I answer you just the same. I would say that I would
assume a statistics of model states. Model states being a state
which in part, which is a state similar to the analog--parent
state--similar to the isotopic spin state which then distributes
its power and strength among its children. That is the resonances
nearest to it, not in the spiritual sense in this case, but in the
energy sense. And I would assume that there is a wealth of such
states with varying wealth (endowed with a varying wealth) and
also endowed with varying strengths of interactions with average
state in its surroundings. Then I would carry out a calculation
as has been carried out by several people in particular--I mention this because you are familar with it in particular also, on
the analog state as it distributes its wealth and that would give
a new type of statistics. I don't know whether in the long run
over an area in which there are several parents available. But,
in the whole sum it will give a distribution different from the
Porter-Thomas distribution.   But it would give correlations between strengths, and in this case, between transition probabilities
or reduced widths--and in this case, I would say a positive correlation, clearly, between transition probabilities as it has been
observed and as it is the essence of the doorway or giant resonance
state, or Isotopic Spin, Parent State, or whatever is available.
Does that more or less answer your question?

GARG

     If I may be able to ask another question and that of course
concerns with the distribution of neutron widths and energy level
spacing. Dr. Rainwater has pointed out in his summary that some
older data, of course on thorium and uranium has been somewhat
unsuitable for a comparison with the theoretical prediction of
statistical distributions. My own personal feeling is that for a
comparison of the experimental data with the theoretical results,
you do require a finite sample of the lavels which belong to the
same symmetry character. And as you improve the resolution or
the precission of your measurements, it is quite natural to be
able to observe the finer resonances which perhaps are due to
higher orbital angular momentum (p and d wave captures). Of
course there are exceptions where you expect S-wave strength
function maxima to be much larger than p wave and so on. And
since you are not able to really get a full sample of the p wave
levels in any higher precision and extensive measurement of these
things, I think what you are actually doing--you are mixing up
levels of different symmetry character into your sample that you
are trying to compare with your theoretical predictions. And
from that point of view I would like to feel that the data which
is not really as sophisticated or as precise as, in fact, reminds
me of the beautiful work done by Dr. Cohen, where he saw these
single particle effects in the measurements done with somewhat poor
resulution. I think that it is a very similar problem. As you

improve the resolution and the quality of the data, you are
mixing up different types of resonances and the question is how
much then, can you rely on the deviation of the experimental data
from the theoretical prediction of either the level spacing distribution or the Porter-Thomas distribution for the transition
probabilites.

WIGNER
    Why don't you say something toward the question.

RAINWATER
    What you're saying is that the measurements of the type that
we have made over the years, where you look at individual
resonances at low energies. You have difficulty including only
a single population and this is a feature where I've tended to
feel that the pertinent use of the theory is more the other way
around. That is, if you believe the Porter-Thomas and look at
the distribution of reduced widths, it tells you the extent to
which other populations are present. And if you believe the
single spacing distribution of Wigner which seems to imply in
good cases it tells you failings of your data, and to some extent,
for example, we have been extracting p-wave strength functions by
calculation of the probability for given sensitivity of detection
that you see p-waves. You can calculate as a function of the
strength function what you would get. But you are, I believe,
talking about the higher energy region where the averaging tends
to be over large numbers of levels and this I think, should be
really answered by one of the people who have been most directly
involved with this type of research and analysis.

WIGNER
    Try to put the blame on the theorists. I would like to ask
Dr. Bollinger who also has great experience in this area, perhaps
he would make a comment. It's a very experimental question really.

BOLLINGER
    Well I believe I understand what Garg was talking about. I'm
not sure if I understand Rainwater's last remarks, so I'll address
myself to Garg's remarks. I agree with him that one worries
about the problem he raised, namely how can you test a model when
you can't assure yourself that you have a pure sequence. The
puzzling thing about this is that the existing data appear in some
cases to predict statistics, such as the Δ_3 statistic that are
actually smaller than are expected from the Random Matrix Theory,
which certainly in itself suggests that the data are very pure
indeed. And yet one at the same time cannot, in a completely
objective way be sure of it, it seems to me. But this allows me
to lead then into a question I would have wanted to ask of someone.
Now to me being totally ignorant of all the theoretical aspects of
this, it appears that the approach that's been introduced by Wong

and French, and discussed in a couple of papers here, is heading
in a direction that seems more realistic than the assumption of
an ensemble and then going ahead with the consequences. And yet
it also seems that the data, the best recent data, are in much
better agreement with the older theories. I would find it useful
at least to have someone summarize for me what might be wrong
with the newer approaches.

WIGNER
     I think what the theorists, like myself--old fashioned
theorists--or call us what you like but the thing is this, the
ideas that unquestionable that the bulk of the forces is two body
forces. However, if you put together, not two, three configurations, but dozens of configurations, and if you introduce reasonably strong interactions between them, you will recover a system
similar to the old fashioned system because as you have strong
interactions between several configurations, the meaning of the
individual particle picture will degrade and degrade until
finally it will be just as bad or just as good as an old fashioned
random matrix. I shouldn't probably have said this, but I should
have asked Dr. French if he is around to say something and maintain
the opposite point of view.

FRENCH
     Thank you very much. I don't actually agree with the last
remark, I'm sorry to say, because I believe that if you have two
body interactions, and if you believe in the shell model then
there are many consequences which will be very very different for
the two body case than for the random than for the more general
case. But there are some other remarks. Now if one actually believed that, if one could now believe that there is a theoretical
rule for the distant neighbor of spacings and that which has been
found by Flores and Boghias and if one could believe that the data
is good and that the agreements are as indicated, then I believe
that it will turn out that one will have to say--let us say there's
a good chance that it will turn out that one will have to say that
the shell model in fact is not really in operation when you get in
this general domain. You could alternatively say that there are
just ten body interactions. It isn't enough, I think, to have
three body or four body. All the active particles have to be re-
interacting together, I think. Now, I don't think the shell model
could survive if the interaction were dominantly ten body, so it
perhaps boils down to the same thing. However, in my mind there
are still lots of uncertainties. One speaks about the rule for
distant neighbors, either the Gaussian orthogonal rule or the
Two Body Random Ensemble Rule, but it is not absolutely demonstrated
I believe, that there is a rule in either case for the force nearest neighbor, say. A rule in my mind means that you have an ensemble average on the one hand and which is such that and you must
have a statistic or a figure of merit with a small variance. That

is to say, besides having an esemble average, it must be that the
members of the ensemble must come close to this. Because of
course, if the variance is too big then there is no rule and then
we can tolerate everything. So I'm sure, that at least for two
body random things, it has not been demonstrated that there is a
rule, as yet. Now another aspect of this is that people who are
unable to do the thing analytically (and these two body random
things are horribly complicated) they actually go about solving
small matrices. Now, they produce a thing that may look like a
rule, but as I say, they haven't checked the variances yet and
besides they may be terribly fooled because one can only diagonalize
small matrices. And I myself, have a terrible fear, that if one
could diagonalize larger and larger matrices, the thing, that one
now has, would actually change. Now I think a great deal of work
has to be done along that line in other words. Therefore, before
one would be sure that he has found a form, namely the forms that
were talked about by Bohigas and Flores and I think a lot more
work before one would be sure he has a rule. I don't think that
the suggestion that was made by Herman the other day that the
spaces are small so you should put in three and four body parts
and then you'd be o.k. I don't think that's really the right way.
I think the right way to look at what Bohigas and Flores are doing,
is that they are attempting to convince themselves about the assymptotic rules for large matrices. But they have to go about it
in this rather poor kind of way of diagonalizing small matrices.
If they could be convinced, and could convince us, that they had
found the assymptotic rules and that the variances were small,
and that the data agreed with the old rules, and disagreed with
them, then I think it would be a most remarkable conclusion. But
I think we are a long ways from that yet, but I find the whole
thing extraordinarily interesting. I'm sorry as I said so much
about it.

BOLLINGER
     I didn't hear your last phrasing...if this all came about
then what?

FRENCH
     I said that if in point of fact, the forms which they have
found are stable. That is to say they do not change in largest of
matrices which are diagonalized, and if you believe that they
would not change if you went to the limit. That's the first
condition. And if you also have demonstrated that the variances
are small and thirdly, if the data really fits the old guassian
orthogonal ensemble then I think that the conclusion is inescapable,
that the shell model does not apply in that domain. I mean, perhaps other people can find other ways of rationalizing that.

WIGNER
     The same with the level density, it does not apply. I mean

in these regions. I think that's Greek isn't it, because
there are so many states mixed that you can't identify--I'm sure
that it is not possible to identify the 1.15 ev neutron resonance
in indium with any configuration. It surely is a mixtrue of fifty-
five configurations. Well, thirty-five configurations.

FRENCH
    When I speak of the shell model..what I really mean is that
there is an underlying set, a finite set of single particles states,
with which you can construct many particle states. And, the whole
outcome being such that you produce a space in which you can
describe what goes on. I mean there may be enormous mixtures,
that's obvious.

WIGNER
    That's a mathematical identity--however, I shan't argue with
you, I shall ask Dr. Bloch who is, as far as I know, the only one
of us who made the calculations one end of which was the new model
and the other end was the old model, to say something on it.

BLOCH
    It was not my calculation, it is Gervois' calculation.
    (Reference)
WIGNER
    I accused you unjustly.

BLOCH
    I would, if I may first, ask for a different question. What
do you mean when you say that the shell model does not work? Of
course when we say the shell model works we do not expect that
nuclear states are really pure shell model. We mean that one can
take the shell model state as a basis, then introduce residual
interaction, and by taking not too many, not too large a set of
these things and make of these states, and make a diagonalization--
we would get the exact state. But we do not expect that...

WIGNER
    Not the exact state but a very good approximation.

BLOCH
    ...a very good approximation.

WIGNER
    But you see, that is sort of an identity, that's a mathe-
matical identity. You can represent every wave function, as a
linear combination of shell model wave function.

BLOCH
    An infinite -

WIGNER

It is an infinite, yes. And if this is infinite, the question is how well does it converge? How many terms are necessary? Now on this we can differ, but I would believe that if I think for instance about the 6.8 electron volt neutron resonance of Uranium 239. Then this has well I said before, fifty-five states... no I would maintain fifty-five states. Is fifty-five a small number or a large number?

BLOCH

It's not very large.

WIGNER

In that case the shell model will work. But if that's what you call working of the shell model, that's a different story. But I would think that if you have such a situation, and this is only I believe, an opinion, and evidently not shared by Dr. French, then the total distribution of second nearest neighbors, third nearest neighbors, and so on, will be very much the same as in at least several other situations including the guassian, including the new Dyson's distribution and probably several others. I speak too much, Dr. Rosenzweig wanted to say something.

ROSENZWEIG

You meant fifty-five configurations, not states.

WIGNER

Yes, yes, that's what I meant.

ROSENZWEIG

So I draw the conclusion from all of this is that if you did construct the matrix of the Hamiltonian in a shell model basis, as Professor French and many others have done, and if you took the dimensionality of the space sufficiently large, you would in fact get the results of the orthogonal ensemble. This is your opinion.

WIGNER

Yes, local properties of the orthogonal ensemble.

ROSENZWEIG

Yes, local properties. Of course, we know that the global properties would presumedly be described better by the shell model approach.

WIGNER

Yes, they will reasonably be well described by a shell model approach. But as I constantly emphasize, everybody realizes that the semicircle distribution does not give the nuclear level density. It's not necessary to repeat this.

FESHBACH

I'm confused by this discussion. I just want to make the point, we spent a lot of time talking about intermediate structure. Doesn't that affect the spacing distributions and so on and don't the two body forces do a better job of representing them than many body forces? Representing the fact that there are non statistical pieces here. I mean I'm not quite sure what the debate is about when we know that statistics does not apply.

WIGNER

We talked about level spacing.

FESHBACH

Well, it affects level spacing, too.

WIGNER

It affects level spacings very little as long as the number of model states, or contributing states, parent states is very much smaller than the number of...

FESHBACH

Within the region in which there is an isobaric analog state, all the levels get shifted around by the presence of the analog.

WIGNER

Yes, they shifted around one level spacing--which is nothing.

FESHBACH

Well I don't see why the distribution can't be changed by that.

WIGNER

I don't know but if you calculate it, and I'm sure you have made the calculation also, the theory is that it is the reduced widths of the levels which change, to which a contribution is made and not the density of the levels.

FESHBACH

I don't agree.

WIGNER

Well, this is a question of mathematics.

FESHBACH

We can disagree, it's allowed. (Laughter)

WIGNER

But I don't think we would disagree if somebody had produced the mathematical proof.

(Laughter)

BLOCH
    Did I understand you, somebody said that there are fifty-
five single particle states in a resonance?

WIGNER
    Configurations...excuse me...and of course the statement is
meaningless because we should have said that these are fifty-five
configurations, if you want to represent the wave functions of the
state vector accurately to let us say five percent. That is a
meaningful statement. And it may be wrong.

BLOCH
    I would think, well, who am I to talk, but it sure seems to
me like there'd be alot more than fifty-five

WIGNER
    You think it will be more? Then you argue with...

BLOCH
    For compound nucleus it would be many more.

WIGNER
    I am perfectly happy to accept it. I did not want to over-
estimate it.

BLOCH
    But certainly the number of single particle states is very
restrictive.

WIGNER
    It's very restrictive.

BLOCH
    Within the imaginary potential "W," nothing beyond that...

WIGNER
    I think somewhat less restrictive than is generally believed.
However, I speak too much...

BLOCH
    I just wanted to say that I was surprised at the modesty of
the figure proposed for the figure fifty-five in which you proposed
for an infinite number--I think was a very modest estimate.

WIGNER
    Well...

BLOCH
    Actually, people working in shell models now-a-days make
diagonalization of very large space.

WIGNER

But I don't mean to say that you can't calculate it on the
basis of shell model and I'm sure you could calculate it on the
basis of fifteen other models. What I am saying is that the end
result is not something that I consider a shell model wave function.
Because I consider something a shell model wave function if there
are two configurations in it. If I want to be very generous,
three configurations. (Laughter) Excuse me, but that is a matter
of definition and to redefine it--it's fifty-five.

BLOCH

But then how should we call a model in which one uses a
rather large number of shells? How should we call a model in
which we take a large number, let's say of the order of fifty-five,
or perhaps a few more shell model configuration, and then start
working from that, either by taking a realistic inter-action or
by taking random matrices-random two body matrices? How should
we call such a model?

WIGNER

Well, let's call it a Bloch Model... (Laughter)

BLOCH

Which Bloch?

WIGNER

I don't want to argue about it and everybody will admit that
if you put in sufficiently many configurations, you get sufficiently
accurate   answer. The question is what we call such a model.
And that is another question. But should we, I don't know, we
don't have terribly much time left. Somebody else has some observation to make or some question to ask in the audience.

NEWSTEAD

Well, just a quick question on a different topic. I wonder
how valid it is to try and analyse the intermediate structure one
sees by putting spreading and damping widths into the optical
model potential, as we've seen in one or two contributed papers!
What scope might there be for that sort of calculation? If
Dr. Feshbach could comment on that.

FESHBACH

Well, it depends of course on the condition of the data.
In other words, I have the feeling that, to do that, let's say
do an optical model and simply look at the angular distribution
and the total cross-seciton--might not be sufficient. In other
words, one might need polarization data as Monahan and Elwyn used
and things of that kind. In other words, I am perfectly willing
to describe the average wave function in terms of coupled channels
of which one of the channels to which your coupling is the Doorway

State and have that responsible for the intermediate structure
that you see. But, I think you have to have enough data before
you guarantee the validity of that description. That would be
the way I would set it up.

NEWSTEAD
     So it's a question of data really rather than formalism.

FESHBACH
     I think so.

WIGNER
     Dr. Moldauer wanted to say something. No he doesn't.

MAHAUX
     I have a question to Professor Feshbach. I would like to
know how optimistic you are about the possibility of making
microscopic calculations of the strength function in a wider
range of nuclei than those you consider--with Barry Bloch--a
number of years ago, or I think equivalently how optimistic are
you about the possibility of making now, microscopic calculations
of the absorptive part of the optical model potential.

FESHBACH
     Optimistic I can always be, as long as I don't have to do it.
Actually some attempts are being made in this direction as you
surely must know--at higher energies, I think the place in question is Michigan State where the optical potentials are calculated
on the assumption of intermediate states of a particular character.
I think I'll answer the question in the same way I answered
Dr. Newstead's question. I think the answer, in a way, is the
same. If you are in a mass and energy region where you know what
the important states are, for one reason or another, then you can
replace the simple optical model by calculation of coupled channels. And in that way get yourself to a microscopic calculation
or closer to it. I think, however, it's really premature to say
that one should be optimistic about it.

WIGNER
     Thank you.

PREIS(R.P.I.)
     I have a question for Dr. Cohen. In view of the fact that
alot of the data that we have seen today, and this week, look
like compound nuclei but might be direct interaction and vice
versa--Could you give us, maybe a new definition for direct
interaction.

COHEN
     Of course there is a classical definition. I mean, you know

just in terms of times; but I think that the thing that is clear
is the definition of a compound nucleus. Now, I would like to
say that a compound nucleus--well, then, there's Griffin's stuff.
Well, I need somebody else.

MOLDAUER

    I think I would like to turn the question around and ask some-
body to give me a good definition of a compound nucleus.

WIGNER

    I think these are very pertinent questions with which we are
not completely in equilibrium. (Laughter.) No, we are not, we
are not. The definition as Dr. Cohen made it, that it takes short
time is very much the same thing as the--well, the delay time is
directly expressible in terms of the derivitive of the "S" of the
Collision matrix element with respect to energy. So that if the
time is short this derivitive should be small, which means that
the cross-section should not vary with energy very strongly. Now,
of course, there is no sharp division between quickly varying cross-
section or short delay time and long delay time; and there is an
intermediate region. It is sort of a work for us--it has been
sort of a work for us--theorists, that the two regions were so
separated that we could find an approximation for one, the distorted
wave Born approximation--an approximation for the other relative
with few compound states and worked that way. It would be very
desirable to develop a theory which has the transition region
between the two types of reactions. I have another question to
Dr. Cohen; and if you will forgive me, I will ask him that. Did
the data which he told us about appear anywhere in the literature?
It's a question he probably can answer. (Laughter.)

COHEN

    Well, some of it has, and some of it hasn't. (Laughter.) On
your question about the separation--now, in the Griffin Model,
which incidentally I don't like so much, but there is, I mean,
the probability of something happening after one collision is well
that's direct reaction. But something happening after two collisions is high, after three collisions is much less, after four
collisions is much less--so it is something which drops down very
rapidly and then you have the statistical region. So that's a
reasonably clean separation, I would say, in some energies. Of
course, in other energies there is no separation.

WIGNER

    I think that's a rather over-classical picture.

COHEN

    Well, I'll agree with that, yes.

MOLDAUER

Well, I just would like to make this remark about the last
comments which seem too flippant. One must, I think, keep in
mind that statements about models--what kinds of classes go on--
are statements about amplitudes. And it's not quite clear that
you can make unique statements about amplitudes by looking at
their squares and maybe the cross-sections, and separate how much
is what in any unique way. And I think that is in part the difficulty.

WIGNER

A very important remark, really. I don't know--our time is
up. We can do two things. I think we should probably close. We
can have private discussions for considerable lengths of time, I
think, in the corridors, but before we adjourn, I would like
Dr. Garg to make an announcement. Can he do it? Or is there
somebody who would like very much to ask a question or make a
comment? No, it doesn't seem so.

GARG

Well, on behalf of the organizing committee, I would like to
thank all the delegates again, and in particular the invited
speakers and the distinguished panelists,whose presence here has
been so essential for the success of this conference and who
have taken the time and effort to present a clear elucidation of
so many complex but still unsolved problems. You have been
informed earlier in one of the Conference circulars that the
proceedings will be published by the Plenum Press and a copy of the
proceedings will be sent to each registered delegate upon publication, free of charge. We will do our very best to have the proceedings published before the end of the year. However, in order
to accomplish this difficult task, we would like all the speakers
and the participants in the discussions to return their edited
manuscripts before leaving the Conference this afternoon. The
Conference on Statistical Properties of Nuclei is now closed, and
I would like to thank you all again.

INDEX

Accelerators
  electron linac 173, 233, 237, 251,
  263, 306, 367, 447, 601, 611, 622
  heavy ion accelerators 490, 498,
  508
  isochronous cyclotron 455
  research reactors 233, 257, 263
  synchrocyclotrons (Nevis) 83, 205
  Tandem Van-de-Graaffs 483, 527,
  536
  Van-de-Graaffs 299, 631

Absorption 476
  strength 589
  of photons 589, 601, 604

Analogue states 8, 32, 99-108, 147,
  271, 275, 299, 312, 428-29, 521,
  542, 549, 557, 567, 570, 572-73,
  591, 593, 616, 636, 642

Angular correlations 243, 474

Angular distribution 57, 64, 352-
  53, 438, 471, 477, 492, 522, 525,
  535, 565, 607, 628, 630, 644
  of compound nucleus process 354
  of elastic scattered radiation 258
  in fragmentation process 89, 173,
  437, 493
  of neutrons 57, 573, 604
  of inelastic scattering 351

Angular momentum 9, 82, 131-141,
  354, 384-87, 406-09, 426, 429-30,
  434, 473, 480, 484, 487, 503, 505-
  514, 520-22, 531-33, 552, 581, 602

orbital 335, 349, 369, 507,
  636
total 101, 195, 364, 573
transfer in direct reactions
  479
in the gamma decay 483

Area analysis 57, 149

Atomic number 133

Auto-correlation function 132-
  138, 167, 346, 351, 458

Average (mean)
  cross section 139, 153, 342,
  345-46, 359, 369, 546, 552-53,
  556
  level width 139, 141, 148, 342,
  456, 458
  level spacing 341, 426, 434,
  447, 545

Beta decay 298, 631

Binding energy 224, 426-29, 470,
  497, 509, 601, 612

Blocking dip 536-37

Blocking effect 410, 434, 535,
  538, 565, 631

Bohr's channel theory 67, 77,
  156-57, 177

Bohr's model 272

LIST OF CONTRIBUTED PAPERS

SESSION 4: STATISTICS OF RESONANCE PARAMETERS (PART II)